AF855964

BIOGEOGRAPHY AND ECOLOGY OF NEW GUINEA

MONOGRAPHIAE BIOLOGICAE

Editor

J. ILLIES

Schlitz, F.R.G.

VOLUME 42

Dr W. Junk Publishers The Hague–Boston–London 1982

BIOGEOGRAPHY AND ECOLOGY OF NEW GUINEA

Edited by

J. L. GRESSITT

VOLUME ONE
PART ONE: GENERAL AND PHYSICAL BACKGROUND
PART TWO: MAN AND HIS IMPACT ON THE ENVIRONMENT
PART THREE: VEGETATION AND FLORA

Dr W. Junk Publishers   The Hague–Boston–London 1982

Distributors:

for the United States and Canada

Kluwer Boston, Inc.
190 Old Derby Street
Hingham, MA 02043
USA

for all other countries

Kluwer Academic Publishers Group
Distribution Center
P.O. Box 322
3300 AH Dordrecht
The Netherlands

This volume is listed in the Library of Congress Cataloging in Publication Data

ISBN-13:978-94-009-8634-3 e-ISBN-13:978-94-009-8632-9
DOI: 10.1007/978-94-009-8632-9

Cover design: Max Velthuijs

Copyright © 1982 Dr W. Junk Publishers, The Hague.
Softcover reprint of the hardcover 1st edition 1982

All rights reserved. No part of this publication may be reproduced, stored in a retrieval system, or transmitted in any form or by any means, mechanical, photocopying, recording, or otherwise, without the prior written permission of the publishers
Dr W. Junk, Publishers, P.O. Box 13713, 2501 ES The Hague, The Netherlands.

Contents

PART FOUR: INVERTEBRATE FAUNA

PART FIVE: VERTEBRATE FAUNA

PART SIX: BIOGEOGRAPHICA SUMMARIES

PART SEVEN: CONSERVATION

PART ONE

General and physical background

1. General introduction

J. L. Gressitt

New Guinea is a fantastic island, unique and fascinating. It is an area of incredible variety of geomorphology, biota, peoples, languages, history, traditions and cultures. Diversity is its prime characteristic, whatever the subject of interest. To a biogeographer it is tantalizing, as well as confusing or frustrating when trying to determine the history of its biota. To an ecologist, and to all biologists, it is a happy hunting ground of endless surprises and unanswered questions. To a conservationist it is like a dream come true, a "flash-back" of a few centuries, as well as a challenge for the future.

New Guinea is so special that it is hard to compare it with other islands or tropical areas. It is something apart, with its very complicated history (chapters I: 2–4, II: 1–4, III: 1, VI: 1, 2). It is partly old but to a great extent very young, yet extremely rich and complex. It has biota of different sources — to such a degree that it is still disputed in this volume as to what Realm it belongs to: the Paleotropical or Notogaean (Australian); or what Region: Oriental, "Oceanic," Papuan or Australian. The terms Papuasian, Indo-Australian and Australasian also have been applied to the area.

New Guinea is the largest tropical island, the highest island, one of only three tropical areas with glaciers, as well as a land of great variety of vegetation types, and of most kinds of environments except deserts (chapters I: 3, 4; III: 3).

New Guinea, or the Papuan Subregion (Moluccas to Solomon Islands) has many characteristic groups of biota, limited to the area or barely spilling over into adjacent areas: the tree kangaroos, the kuskuses, giant spiny ant-eaters, birds of paradise, cassowaries, green tree pythons, long-necked turtles, bird-winged butterflies and many other special groups of insects and others — endemic or nearly endemic. It has many groups of plants and animals with relationships extending in all directions, to the Indonesian, Philippine, Malaysian areas on to E Asia, S Asia or tropical Africa on one hand; to Micronesia or Polynesia to north and east; and to Australia to the south.

New Guinea boasts the world's largest and smallest parrots, the largest doves, the longest lizard, some of the smallest frogs, the largest butterflies and moths, some of the longest stick insects, the widest-headed (stalk-eyed) flies, the tallest tropical trees, the largest rhododendron flowers, the richest mangrove and sea grass floras, and many other forms and habitats unique and fascinating if not extreme.

Monographiae Biologicae, Vol. 42, ed. by J. L. Gressitt.
© 1982, Dr W. Junk Publishers, The Hague.

New Guinea is largely covered with lush rainforest of many types, some of the lowland types being nigh impenetrable. Many of the extensive swamps and wide shifting rivers are great barriers to road construction. Much of the mountainous areas are so steep and so rapidly eroding (even though covered by forest) that roads have to be constantly repaired and bridges replaced. In the high mountains there are areas of grassland and of tundra (chapters III: 2, 9), and in the far south some areas of savanna woodland (chapter III: 8) unique for New Guinea.

New Guinea is the only area where oaks occur together with southern beaches, or with *Araucaria* and other "trans-Antarctic" elements in several groups of plants.

The biogeography of New Guinea is an intriguing problem, and one difficult of resolution. Not only did the biota come from different directions, such as from SE Asia, the Philippine and Indonesian islands and from Australia, but even from S America, or Gondwanaland, via Australia (see especially chapters VI: 1, 2).

The ecology of New Guinea is a vast field barely touched, which will certainly offer fascinating scientific discoveries as well as disclose economic potential. It is clear that information on ecological processes in any other tropical region cannot be assumed to apply equally to New Guinea. One example of the uniqueness of New Guinea ecology is the "epizoic symbiosis" with a varied community of plants and animals living on the backs of large moss-forest weevils (chapter IV: 13). No equivalent of this association has been reported even from other humid tropical areas. Thus the ramifications of ecology in the humid forests which cover much of the island will certainly prove complex. For most groups of animals, so little is known about feeding habits that rather little charting of basic food-chains can be done at this point. Certain birds, just a few of the mammals (chapter IV: 5), butterflies, mosquitoes, more serious agricultural insect pests, and very few others have received some special attention. A few major forest trees and other economic plants have been studied in more or less detail; shallow-water marine biota, such as at the Motupore I. marine station of the University of Papua New Guinea (PNG), as well as fisheries resources, are being studied in just a few areas. At Wau Ecology Institute some basic ecological and behavioral work is being done, mostly on birds (chapters V: 3, 4), lizards (chapter V: 2), spiders (chapter IV: 3) and insects. There is a great need for basic studies in ecosystem ecology in different zones, as well as on potential food and medicinal plants and animals from land and sea. Some general statements in ecology textbooks, even in some directed toward the tropics, do not relate accurately to situations in New Guinea. For instance, some texts say that it takes 700 years for climax moss forest with epiphytes to regenerate, whereas some in New Guinea have regenerated in 200 years following complete destruction from a volcanic island explosion (Diamond 1974, Science 184:803–806).

The scope of this volume, geographically, is the entire island of New Guinea, plus immediately neighboring islands with essentially similar biota, such as Waigeo, Salawati, Aru, Biak, Japen, and the d'Entrecasteaux and Louisiade groups; also the Bismarck Archipelago, including the Admiralty Is (Manus and others), New Hanover, New Ireland, New Britain and nearby islands, with their related biota reduced and partly different from mainland New Guinea's. In subject matter this book attempts to give a general picture of the geology, geography, climate, scientific history, peoples, crops, vegetation, and terrestrial biota of the area. But in spite of the paucity of detailed information on so many aspects of these subjects, the scope is far too great for a volume necessarily limited to this size. Many groups of plants and animals are not discussed at all. The marine environment has been excluded from treatment except for a very little on plants essentially in the brackish fringe. The fresh-water environments are sadly neglected, and no mention is made of their fish, nor of most of their other biotic elements. Most of the reptiles are not treated here, and most invertebrate groups, other than a few of the arthropods, are also omitted. Only a small fraction of the insect fauna is discussed in any way and here there is very little ecological information to present. These serious gaps are regretted, but these, and the paucity of ecological data, clearly demonstrate the need for much more research. It is hoped that the many needed separate studies, and the ecological coordination, may materialize and be followed by a larger and more balanced summary work after a few years. If this book serves to inform students and specialists of some of the basic aspects of New Guinea biogeography and to give a little insight into the vast biota and the intriguing ecology of the area, and particularly to offer the challenge to help fill the great gaps in our knowledge, then it will have served its purpose.

New Guinea has had a complicated political history, the impact of which on geographical and scientific discovery and biological investigation is discussed briefly in Chapter II: 1. Only a brief review of the chronology of colonial and current independent administrations is presented here. The complicated subject of local and foreign-applied place-names and their homonyms is unfortunately omitted from this volume, except as individual authors have chosen to mention such in presenting their information.

The first European discoveries of land in this area were made by Portuguese (starting from 1527) and then Spanish, although there had been much earlier contacts with the Moluccas in the west, and at least indirectly with China. However, European claims on the area were by Netherlands, for the west half, in 1875 (and earlier) and by Germany for the NE quarter (also for the Bismarks and Bougainville I. of the Solomons) and Great Britain for the SE quarter, at the same time. Netherlands New Guinea (Dutch New Guinea) became the Indonesian Province of Irian Barat, now Irian Jaya, in 1963 (with an interlude during World War II when it came under Japanese control). German New Guinea (Deutsch Neu Guinea or Kaiser Wilhelmsland) was taken by Australian forces in 1914, became a League of Nations Mandate to Australia till WW2, was temporarily occupied by Japan, and then became a UN Trust Territory assigned to Australia. British New Guinea was turned over to Australia in 1901 and continued to be a colony. Only parts of it, mostly north coastal and insular, were occupied by Japan in 1942–44. From 1946 Australia governed NE New Guinea and Papua as a single territory. In 1975 this combined area, all of the eastern half of New Guinea plus Bismarcks and Bougainville I., became the independent nation of Papua New Guinea. This sequence is summarized in

Table 1 and partly shown by map in Fig. 1. Thus this volume treats the Indonesian province of Irian Jaya and the country of Papua New Guinea minus Bougainville and Buka which are geographically and biologically part of the Solomon Is.

New Guinea forms the western end of Melanesia. This is an area based on ethnology, but it nearly coincides biogeographically, at least for some groups of insects, with a tapering extension to eastward of the Papuan Subregion (Moluccas, New Guinea, Bismarcks and Solomons). The Melanesian area, which extends to New Hebrides, New Caledonia and Fiji, comprises the richer portions of terrestrial fauna of Pacific islands and the source areas or migration routes for fauna of the outer islands — for part of the Micronesian fauna and much of the Polynesian fauna. For land biota the New Hebrides is the poorest part of the Melanesian area, being especially low in endemics. This reflects the younger nature of the purely volcanic and coralline New Hebrides. Melanesia is occupied by the dark Melanesian peoples, though much of the highlands of New Guinea is inhabited by the shorter and stockier Papuan race. These, however, have mixed in many areas, and there are additional non-Melanesians, strictly speaking, to make a very complex situation (chapters II: 2, 3).

The most conspicuous feature of New Guinea is the great central spine of mountain ranges, extending for the length of the island, with few gaps below

Table 1. Diagram of periods of political administration.

	West	Northeast (+Bismarcks, Bougainville)	Southeast (w. isles)
1875	Netherlands New Guinea (Dutch New Guinea Nederlands Nieuw Guinea)	German New Guinea (Kaiser Wilhelmsland, Deutsch Neu Guinea)	British New Guinea
1901			Papua (Australian Territory)
1914		Australian occupation	
1919		League of Nations Mandate: Australia	
1941–45	(Japanese occupation) (U.S., Au., Dutch reoccupation)	(Japanese occupation) (U.S., Au., Dutch reoccupation)	(Japanese invasion of north)
1946		UN Trust Territory assigned to Aust.	
1948		Territory of Papua and New Guinea (also "Papua-New Guinea") Administered by Australia	
1963	Indonesia province of Irian Barat, now Irian Jaya		
1975		Independent nation of Papua New Guinea	

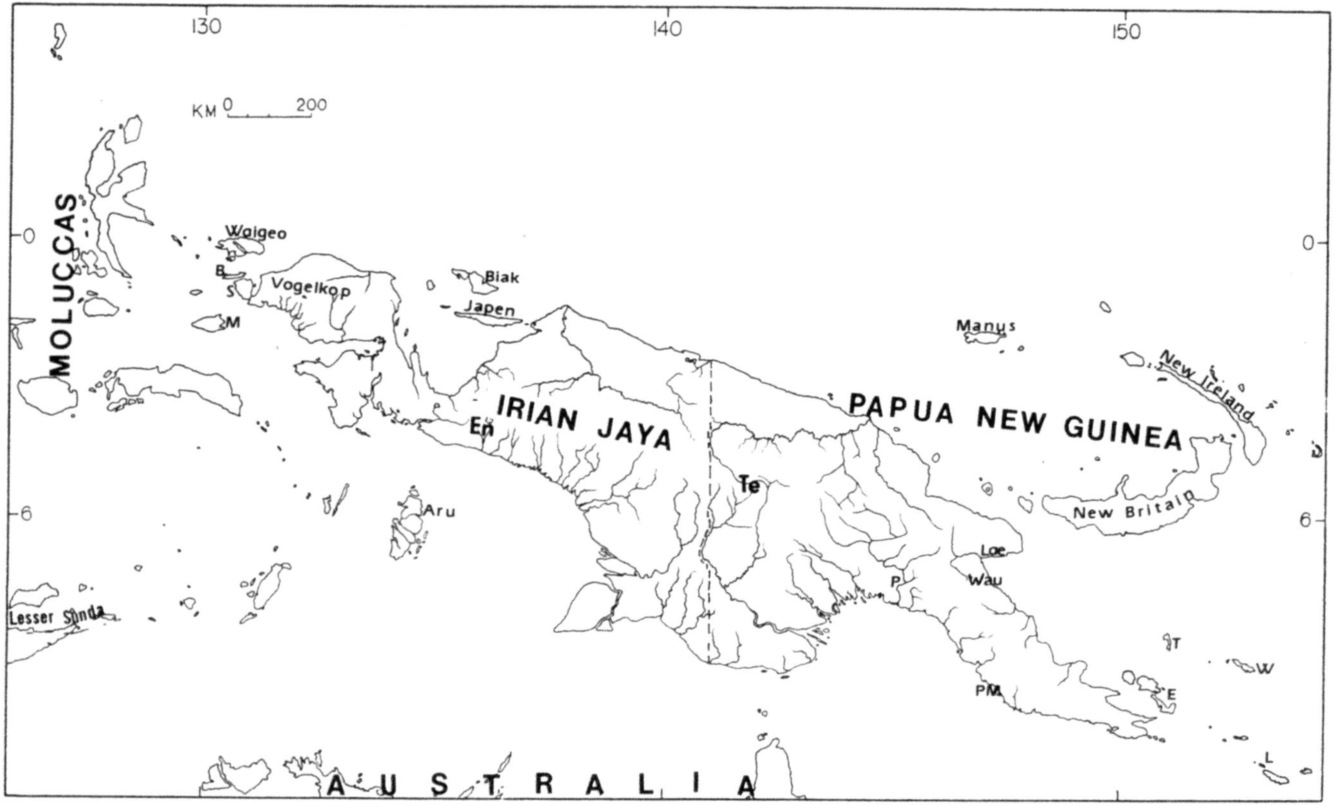

Fig. 1. Map of New Guinea area with islands labelled, from Waigeo to the Louisiades, which are essentially part of the New Guinea biotic area. B = Batanta I.; S = Salawati; M = Misoöl; T = Trobriands; W = Woodlark; E = d'Entrecasteaux; L = Louisiades. Bougainville I. of the Solomons, not shown (SE of New Ireland), is also part of the nation of PAPUA NEW GUINEA. Irian Jaya, Moluccas and Lesser Sundas are provinces of Indonesia. Also showing climatological sites (Fig. 2, 3): P = Wabo, lower Puraru R.; PM = Port Moresby; Lae; Wau; Te = Telefomin; En = Enarotali, Wissel Lakes (see text for latter).

2000 m for much of its length. This contrasts with low plains to the south in the wider portion and extensive low areas to north of the central cordillera. North of these, except at some river mouths, are the north coastal ranges, which in the west jump to Japen and Biak islands and then to the NE part of the Vogelkop (bird's head), and in the east terminate on the Huon Peninsula which points both to New Britain and to the D'Entrecasteaux and associated islands and then the Louisiades beyond the east end of the SE peninsula. As pointed out in various chapters in this volume, these various major ranges, portions and islands have had partly differing origins and histories, and some parts of the mainland were separate islands at certain periods. All this very complicated geomorphology and history has contributed to the great diversity and various aspects of the uniqueness of the biota and ecology of New Guinea.

The unique features of being the second largest island, highest island and having glaciers, have been pointed out above. The fact that the island is to a great extent covered by lush rainforest reflects its tropical location, benign climate with ample rainfall (see below) and the fact that human population density is low and for the most part rather recently emerged from the stone age era with its warfare, superstition, black magic, cannibalism or other practices, all of which, aided by disease, tend to keep population low. The lack of metal tools made elimination of forest difficult, except when dry enough to make extensive use of fire. The fact that most of the island lacks a true dry season again reduced potential destruction of forests by fire. The tendency for grassy areas to gradually be extended relates to the common custom of burning grass when it is dry, by ingrained habit or to watch slopes burn, if not to kill (and cook) rodents or marsupials living in the grassland, or less often to re-use the land for agriculture. The traditional custom of swidden or slash-burn agriculture, involving the cutting and burning of mature or submature forest for vegetable gardens and abandoning them when fertility is depleted after two years or so, was adequate under the earlier low population situation, but with current rapid population growth it menaces the forests, especially on slopes. As discussed in chapters II: 5 and VII: 1, Wau Ecology Institute is demonstrating a system called contour-mound composted agro-silviculture, to counter the negative effects of the old system and to conserve forest and biota as well as to reduce erosion and improve nutrition.

Studies on Mt Kaindi at Wau, NE New Guinea (Gressitt and Nadkarni, WEI Hdbk 5, 1978) have shown that only a modest rainfall for New Guinea (of little over 2200 mm per year) can maintain forest and allow natural regeneration of native forest, essentially because rainfall is somewhat evenly distributed through the year, there is almost daily fog in the mountains, and high humidity much of the time. Thus even frequent fires there, when grass dries in less-wet periods, impose rather slow extension of the grassy areas unless active tree-felling and agriculture is involved. It is this same highly humid and foggy regime which has permitted the development of luxuriant moss forest as well as the "epizoic symbiosis" mentioned above and discussed in chapter IV: 13. The studies on Mt Kaindi have also shown that the highly benign environment of adequate rainfall, lack of dry season, lack of extremes of temperature, general high humidity, frequent fogginess, rarity of strong winds, and the zonation with altitude, plus partial isolation with deep valleys and differing geological history and substrates, have combined to promote amazing diversity of forms in many groups, and local endemism. Certainly on different mountains or ranges with barriers such as low passes, deep ravines or wide valleys between them, much local speciation has taken place in the New Guinea mountains. This is strikingly so for some groups of insects, especially for some restricted to higher altitudes (see chapter IV: 13 and other chapters).

Conservation takes on an interesting light in New Guinea because in contrast to so many other countries in the world, much of the island is still covered with rainforest. Limited areas have been deforested and become grassland or agricultural land, and only a small fraction of the island has savanna woodland, in the south. Some portions of the island are under serious threat of extension of the grasslands by fire; erosion and landslides are a constant menace; and human population growth and timber export threaten much of the tall forests in the lowlands and on the less-steep slopes (see chapter VII: 1).

Acknowledgements

I am deeply indebted to all the authors of chapters in this book. Most of them have had rather difficult assignments considering the inadequacy of information, especially ecological. I am much indebted for helpful advice or reading of some chapters to A. Allison, D. I. Axelrod, F. R. Fosberg, D. G. Frodin, W. C. Gagné, G. H. Hope, F. G. Howarth, H. S. Howden, R. Johns, Z. Kaszab, H. Levi, E. Löffler, Barbara York Main, G. B. Monteith, P. H. Raven, M. H. Robinson, P. van Royen, J. M. B. Smith, J. J. H. Szent-Ivany, D. Walker and others, including staff-members of Bishop Museum, Bulolo Forestry College and University of Papua New Guinea. Assistance with illustrations has been given by Mark Thomas and Margaret Gressitt. For various other advice and support I am in debt to Professor J. Illies, editor of this series, and to the publishers.

ADDENDUM

Since the climate chapter contracted for this volume did not materialize, I have prepared the following brief discussion which has benefited by comments from Drs Roger G. Barry, Allen Allison and Francis G. Howarth.

The climate of New Guinea

New Guinea is a humid tropical island with moderate to very high rainfall. The environment is, in general, benign with moderate temperatures, high humidity, adequate rainfall through most or all of the year, and feeble wind much of the time in many areas. Only an interrupted coastal strip in the far southcentral portion and parts of the southeastern portion have a low rainfall, distinctly seasonal, of only about 1000–1300 mm per year. There are a few other limited rainfall areas. These seasonal or low rainfall areas are characterized by savanna vegetation of primarily Australian floral types, whereas the natural vegetation of most of New Guinea is rainforest. A characteristic feature of New Guinea climate is very high cloudiness. Solar radiation averages 400 langleys/day in the tropics with little seasonality in areas with even rainfall distribution (see Kalma 1972).

Wind and rain. New Guinea is close to the equator where winds from both hemispheres converge in the Inter-tropical Convergence Zone. General movement of air in this zone is westward. For most of the year the SE trade winds (or 'SE variables') come off the Coral Sea bringing rain to the coasts or mountains, or winds come from other directions, likewise from over sea, often bringing rain. New Guinea is affected by equatorial westerlies — the winds at 3000 m above Lae are westerly from December to March and easterly from May to October. Surface winds are usually weak and are not usually reliable indicators (Barry 1978). It is not fully understood why some areas in the west of the island have very heavy rainfall throughout the year while others nearby (like Waigeo I.) have low rainfall. The SE trades ('variables') partly originate from 25–30°S and receive occasional high latitude air, especially in mid-year. The temperature inversion rises as the equator is approached, reaching 4000–4500 m at Lae. The southeasterlies are deflected westward and may alternate with easterly currents of equatorial air (which are less moist) (Brookfield and Hart; Wittwer). Even distribution through the year rather than total rainfall is significant for plant growth.

Seasons. Temperature seasonality is minimal but the diurnal cycle is quite important. It usually rains on both sides of the main cordillera of New Guinea through much of the year. In general there is less rain from May to August or September in the eastern mountains (Fig. 2b) and especially on the S coast (Fig. 2d). In general, in the E, the very high rainfall areas (Fig. 2a, c; 3) have a very wet season June–September at the same time that many moderately high to moderate rainfall areas in the mountains (Fig. 2b) and the low rainfall areas (Fig. 2d) have their seasons of less rain (McAlpine et al. 1975). In the W the regime at Enarotali, Wissel Lakes, is not greatly different from that at Telefomin (Fig. 3), though rainfall is just over 2000 mm, with maximum in February, at Wamena (Balim Val.) which is between them (Barry 1980, after Tsuchiya). In the mid-year months the southeasterlies penetrate beyond the Bismarck Archipelago, while in January–February similar air masses may enter from near New Caledonia. From January to April tropical cyclones may develop, originating in a triangle between Tuvalu (Ellice Is.), Fiji and Vanuatu (New Hebrides), more rarely in the Solomons. These travel SW, passing over, or near, New Caledonia (av. 3 cyclones/year) and Vanuatu. Few of these cyclones reach New Guinea. In general the highest rainfall areas are a long area stretching westward from the Gulf of Papua (Fig. 2c) south of the main divide, and the S coasts of Huon Peninsula (Fig. 2a) and New Britain. Areas of double maxima of rain occur in some southern valleys, (Fig. 3), the Schouten Is. (Biak, Japen) and Admiralty Is. (Manus). (See rainfall map, Fig. 1, in chapter V: 3).

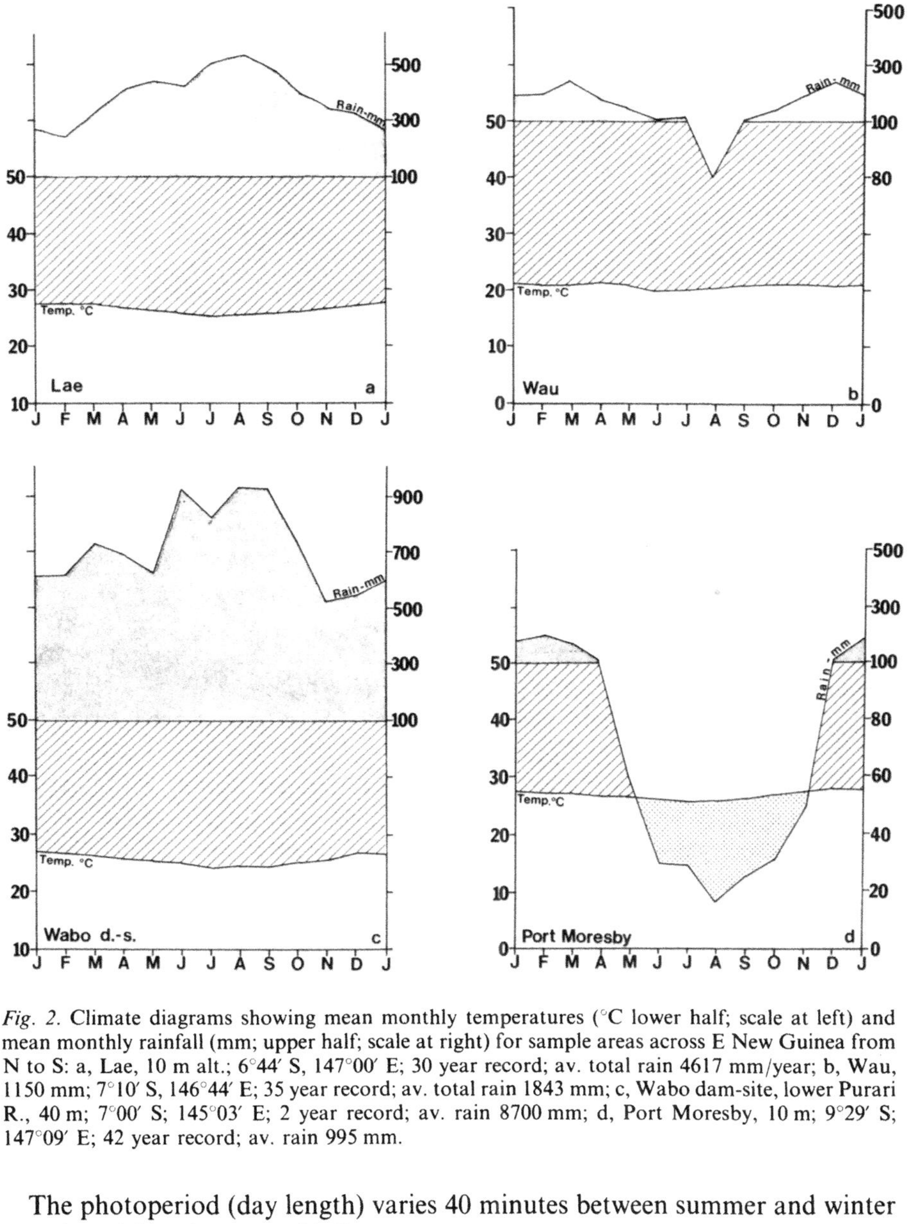

Fig. 2. Climate diagrams showing mean monthly temperatures (°C lower half; scale at left) and mean monthly rainfall (mm; upper half; scale at right) for sample areas across E New Guinea from N to S: a, Lae, 10 m alt.; 6°44′ S, 147°00′ E; 30 year record; av. total rain 4617 mm/year; b, Wau, 1150 mm; 7°10′ S, 146°44′ E; 35 year record; av. total rain 1843 mm; c, Wabo dam-site, lower Purari R., 40 m; 7°00′ S; 145°03′ E; 2 year record; av. rain 8700 mm; d, Port Moresby, 10 m; 9°29′ S; 147°09′ E; 42 year record; av. rain 995 mm.

The photoperiod (day length) varies 40 minutes between summer and winter at 5°S and 70 minutes at 10°S.

Topography and climate. Topography is of great importance because winds are stopped or deflected upwards by mountain ranges, where clouds drop most of their moisture. This is also strongly evident on small islands which may be wet on one side (usually SE) and dry on the other. The strip of very high rainfall area S of the main mountain ranges W of the lower Purari River (Fig. 2c) has been explained as 'air stagnation' caused by the very high range which is

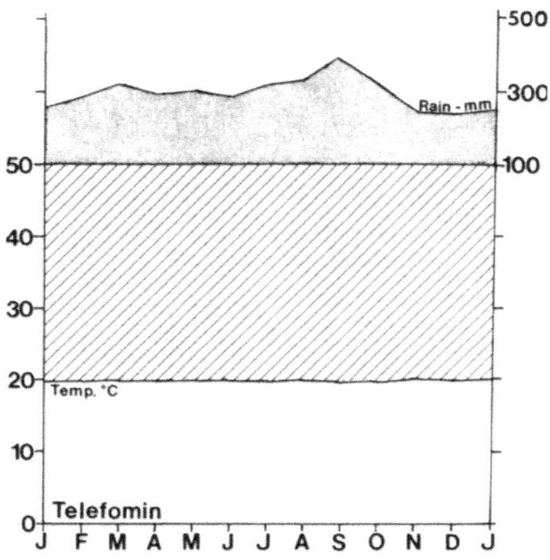

Fig. 3. Climate diagram for Telefomin, W. Sepik Province, Papua New Guinea, 1520 m alt; 5°09′ S, 141°38′ E; 16 year record; av. rain 3527 mm/year.

particularly steep on the S side. The fact that in general the SE sides of islands or ranges are wetter and NW sides often drier is evidence that there is no NW monsoon in the general area of New Guinea (Brookfield and Hart 1966).

Temperatures vary from about 35°C at sea level to below freezing at high altitudes. Average temperature declines 0.6°C with each 100 m increase in altitude. Occasionally there is frost at the highest altitudes where people normally live and cultivate crops (about 2500 m), and also more rarely, in dry periods, down to about 2000 m or lower (Brown and Powell 1974). Above 3600 m there may be freezing every night, and above 4600 m (W New Guinea) there are some glaciers. Snow falls at different times of year above 3800 m. There is little manifestation of annual temperature cycle, and local minima may relate to topography, such as with cold air descending slopes from high altitudes. The diurnal temperature cycle spans more than the annual cycle of monthly means at any given locality. On the mainland about 34% of land area is below 300 m, 40% is from 300 to 1500 m, 13% from 1500 to 3000 m and 1% above 3000 m. Topography also has a great bearing on microclimate.

Local patterns. Again to a great extent related to topography and wind directions, inland valleys often have reduced rainfall (rain shadows). Examples are inland areas of the Vogelkop, the Lake Sentani area, the lower Sepik plains, the Markham Valley, lower Asaro Valley and other small areas such as the middle of Wau, Bulolo, Watut and Bulowat valleys. Annual rainfall varies from under 1000 mm/year in the far S (Fig. 2d), to under 2000 in other drier areas (Fig. 2b) to 5000 near Lae (Fig. 2a), more on S coast of New Britain and 10,000, more or less, SE of the main central ranges (Fig. 2c). The diagrams (Figs. 2, 3) also show mean temperatures for the sample areas: Lae, Wau, lower Purari R., Port Moresby and Telefomin. As indicated above, highland rainfalls vary considerably, and data are very incomplete. In many highland valleys fog fills the bottoms of the valleys at night and gradually dissipates or rises to the sides or tops of the surrounding mountains in the mid-morning or earlier. Wind in these protected valleys is usually insignificant.

Humidity. This varies but is usually very high in the mountains and moderately high in most other areas, except in very dry periods, such as the severe drought in PNG highlands of 1971–72. In many montane areas humidity is at saturation all night every night of the year as well as often in foggy day-time periods. In these areas on sunny days it may drop to 50% or a bit less in early afternoons.

References

Allison, I. and J. Bennett. 1976. Climate and microclimate. In: Hope et al. 1976, The equatorial glaciers of New Guinea, 61–80.

Barry, R. G. 1978. Aspects of the precipitation characteristics of the New Guinea mountains. J. Trop. Geogr. 47: 13–30.

—— 1980. Mountain climates of New Guinea. In: van Royen, P., 1980. The alpine flora of New Guinea. vol. 1: 75–109.

Brookfield, H. C. and D. Hart. 1966. Rainfall in the tropical southwest Pacific. Australian National Univ. Res. Sch. Pac. Stud., G/3, 25 p.

Brown, M. and J. M. Powell. 1974. Frost and drought in the highlands of Papua New Guinea. J. Trop. Geogr. 38: 1–6.

Kalma, J. D. 1972. Solar radiation over New Guinea and adjacent islands. Austr. Met. Mag. 20: 116–27.

McAlpine, J. R., G. Keig and K. Short. 1975. Climatic tables for Papua New Guinea. C.S.I.R.O., Div. Land Use Res., Tech. Pap. 37, 177 p.

Tsuchiya, I. 1971. Fluctuations of rainfall in southeast Asia — Pacific and low and middle latitude circulations in the Southern Hemisphere. In: M. M. Yoshino, 1971, Water balance of Monsoon Asia, 217–39.

Wittwer, E. L. 1973. Seasonal changes in wind regimes in northern Australia and adjoining intertropical area. Austr. Geogr. 12: 340–62.

2. Geology of New Guinea

P. E. Pieters

1. Introduction

New Guinea Island is politically divided along 141°E long. into Papua New Guinea (PNG) to the east and the Indonesian province of Irian Jaya to the west. The western part of the island is one of the few remaining land areas in the world where topographical and geological maps are marked by blank spots.

Although the earliest geological exploration dates back to the mid 1800's, only in the 1960's was a start made with the systematic geological mapping of PNG by the Bureau of Mineral Resources of Australia (BMR) and the Geological Survey of Papua New Guinea (GSPNG) when airphotos and helicopter transport became easily available. For the same reasons, the systematic mapping of Irian Jaya was initiated as late as 1976 by a geological and geophysical team of the Geological Survey of Indonesia (GSI) and BMR.

Reconnaissance mapping supplemented by onshore and offshore geophysical work is now completed in PNG and resulted in the issue of a series of 1:250,000 maps and accompanying explanatory notes, and map compilations on 1:1,000,000 (Bain *et al.* 1972) and on 1:2,500,000 scale (d'Addario *et al.* 1976). These maps and notes supplied the basic information for numerous detailed and regional geological publications. The first 1:250,000 geological maps of Irian Jaya are still in preparation for publication.

In pioneer days geological information of New Guinea was collected by expeditions and surveys instigated by the Dutch, German and Australian governments and also by private organisations (for example the Archbold Expeditions).

Until the 1960's petroleum exploration in Irian Jaya was controlled by the Nederlandsche Nieuw Guinee Petroleum Maatschappij N.V. (NNGPM) and in PNG by the Australasian Petroleum Co. (APC), but subsequently other companies — in Irian Jaya always in joint venture with the state controlled Pertamina — got a foothold in the area and also started offshore exploration. In this work I have drawn heavily from the 1:250,000 maps and explanatory notes covering PNG and from publications dealing with the regional geology of New Guinea. The most important geological compilations of PNG were prepared by Australasian Petroleum Co. (1961), Thompson and Fisher (1965), Dow *et al.* (1972), Bain *et al.* (1975). Brown *et al.* (1975), Smith and Davies (1976) and

Monographiae Biologicae, Vol. 42, ed. by J. L. Gressitt.
© 1982, Dr W. Junk Publishers, The Hague.

Dow (1977). The only up-to-date compilation of the geology of Irian Jaya was carried out by Visser and Hermes (1962) but left untouched large areas in the central range and its northern flank where geological information is sparse.

2. Physiography

New Guinea is the second largest island in the world; it is 800,000 km^2 in area, 2400 km long, up to 720 km wide and the highest mountain (Mt Jaya) reaches to almost 5000 m (Fig. 2). The island's bird-like shape (from W to E: Birdshead, "neck", "main body", and "tail") and its coincidental, almost unique habitat for the famous "birds-of-paradise" are both the result of tectonic movements which rocked the SE Asian and SW Pacific region during the Cenozoic.

A rugged mountain range from the "tail" to the "neck" forms the spine of the island (Fig. 1). Mountains over 3000 m are common and tower over extensive areas of lowlands and swamps. The young topography of the country is indicated by ungraded rivers, numerous waterfalls and rapids, stream-piracy, steep and deep valleys and sharp cliffs. Present-day tectonism is accompanied by earthquakes, active volcanism and land instability. The three highest peaks in Irian Jaya are capped by snow and ice; however the snowline is rapidly receding (Fig. 2). Glacial landforms, such as moraines and cirques, are widespread in the central mountain range and were formed during the Pleistocene.

The rugged terrain of the central range is at places interrupted by large, elongate intermontane valleys. In the wide middle part of the island the mountain range falls steeply into the southern lowlands and alluvial and coastal plains which form a composite delta. Northward the mountains give way to

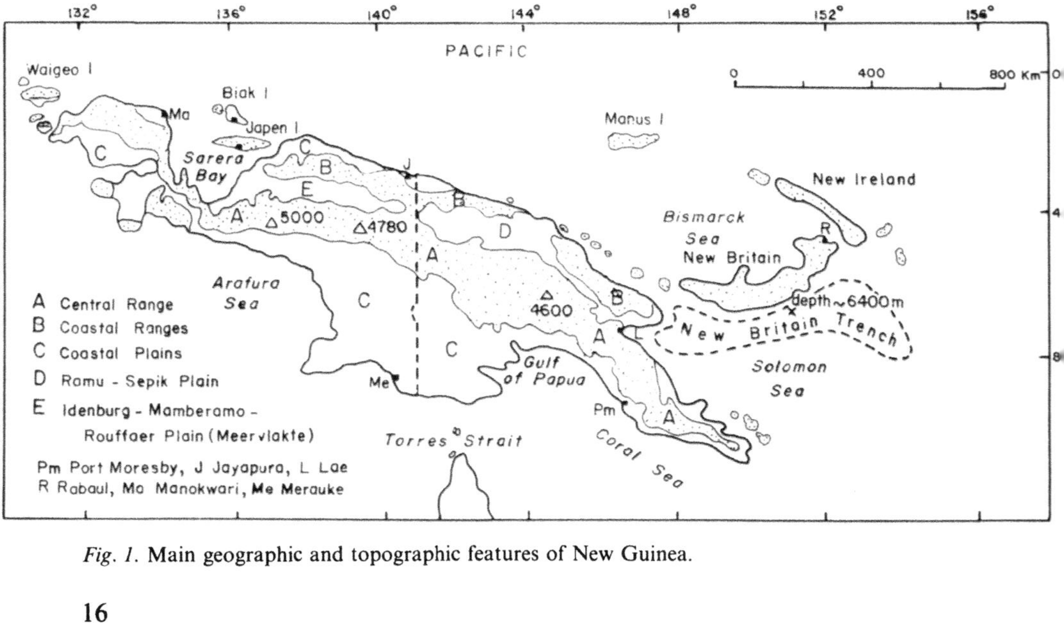

Fig. 1. Main geographic and topographic features of New Guinea.

wide, east-west trending plains bordering the Ramu, Sepik, Idenburg, Mamberamo and Rouffaer Rivers. These swampy, mosquito infested plains are flanked to the north by a discontinuous coastal range which gradually changes from rugged, steep mountains with peaks up to 4160 m in the E (Huon Peninsula) to hilly country in the W.

The "tail" (E Papua Peninsula) is mostly occupied by the central range with the foothills reaching the coast at many places. Young volcanic landforms occur at places along the north coast. A low mountainous fold belt with karst topography dominates the "neck" of the island. The mountain massif of the NE part of the Birdshead with peaks up to 3000 m tapers south and westward into narrow coastal ranges; these Birdshead highlands drop down to an extensive area of lowlands and alluvial and coastal plains locally intercepted by short belts of karst hills or low mountains.

The islands of the Bismarck Archipelago have a mostly rugged topography with mountain ranges rising directly from the sea or from narrow coastal plains. Quaternary volcanic landforms dominate N New Britain. Japen I. in Sarera Bay is also mountainous and its remarkable straight, east-west striking coastlines are fault controlled. Biak I. and Waigeo I. consist of low limestone plateaux and hilly or low mountainous country.

A deep sea trough (with depths to 6400 m) parallel to the S coast of New Britain and swings sharply to SSE following the trend of the Solomon Is. The Coral Sea, Bismarck Sea and Banda Sea are deep basins underlain by Tertiary oceanic crust and sediments; the Gulf of Papua, Torres Strait and Arafura Sea are shallow shelf seas. During Pleistocene ice ages the Torres Strait was largely dry forming a landbridge between New Guinea and Australia. Coral reefs are common throughout the region.

Fig. 2. View west over the highest part of the central range in Irian Jaya. Mt Jaya is flanked by short glaciers to the north. The anticlinal structure is developed on Miocene stable shelf limestone of the Platform.

3. Regional geological setting

Geologically New Guinea may be subdivided into three provinces (Fig. 3) which differ characteristically in stratigraphic, tectonic and magmatic history (Dow 1977).

An outer zone of *Ophiolite and Volcanic Arc* on the edge of the Pacific plate is separated from the *Platform* underlain by continental crust of the Australian plate by a medial up to 150-km-wide deformed *Mobile Belt*. The distribution and contrasting geology of the provinces is the result of interaction between the Australian and Pacific plates which probably dates back to the Middle Mesozoic and culminated into a major orogeny during the Early Tertiary when the two plates collided.

In the Eocene and Oligocene, following rifting of Australia from Antarctica, Australia rafted northward over a distance of about 30° lat. (see Chapter 00).

Compression between the Australian and Pacific plates may have been relieved by the formation of a northerly dipping subduction zone within the Pacific plate some distance off NE Australia. A volcanic arc developed over the hanging wall while oceanic crust and overlying deep marine sediments and volcanics subsided. Movement was immobilized by the arrival of buoyant crust of the Australian continent in the subduction zone. Further relief of stress resulted in folding and thrusting at the continental margin, low-angle overthrusting of segments of oceanic crust and volcanic arc onto the continent, and uplift and fracturing of the volcanic arc. Post-collision plate interaction continues over a wide zone covering the Mobile Belt and Ophiolite and Volcanic Arc; it includes rifting, transcurrent (transform) faulting, differential uplift, and the formation of a Neogene subduction zone S of New Britain and W of the Solomon Is. New Guinea N of the mobile belt forms the youngest accretion to the Australian continent.

The *Platform* is developed on Precambrium basement in the W and on basement of the Tasman Orogen in the E. In N Australia the boundary between these crustal blocks skirts the W coast of Cape York Peninsula and probably trends into New Guinea along 141° long. (Fig. 3).

Outcrops of Precambrian rock have not been found, but gently dipping probably Cambrian volcanics and sediments are exposed along the southern slope of the central range in E Irian Jaya and were intersected by petroleum exploration wells near Merauke (Visser and Hermes, 1962). On the other hand deformed metasediments and Carboniferous to Triassic granites are exposed at several localities in the C and W highlands of PNG and were penetrated by wells in the lowlands further S. The age of metamorphism is Carboniferous-Permian, but the age of deposition is unknown and could be much older. In the Birdshead occurs an isolated fragment of continental crust which consists of folded Silurian to Devonian metasediments and Permian-Triassic granites.

The basement rocks, including the suspected Cambrian deposits, are unconformably overlain by an almost continuous sequence of transgressive-regressive, open marine, marginal marine and minor continental sediments with an age range from Late Triassic to Holocene in PNG, and from Late Carboniferous to Holocene in Irian Jaya.

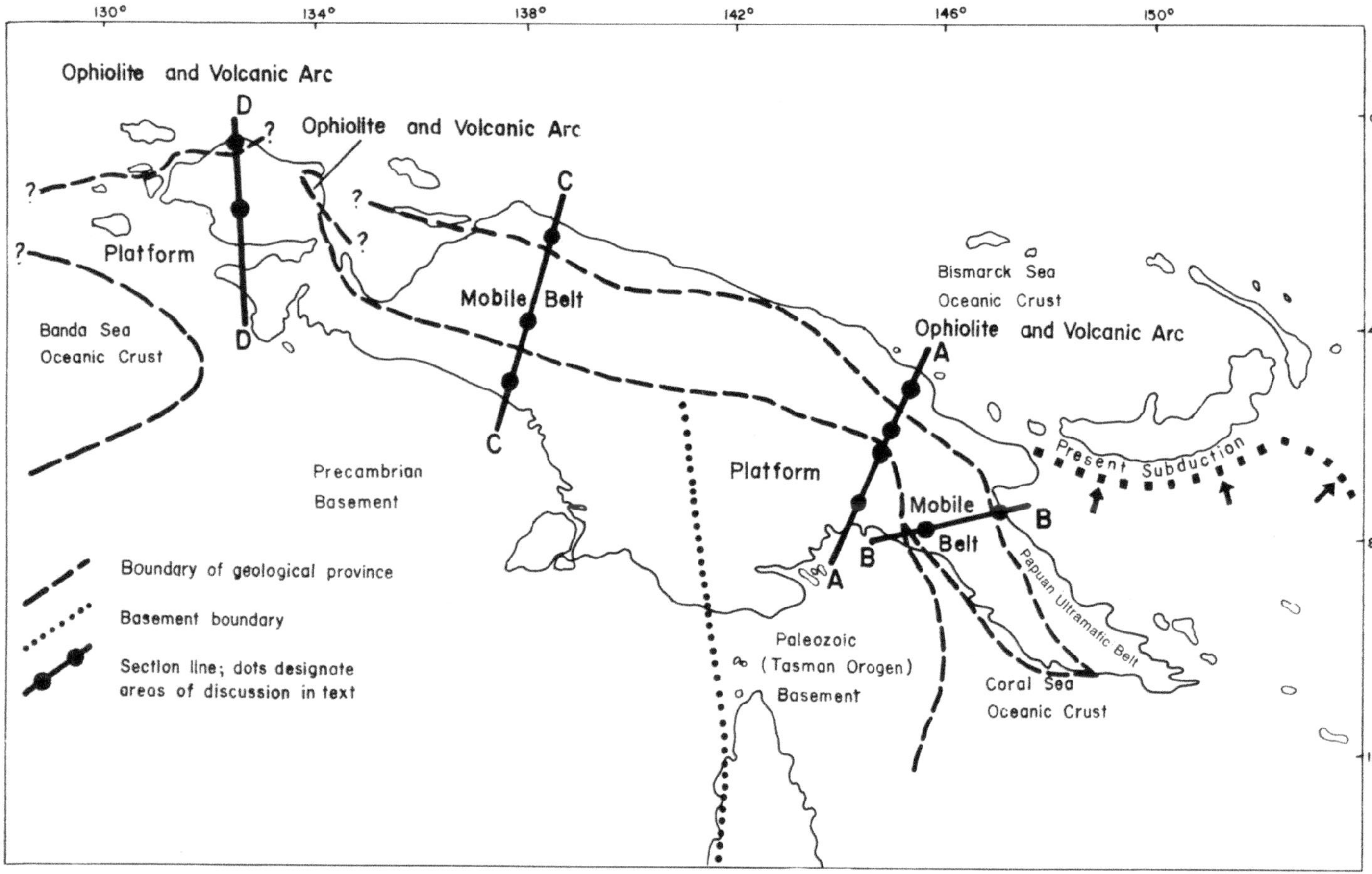

Fig. 3. Major geological provinces of New Guinea.

The Late Paleozoic and Mesozoic consist of shale, siltstone, sandstone, conglomerate, limestone and volcanics; the Tertiary rocks comprise a large thickness of shelf and basinal limestone blanketed by terrigenous sediments. The Platform cover is only locally deformed by block-faulting and gravitational folding and thrusting.

The *Mobile Belt* originated in the Mesozoic as a deep basin along the margin of the Australian continent (Fig. 3); it may be underlain by oceanic or continental crust. The rocks comprise a thick pile of metasediments, and minor shale, greywacke sandstone and siltstone, metavolcanics, altered volcanics, limestone and conglomerate.

Sedimentation continued until late Eocene, when the sedimentary-volcanic pile was folded and metamorphosed and subsequently uplifted during the major orogeny which lasted until the middle Miocene. Metamorphism ranges from lowest greenschist to amphibolite facies, but at several places occur lenses of metavolcanics and metasediments with mineral assemblages typical for the high pressure/low temperature blueschist facies. Fossils are rare and the oldest dated rocks are Cretaceous. The bulk of the sediments derived from the Australian continent, but the volcanic intercalations and the composition of some metamorphics indicate, at times, a nearby provenance of intermediate to basic igneous rocks, possibly associated with a volcanic arc.

Following emergence of the orogen, flysch and molasse-type deposits were deposited in adjacent troughs; erosion of the new landmasses and widespread volcanic activity provided the main source of sediment. Renewed uplift during the Pliocene and Pleistocene was accompanied by subaerial volcanism and gave rise to regression and deposition of fringing shallow marine and continental clastics. Clusters of volcanos, some active or dormant, others with spectacularly preserved original form, but most eroded to various degrees, mark the central range in PNG. Quaternary volcanic landforms are conspicuously missing in Irian Jaya and young small intrusions are the only evidence of this late magmatic phase.

The *Ophiolite and Volcanic Arc* comprise Mesozoic and Tertiary ultramafic rocks, gabbro and basalt representing upper mantle and oceanic crust, at places overlain by Late Mesozoic or Tertiary volcanics.

The largest ophiolite bodies are exposed in the northern part of the "tail" (Papuan Ultramafic Belt; Davies, 1971) and on Waigeo I., NW of the Birdshead. Smaller bodies occur as outliers in the Mobile Belt along the N slope of the central range, and also sporadically in the coastal ranges along, and on some islands off, the N coast. The Late Mesozoic rocks of the Papuan Ultramafic Belt are interpreted as a slice of Jurassic or Cretaceous oceanic (Pacific) crust and mantle which was thrust over the Mobile Belt during the Early Tertiary.

The "tail" and Birdshead are separated from the Australian continent respectively by the Coral Sea and Banda Sea which are underlain by oceanic crust. Rifting in the Coral Sea may have started in the Late Cretaceous and was followed by triangular spreading in the Paleocene which progressively moved NW-ward. There is some geological and geophysical evidence of Mid- and Late Tertiary rifting or spreading in the Woodlark, Solomon and Bismarck Seas.

The volcanic arcs are composed of a wide range of late Eocene to early Miocene intermediate to basic volcanics and volcanically derived sediments. They are exposed in the north coastal ranges and on the islands of the Bismarcks. The volcanics are overlain by Miocene and younger limestone and clastic sediments. Pliocene to present-day volcanic arcs occur along the N coast of New Britain and on small islands NW of New Ireland; they are absent in Irian Jaya.

4. Stratigraphy along representative sections

In order to spare the reader a jumble of formation and topographic names and to avoid sweeping correlations of rock units, the stratigraphy is described along 4 representative sections about perpendicular to the regional structural strike (Fig. 3 and 4). Section A is described in more detail than the others, as it intersects the 3 provinces where the geology is best known.

4.1 Section A (Central Papua New Guinea)

4.1.1 Inner Platform. In the Triassic and Early to Mid-Jurassic the Paleozoic basement rocks of the inner Platform formed a block-faulted landmass with local accumulations of continental sediments. During the Middle and Late Jurassic the sea advanced from the N and E, and thin deposits of marginal and continental facies conglomerate and sandstone were overlapped by marine fine clastics. A NNW trending tectonic downwarp which partially underlies the Section, was the center of a thick pile of shale with intercalations of sandstone reflecting minor regressions.

The Early Cretaceous started with a regional regression; uplift and basin shallowing were accompanied by deposition of quartz sandstone and conglomerate. In the Late Cretaceous fine clastics transgressed over the shallow marine sediments, but at the close of the Cretaceous tilting and warping caused uplift and erosion in the easternmost part of the Platform. The Jurassic and Cretaceous sediments were dated by pollen, microfossils and macrofossils including ammonites, belemnites and bivalves.

There is no record of widespread Paleocene and early Eocene sedimentation over the Platform. Late Paleocene to Eocene interbedded quartz-feldspar sandstone and mudstone rest locally unconformably over Mesozoic sediments and were deposited in restricted, shallow marine basins. Thin deposits of mid- and late Eocene shallow shelf detrital and micrite limestone and calcareous sandstone overlap the Mesozoic from E to W. Although the close of the Eocene was followed by a general retreat of the sea, limestone sedimentation continued locally until early Oligocene along the Platform margin.

A more widespread phase of limestone deposition began in late Oligocene and lasted until middle Miocene. Reefs were growing near the shoreline of a stable shelf sea and around upfaulted segments of the basement and supplied detritus for an extensive sheet of forereef bioclastic limestone. Farther offshore,

SECTION A (CENTRAL PAPUA NEW GUINEA)

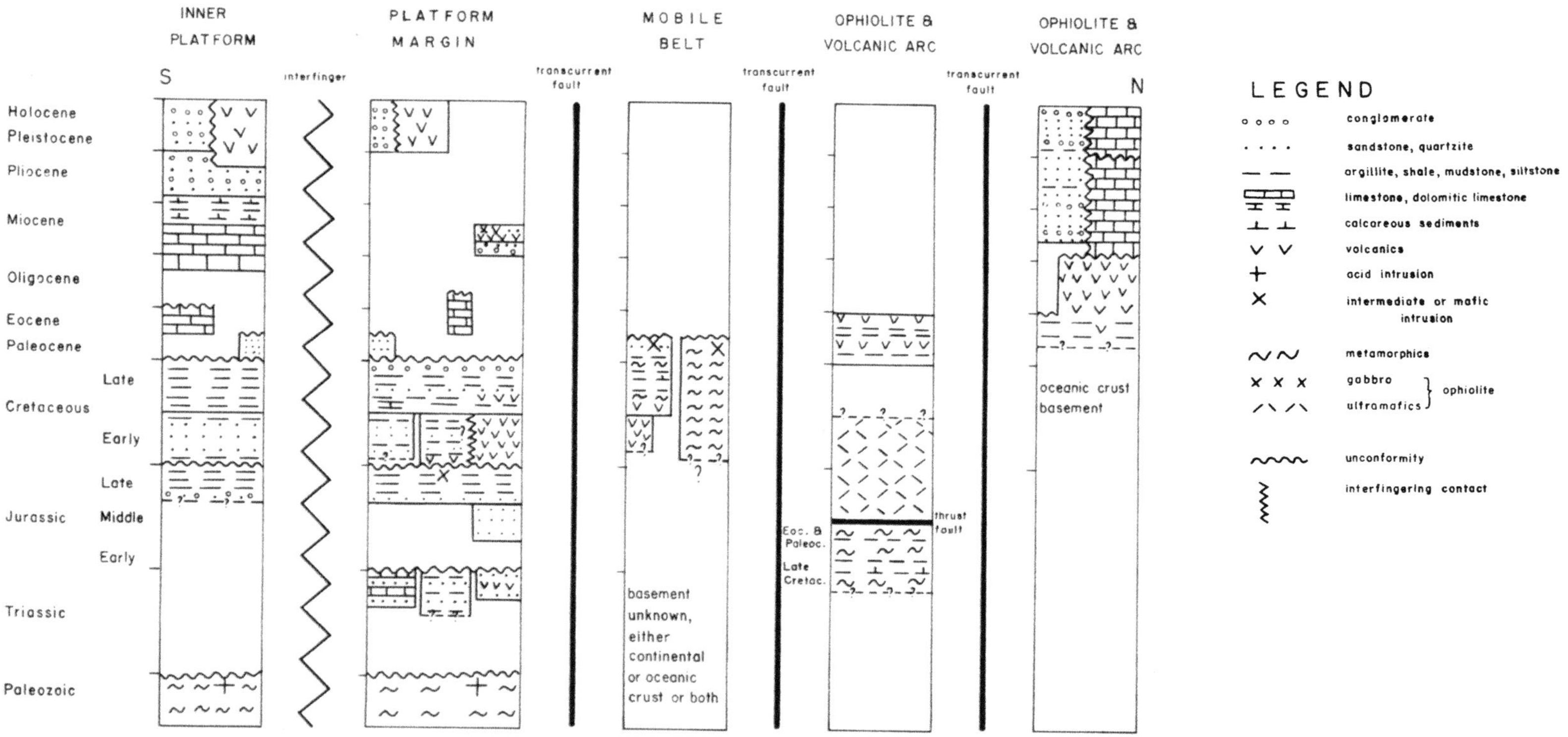

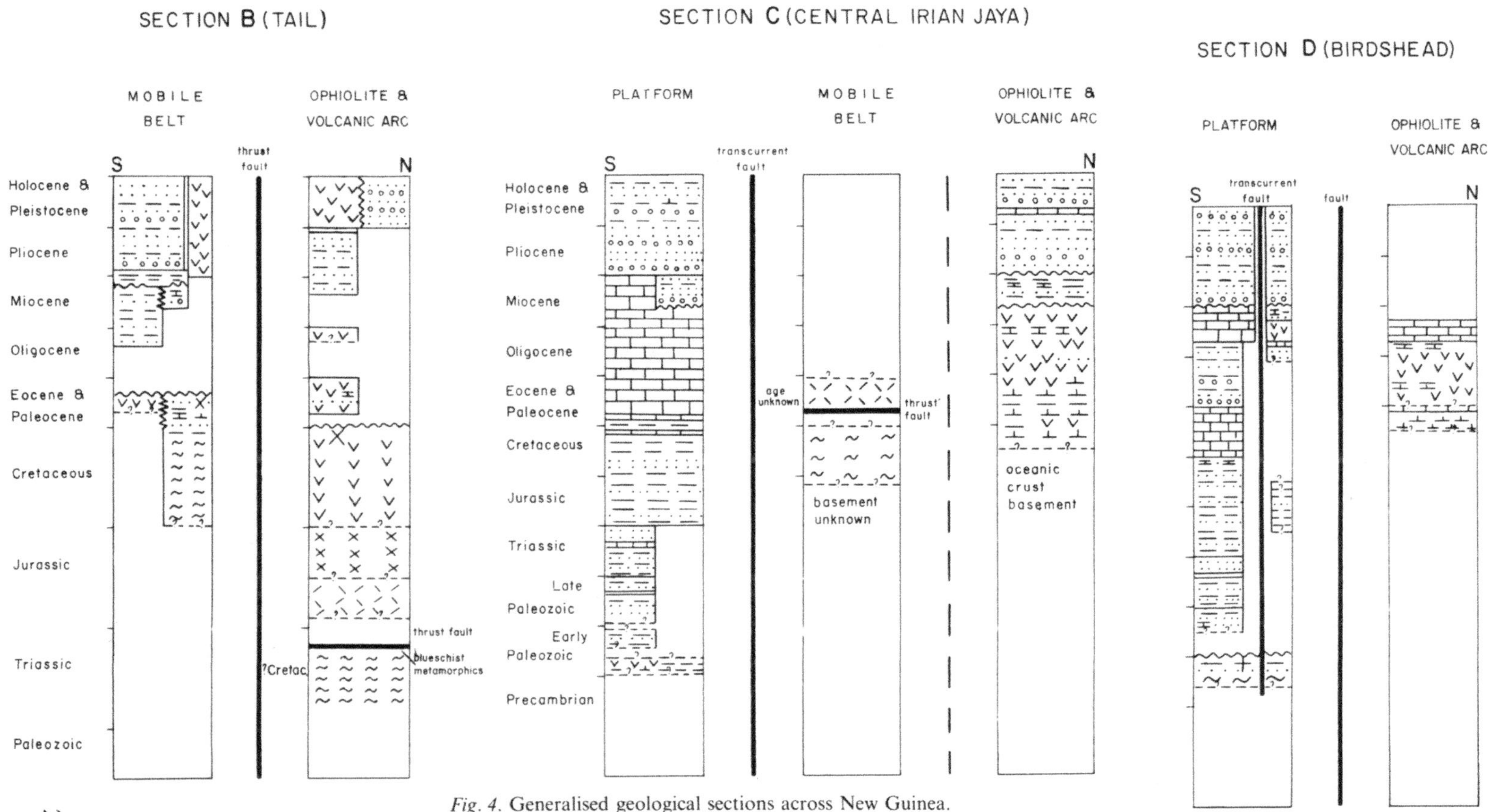

Fig. 4. Generalised geological sections across New Guinea.

fine-grained argillaceous basinal limestone was deposited. The shelf limestone is packed with benthonic and planktonic foraminifera, algae, corals, molluscs and echinoderms.

Widespread open marine deposition of mudstone, commonly calcareous, during the late Miocene was followed in the Pliocene and Quaternary by a pronounced regression as the result of uplift of the Mobile Belt, and accompanying volcanism. A sheet of coarse and fine-grained continental and deltaic sediments progressively spread out over the southern foreland, and lahar and terrestrial deposits were deposited around the lofty central-type stratovolcanos. The volcanic rocks include basalt (shoshonite) and andesite lava and pyroclastics. None of the volcanos of the central range is at present active.

4.1.2 Platform margin. The sedimentation record along the Platform margin dates back to Late Triassic. The basement is unconformably overlain by marginal marine and fluviatile reefal limestone, sandstone and arkose with a varied fauna of brachiopods, gastropods, ammonites, corals and foraminifera (Skwarko *et al.* 1976). Farther offshore, on a marine shelf and possibly on the continental rise, accumulated shale, greywacke, siltstone and feldspathic sandstone containing ammonites, nautiloids, pelecypods, gastropods, brachiopods and crinoidal stems. The base of these sediments is not exposed but they probably also overlie basement of the Australian continent.

At the close of the Triassic the continental margin was the site of volcanic activity and associated sedimentation. The mostly marine volcanics include altered andesite and some basalt, and the sediments consist of greywacke, coral limestone, calcarenite, and volcanically derived sandstone with rare bivalves.

The Late Triassic volcanic phase was followed by a period of uplift and erosion until the Mid-Jurassic, when shallow seas invaded parts of the Platform margin. The continental fluvio-deltaic and marginal marine sediments in these restricted basins are composed of greywacke, feldspathic sandstone and siltstone with ammonites, belemnites and bivalves; they are overlain by a thick sequence of Late Jurassic open marine carbonaceous shale and siltstone.

During the Early Cretaceous a thin regressive deposit of shallow marine reworked clean quartz sandstone, and mudstone is restricted to the area of the Section, but farther westward, and extending into Irian Jaya, deposition of Late Jurassic fine clastics continued until Early Cretaceous. A volcanic arc developed some distance off the Platform margin and supplied the detritus of a thick pile of sediments in the intervening basin. The volcanic arc was probably founded on oceanic crust; it is composed of altered pillow lava and lava breccia, basalt and andesite agglomerate and tuff, and volcanically derived sediments. The basinal sediments onlap the Platform margin and comprise greywacke, tuffaceous sandstone, shale and siltstone with ammonites, pelecypods, gastropods, belemnites, carbonized wood fragments and leaf impressions.

These sediments grade upward into Late Cretaceous shallow marine calcareous shale, sandstone and siltstone with minor calcarenite and conglomerate towards the top. Rare and local intercalations of altered volcanics and volcanically derived greywacke indicate that nearby volcanic activity continued until the Late Cretaceous.

Paleocene and early Eocene clastics and Eocene limestone are similar to those of the inner Platform, but limestone deposition locally continued until the early Oligocene. Orogenesis of the Mobile Belt culminated in an almost total retreat of the seas over the Platform during the Oligocene. In the late Oligocene, uplift of the orogen was accompanied by volcanic activity; the new emerging landmasses replaced the Australian continent as the main source area for detritus. Sedimentation at the Platform margin resumed in the early Miocene with shallow marine calcareous siltstone, shale, sandstone, limestone and conglomerate in a basin which is thought to be a western arm of a deep unstable Tertiary trough farther E (Aure Trough; see *Mobile Belt* below). The sequence contains abundant early to mid-Miocene foraminifera. Widespread volcanic activity along the Platform margin started in the mid-Miocene with andesite and basalt (shoshonite) agglomerate, tuff and subordinate lava interbedded with volcanically derived sediments and limestone. They form part of a long belt of middle Miocene volcanics and sediments which stretch out from the "tail" to 141°E, but, with the exception of the Birdshead, there is no record of similar volcanics in Irian Jaya. This phase of volcanism was accompanied by intrusion of basic to acid plutons.

A small number of Quaternary volcanic centers extend from the inner Platform onto the Platform margin. At places the volcanic rocks are interbedded with Pleistocene lake beds and alluvium in intermontane valleys.

4.1.3 Mobile Belt. The sediments of the Mobile Belt were deposited on the continental slope and rise marginal to the Australian continent. The rocks consist of shale, siltstone and greywacke with minor limestone and conglomerate and intercalations of marine, altered volcanics; lithologies and sedimentary structures suggest a turbidite sequence. Foraminifera range in age from Late Cretaceous to Eocene, but westward similar sediments contain Early Cretaceous ammonites. Locally the sequence overlies conformably Early Cretaceous volcanics (see *Ophiolite and Volcanic Arc* below). The sediments are deformed and metamorphosed to various degrees; on one hand they are schistose or phyllitic without much recrystallization, on the other hand they are metamorphosed to slate, phyllite, schist or gneiss containing minerals such as biotite and andalusite. The age of regional deformation and metamorphism was probably late Eocene to early Miocene. The Miocene sediments and volcanics covering the Mesozoic and Early Tertiary sequence of the Platform margin probably extended over parts of the orogenic belt but were subsequently eroded.

4.1.4 Ophiolite and volcanic arc. An ophiolite complex of gabbroic and ultramafic rocks, and basalt lava with pelagic sediments crops out along the N slope of the central range; the fault-bounded body forms a low-angle thrust sheet (Jaques and Robinson 1977).

At the base of the thrust sheet is harzburgite with some dunite overlain by cumulus peridotite, pyroxenite and layered norite, gabbro and anorthosite gabbro. The rocks are serpentinized to various degrees. Pelagic sediments including radiolarites of probable Eocene age overlie and are intercalated with pillow basalt and pass upwards into argillite, greywacke and tuff. The ophiolite complex is thrusted over Late Cretaceous and Eocene metasediments which are exposed in windows. Emplacement of the complex postdates early Eocene and was before mid-Miocene. The age of the plutonics is unknown but they represent a segment of oceanic crust and mantle of the Pacific plate.

The coastal range is separated from the central range by a fault controlled basin which is filled with a thick sequence of Oligocene to Holocene marine and continental clastics and minor limestone, and volcanics near the base. Basement of Jurassic gabbro was penetrated by a petroleum exploration well. The coastal range is developed on a volcanic arc which is underlain by pelagic sediments and covered by limestone. The pelagic sediments include Eocene siliceous argillite which, towards the top, becomes tuffaceous before it grades into volcanics. The base of the sediments is not exposed, but they are probably deposited on oceanic crust. Widespread volcanic activity burst forth in early Oligocene and lasted until early Miocene. An island arc was formed on a 4500 m thick pile of basalt and andesite flow breccia, volcanically derived sediments, tuff, agglomerate, lava, pillow lava, argillite and limestone lenses with foraminifera. In early Miocene, at waning stages of volcanism, limestone encroached onto the volcanic edifice, and was deposited on faulted step blocks until the mid-Miocene. Limestone sedimentation continued over the northern arc flank until the Holocene, but was locally interrupted by deposition of coarse clastics in the Pleistocene. The interfingering forereef and basinal facies limestones contain abundant fossils such as benthonic and planktonic foraminifera, algae, corals, bryozoans, bivalves and gastropods. Raised coral reefs fringe the coast E of the Section indicating rapid present-day uplift. A thick clastic wedge flanks the arc to the S. Greywacke, lithic arenite, conglomerate, siltstone and mudstone, and minor limestone overlie unconformably the Paleogene argillite and volcanics from which they are derived. The clastic wedge is covered by Quaternary sediments of similar derivation and composition.

The volcanic islands off the N coast form part of a Quaternary volcanic chain which extends eastward towards N New Britain (Johnson 1972). The dissected stratovolcanos rest on basalt and andesite pyroclastics and minor lava, and probably overlie Miocene or Pliocene offshore clastic sediments.

4.2 Section B ("tail")

4.2.1 Mobile Belt. The Section crosses an arcuate belt of metamorphic and deformed sedimentary rocks which occupies the larger part of the central range of the "tail". During the Cretaceous and up to mid-Eocene a trough, marginal to the Australian continent, was the repository of a thick pile of sediments and minor volcanics which were deformed and metamorphosed between late Eocene and early Miocene. Folded argillite, shale, siltstone, lithic and feldspathic sandstone, greywacke, fine-grained limestone and conglomerate and intercalated spilitic volcanics grade eastward into slate, phyllite, schist, metasandstone, metaconglomerate and metavolcanics. Fossils are poorly preserved Late Cretaceous and Eocene foraminifera, and Cretaceous bivalves and ammonites. Near the eastern fault boundary the felsic metamorphics interfinger with or grade into metamorphics of a more mafic composition, and both rock types have local mineral assemblages which are typical for the high pressure/low temperature blue-schist facies (assumed to be related to metamorphism in a subduction zone).

Following orogenesis of the Mobile Belt, flysch and molasse basins developed adjacent to the new landmasses from late Oligocene onwards.

A prominent basin, the northerly trending Aure Trough (Brown *et al.* 1975), in the S, formed over a zone of crustal thinning (St John 1970; Mutter 1975) which may have resulted from the opening of the Coral Sea. The Aure Trough is an asymmetric downwarp with more rapid sedimentation towards the steeper eastern flank. Eastward the trough sediments grade abruptly into unstable shelf, whereas to the W they are transitional into (calcareous) mudstone of the Platform margin. The predominantly turbiditic sequence of the Aure Trough has an apparent thickness of 9000 m made up of lithic greywacke with increasing mudstone and fine-grained limestone towards the W. The sediment is largely volcanically derived with variable amounts of metamorphic and sedimentary clasts from the orogen. Planktonic foraminifera are restricted to the upper fine-grained calcareous laminae of the graded beds. Late Miocene and early Pliocene open marine mudstone overlies locally unconformably the turbidites, and the mudstone in turn is conformably covered by Pliocene to Recent shallow water marine, deltaic and fluviatile sediments.

In the Miocene, Pliocene and Quaternary intermittent uplift of the orogen and adjacent unstable shelf was accompanied by submarine and subaerial volcanism and the emplacement of related intrusives.

4.3.2 Ophiolite and volcanic arc. Ophiolites are exposed in a 400 km long by 40 km wide belt, the Papuan ultramafic belt (Davies 1971) which is faulted against the Mobile Belt (Fig. 5). The ophiolite body consists of 4–8 km thick ultramafic rocks overlain by 4 km thick gabbroic rocks which are overlain by 4–6 km thick basalt volcanics. Most of the ultramafic rocks are harzburgite, dunite and enstatite pyroxenite with allotriomorphic metamorphic textures. Near the contact with the gabbro is a discontinuous layer of ultramafic rocks with cumulus textures, which contains olivine, orthopyroxene, chromite and clinopyroxene. The gabbro layer consists of gabbro and norite, some with layering and cumulus textures. The cumulus gabbroic rocks grade downward into the cumulus ultramafic rocks. Elsewhere the contact between the two layers is intrusive, with gabbro into peridotite. The basalt layer consists of massive basalt and submarine basalt lava and pillow lava. The contact between the gabbro and basalt layers is in places transitional through a dolerite phase.

The age of crystallization of the gabbro and basalt portions of the ophiolite body is Jurassic or Cretaceous. The ophiolite belt is intruded by Eocene tonalite and overlain by consanguineous andesite basalt and dacite lava, pyroclastics and limestone which form the vestiges of a volcanic arc. Late Oligocene lava and tuff are exposed SE of the Section not far from the ophiolite body, and in the same area is a thick sequence of mid-Miocene to Pliocene clastics composed of predominantly basalt detritus.

Pliocene and Quaternary volcanics and volcanically derived sediments overlie

Fig. 5. View SE over central range in "tail" end of Papua New Guinea. The fault-line valleys separate ophiolites to the N from metavolcanics and metasediments to the S. The metamorphic rocks in center and right foreground show northeasterly dip-slopes.

the ophiolite belt. The volcanics range from basalt to rhyolite and are probably related to Late Tertiary and Quaternary uplift of the ophiolite belt.

4.3 Section C (Central Irian Jaya)

4.3.1 Platform. The oldest Platform rocks, tentatively assigned to the Cambrian, are mapped locally along the S slope of the central range. Their base is not exposed, and they are unconformably overlain by Jurassic or Cretaceous basal conglomerate. The rocks show some resemblance to Cambrian in N Australia (Visser and Hermes 1962) which overlies a Precambrian sedimentary series. The rocks include silicified basalt, volcanic breccia and pyritic slaty mudstone and dolomite with thin intercalations of chert. The sediments are intruded by dykes of quartz diabase.

The S side of the central range is mainly developed on a thick sequence of sediments with ages ranging from possibly Silurian or Devonian to Permian. Early Paleozoic brachiopods and cephalopods were only found in river pebbles, but may have derived from very hard slate, shale, quartzite and dense limestone which occur in fault contact with Carboniferous and Permian sediments.

The Carboniferous and Permian sediments consist of shale, sandstone or quartzite, calcareous sandstone, limestone and conglomerate with a varied fauna of brachiopods, corals, fusulinids, bryozoans, crinoid stems, gastropods, pelecypods and cephalopods, and also with plant remains. The depositional environment graded from stable shelf marine to paralic and continental, fringing a low-relief southerly hinterland.

Sedimentation continued uninterrupted into the Triassic with red and mottled fine to coarse-grained clastics and dolomitic limestone. The different character of the sediments may be explained by a change of climate from warm and humid to dry.

The Jurassic, Cretaceous and Paleocene interval began with a transgression, and shallow and open marine conditions prevailed during most of this time. Shale, glauconitic sandstone and aphanitic limestone were deposited near the edge of the Australian landmass and passed into shale farther offshore. The sequence is up to 2000 m thick and contains a rich fauna of ammonites, belemnites, bivalves, pelagic foraminifera and minor benthonic foraminifera.

Towards the close of the Mesozoic a regional regression was accompanied by a diminished supply of clastics from the southern hinterland. The Late Mesozoic and Paleocene beds grade upwards into a thick series of stable shelf, reef and forereef, and minor basinal limestone. Limestone deposition continued uninterrupted until late Miocene but fluctuations of sealevel are reflected by changes in limestone lithology and faunal contents. The limestone is widely distributed in the central range and occupies most of the higher peaks. The up to 1600 m thick unit consists of coarse and fine-grained detrital, aphanitic and biolithic limestone with abundant bryozoans, algae, corals, benthonic foraminifera, echinoid fragments and spines, and bivalves.

During the mid-Miocene parts of the Mobile Belt and possibly also the Platform margin became emergent and formed the source area of coarse and fine-grained clastics which accumulated in adjacent basins. The basal beds of these deposits commonly overlie the Mesozoic clastics and Tertiary limestone with an erosive contact. Renewed regional uplift during the Pliocene and Quaternary was followed by rapid sedimentation of coarse clastics along the flanks of the new landmasses. The uplifted coalescing alluvial fans along the steep S side of the central range were deposited during the Pleistocene.

4.3.2 Mobile Belt. The N part of the central range, where the Mobile Belt crops out, is one of the least explored areas of New Guinea. Scattered observations indicate the occurrence of an elongate belt of metamorphics in fault contact with ophiolite bodies. The metamorphics probably range in age from Jurassic to Cretaceous; they include metasediments such as slate, phyllite and quartz-mica schist, and mafic rock types with amphibole, chlorite, epidote and feldspar. Some mafic metamorphics contain glaucophane and lawsonite typical for high pressure/low temperature metamorphism (blueschist facies) in a subduction zone. These rocks, and also closely associated eclogites, are quarried by the local highland Papuans as material for high-quality adzes. The age of deformation and metamorphism is probably the same as in PNG, late Eocene to early Miocene.

4.3.3 Ophiolite and volcanic arc. During Late Cretaceous, Paleocene and Eocene pelagic limestone, calcareous argillite, radiolarite and submarine altered andesite and basalt volcanics were deposited in a deep marine environment. These rocks are closely associated with serpentinite, ultramafics and diorite or gabbro, and, therefore, probably overlie oceanic crust. Upwards in the sequence the character of the sediments changes to unstable shelf, shallow-water limestone with corals, algae and foraminifera, and volcanically derived clastics which form intercalations in a thick pile of andesite and basalt pyroclastics and lava. This suite of rocks is typical for a volcanic island arc which probably developed some distance away from the Australian continent in late Eocene to early Miocene. The waning stages of volcanism in early Miocene were followed by a large influx of clastics and turbidite sedimentation in deep basins. The lithologies include greywacke, shale, siltstone and argillaceous limestone, which show graded bedding and contain rich faunas of pelagic foraminifera. The detritus of these mid-Miocene sediments was mostly derived from the Mobile Belt, which was then emergent. During late Miocene, Pliocene and Quaternary renewed uplift of the Mobile Belt was accompanied by rapid deposition of up to 4000 m thick clastics, and shoal and basinal limestone.

4.4 Section D (Birdshead)

4.4.1 Platform. A distinct unconformity marks the contact between deformed and regionally metamorphosed basement of Silurian to Devonian age, and an almost continuous sequence of little disturbed Late Paleozoic to Holocene sediments. The metasediments consist of shale, slate, metagreywacke, quartzite, schist, and minor metaconglomerate and metalimestone commonly with well-preserved graded bedding and cross-lamination. At a few places graptolites and ostracods occur in the shale. Higher grade metamorphics locally contain biotite, andalusite, sillimanite and garnet porphyroblasts which formed during a phase of thermal metamorphism accompanying the emplacement of granite batholiths and pegmatite veins in Late Permian to Early Triassic.

The late Paleozoic and Mesozoic stratigraphy of the central Birdshead and along the S flank of the central range show a remarkable resemblance, and the widespread uniform depositional environment and provenance along the Platform margin indicate a stable tectonic regime during the Late Paleozoic and Mesozoic.

Platform sedimentation continued in the Eocene with reef and forereef limestone which conformably overlies Mesozoic clastics. The shoal limestone grades laterally into pelagic limestone towards the "neck", and southward into a dolomite-clastic-evaporite sequence (backreef). The main reef-builders were algae and bryozoans. The Eocene limestone is overlain by Oligocene and early Miocene paralic sandstone, shale and conglomerate with plant imprints and coal films. This clastic interval is possibly related to uplift farther N where the Late Paleozoic, Mesozoic and Early Tertiary sedimentary cover is partly removed and mid-Miocene limestone rests on rocks as old as the Early Paleozoic basement. This mid-Miocene limestone is part of a very extensive mid- to late Miocene carbonate sheet which extends E along the central range. A long belt of coral pinnacle reefs supplied the detritus for the forereef and basinal limestone. At the same time fine-grained clastics derived from the N and fine carbonate detritus accumulated in local basins.

In late Miocene, Pliocene and Quaternary renewed emergence of the landmasses to the N and E was followed by rapid deposition of fluviatile, paralic, deltaic and shallow marine fine and coarse clastics in adjacent basins.

Farther N, and separated from the main part of the Birdshead by a major transcurrent fault, is a sliver of Early Paleozoic basement overlain unconformably by remnants of the Late Paleozoic, Mesozoic and Early Tertiary sedimentary cover. These sediments are overlain by a sheet of mid-Miocene andesite and dacite pyroclastics, lava, volcanically derived sediments and limestone. Deposition of calcareous clastics continued into late Miocene, and in the Pliocene and Pleistocene coarse and fine clastics were draped around uplifted landmasses.

4.4.2 Ophiolite and volcanic arc. Basalt and andesite lava breccia, pillow lava, pyroclastics, volcanically derived sediments, intercalated limestone, and diorite and gabbro intrusions built up an island arc from Eocene to early Miocene. The volcanics rest on Eocene argillaceous fine-grained limestone, and are overlain by early or mid-Miocene reefal limestone.

5. Structure

The fundamental differences between the stratigraphy of the Platform, Mobile Belt and Ophiolite and Volcanic Arc provinces are also shown by the structure. The province boundaries are the sites of deep seated crustal dislocations which controlled the geology of the region since at least as long ago as Late Mesozoic (Fig. 6).

5.1 Platform

The metamorphic and granitic basement underlying the Platform in PNG is part of the Tasman Orogen which became accreted to the Australian continent in the Carboniferous and Permian. The continental crust reacted to stress by block-faulting, differential uplift, tilting and broad warping, and the distribution of sedimentary basins and land areas was largely controlled by these tectonic processes. The Late Paleozoic, Mesozoic and Tertiary sedimentary cover is generally little disturbed, but Late Tertiary and Quaternary uplift of the central range involved the Mobile Belt as well as parts of the Platform margin.

North of the Gulf of Papua occurs a belt of parallel, subhorizontal, superficial asymmetrical folds which are commonly overturned and broken by northerly dipping thrust faults. The structures are probably the result of gravity sliding off the S flank of the emerging Platform margin in the Pliocene (Jenkins 1974).

The "neck" area is occupied by a belt if irregularly folded, and faulted limestone. The fold belt may have formed as a response to compressional forces, but there is also evidence of gravitational sliding.

5.2 Boundary between Platform and Mobile Belt

The Platform and Mobile Belt sediments are separated by a system of major strike-slip faults. The faults have steep dips and consist of zones of cataclasite

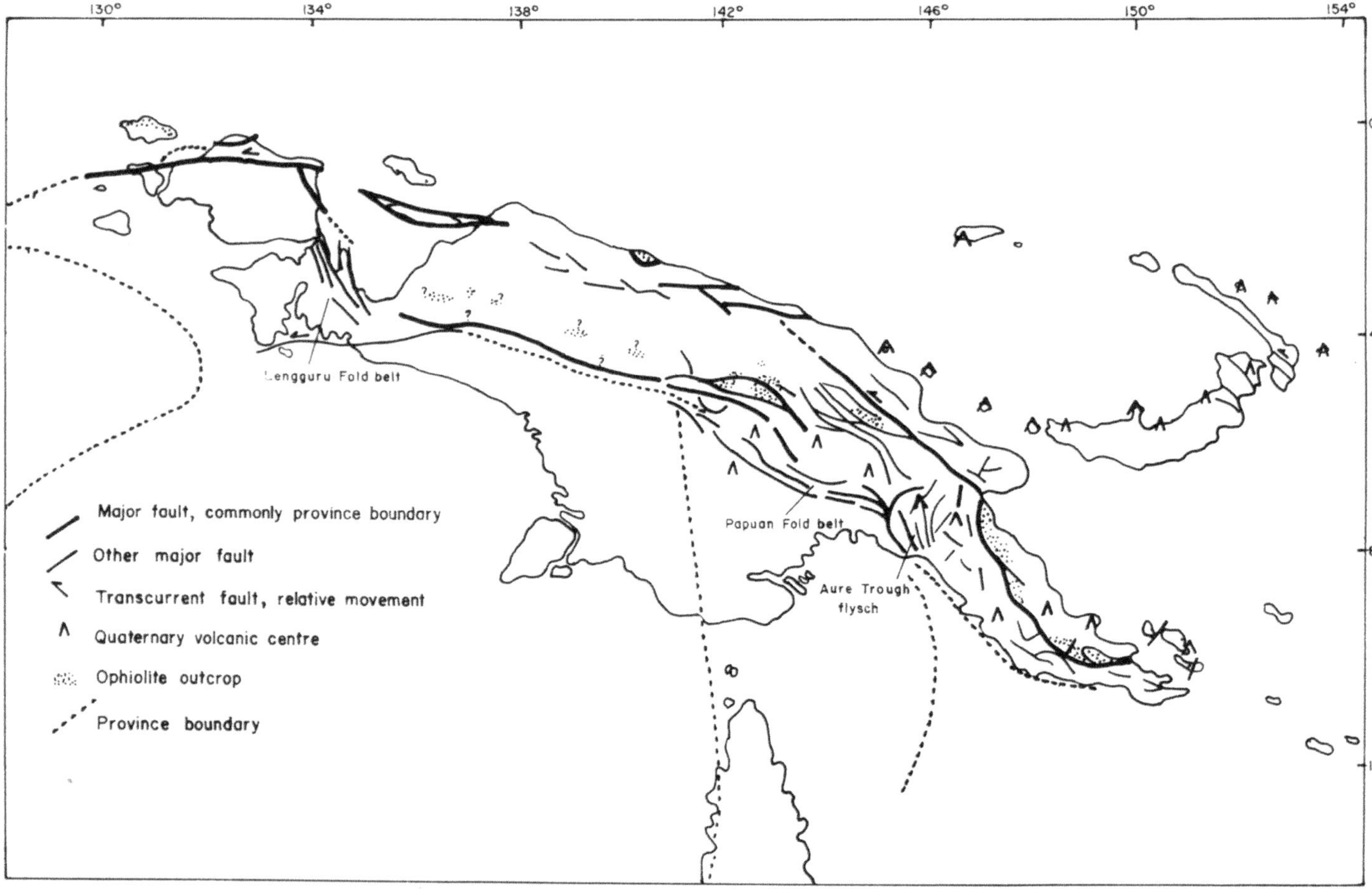

Fig. 6. Major structural elements of New Guinea.

and mylonite up to several hundred meters wide. In the widest part of the central range in PNG fragments of the Platform basement were sheared off from the Australian continent by transcurrent faulting and emplaced as fault wedges within the Mobile Belt. Most strike-slip faulting probably dates from early Miocene to Recent subsequent to orogenesis of the Mobile Belt.

5.3 Mobile Belt

The sediments and volcanics of the Mobile Belt were deformed and metamorphosed during a late Eocene to early Miocene orogeny when the Australian and Pacific plates collided along a northerly dipping subduction zone. Deformation produced large-scale folds in the sediments, and a well-developed pervasive foliation, parallel to the axial plane of small-scale isoclinal folds, in the metamorphics. The primary folding and foliation are commonly deformed by minute folds and locally associated crenulation or fracture cleavage.

From early Miocene to Quaternary intermittent uplift and transcurrent faulting chopped up the Mobile Belt into an intricate system of fault wedges. The major faults are near vertical and accompanied by wide zones of cataclasite and mylonite which may be traced over ten's of kilometers. In W PNG and C Irian Jaya the faults are associated with bodies of ultramafic rock.

5.4 Boundary between Mobile Belt and Ophiolite and Volcanic Arc

The boundary between these two provinces is at places defined by low-angle thrust faults which mark the site of collision between the Australian and Pacific plates. In the "tail" and NE Irian Jaya ophiolites are thrust over metamorphics but in other areas the thrust contact is masked by more recent normal or transcurrent faulting, or by downwarping into large elongate alluvial basins between the coastal and central ranges.

5.5 Ophiolite and Volcanic Arc

The rocks of this province are only mildly deformed and uplift was accompanied by block-faulting, tilting and broad folding. Although broken up by many faults, the gross layering of the ophiolite complex in the "tail" dips to the NE, and limestone sheets overlying the volcanic arc in the coastal ranges in PNG are moderately tilted to the N.

6. Economic geology

6.1 Petroleum

The discovery of oil seepages in 1911 in the Papuan Basin (Fig. 7) around the Gulf of Papua instigated an extensive search for petroleum in PNG which is still continuing. Minor exploration was carried out in the N New Guinea and Cape Vogel Basins (Fig. 7). As yet commercial oil or gas fields have not been detected in PNG.

Petroleum exploration in Irian Jaya started in 1935 and was concentrated on the W part of the Birdshead (Salawati Basin; Fig. 7). During the early work oil was drilled in a Miocene reef structure and more recently, with the application of detail seismic techniques, a belt of pinnacle reefs was delineated. At the moment several wells of this field are in production.

6.2 Coal

The several lignite deposits in New Guinea are of low-grade quality and prohibitively remote, and therefore not suitable for exploitation.

6.3 Copper, lead and zinc

New Guinea is well endowed with copper mineralization and several major porphyry-type deposits have been discovered. Despite the numerous difficulties of operating in remote and mountainous areas copper mining has started at Panguna on Bougainville (Solomon Is) and at Ertsberg in the W part of the central range in Irian Jaya (Fig. 7). Other major copper prospects occur in the W and middle part of the central range in PNG, on New Britain and on Manus I. (Fig. 7). A small quantity of copper has also been mined from sedimentary deposits near Port Moresby.

Only insignificant lead-zinc mineralization is known from New Guinea.

6.4 Gold

Gold (and alloyed silver) is widely distributed in PNG and has been mined since 1888; gold prospecting in Irian Jaya never reached the same level of activity as in PNG and only insignificant prospects have been reported.

Initially most gold was won from alluvial deposits and to a less degree from primary lodes. However, gold is commonly an important by-product of porphyry copper deposits and with the opening of the porphyry copper mines the production of gold has increased significantly.

6.5 Nickel, cobalt and chromium

Nickel, cobalt and chromium are associated with ultramafic rocks which are widely distributed on New Guinea. At many places these basemetals are concentrated by lateritic weathering. With the exception of Gag I. near Waigeo I. in W Irian Jaya, testing of several nickel laterite deposits failed to find any economic prospects. An international consortium of companies is investigating the possibility to open an open-cut mine on Gag I.

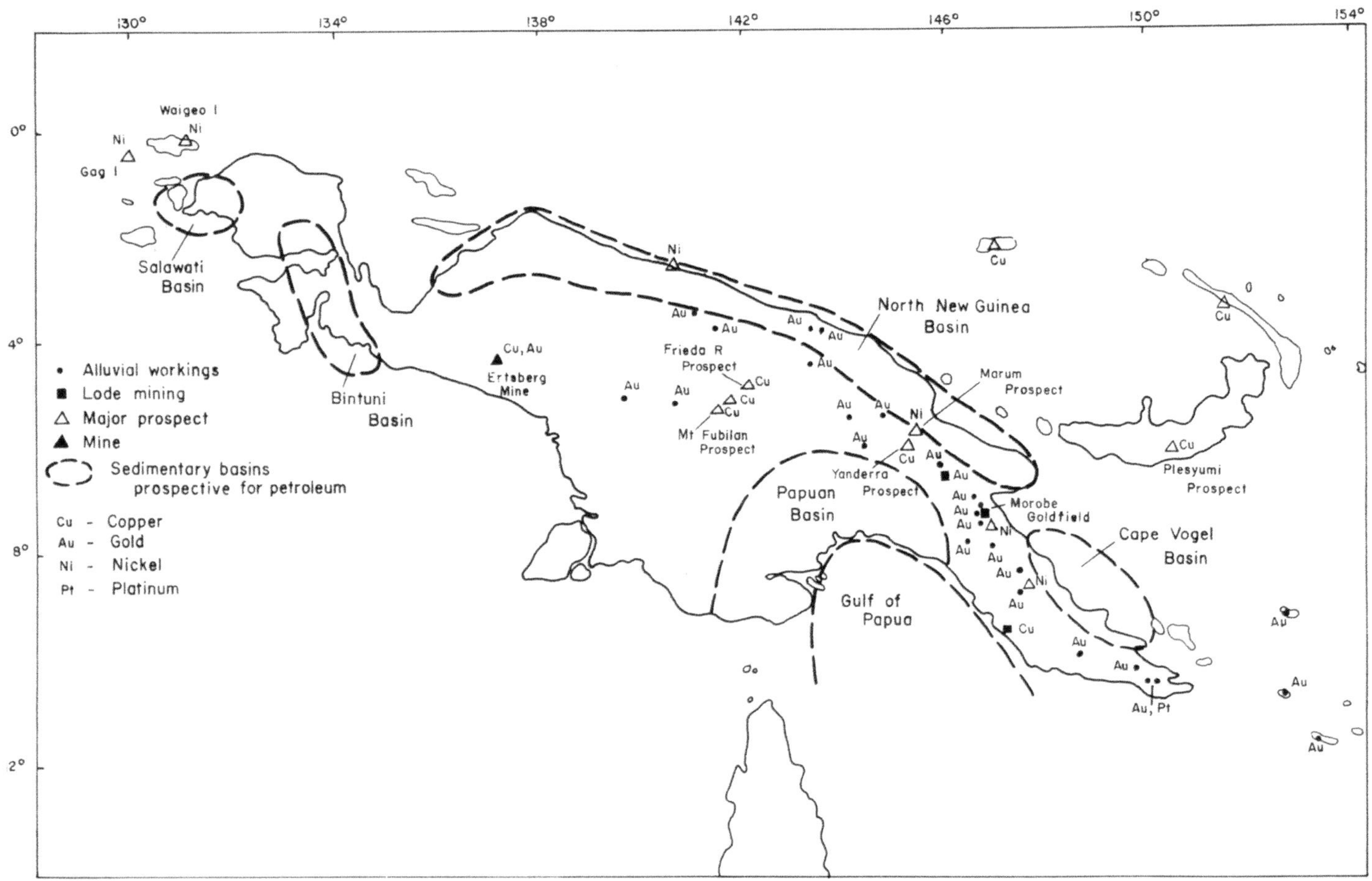

Fig. 7. Main mineral occurrences of New Guinea.

6.6 Platinum

In PNG alluvial platinum was recovered from most alluvial gold workings in areas where ultramafic rocks are exposed.

7. Summary

New Guinea straddles the boundary between the Australian and Pacific plates and the interaction between these 2 major crustal elements played an important role in the eventful geological history of the area.

The island is subdivided into 3 geological provinces which conform to the configuration of the Australian and Pacific plates and the intervening zone of interaction.

The *Platform*, the southernmost province, is developed on stable continental crust of Precambrian and Paleozoic age which is overlain by Paleozoic, Mesozoic and Cenozoic sediments and some volcanics. The sequence is flat-lying, or locally tilted and warped.

The Ophiolite and Volcanic Arc occupy the N part of the island where they include Mesozoic oceanic crust and mantle overlain by Eocene to early Miocene basalt and andesite volcanics. Between these 2 provinces occurs the *Mobile Belt* which consists of folded and metamorphosed oceanic sediments of Cretaceous and Eocene age.

A major orogeny, as the result of collision between the Australian and Pacific plates from late Eocene to early Miocene, caused deformation and metamorphism of the Mobile Belt and was accompanied by vigorous volcanic activity along an island arc. Post-collision plate adjustments include regional uplift, extensive transcurrent faulting about the plate boundary and thrusting of the Ophiolite and Volcanic Arc against the Mobile Belt.

In the Miocene erosion of the emergent Mobile Belt and volcanic activity provided the main source of sediment in adjacent flysch and molasse basins, but shoal and basinal limestone deposition prevailed over the stable Platform.

In the Pliocene and Quaternary renewed emergence of the Mobile Belt, and uplift and tilting of the Platform margin and the Ophiolite and Volcanic Arc, were accompanied by deformation of molasse, flysch and stable shelf sediments, further widespread subaerial volcanism, regional regression and continuing deposition of unmature clastics.

References

Adams, B. W. 1973. The Ertsberg project. Mining Magazine, 1973: 310–323.

APC (Australasian Petroleum Co.). 1961. The geological results of petroleum exploration in western Papua. J. Geol. Soc. Aust., 8(1): 1–133.

Bain, J. H. C., H. L. Davies, P. D. Hohnen, R. J. Ryburn, I. E. Smith, R. Grainger, R. J. Tingey and M. R. Moffat. 1972, Geology of Papua New Guinea. Bur. Miner. Resour. Aust. 1:1,000,000 map.

Bain, J. H. C., D. E. Mackenzie and R. J. Ryburn. 1975. Geology of the Kubor anticline, central highlands of Papua New Guinea. Bur. Miner. Resour. Aust. Bull. 155.

Blake, D. H. and Y. Miezitis. 1967. Geology of Bougainville and Buka Islands, New Guinea. Bur. Miner. Resour. Aust. Bull. 93.

Brown, C. M., P. E. Pieters and G. P. Robinson. 1975. Stratigraphic and structural development of the Aure Trough and adjacent shelf and slope areas. APEA J. 15(1): 61–71.
D'Addario, G. W., D. B. Dow and R. W. Swoboda. 1976. Geology of Papua New Guinea, Bur. Miner. Resour. Aust. 1:2,500,000 map.
D'Audretsch, F. C., R. B. Kluiving and W. Oudemans. 1966. Economic geological investigations of NE Vogellop (Western New Guinea). Koninklijk Nederlands Geologisch Mijnbouwkundig Genootschap, Verhandelingen, Geologische ser. 23: 1–151. Staatsdrukkerij en Uitgeverijbedrijf, 's-Gravenhage.
Davies, H. L. 1971. Peridotite-gabbro-basalt complex in eastern Papua: an overthrust plate of oceanic mantle and crust. Bur. Miner. Resour. Aust. Bull. 128.
Davies, H. L. and I. E. Smith. 1971. Geology of eastern Papua. Bull. Geol. Soc. Amer. 82: 3299–312.
Denham, D. 1969. Distribution of earthquakes in the New Guinea-Solomon Island region. J. Geophys. Res. 74: 4290–9.
Dow, D. B. 1977. A geological synthesis of Papua New Guinea. Bur. Miner. Resour. Aust. Bull. 201.
Dow, D. B., J. A. J. Smit, J. H. C. Bain and R. J. Ryburn. 1972. Geology of the south Sepik region, New Guinea. Bur. Miner. Resour. Aust. Bull. 133.
Grainger, D. J. and R. L. Grainger. 1974. Explanatory notes on the 1:2,500,000 mineral deposits map of Papua New Guinea. Bur. Miner. Resour. Aust. Bull. 148.
Hohnen, P. D. 1978. Geology of New Ireland. Bur. Miner. Resour. Aust. Bull. 194.
Hutchison, D. S. and M. Norvick. In prep. Geology of the north Sepik region, Papua New Guinea. Bur. Miner. Resour. Aust. Bull.
Jenkins, D. A. L. 1974. Detachment tectonics in western Papua New Guinea. Bull. Geol. Soc. Amer. 85: 533–548.
Jaques, A. L. and G. P. Robinson. 1977. The continent/island-arc collision in northern Papua New Guinea. BMR J. Aust. Geol. and Geophys. 2, 289–303.
Johnson, R. W., G. A. M. Taylor and R. A. Davies. 1972. Geology and petrology of Quaternary volcanic islands off the north coast of New Guinea. Bur. Miner. Resour. Aust. Rec. 1972/21 (unpubl.).
Johnson, T. and P. Molnar. 1972. Focal mechanisms and plate tectonics of the southwest Pacific. J. Geophys. Res. 77, 5000–32.
Mackenzie, D. E. 1973. Quaternary volcanoes of the central and Southern Highlands of Papua New Guinea. Bur. Miner. Resour. Aust. Rec. 1973/89 (unpubl.).
Manser, W. 1974. Earth science abstracts, Papua New Guinea, to 1971. Bur. Miner. Resour. Aust. Bull. 143.
Mutter, J. C. 1975. A structural analysis of the Gulf of Papua and northwest Coral Sea region. Bur. Miner Resour. Aust. Report 179.
Page, R. W. 1978. Geochronology of igneous and metamorphic rocks in the New Guinea highlands. Bur. Miner. Resour. Aust. Bull. 162.
Reynolds, C. D., I. Havryluk, S. Bastaman and S. Atmowidjojo. 1973. The exploration of the nickel laterite deposits in Irian Barat, Indonesia. Geol. Soc. Malaysia, Bull. 6: 309–323.
Skwarko, S. K., R. S. Nicoll and K. S. W. Campbell. 1976. The Late Triassic molluscs, conodonts and brachiopods of the Kuta Formation, Papua New Guinea. BMR J. Aust. Geol. and Geophys. 1(3), 219–30.
Smith, I. E. 1973. Late Cainozoic volcanism in the southeast Papuan islands. Bur. Miner. Resour. Aust. Rec. 1973/67 (unpubl.).
Smith, I. E. and H. L. Davies. 1976. Geology of the southeast Papuan mainland. Bur. Miner. Resour. Aust. Bull. 165.
St. John, V. P. 1970. The gravity field and structure of Papua New Guinea. APEA J. 10(2): 41–55.
Thompson, J. E. and N. H. Fisher. 1965. Mineral deposits of New Guinea and Papua and their tectonic setting. Proc. 8th Comm. Min. metall. Cong. 6: 115–48.
Trail, D. S. 1967. Geology of Woodlark Island, Papua. Bur. Miner. Resour. Aust. Report 115.
Valk, W. 1960. The mining potential of Netherlands New Guinea. Nova Guinea, Geology 2: 5–12.
Vincelette, R. R. and R. A. Soeparjadi. 1976. Oil-bearing reefs in Salawati Basin in Irian Jaya, Indonesia. Am. Assoc. Petrol: Geol. Bull. 60(9): 1448–1462.
Visser, W. A. and J. J. Hermes. 1962. Geological results of the exploration for oil in Netherlands New Guinea. Koninklijk Nederlands Geologisch Mijnbouwkundig Genootschap, Verhandelingen, Geologische ser 20: 1–265. Staatsdrukkerij en Uitgeverijbedrijf, 's-Gravenhage.
Wegen, G. van der. 1971. Metamorphic rocks in West Irian. Netherlands National Museum of Geology and Mineralogy, Scripta Geologica 1:1–11.

In addition to the above references the following 1:250,000 Geological Series maps and explanatory notes have been published by BMR.

Bain, J. H. C. and D. E. Mackenzie. 1974. Karimui, Papua New Guinea–1:250,000 Geological Series. Bur. Miner. Resour. Aust. explan. Notes SB/55–9.
—— 1975. Ramu, Papua New Guinea–1:250,000 Geological Series. Bur. Miner. Resour. Aust. explan. Notes SB/55–5.
Blake, D. H. 1967. Bougainville Island North and South, Papua New Guinea–1:250,000 Geological Series. Bur. Miner. Resour. Aust. explan. Notes SB/56–8, 12.
Brown, C. M. 1978. Yule, Papua New Guinea–1:250,000 Geological Series. Bur. Miner. Resour. Aust. explan. Notes SC/55–2.
Davies, H. L. 1973. Gazelle Peninsula, New Britain–1:250,000 Geological Series. Bur. Miner. Resour. Aust. explan. Notes SB/56–2.
—— 1973. Fergusson Island, Papua New Guinea–1:250,000 Geological Series. Bur. Miner. Resour. Aust. explan. Notes SC/56–5.
Davies, H. L. and M. Norvick. 1974. Blucher Range, Papua New Guinea–1:250,000 Geological Series. Bur. Miner. Resour. Aust. explan. Notes SB/54–7.
Davies, H. L. and I. E. Smith. 1974. Tufi-Cape Nelson, Papua New Guinea–1:250,000 Geological Series. Bur. Miner. Resour. Aust. explan. Notes SC/55–8, 4.
Dow, D. B., J. A. J. Smith and R. W. Page. 1974. Wau, Papua New Guinea–1:250,000 Geological Series. Bur. Miner. Resour. Aust. explan. Notes SB/55–14.
Pieters, P. E. 1979. Port Moresby-Kalo-Aroa, Papua New Guinea–1:250,000 Geological Series. Bur. Miner. Resour. Aust. explan. Notes SC/55–7, 11, 6.
Robinson, G. P. 1974. Huon-Sag Sag, Papua New Guinea–1:250,000 Geological Series. Bur. Miner. Resour. Aust. explan. Notes SB/55–11, 7.
Robinson, G. P. and A. L. Jaques. 1978. Karkar Island, Papua New Guinea–1:250,000 Geological Series. Bur. Miner. Resour. Aust. explan. Notes SB/55–2.
Robinson, G. P., A. L. Jaques and C. M. Brown. 1976. Madang, Papua New Guinea–1:250,000 Geological Series. Bur. Miner. Resour. Aust. explan. Notes SB/55–6.
Ryburn, R. J. 1974. Pomio, New Britain–1:250,000 Geological Series. Bur. Miner. Resour. Aust. explan. Notes SB/56–6.
—— 1975. Talasea-Gasmata, New Britain–1:250,000 Geological Series. Bur. Miner. Resour. Aust. explan. Notes SB/56–5, 9.
—— 1979. Cape Raoult-Arawe, New Britain–1:250,000 Geological Series. Bur. Miner. Resour. Aust. explan. Notes SB/55–8, 12.
Smith, I. E. and H. L. Davies. 1973. Abau, Papua New Guinea–1:250,000 Geological Series. Bur. Miner. Resour. Aust. explan. Notes SC/55–12.
—— 1973. Samarai, Papua New Guinea–1:250,000 Geological Series. Bur. Miner. Resour. Aust. explan. Notes SC/56–9.
Tingey, R. J. and D. J. Grainger. 1976. Markham, Papua New Guinea–1:250,000 Geological Series. Bur. Miner. Resour. Aust. explan. Notes SB/55–10.
Willmott, W. F. 1972. Daru-Maer, Papua and Queensland–1:250,000 Geological Series. Bur. Miner. Resour. Aust. explan. Notes. SC/54–8, SC/55–5.

3. Pleistocene and present-day glaciations

Ernst Löffler

Introduction

The presence of Pleistocene and present-day glacial features in the humid tropics has fascinated researchers for a long time. Pleistocene glacial phenomena are the most obvious manifestations of climatic change and much of our knowledge on Pleistocene climatic variations in the tropics is based on research into the glacial geomorphology of the high mountains. Other disciplines like palynology have also taken advantage of the high mountains because preservation of plant and pollen material is usually much better in this cold environment than in the tropical lowlands, and these studies have substantially added to the reconstruction of the Pleistocene history of these areas. Access has also been an important factor. Although initially access to the mountains may be difficult, once one has left behind the forested slopes, movement and field observations are much facilitated by lack of forest cover.

In New Guinea detailed research into the Pleistocene and present-day glaciations only started about two decades ago because many of the high mountains were situated in areas discovered by Europeans only a few decades earlier. However, the presence of Recent glaciers and an extensive Pleistocene glaciation has been known for much longer.

Early work was concentrated on the highest mountains of New Guinea situated in the former Dutch Territory of West New Guinea, the present Indonesian province of Irian Jaya. The ice cap on Mt Carstensz was in fact already sighted in 1623 by the man whose name the mountain still bears but it was not until the beginning of this century in 1912 that the first European, A. F. R. Wollaston reached the ice (Wollaston 1914).

The next expedition to visit the glaciers was undertaken in 1936 and led by A. H. Colijn. He and his colleague J. J. Dozy undertook the first scientific work on the glacial geomorphology of the mountain. The extent of the glacier tongues was mapped and marked with cairns and the extent of the Pleistocene glaciation was studied and it was claimed that one valley glacier descended as low as 2000 m, a surprisingly low altitude for a tropical glacier (Dozy 1938).

Several expeditions visited the Carstensz glaciers after World War II but mostly for mountaineering purposes. The last two expeditions, however, undertaken as joint research projects by several Australian universities were to

Monographiae Biologicae, Vol. 42, ed. by J. L. Gressitt.

© *1982, Dr W. Junk Publishers, The Hague.*

provide some of the most detailed and best researched account for any glaciated mountain in the tropics (Peterson and Hope 1972; Peterson *et al.* 1973; Hope *et al.* 1976).

Two other presently glaciated mountains in West New Guinea, Mt Wilhelmina (now Mt Trikora) and Mt Juliana (now Mt Mandala), were also visited and their ice cover photographed before World War II.

In the eastern part of the island no work was done on the mountains until the 1960s, but there was one report of a possible Pleistocene ice cover on the Saruwaged Range by Detzner (1933) a captain in the Imperial German army whose fame however rests not on this discovery but on having survived the four years of the first world war in hiding in the interior of New Guinea.

In the last decade much research effort has been concentrated on the mountains of Papua New Guinea and it is probably no exaggeration to claim that today the formerly glaciated mountains of New Guinea belong to the best researched tropical mountains in the world and there are numerous accounts not only on the glacial geomorphology but on vegetation, vegetation history, palynology and climate (Reiner 1960; Wade and McVean 1969; Blake and Löffler 1971; Bik 1971; Löffler 1970, 1971, 1972, 1977; Hope 1973; Hope and Peterson 1975; Hope *et al.* 1976; Peterson *et al.* 1973; Galloway *et al.* 1972; Smith 1975, 1977; Walker 1970).

Pre-Würm glaciations

Evidence for glaciations pre-dating the last (Würm or Wisconsin glaciation) is notoriously difficult to find in areas where only cirque and valley glaciers were developed and the ice never developed extensive ice sheets. The reason for this lack of evidence is usually sought in the high rates of erosion in the mountains obliterating traces of a former ice cover. As will be shown later the traces of the last glacial cover are exceptionally well preserved despite the twelve to fourteen thousand years that have passed since the ice has disappeared and it would be surprising if older traces were not preserved if previous glaciations had indeed covered larger areas than the last. It is therefore probably fair to assume that the last glaciation was the most extensive on most mountains.

In New Guinea there are only two areas where evidence of older glaciations has been found. In West Irian weathered moraine material was found beyond the moraines of the last glaciation on the Kemabu Plateau north of the Jaya Mountains (Dow 1968; Hope and Peterson 1975). No dating is available but the advanced weathering status of the deposits as compared to the virtually unweathered tills of the last glaciation suggests that they are considerably older and do not merely represent an earlier stage of the last glaciation.

The other area of an earlier glaciation in Papua New Guinea is on Mt Giluwe and is of an entirely different nature. Mt Giluwe is a dome-shaped volcanic cone built up of innumerable lava flows. Intermittent with the eruptions of the lavas however were periods of ice cover. If lava erupts under ice it is drastically chilled and forms a palagonite or palagonitic breccia. Two sites of palagonitic breccia have been found on Mt Giluwe one in the upper Gogon valley and one on the east peak (Blake and Löffler 1971). Dating of lavas associated with the breccias has shown that they formed at approximately 290 000 years BP and also possibly at about 700 000 years BP (Löffler *et al.* 1980). The latter date is not as reliable as the first and only gives a rough estimate. The first date is supported by dates from another finding in the upper Tamal valley where moraine material is sandwiched between two lava flows which gave dates of between 290 000 and 319 000 years BP. All findings are situated within the areal limits of the last glaciation.

There can be no doubt that New Guinea mountains experienced earlier glaciations. The 290 000 years date is consistent with findings in the mountains of Hawaii (Porter *et al.* 1977) and also with the paleotemperature curve of Emiliani and Shackleton (1974). The result also shows that the mountains must have reached altitudes similar to the present ones relatively early in the Pleistocene.

Last glaciation

The last glaciation had a profound effect on the New Guinea mountains which is still clearly visible today. It is not known how much the older glaciations had already modified the mountain terrain and whether the glaciers of the last ice age simply occupied the glacial landscape carved out by earlier glaciers. However, what we see today is the imprint of the last glaciation.

Even to the casual observer the formerly glaciated terrain stands in striking contrast to the non-glaciated landforms below it. Whenever one moves into a glacial valley one leaves behind the V-shaped, steep-sided, forest-covered fluvially formed valleys and enters an open U- or saucer-shaped valley, the wide and flat floor of which is swampy and covered with grass. The entrance into the glacial valley may be accentuated by a terminal moraine crossing the valley floor, but cut into by the swiftly flowing river draining the valley. The terminal moraines often link up with lateral moraines forming sharp-crested, steep-sided ridges or perching on top of pre-existing crests. The excellent preservation of terminal and lateral moraines and their great height, which sometimes exceeds 100 m, is a typical feature of the New Guinea mountains and seems to be characteristic for tropical mountains in general. It is probably due to the diurnal regime of melting evenly spread over the year and consequently the relatively constant rate of discharge, even during periods of glacial retreat, as compared with high-latitude glaciers — which produce very high discharges during spring and summer and therefore erode more severely.

Traces of glacial erosion are equally well preserved, and rock thresholds, roches moutonnes, over-deepened basins, and precipitous cirque backwalls are widespread. The basins have proved particularly useful for the study of the glacial history of the area since they contain datable peat material and pollen.

The total area covered by Pleistocene glaciers was about 2000 km^2 of which 1400 km^2 are situated in Irian Jaya (Hope and Peterson 1976) and the remaining 600 km^2 in Papua New Guinea (Fig. 1). In Irian Jaya there was a nearly continuous ice field from Mt Idenburg in the west to the Star Mountains in the east only interrupted by a few deeply incised passes. In Papua New Guinea the glaciated mountains remained isolated ice-covered peaks scattered through the highland areas, along the Owen Stanley Range and the eastern parts of the northern coastal ranges (Fig. 1).

The Late Pleistocene glacial snowline was at about 3550–3600 m on most mountains but there were some variations due to local climatic or tectonic conditions. This snowline corresponds to a temperature depression of about 5–6°C in the mountain areas assuming that the lapse rate was similar to the present. Because of the likelihood of an increased lapse rate due to cooler, drier air masses in the lowlands (Nix and Kalma 1972) the temperature depression may have been less at lower altitudes where it probably amounted to only 3–4°C. The temperature depression must have resulted not only in the formation of glaciers but also in a general lowering of the vegetation zones and climatic geomorphological belts such as the zone of periglacial solifluction. However there is little evidence that this shift of altitudinal zones had any significant geomorphic effect other than glaciation. All evidence for Pleistocene periglacial activity has been found inside the formerly glaciated area and such activity must have taken place shortly after the ice had receded. This observation is not surprising if one looks at the present-day distribution of the periglacial zone. Periglacial solifluction as an important landforming process occurs only from about 4350 m upwards (Löffler 1975). The zone is thus compressed to a mere 200–300 m altitude range which is surely the result of the great uniformity of the climate with its lack of diurnal temperature extremes. During the Pleistocene the ice on most mountains extended well below the snowline and thus covered most of the areas that were climatically suitable for periglacial activity which is the zone of diurnal frost and thaw action.

The Maoke Range (Snow Mountains)

This range is the highest and most rugged part of the central ranges and its divide is nearly continuously above 3000 m. The Maoke Range contains the highest mountains of the island with two peaks Mt Jaya (Carstensz) and Mt Idenburg still supporting ice caps (Fig. 2). The area used to be the most remote, most inaccessible and least well known high mountain area of New Guinea but thanks to the efforts of the two Australian Universities Expeditions (Hope *et al.* 1976) it has become one of the best researched. The following account is based on this work.

Mt Jaya also known as Mt Carstensz is the highest mountain of New Guinea reaching 4884 m above sea level. The mountain is part of a range that is capped by a massive limestone block which is bounded by cliffs up to 1000 m high on all sides except the west where the limestone dips down gently to the general level of the range. The limestone sheet is folded into two narrow anticlines and two wider synclines that have been opened up by glacial erosion.

To the south the Carstensz Mountains fall down steeply (Fig. 3) to low foothills but to the north they abut against a rolling plateau area, the Kemabu Plateau, that extends north for 25 km with altitudes between 2400 and 3300 m.

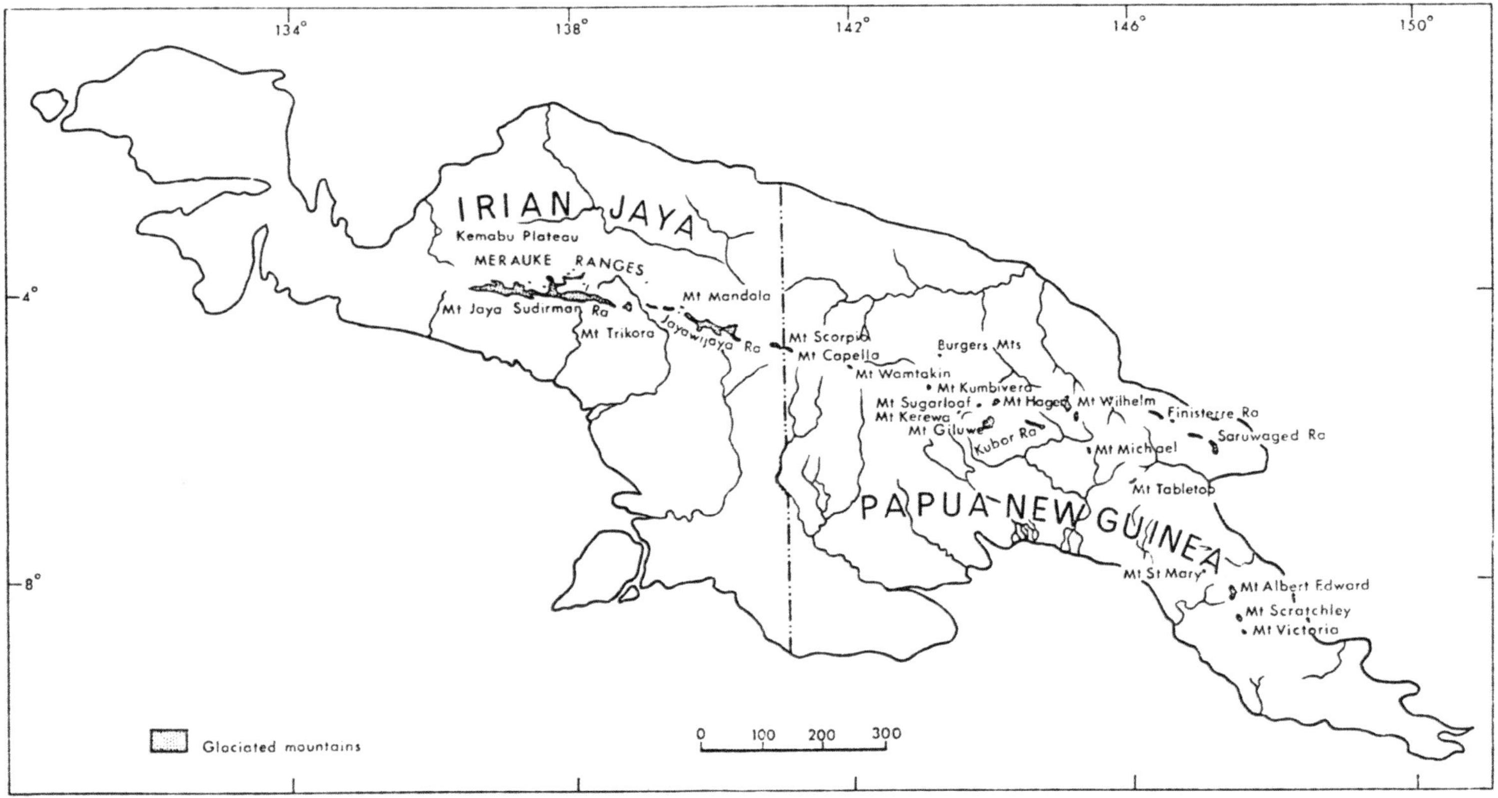

Fig. 1. Distribution of glaciated areas in New Guinea.

This plateau area has been the site of extensive glacial deposition and huge moraines up to 200 m high form prominent ridges on this plateau. While on the plateau the moraines lie at about 3400 m altitude on the southern fall the moraines descended as far down as 2300 m obviously because of the different topographic situation.

The total area covered by ice during the last glaciation amounted to nearly 1000 km^2 and the snowline has been estimated at 3600–3700 m (Hope and Peterson 1976). The reason for the relative high snowline in comparison with the mountains further east could be somewhat drier conditions caused by the closure of Torres Strait and drying up of the Arafura Shelf. The southeast trades would then not have blown over warm ocean waters but over dry and probably semi arid land.

The recession of the Pleistocene glaciers coincided with the recession on the other New Guinea mountains at about 15 000 years BP or possibly earlier since these dates usually give minimum ages.

As on other New Guinea mountains the recession of ice was not a continuous melting away but was interrupted by minor readvances which are manifested in numerous recessional moraines. It appears that the glaciers retreated from their maximum extent from about 15 000 years BP onwards until they had entirely disappeared at about 7000 years BP. However, glacial ice formed again on the high peaks at about 5000 years BP (see p. 49).

The Star Mountains

The Star Mountains are part of the central ranges and extend across the international border from about 141° 15′ E to 140° E. Most of the glaciated areas are situated in Irian Jaya as are the highest elevations, but small areas of glaciated terrain are also to be found on the Papua New Guinea side of the range. The highest peaks are Mt Mandala (formerly Juliana) (4640 m) in Irian Jaya and Mt Capella (4000 m) and Mt Scorpio (3830 m) in Papua New Guinea.

The Star Mountains are predominantly formed of massive limestone that is tilted to the north thus giving the mountains a markedly asymmetric cross profile with cliffed faces to the south and smoother gentler slopes to the north. This has of course also influenced the distribution of the glacial ice cover which is most extensive on the north facing slopes. The Pleistocene glaciation of the Western Star Mountains has been investigated by Verstappen (1964) and that of the eastern Star Mountains by Shepherd (unpubl.). Cirques, glacial valleys and moraines are well developed on both the eastern and western parts of the mountains but the extent of glaciation was considerably greater on the Irian Jaya side. Verstappen (1964) estimates an area of 750 km^2 for the glacial cover but a more recent estimate of Hope (1976) is only about half that figure. The glaciers descended to an average altitude of 3200 m with some larger glaciers extending as low as 2940 m. The Pleistocene snowline of 3550–3600 m given by Verstappen agrees well with the estimates from the Papua New Guinea highlands (Löffler 1972).

The greatest number of glaciated mountains and the largest ice covered areas in eastern New Guinea are to be found in the so called highland areas (Fig. 1). Mt Wilhelm and Mt Giluwe the two highest mountains in Papua New Guinea had glaciers covering well over 100 km^2 while glacial ice on Mt Hagen and the Kubor Range extended over 20 and 30 km^2. Apart from these main areas of Pleistocene glaciation there were several isolated mountain peaks that just extended above the snowline such as Mts Michael, Tabletop, Kumbivera, Sugarloaf, Kerewa and the Burgers Mountains. Each of these peaks had only a few square kilometres of ice.

Mt Giluwe (4368 m) is a huge volcanic dome built up of a succession of thin outwardly dipping lava flows most of which erupted between 300 000 and 220 000 years BP, (Fig. 4) surrounding a considerably older central core. It rises from about 2000 m at the base to a summit height of 4368 m and measures over 100 km in circumference. The summit area above 3200 m was covered by a more or less continuous ice cap during the last glaciation and several valley glaciers descended further down to between 2750 and 3000 m. The glacial landforms have been very well preserved.

The maximum extent of the ice cap during the last glaciation was 188 km^2 and this can easily be traced because of the excellent preservation of the terminal and lateral moraines that form a nearly continuous belt around the entire mountain. Another striking feature is the multitude of recessional moraines present (Fig. 5). There are over 20 individual moraines forming low ridges 0.5–5 m high and they occur on the valley floors of U-shaped valleys, on the inner sides of lateral moraines and in groups of 2 to 7 ridges on the eastern and south-western slopes. These recessional moraines represent minor re-advances during the general ice retreat which started at about 13 000 years BP and was completed at about 10 000 years BP.

Mt Wilhelm (4509 m) is an extensive granodiorite massif and is the highest mountain of Papua New Guinea. Although higher than Mt Giluwe the glaciated area is only little more than half that of Mt Giluwe. However the glaciation was more truly alpine in character with deep cirques, precipitous cirque backwalls and over-deepened floors that contain large lakes, deep glacial troughs with shoulders and stepped valley profiles and extensive terminal and lateral moraines (Fig. 6).

Most of the cirques are situated west and southwest of the summit ridge. This asymmetry in cirque distribution has been attributed to the preglacial asymmetry of the massif favoring extension of the ice in a westerly and southwesterly direction (Reiner 1960). This asymmetry is, however, not restricted to cirques and glaciers but is also apparent in general slope profiles with smoother and gentler east-facing slopes and more irregular and steep west-facing slopes.

A similar asymmetry has been observed in mountains of subtropical latitudes but here it is a north-south asymmetry due to the difference in sun insolation between north and south aspect favoring glacial activity on the northern slopes but periglacial solifluction on the south sides. A similar explanation could be sought for Mt Wilhelm except that it is not the difference in insolation between

Fig. 2. Carstensz area. View from Meren Valley west to Grasberg (middle) and the ice capped peak of Mt Idenberg. The section of the Meren Valley seen in the foreground was ice covered during a neoglacial advance at about the middle of last century. The sharp linear boundary between grassland and shrubland on the right indicates the extent of the glacier. (Photography by G. S. Hope.)

Fig. 3. Oblique aerial photograph showing Carstensz ice fields looking west. Foreground Carstensz and Meren glaciers and their extensive firn areas. Background right the Northwall Firn. (Photograph by Australian Universities Expeditions to Irian Jaya.)

Fig. 4. Summit area of Mt Giluwe with series of thin, flat lying lava flows (middle right) and two castelated peaks of steeply dipping lava flows and pyroclastics rising above the level of the flat lying lava flows. The flat lying lava flows erupted at between 300 000 and 220 000 years BP while the lavas of the two peaks erupted at about 700 000 years BP.

Fig. 5. One of the outstanding features of Mt Giluwe is the presence of a great multitude of recessional moraines and associated tarns and peat bogs on the broad plateau-like eastern (shown) and western slopes.

north and south but between east and west caused by the fact that in the tropics east facing slopes receive more sun insolation than west facing slopes because of the rapidly increasing cloud cover during the day. This has also been used to explain the asymmetry in plant distribution on Mt Wilhelm which extend further up on east aspects than on west aspects (Smith 1977).

One difficulty with this climatic explanation is that this kind of asymmetry is not present on many other New Guinea mountains but it may be that there structural control overrides the climatic factors (Löffler 1977).

Mt Hagen is situated some 40 km north of Mt Giluwe and is also an extinct volcano somewhat older than Mt Giluwe with a remnant cone that was formed at about 1.1 million years BP, but volcanic activity has continued until 200 000 years BP. Glacial features are restricted to the small summit area and include cirques and moraines none of which are particularly well developed. During the maximum extent of the glaciation about 21 km^2 were covered by an ice cap from which short lobes descended down to about 3400 m.

The Kubor Range a westnorthwesterly trending ridge formed of old Paleozoic rocks also shows clear evidence of glaciation along the main summit ridge where small cirques, rock thresholds and short glacial valleys with low moraines are present. The peak most extensively glaciated is Mt Digini at the eastern end of the range.

All the other peaks mentioned only had minor ice covers never exceeding a few square kilometres in size. Mt Kumbivera represents a limestone syncline with a broad central basin that was partially filled with glacial ice. Mts Kerewa and Sugarloaf are both volcanic and may have had small ice caps but this needs verification in the field. The Burgers Mountains and Mts Udon and Tabletop show evidence of glacial cover on air photographs but only small areas were affected.

The Owen Stanley Range

The Owen Stanley Range forms the rugged backbone of eastern New Guinea and consists of low grade metamorphic rocks in the northwest and basaltic rocks in the southeast. A distinctive feature of the Range is the presence of plateau-like surfaces along the main watershed and some off-branching divides which provided excellent areas for snow accumulation once the snowline had descended to their level. Glacial activity was most pronounced on Mt Albert Edward and Mt Scratchley where areas of 90 km^2 and 28 km^2 were covered by glacial ice.

The most outstanding features on Mt Albert Edward are two asymmetric summit ridges that rise some 200–300 m above the surrounding plateau and the numerous picturesque cirque lakes that fill the overdeepened rock basins (Fig. 7). During the maximum extent of the glaciation the plateau area was covered by a nearly continuous ice cap down to about 3400 m. The snowline was at 3600–3650 m.

Mt Scratchley is a smaller plateau area right on the main watershed and was also covered by a continuous ice cap. Well-developed moraines and glacial valleys with stepped profiles and broad valley floors are prominent glacial features and form a marked contrast to the surrounding non-glaciated area.

Mt Victoria (4036 m) a narrow, steep sided mountain ridge and the highest peak of the Owen Stanley Range shows only minor evidence of glacial activity in form of two small cirques and some low moraines because the area available for snow accumulation was very small.

The Saruwaged and Finisterre Ranges

The Saruwaged and Finisterre Ranges are the easternmost and highest parts of the northern ranges of New Guinea and the only glaciated areas outside the central ranges. They are also the most unstable and tectonically most active areas of New Guinea and rates of uplift have been measured from raised coral terraces along the adjacent coast to be in the order of 3 m/1000 years (Chappel 1974). The highest peaks of the Ranges are Mt Bangeta (4121 m) on the Saruwaged Range and Mt Finisterre (3917 m) on the Finisterre Range.

The summit areas of the Finisterre and western Saruwaged Ranges consist of a single narrow steep-sided ridge of limestone but the central and eastern Saruwageds are formed of irregular plateau surfaces which show clear evidence of a former ice cover in their higher parts (Fig. 8).

Glacial features on the Bangeta Plateau which is the highest part of the Saruwaged Range include cirques, overdeepened basins, short glacial valleys and low moraines (Fig. 8). The type of glaciation was similar to that on Mt Albert Edward with a continuous ice cap covering the plateau and short lobes of ice extending from it.

The smaller and lower Uruwa Plateau west of Bangeta supported mainly cirque glaciers and short valley glaciers as did the highest peaks of the Finisterre Range further west.

The Pleistocene snowline has been estimated at 3650–3700 m. This relatively high snowline in comparison to the values of the highland areas is probably due to postglacial uplift of the Ranges.

Deglaciation and present-day glaciation

From about 15 000 years BP the glaciers receded rather rapidly from their maximum extent. By 9000 years BP most New Guinea mountains were ice free and by about 7000 years BP even those that have a present day ice cover were bare (Hope and Peterson 1975). The recession did not proceed in a continuous ice retreat but was interrupted by numerous minor readvances and these are manifested in low recessional moraines that occur in great numbers between the main moraines of the last glaciation and the cirques. The number of recessional moraines varies from mountain to mountain, an indication that local orographic and climatic conditions were important. Most numerous are the recessional moraines on the eastern and western flanks of Mt Giluwe where over 20 single moraine ridges can be distinguished. These ridges, many of which

Fig. 6. Pindaunde Valley, Mt Wilhelm with two cirque lakes and high lateral moraine forming the left margin of the valley. The smooth evenly sloping crest of the moraine contrasts sharply with the irregular high ridge crest further to the left and is a clear indication of its depositional character. (Photograph by G. S. Hope).

Fig. 7. Mt Albert Edward with prominent summit ridge and deep glacial basins occupied by lakes. Note the forest remnant to the right of the large lake at between 3650 and 3850 m.

Fig. 8. Saruwaged Range with summit area formed of massive limestone plateau bounded by steep escarpments. The plateau was covered by an ice cap during the last glaciation and short valley glaciers developed in the valleys at the margin of the plateau. The ice feeding these short valley glaciers may have originated largely from ice avalanches from the plateau above.

are only a metre high are continuous for several hundred metres and follow the local relief of the broad plateau-like slopes like reversed contour lines, curving downslope in the shallow valleys and upslope on the ridges. The moraines can be grouped into 5 major stages numbered 1–5 (Löffler 1972). Dates of peat bogs behind these moraines have given minimum dates of 13 050 ± 700 years BP for the second oldest moraine group at 3510 m (Stage 2), 11 250 ± 550 years BP for the second youngest group at 3710 m (Stage 4) and 9980 ± 280 years BP for a peat bog on the western summit at 4160 m. These samples were collected by G. S. Hope and the author in 1973 and supercede the dates given by the author in a previous publication (Löffler 1972). The dates show that by 13 050 years BP the ice already had receded from the second oldest stage and that by 9980 years BP Mt Giluwe must have been free of ice. The rate of ice retreat was in the order of 1 m/year if one does not take into account the readvances, but because of the oscillations the rate must have been considerably larger during certain periods.

The dates on Mt Giluwe correspond well with deglaciation dates from Mt Wilhelm which are supported by pollen data. The estimated rate of retreat on Mt Wilhelm was about 0.24–0.31 m altitude per year (Hope and Peterson 1975) which would correspond to retreat of about 0.8–0.9 m per year. There are only a few recessional moraines present on Mt Wilhelm and this may reflect the completely different nature in the type of glaciation here as compared to Mt Giluwe. On Giluwe the glaciers were broad, probably relatively thin plateau glaciers reacting quickly to minor climatic oscillations, while the glaciers on Mt Wilhelm were thick valley glaciers partly protected from insolation by steep ridges towering above the ice and consequently were much less likely to react to small climatic oscillations. The retreat on Mt Wilhelm was therefore more uniform reflecting the major climatic change to warmer conditions.

Deglaciation dates from most other mountains are in complete agreement with the Wilhelm and Giluwe results. G. S. Hope (unpubl.) collected peat from glacial valleys on Mt Albert Edward and Mt Scorpio which gave ages of 12 850 ± 300 years and 13 250 ± 395 years respectively.

The only deglaciation date not fitting the series of dates reported above has been reported by Peterson and Hope (1972) from the Carstensz area. In one of the major valleys draining the Carstensz Mountains they found moraine-like material containing wood at 1750 m altitude and obtained a date of 10 300 years BP. The initial interpretation of this anomalous result was that the supposed moraine originated from a major readvance of the Carstensz glacier possibly caused by the sudden influx of moist air masses from the flooding of the Arafura Sea and Torres Strait following the post glacial rise in sea level (Peterson and Hope 1972). This interpretation is however not upheld by the authors in a more recent publication (Hope and Peterson 1975) where they point out that the material probably originated from a mud flow type deposit.

The relatively high tree line indicated by pollen records between about 7000 and 5000 years BP suggests that all New Guinea mountains were ice free but after 5000 years BP new ice was again developing on the highest peaks in Irian Jaya (Hope and Peterson 1975). According to these authors ice advanced down to about 4200 m in the Yellow Valley, Mt Carstensz area at about 3500 years BP and there were three further advances interrupted by retreats at 3000 to 2400 years BP, after 2400 years BP and at 1350–1500 years BP. These are the only glacial dates for this time period and it is probable that these fluctuations did not occur in the presently unglaciated areas although there may have been small snow caps on the higher peaks.

Present day glaciation is restricted to Mt Carstensz where it is most extensive (Fig. 2), and to three isolated summits Mt Juliana (Mandala Pk) Mt Wilhelmina (Trikora Pk) and Idenburg Top. With the aid of aerial photographs, written records and dated cairns placed in front of the glacier tongues the recent fluctuations of the Carstensz glaciers and the other smaller ice caps have been investigated by Peterson *et al.* (1973) and Allison and Peterson (1976) and the following account is based on this work.

The glacial ice in the Carstensz area, covering 6.9 km^2, consists of several distinct ice masses. The two largest, the Carstensz and Meren glaciers are short valley glaciers with well-developed glacier tongues that have average surface slopes of 20 and 25% and descend to 4260 and 4380 m respectively. The estimated thickness of the glacial ice is 40 m. The glaciers are dynamically active and their surface velocities are 25 m/year for the steeper Meren glacier and 15 m/year for the Carstensz glacier.

These glaciers have been retreating relatively rapidly over the past 35 years at rates of 33 m/year for the Meren glacier and 16 m/year for the Carstensz glacier. The total loss of ice covered area since the last neoglacial advance, which is thought to have occurred at the middle of last century, is 16.4 km^2.

All other New Guinea mountain peaks had only relatively small ice covered areas. Mt Juliana (Mandala Pk) has been described by Verstappen (1964). In 1959 the ice cap covered about 0.29 km^2 and the maximum thickness of the ice amounted to about 50 m, while the snowline was at 4580 m. Verstappen estimated that the ice cap had been reduced by 15% between 1937 and 1959. Peterson *et al.* (1973) have documented a further reduction between 1959 and 1973.

Mt Wilhelmina (Trikora Pk) was capped by ice in 1905–1939+ but the ice has since completely disappeared. The Idenburg Top west of Carstensz was covered by three small ice bodies in 1943 but only the largest of these has survived, greatly reduced in size.

The ice retreat in the New Guinea mountains is broadly in sympathy with the world wide glacier fluctuations. This is probably due largely to changes in temperatures as tropical glaciers tend to react much more sensitively to temperature changes than to changes in precipitation because precipitation is usually not the limiting factor in ice development in these highly humid areas.

Conclusion

The Pleistocene glacial history of New Guinea may date back to about 700 000 ± 100 000 years BP when lava erupted under ice on Mt Giluwe. There is no evidence on earlier glacial events and it is unlikely that it ever will be found. At 290 000–300 000 years BP Mt Giluwe was again covered by ice but the extent of the ice cover is unknown. This glacial event was probably synchronous with glaciations in other parts of the world. No evidence has as yet been discovered for glaciations following this period apart from the last major glaciation that covered all New Guinea mountains exceeding 3600 m in altitude. The onset of this last glaciation is not known although vegetation histories indicate (Williams *et al.* 1972; Walker 1970; Hope and Peterson 1975) that cold conditions existed before 37 000 years BP and again from about 30 000 years BP. This last glaciation lasted until about 15 000 years BP when the ice receded rather rapidly to completely disappear by about 9000 years BP. These dates are consistent over the entire island and also with evidence of climatic change in Australia where conditions changed from arid to less arid (Bowler *et al.* 1976). The dates are also broadly synchronous with world wide glacial events. In detail there are however some differences like the lack of glacial advances at 9000 years BP which occurred in the European Alps (Heuberger 1974).

The type of glaciation in New Guinea varied from alpine type valley and cirque glaciation to plateau glaciation. Aspect was mostly not a significant factor in the development of glaciers but there may have been a slight east-west asymmetry in glacial development caused by the difference in insolation between east and west facing slopes due to increase in cloudiness during the day. However this only seems to have affected deeply incised valley and cirque glaciers but not the plateau glaciers. It is also not apparent in the present day glacier distribution.

The snowline during the last glaciation was at about 3550–3600 m altitude indicating a lowering of the snowline by about 1000–1100 m. This would correspond to a temperature depression of about 5°–6°C assuming that precipitation was similar to the present. If precipitation was significantly less than today the temperature depression would have been greater but there is no climatic evidence to assume this (Nix and Kalma 1972). The relatively low Pleistocene snowline is therefore seen as the result of a temperature depression in the order of 5–6°C and of high rates of precipitation similar in magnitude to the present.

References

Allison, I. and J. A. Peterson. 1976. Ice areas on Mt Jaya: their extent and recent history. In: Hope et al. 1976. The equatorial glaciers of New Guinea. A. A. Balkema, Rotterdam.

Blake, D. H. and E. Löffler. 1971. Volcanic and glacial land forms on Mount Giluwe, Territory of Papua and New Guinea. Bull. Geol. Soc. Am., 82: 1605–1614.

Bowler, J. M., G. S. Hope, J. N. Jennings, G. Singh and D. Walker. 1976. Late Quaternary climates of Australia and New Guinea. Quat. Res. 6: 359–394.

Chappell, J. 1974. Geology of coral terraces, Huon Peninsula, New Guinea. Bull. Geol. Soc. Am. 85: 553–570.

Detzner, H. 1933. Der Saruwaged und seine östlichen und südöstlichen Anschlussgebirge. Kolon. Rdsch. Jg. 1933: 209–221.

Dow, D. B. 1968. A geological reconnaissance in the Nassau Range, West New Guinea. Geol. en. Mijnbouw 47.

Dozy, J. J. 1938. Eine Gletscherwelt in Nederländisch Neuguinea. Z. Gletscherkd. Glazialgeol. 26: 45–51.

Emiliani, C. and N. J. Shackleton. 1974. The Brunhes Epoch: isotopic paleotemperatures and geochronology. Science: 183: 511–514.

Galloway, R. W., G. S. Hope, E. Löffler and J. A. Peterson. 1973. Late Quaternary glaciation and

periglacial phenomena in Australia and New Guinea. In: E. M. van Zinderen Bakker (ed.). Palaeoecology of Africa and Antarctica. Vol. 8. Cape Town, Balkema.
Heuberger, H. 1974. Alpine Quaternary glaciation. In: J. D. Ives and R. G. Barry (ed.). Arctic and Alpine Environments. London, Methuen.
Hnatiuk, R. J., J. M. B. Smith and D. N. McVean. 1976. Mt Wilhelm Studies 2. The climate of Mt Wilhelm, Dept. Biogeogr. and Geom. Publ. BG/4. Aust. Natn. Univ., Canberra.
Hope, G. S. 1976. The vegetational history of Mt Wilhelm Papua New Guinea. J. Ecol. 64: 627–664.
Hope, G. S. and J. A. Peterson. 1975. Glaciation and vegetation in the high New Guinea mountains. Bull. Roy. Soc. N.Z. 13: 155–162.
Hope, G. S., J. A. Peterson, U. Radok and I. Allison. 1976. The equatorial glaciers of New Guinea. A. A. Balkema, Rotterdam.
Löffler, E. 1970. Evidence of Pleistocene glaciation in East Papua. Aust. Geogr. Stud. 8: 16–26.
—— 1971. The Pleistocene glaciation of the Saruwaged Range, Territory of New Guinea. Aust. Geogr. 11: 463–472.
—— 1972. Pleistocene glaciation of Papua New Guinea. Z. Geomorphol., N.F. Suppl. 13: 32–58.
—— 1975. Beobachtungen zur periglazialen Höhenstufe in den Hochgebirgen von Papua New Guinea. Erdkunde 28: 13–18.
—— 1976. Potassium-argon dates and pre-Würm glaciations of Mount Giluwe volcano, Papua New Guinea. Z. Gletscherkd. Glazialgeol. 12: 55–62.
—— 1977. Geomorphology of Papua New Guinea. Aust. Nat. Univ. Press.
Löffler, E., D. E. Mackenzie and A. Webb. 1980. Potassium Argon dates from New Guinea highland volcanoes and their relevance to the geomorphic history. J. Geol. Soc. Aus. 26: 387–397.
Nix, H. A. and J. D. Kalma. 1972. Climate as a dominant control in the biogeography of northern Australia and New Guinea. In: Bridge and Barrier: the Natural and Cultural History of Torres Strait, ed. D. Walker. pp. 61–91. Dept. Biogeogr. and Geomorph. Publ. BG/3, Aust. Natn. Univ., Canberra.
McVean, D. N. 1968. A year of weather records at 3480 m on Mt Wilhelm, New Guinea. Weather 23: 377–381.
Nogami, M. 1976. Altitude of the modern snowline and Pleistocene snowline in the Andes. Geogr. Reports, Tokyo Metrop. Univ. 11: 71–86.
Peterson, J. A. and G. S. Hope. 1972. A lower limit and maximum age for the last major advance of the Carstensz glaciers, West Irian. Nature, 240: 36–37.
Peterson, J. A., G. S. Hope and R. Milton. 1973. Recession of snow and ice fields of Irian Jaya, Republic of Indonesia. Z. Gletscherkd. Glazialgeol. 9: 73–87.
Porter, S. C., M. Stuiver and I. C. Yang. 1977. Chronology of Hawaiian glaciations. Science 195: 61–63.
Reiner, E. 1960. The glaciation of Mt Wilhelm, Australian New Guinea. Geogr. Rev. 50: 491–503.
Smith, J. M. B. 1975. Mountain grassland of New Guinea. J. Biogeogr. 2: 27–44.
—— 1977. Vegetation and microclimates of east-and-west facing slopes in the grasslands of Mt Wilhelm, Papua New Guinea. J. Ecol. 65: 39–53.
Verstappen, H. Th. 1964. Geomorphology of the Star Mountains. Nova Guinea, Geology 5: 101–155.
Wade, L. K. and D. N. McVean. 1969. Mt Wilhelm Studies 1, The alpine and subalpine vegetation. Dept. Biogeogr. and Geom. Publ. BG/1. Aust. Nat. Univ. Canberra.
Walker, D. 1970. The changing vegetation of the montane tropics. Search, 1: 217–21.
Williams, P. N., I. McDougall and J. H. Powell. 1972. Aspects of the Quaternary geology of the Tari-Koroba area. J. Geol. Soc. Aust. 18: 333–347.
Wollaston, A. F. R. 1914. An Expedition to Dutch New Guinea. The Geogr. J. 43: 248–273.

4. Landforms and landform development

Ernst Löffler

Introduction

New Guinea is situated between the stable land mass of Australia and the deep ocean basin of the Pacific and is thus part of one of the most mobile zones of the earth's crust, the circum Pacific Mobile Belt. It is characterized by high seismic activity, widespread volcanism, young folded and faulted mountain chains and curved chains of islands and oceanic rises, so called island arcs. Typical New Guinea landforms are high mountain ridges with sharp narrow crests separated by deeply incised V-shaped valleys. Structurally controlled landforms like homoclinal ridges, hogback ridges are also important in certain areas, as are large areas of limestone karst.

The mountain areas are flanked by extensive alluvial plains particularly to the south, where a stable foreland — the extension of the stable Australian Platform — provided favourable conditions for continuing deposition.

Because of the youthfulness of the New Guinea landscape the major landform regions or physiographic regions closely follow well-defined structural regions. These are from south to north the southern plains and lowlands, the central ranges, the intermontane trough and the northern ranges.

Major landform regions

The southern plains and lowlands

The Southern Plains and Lowlands extend from the south coast to the foothills of the central ranges. In the Fly and Digul Rivers area, they cover approximately 230 000 km^2 and extend over 400 km from the coast to the mountains. To the east and west they become progressively narrower. Westward they narrow down rapidly to a small stretch of alluvium, eastward they become divided by the foothills of the central ranges into several embayments consisting of marshy coastal plains at the mouths of the main rivers, merging upstream into relatively stable and well-drained alluvial plains. Most of the Fly-Digul area constitutes a relict alluvial plain, which has been dissected to varying degrees (Fig. 12). In the south the area is composed of gently undulating to low

Monographiae Biologicae, Vol. 42, ed. by J. L. Gressitt.

© 1982, Dr W. Junk Publishers, The Hague.

hilly country, in the north it is formed by closely spaced narrow ridges with an even summit level that indicates the former depositional plain surface. Swamps and lakes are less prominent than is generally believed, and occur mainly along the rivers as blocked valley swamps or oxbow lakes, and along the coast behind beach ridges. In the area west of the Mappi River however swamps and floodplains occupy most of the lowlands.

The coastal plains consist of extensive tidal flats (Fig. 12), such as the Purari and Kikori deltas or those at Flamingo Bay, and narrow bands of beach ridges and swales and mangrove flats. The tidal flats are mostly extensive mangrove swamps built up of recent marine and estuarine sediments and traversed by numerous anastomosing channels. The relatively straight stretch of coastline between Daru and Fredrik Hendrik Island and much of the Papuan coast consists of a narrow belt of tidal flats and low beach ridges and swales and swampy back-plains. The coastal plains are extending gradually seaward.

The recent alluvial plains form bands 5–35 km wide along the present-day

Fig. 1. Blocked valley swamp with floating vegetation in one of the smaller tributaries of the Aramia River, Fly River area, Papua New Guinea.

rivers. They consist generally of a central meander belt including numerous cut-off meanders, scroll complexes, and sand bars flanked by slightly lower back-plains and swamps. These back-plains and swamps are often connected with blocked valley swamps and lakes formed by damming of the tributary valley by the more rapid sedimentation of the main river (Fig. 1). Major changes in stream courses have occurred in recent times like the one of the Angabunga River (Speight 1965).

The older alluvial plains are generally above the flood level of the present-day streams and receive little or no deposits; they are relict depositional features but parts can still be inundated during the wet season. The largest part of the Fly-Digul area is formed by such a relict alluvial plain. The southern part of these lowlands is the Oriomo plain between 20 and 50 m above sea level which extends from the mouth of the Digul River in the west to the mouth of the Fly River in the east. The northern part of the lowlands differs considerably from the Oriomo plain as it is formed of an intricately dissected pattern of low hills and ridges with a local relief of 10–30 m and slopes of between 5° and 30° (Fig. 2). It is thought that the difference in dissection is due to the northerly increasing amount of precipitation.

The central ranges

The central ranges are the dominant physiographic feature of New Guinea and occupy nearly half of the island. This mountain chain is a complex system of ranges and broad, generally strike-aligned, upland valleys and intermontane

Fig. 2. Northern part of the southern plains and lowlands with very low ridges and hills on old alluvium. Geomorphologically the area represents a relict alluvial plain.

plains. The chain varies in width from about 50 km at its western and eastern ends to about 200 km in its central parts. In the area west of the Strickland River the ranges rise abruptly from the foreland to reach altitudes of over 4000 m within a distance of 15 km. East of the Strickland, a zone of low foothills of variable width forms a gradual transition from the plains to the mountains.

There are four major landform types forming the central ranges (Fig. 12). These are mountains without obvious structural control, mountains with prominent structural control, karst landforms, and volcanic landforms.

Mountains without obvious structural control form the most widespread landform type and occur on any kind of rock. These mountains are characterized by narrow crested ridges with long, steep slopes, often irregular in detail. The valleys are V-shaped and the drainage pattern dendritic (Fig. 3). Weathering mantle varies with altitude and rock type but, in general, it is rather shallow and

Fig. 3. Typical ridge and V-valley mountains on soft sedimentary rocks with a dense pattern of dissection in the Chimbu area, Papua New Guinea.

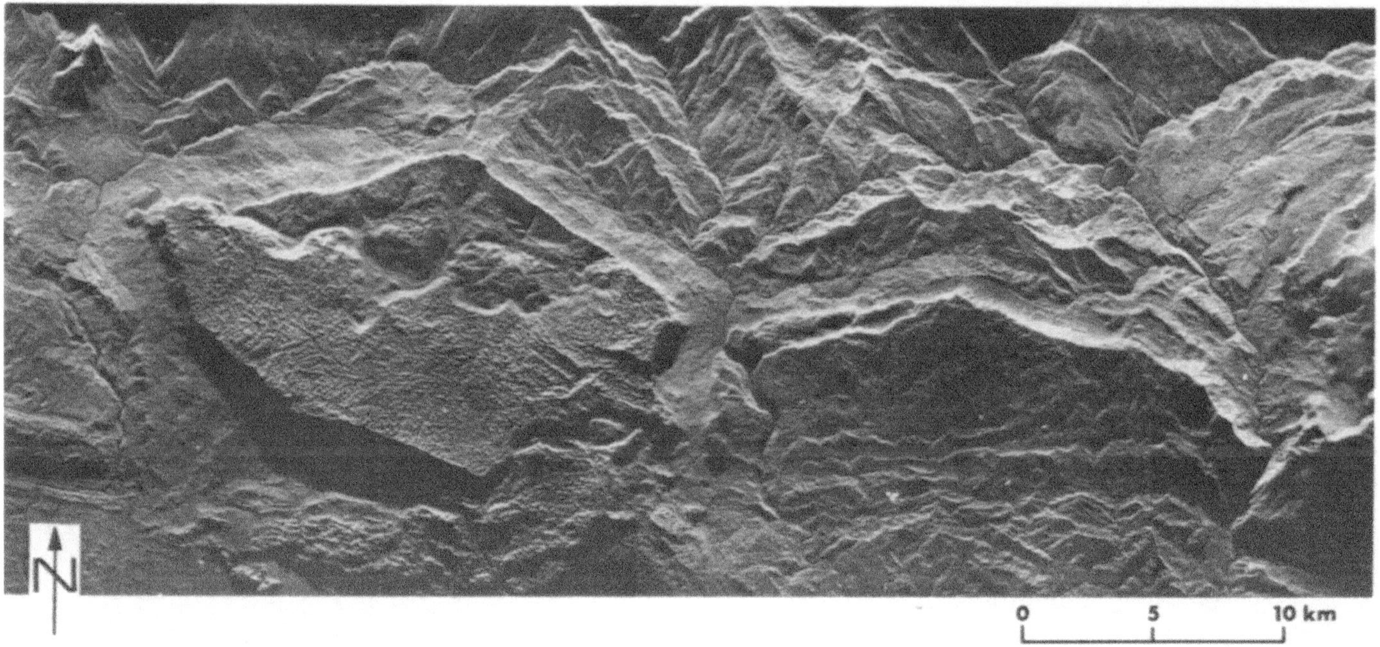

Fig. 4. Radar imagery of the southern part of the central ranges around the Strickland River (eastern margin). The imagery exhibits well the predominant westerly strike and southerly dip of the strata leading to southerly dipping homoclinal ridges. The massive limestone plateau to the west however shows only minor tilting. Its surface is pitted with tropical tower karst.

in spite of the humid tropical climate only immaturely weathered, a fact which is a reflection of the very active erosional processes. In detail ridges and V-valley vary with rock type from massive ridges with straight slopes and a coarse drainage pattern on ultrabasic and other igneous rocks to intricately dissected ridges with a fine grained drainage pattern on sedimentary rocks (Löffler 1977).

Mountains with prominent structural control develop where beds of varying resistance are present and differential erosion causes the less resistant beds to erode more rapidly than the resistant layers (Fig. 4). The rock structure of these resistant layers is thus given surface expression usually in the form of a smooth, gentle dip slope opposing an irregular steep escarpment. The greater the difference in relative resistance to weathering and denudation the more pronounced the structural control tends to be. These landforms are therefore best developed where resistant rocks like limestones or sandstones alternate with mudstones and siltstones. For structural reasons these landforms are most widespread in the southern parts of the central ranges. According to the steepness of the dip slope one distinguishes plateaux and cuestas (gentle slope) homoclinal ridges (moderate slope) and hogbacks (steep slope).

Karst landforms develop where rocks have a high degree of solubility in water and they are therefore most widespread (though not restricted) in limestone. Large areas of the central ranges west of the Erave River are formed on limestone and in most of these spectacular karst landforms are present. The dominant forms are rounded hemispherical hills, steep-sided pyramidal hills (Fig. 5) and spire-crowned towers with precipitous slopes surrounding enclosed depressions of great depth, but plateaux with long narrow crevices and deep sinkholes are also present (Fig. 4). Some of this country is completely inaccessible and uninhabited. The reason for the differences in karst landforms is not well understood but there is probably a complex relationship between climate, type of limestone and geomorphological history.

Volcanic landforms are restricted to the eastern part of the central range where the great strato-volcanoes like Mt Bosavi, Mt Soaru and Mt Karimui, and lava shield volcanoes like Mt Giluwe (Fig. 6), form prominent features in the landscape. The volcanoes are mostly conical in shape and are deeply dissected in their central parts and surrounded by broad, gently sloping footslopes and volcano alluvial fans. The two main volcanic areas in the central ranges are the highlands of Papua New Guinea where there are some 20 major volcanic centres and in Cape Vogel Basin, eastern Papua New Guinea where the Mt Lamington group and the Hydrographers volcano form two complex volcanic centres. The Hydrographers volcano is extinct but Mt Lamington erupted violently in 1951 causing widespread destruction and heavy loss of life. In contrast all the highland volcanoes are extinct and their ages range from 1.8 million years to about 200 000 years. However ash falls continued for some time, but it is not known when this activity ceased.

Interspersed with these mountainous landforms are upland valleys, which, although covering much less area, are more suited to human occupation. The most extensive upland valleys or intramontane plains are the Wahgi (Fig. 7), Goroka, Lai, Andabare, Kandep and Wasumu valleys in Papua New Guinea, and the Baliem valley in Irian Jaya. They are generally strike-aligned and

Fig. 5. Tropical tower karst south of Porema, Southern Highlands, Papua New Guinea. The asymmetry of the towers is due to the northerly (left) dip of the limestone.

Fig. 6. Mt Giluwe volcano, Southern Highlands, Papua New Guinea. Most of the presently extinct volcano was formed between 220 000 and 310 000 years BP. The Pleistocene ice cap roughly coincided with the presently grass covered summit areas.

Fig. 7. Intramontane basin in the highlands of Papua New Guinea at Goroka. This basin is filled with lacustrine and alluvial sediments and was probably formed by damming of the Ajura River (foreground) by faulting at the lowest point of the basin.

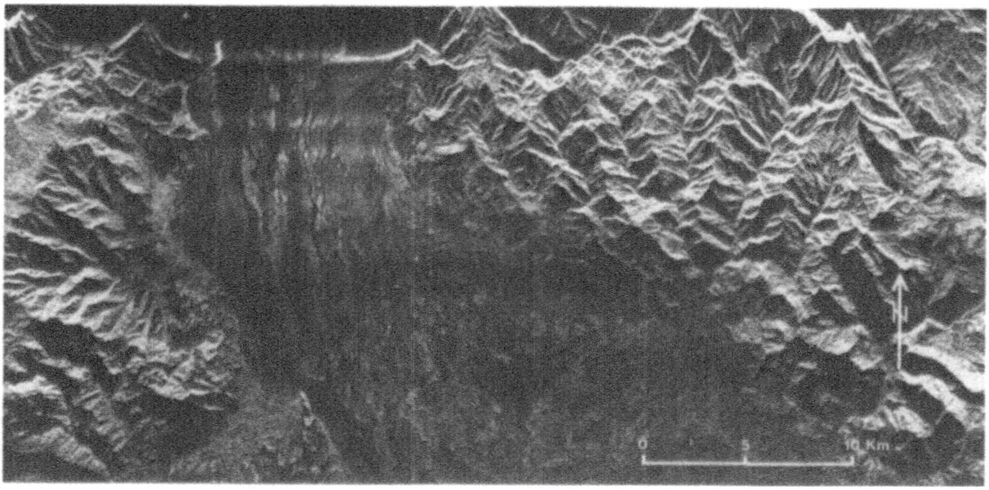

Fig. 8. Radar imagery of the Markham-Ramu graben with steep sided mountains on either side. The floor of the graben is filled with alluvial fan material most of which has been deposited by the streams and rivers from the Saruwaged Ranges (top). The braided floodplain with its numerous distributary channels can be clearly seen on the imagery.

Fig. 9. Massive mountain ridges along the southern fall of the Saruwaged Range in the northern ranges of Papua New Guinea.

covered with alluvium and also with volcanic ash from nearby volcanoes. The history of these intramontane basins is complex but most owe their existence to blocking of river courses by volcanic events.

The geomorphic history of the central ranges is closely related to the orogenic movements of the Pliocene and early Pleistocene and the framework of the present-day landforms is clearly tectonic. Well-preserved erosion surfaces similar to the ones described for Australia are absent but some relict surfaces not in equilibrium with present-day erosional processes, such as Mt Albert Edward, Mt Scratchley and Mt Tabletop, as well as summit accordances in several areas, are evidence that prior to the uplift a landscape of much lower relief existed. Some of the intermontane valleys and plains are probably also remnants of older landscapes, although most are due to ponding by volcanic events (Blong and Pain 1976, Löffler 1977).

The Intermontane Trough

The Intermontane Trough is a young structural depression; extending across the whole island from Geelvink Bay in Irian Jaya to the Huon Gulf in Papua New Guinea. This intermontane trough consists of extensive plains and lowlands, as well as flood plains and swamps. Most extensive are the Sepik plains in Papua New Guinea, and the Meer plain in Irian Jaya, with a meander belt flanked by wide swamps. The upper Ramu and Markham plains, by contrast, represent a young graben system (Fig. 8), with steeply sloping fans, and unstable braided rivers. These fans are formed of coarse material derived from the steep, tectonically active Finisterre and Saruwaged Ranges.

The Northern Ranges

The Northern Ranges run parallel to the central ranges; they consist of several individual ranges, separated by structural lowlands such as the lower Sepik-Ramu plain, and the Gogol plain. In Irian Jaya a major gap is caused by the erosional valley of the Mamberamo River which drains the Meer plain. The northern ranges are generally much lower and narrower than the central ranges, although they reach 4121 m at Mt Bangeta on the Saruwaged Range. The ranges consist predominantly of mountain ridges and V-shaped valleys (Fig. 9), flanked by irregular foothills. On the Huon Peninsula there are prominent structural limestone plateaux, and sloping surfaces with some karst features. The ranges descend steeply to a narrow, discontinuous coastal plain consisting of beach ridges and mangrove flats, alluvial plains and swamps and coral platforms and terraces. In contrast with the southern coast of New Guinea, which is relatively stable, the northern coast is rising at rates of up to 3 m per 1000 years. These earth movements are manifested in a spectacular staircase of coral terraces along the north coast of the Huon Peninsula (Fig. 10).

Islands ranging in size from about 30 000 km^2 to only a few square kilometres are situated off the northern and eastern coasts of New Guinea. The Bismarck Archipelago comprises all the larger islands which form an oval ring about the Bismarck Sea. Structurally these consist of two island arcs, the southern Bismarck Island arc and the northern Bismarck or New Ireland island arc. The southern Bismarck island arc includes New Britain and the chain of volcanic islands off the north coast of New Guinea. It is a typical island arc with an arcuate shape, a deep oceanic trench and a belt of volcanoes along its concave northern coast. Most of these are active or potentially so; they include the Manam, Bam, Karkar, Bagabag, Long I., Pago (Fig. 11), Langila, Ulawun and Bamus volcanoes.

The northern Bismarck island arc consists of New Ireland, New Hanover, the St Matthias Group and the Admiralty Is and represents structurally an extension of the Solomon Is.

New Ireland, is a long thin island trending north-west, with several narrow necks. The south-westerly part of the island is mountainous, while the central and north-western parts are dominated by well-developed tropical karst landforms.

Running parallel to New Ireland are the volcanic islands Feni, Tanga, Lihir and Tabar and numerous coral islets and reefs. Mussau, the main island of the St Matthias group, is a strongly dissected volcanic complex.

The Admiralty Is to the west comprise Manus Island and some very small

Fig. 10. Staircase-like series of uplifted coral terraces along the north coast of Huon Peninsula, Papua New Guinea.

islands grouped south of Manus in a rough semi-circle. The islands around Manus are mostly volcanic or coral. Tuluman volcano, south of Lou I., is an active submarine volcano.

The islands near the eastern tip of Papua New Guinea mainland, the D'Entrecasteaux Islands and the Louisiade Archipelago, are much smaller in area and represent structurally an extension of the mainland. They are formed by rugged mountains and surrounded by raised coral platforms.

The Trobriand Is, to the north of the D'Entrecasteaux group, are raised coral reefs with flat to undulating surfaces, combined often with sharp, rough pinnacles. Woodlark I., further to the east, is also basically an uplifted coral platform, but it has a core of metamorphic rocks rising above the coral at its southern corner.

The islands in Geelvink Bay, Schouten and Japen islands and Waigeo, Batanta and Salawati islands off the Vogelkop Peninsula are like their eastern

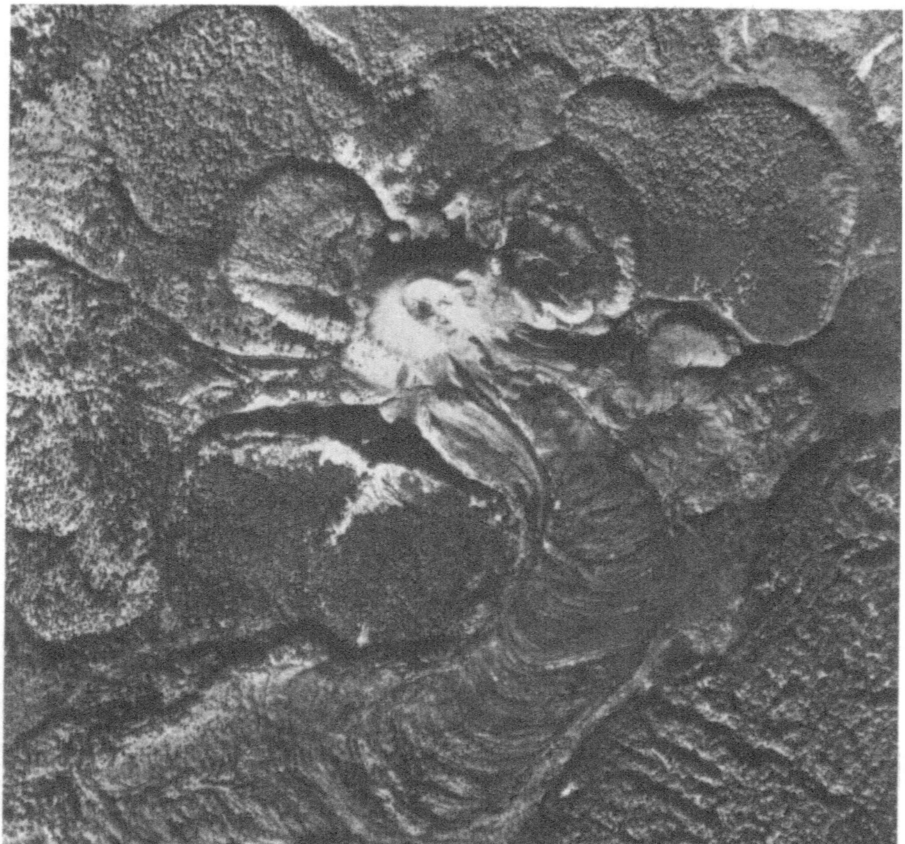

Fig. 11. Pago volcano, New Britain, Papua New Guinea. The volcano has emitted several tongue-shaped lava flows during the last few hundred years. The latest of these occurred in 1914 and the flow can be recognized because of its sparse vegetation cover (bottom).

counterparts sections of the mainland that became separated from it by the post glacial rise in sea level. The dominant landforms are hills and mountains as well as some karst landforms.

Geomorphological evolution

The present land mass of New Guinea started to develop in the lower Miocene with the emergence of the northern half of the central ranges which have remained dry land ever since. However it was not until the Pliocene, some 5–6 million years ago, that the present land mass of New Guinea finally became established. Vertical movements in the order of several thousand metres occurred along major faultlines and created the present mountainous relief. Folding was not as important being largely restricted to the northern margin of the basement block where large amplitude folds such as the Müller Anticline developed. Some tighter folding occurred in the transition zone between the Australian and the Pacific plates where today we see series of steeply dipping razorback like limestone ridges.

Uplift has continued till today but the main centres of tectonic activity appear to have moved from the central ranges to the northern ranges where maximum rates of uplift are in the order of 3 m/1000 years. It is important to realize that even though not noticeable on a human time scale, in geological terms this is a spectacular rate of uplift and the rates of uplift of the New Guinea mountains during times of intense tectonic activity are unlikely to have been much greater.

Closely following the uplift was extensive volcanism in the highland area. The exact date of the start of the volcanism is not known but it appears certain that by mid Pleistocene most of the volcanoes had been in existence for some time. Dating of rocks from the Doma Peaks area gave dates ranging from 1.6 million to 1.8 million years BP, while the oldest rocks so far found on Mt Hagen and Mt Giluwe are 1.27 and 0.86 million years.

The volcanoes formed at or close to their present altitudes since they erupted onto a deeply dissected mountain landscape with a drainage system very similar to the present one. A number of drainage basins were changed or disrupted by the volcanic events and this resulted in the formation of several intramontane basins. Some of these are simply valleys filled in with volcanic debris while others experienced more severe disruptions such as drainage reversals. Not all the intramontane basins are the result of volcanic activity, some are tectonic or structural in origin while others may be relicts of erosional basins formed close to sea level.

Major volcanic activity ceased in the highlands at about 200 000 years before present but eruptions of ash took place until about 50 000 years ago (Pain and Blong 1976). Minor ash showers have occurred intermittently till very recently but from sources outside the highlands, most probably the active volcanoes off the north coast.

Apart from the vigorous fluvial erosion that affected the entire mountain area and created the typical sharp crested ridges, the most important erosional process in the high mountains was glacial erosion and many of the high New

Guinea mountains show traces of a former ice cover. It may sound surprising to find traces of glaciation so close to the equator but it must be kept in mind that temperatures decrease with increasing altitude. During the ice ages when temperatures were generally 5–6°C below the present ones conditions for ice formation existed from about 3600 m upwards. Even today there is occasional snowfall on Mts Wilhelm and Giluwe and in Irian Jaya there are small active glaciers (see chapter I: 3 on Pleistocene and present day glaciations).

Another important landforming process was the so called karst erosion. Karst is a term used for the solution processes taking place in rocks that have a high degree of solubility in natural waters such as limestone. Typical features of karst are the lack of surface drainage and the development of conical and tower shaped residual hills. These landforms dominate much of the southern part of the central ranges.

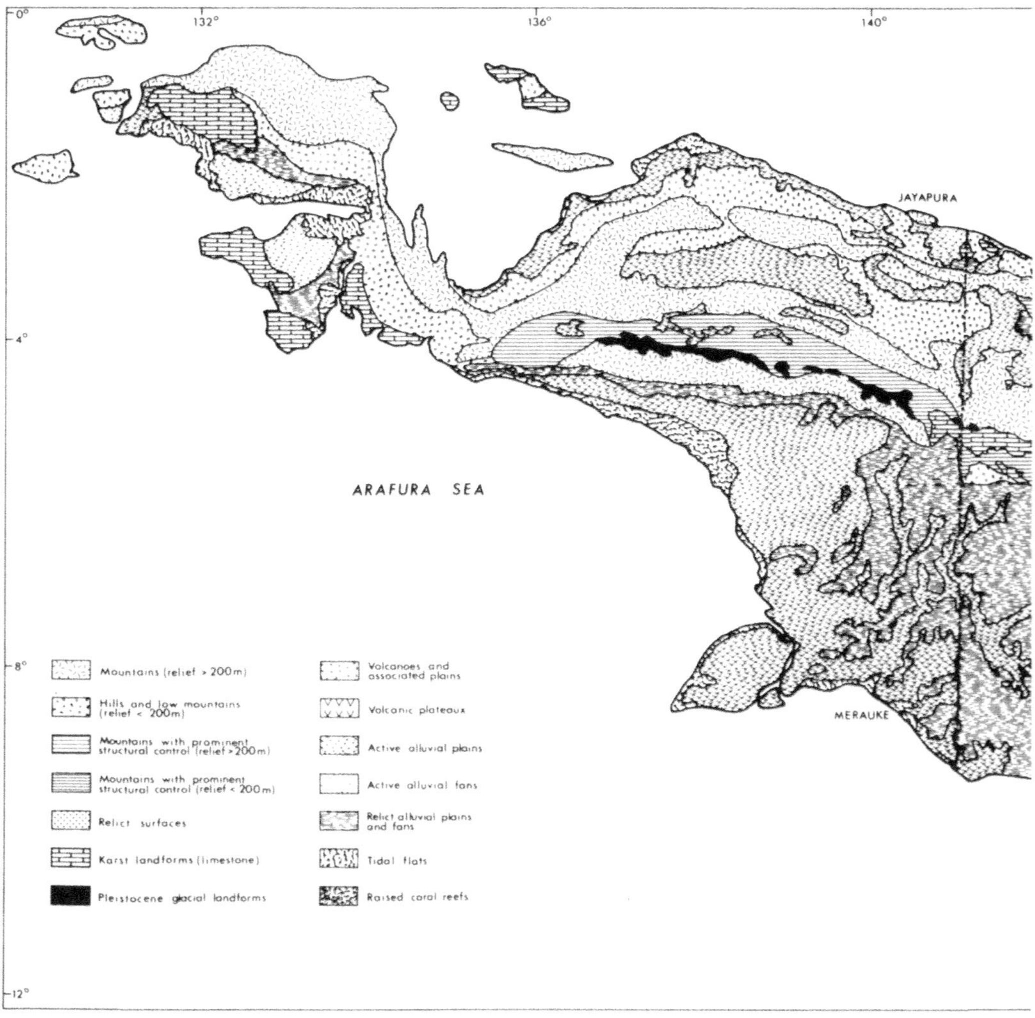

Fig. 12. Map of New Guinea Landforms.

Along the coastline there was another important event that influenced the landform development. This was the world wide lowering of sea level by some 130 m or more caused by the fixing of large quantities of water on the northern hemisphere ice shields. This sea level lowering caused some major changes in the distribution of land and sea in the south of New Guinea where the Torres Strait became dry land and thus Australia and New Guinea formed one land mass. This was of great significance not only for the movement of animals and plants but also of man.

Along the other sections of the coastline the drop in sea level did not result in any significant land gains because of the steep offshore slopes of most of the land. Sea level rose rather rapidly from about 15 000 years before present until about 6000 years ago when sea level reached its present level and this must have caused considerable hardship for the coastal people who must have seen their

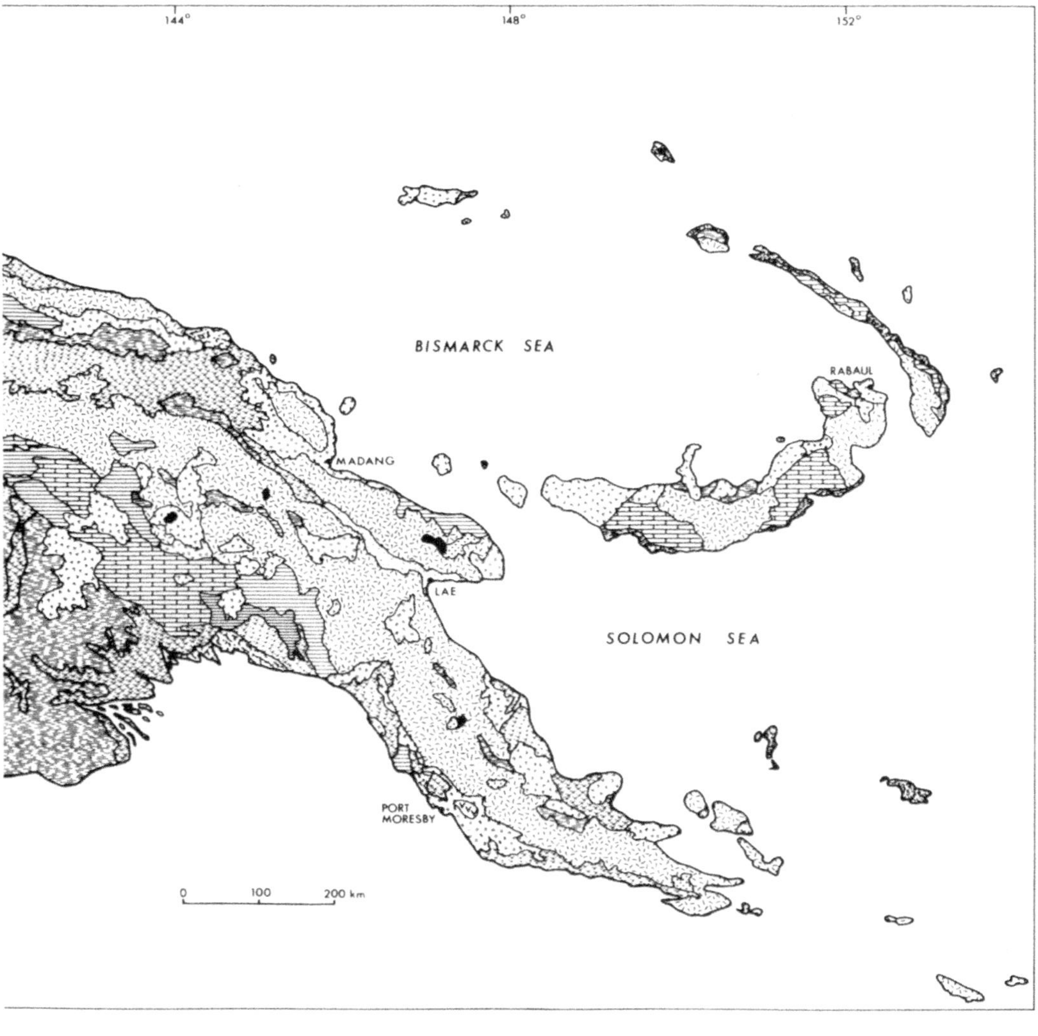

land steadily decreasing. Partly the land loss was offset by creating a new and longer coastline between Australia and New Guinea and by rapid filling in of embayments and other low lying plains with sediment and organic matter.

Landforming processes continue to shape and change the landscape some more readily observable than others. Earthquake induced land sliding, volcanic eruptions, flooding and changing river courses are constant reminders that the New Guinea landscape is actively changing. But other processes much less obvious such as weathering, slope wash, subsurface erosion, solution, and soil creep are constantly at work. Even animals such as earth worms and termites contribute to the processes of erosion and deposition even though their total contribution may be small.

References

Löffler, E. 1977. Geomorphology of Papua New Guinea. Aust. Nat. Univ. Press.

Pain, C. F. and R. J. Blong. 1976. Late Quaternary tephras around Mt Hagen and Mt Giluwe, Papua New Guinea. In: Volcanism in Australasia, ed. R. W. Johnson. pp. 239–251. Elsevier, Amsterdam.

Speight, J. G. 1965. Flow and channel characteristics of the Angabunga River. J. Hydrol. 3: 16–36.

5. The soils of New Guinea

A. W. Wood

1. Introduction

Only two soil maps of the whole island of New Guinea have been produced. Haantjens *et al.* (1967) published a map and report describing the major soil groups of New Guinea, and this resulted from the correlation of soil information from West and East New Guinea for the FAO Soil Map of the World project. The FAO Soil maps were subsequently published and the mapping units are described in FAO-UNESCO (1974).

The FAO project has been the only attempt to correlate soil information from the eastern and western parts of the island and since then survey work and research have proceeded independently. Soil data from Irian Jaya consists largely of the results of several semidetailed surveys of small areas (Reynders 1961; Schroo 1963a, 1963b, 1964), and a reconnaissance survey in the Central Highlands (Reynders 1964), all carried out during the Dutch period. In Papua New Guinea, soil investigations have been of three main types: reconnaissance soil and land resource survey by CSIRO Division of Land Use Research between 1964 and 1971 (Haantjens 1965); semidetailed soil surveys carried out by the Land Utilization Section of the Department of Primary Industry (Hartley, Aland and Searle, 1967; Knight, 1973; Zijsvelt 1977); and investigations by various independent research workers.

Two maps have been published of the soils of Papua New Guinea, one a modification of the 1967 map of the whole island (Haantjens 1970), and the other an updated version of this map using FAO nomenclature (Bleeker 1974).

In view of the limited information available on the soils of Irian Jaya, the soil descriptions in this chapter will deal mainly with Papua New Guinea soils.

2. Soil classification

There is no national system of soil classification in Papua New Guinea, and this has resulted in the use of several different systems in soil survey reports. The various classification systems used in Papua New Guinea have been reviewed by Wood (1976).

The CSIRO Division of Land Use Research has used classification systems of

Monographiae Biologicae, Vol. 42, ed. by J. L. Gressitt.
© 1982, Dr W. Junk Publishers, The Hague.

local application in its earlier reports before 1967. The soil groupings were defined to suit local conditions and were not part of an overall soil classification scheme of wider application. The report on the soils of the Safia-Pongani area (Haantjens 1967a) was the first to use an international system of soil classification, in this case the American 7th Approximation system (USDA 1960), later to become the Soil Taxonomy (USDA 1975).

Haantjens (1967b) has outlined some of the problems encountered in using the American system. This system is not entirely suited to the classification of soils during rapid reconnaissance surveys where there is generally a lack of soil analytical data, and where field soil descriptions may be lacking in detail. Also the precisely defined differentiating criteria of the American system has in some cases led to the creation of more soil groupings than are practicable, some of which have little significance to the major objectives of the survey. In view of these restrictions, Haantjens (1967b) found it necessary to modify and simplify some of the differentiating criteria of the American system.

The Land Utilization Section of the Department of Primary Industry has also used classification schemes of local application in many unpublished reports. A virtually unmodified version of the 7th Approximation was used in the reports on the Markham Valley (Knight 1973), and a modified version of Northcote's classification (Northcote 1965) in a report on the soils of West New Britain (Hartley, Aland and Searle 1967).

The large number of different systems which have been used to classify New Guinea soils make soil correlation between various areas a difficult task. Attempts were made to establish a computer based storage and retrieval system for the soil and landscape data gathered by the CSIRO surveys (Haantjens, Bleeker and Cook 1975). This was hoped to provide a general review of the nature and distribution of Papua New Guinea soils, and to allow extrapolation. This computer based system however has proved to be unworkable.

3. The soil map

A map of New Guinea soils is presented in Fig. 1. Twelve map units have been used and these are largely based on the units shown in the earlier maps of the area by Haantjens *et al.* (1967), Haantjens (1970) and Bleeker (1974). The number of soil mapping units has been reduced and modifications have been made to some units, based partly on my own field observations. The soil nomenclature has been changed to more simple terms that convey information not only about the type of soil present, but also about the geology and geomorphic environment.

4. The soil mapping units

The mapping units are essentially associations of soils, and some contain a wide variety of soil types. Due to the great variability of soils over small areas within New Guinea, only a generalised picture of soil distributions can be presented in map form.

The map units have been divided into two groups: those occurring mainly in highland areas; and those of the lowlands. Map unit 5 occurs in both highlands and lowlands.

4.1 Highland soil associations

4.1.1 Lithosols and Alpine Humus soils. This association occurs mainly on rugged mountain summit areas at 3000–4500 m on a wide variety of parent materials. The majority of the summit areas have been glaciated, and there are extensive areas of bare rock. In Irian Jaya, summit areas above 4500 m still carry permanent ice fields (Hope *et al.* 1976). The soils are generally shallow Lithosols and Rankers, and the cool, wet climatic conditions have encouraged local accumulations or raw organic matter, forming Alpine Humus soils or areas of peat.

Rutherford (1964) has described a sequence of Alpine Humus soils formed on glacial and fluviogacial deposits on the summit of Mt Giluwe. The soils are generally shallow and have a dark brown silty clay to clay surface horizon overlying a well decomposed black organic clay subsoil. The soils are strongly acid, high in organic matter and total phosphorus, and low in base saturation. In another paper, Rutherford (1968) has shown that different soil types are arranged in a series of semi-parallel zones around Mt Giluwe. Soil types range from shallow poorly developed, highly organic Alpine Humus soils on the summit, to poorly drained soil on the fluvioglacial deposits of the Upper slopes, and Humic Brown Ash soils (Map unit 4) on the ash covered lower slopes.

The soils of the Irian Jaya mountains are briefly described in Hope *et al.* (1976: 119, 121). Alpine humus soils are dominant below the limit of recent glacial action, and mineral soils have developed on rock fans.

4.1.2 Rankers and regosols. This association occurs on rugged mountains with steep slopes, 500–3000 m. Slopes tend to be unstable, soil creep is common and occasional landslides occur. Soil profiles are generally shallow due to the instability of most slopes, although deeper soils do occur on more stable hill crests and upper slopes. Rankers and Regosolic Brown soils predominate on the unstable slopes, and Lithosols occur on the steepest slopes. Podzolic soils are included in this association and occur more on acid parent materials. There are more extensive areas of podzolic soil in Irian Jaya than in Papua New Guinea, and Haantjens *et al.* (1967) have suggested that this can be explained by a general decrease in the acidity of rock types from west to east.

4.1.3 Rendzinas and limestone soils. This association occurs only on limestone rocks, 1000–4000 m. Karst features are common, shallow, stony Rendzinas are dominant and occur on steep slopes, with Lithosols on very steep slopes. Deeper red and brown calcareous soils are more common on older surfaces and on more stable slopes.

Rutherford and Haantjens (1965) have described the main soil types occurring on limestone in the Southern Highlands Province of Papua New

Guinea as being either Shallow Dark Clay soils (Rendzinas) on steep limestone slopes with much rock outcrop, or Deep Dark Clay soils on gentle colluvial slopes, flat surfaces, depressions and doline floors. In some of the main valleys in the limestone area, the limestone has been mantled with volcanic ash, and Humic Brown Ash soils have formed (map unit 4).

4.1.4 Humic brown ash soils. This association is found in the Central Highlands of Papua New Guinea, 1000–3000 m. The parent material is mainly weathered andesitic volcanic ash occurring on extensive plains and slopes around extinct Pleistocene volcanoes. The ash overlies a variety of sedimentary rocks which include sandstone, shale, mudstone and limestone. On steeper unstable slopes the ash cover may have been partially or completely removed. Soil profiles are generally deep, with thick dark brown or black humic topsoils overlying yellowish brown to strong brown clay subsoils.

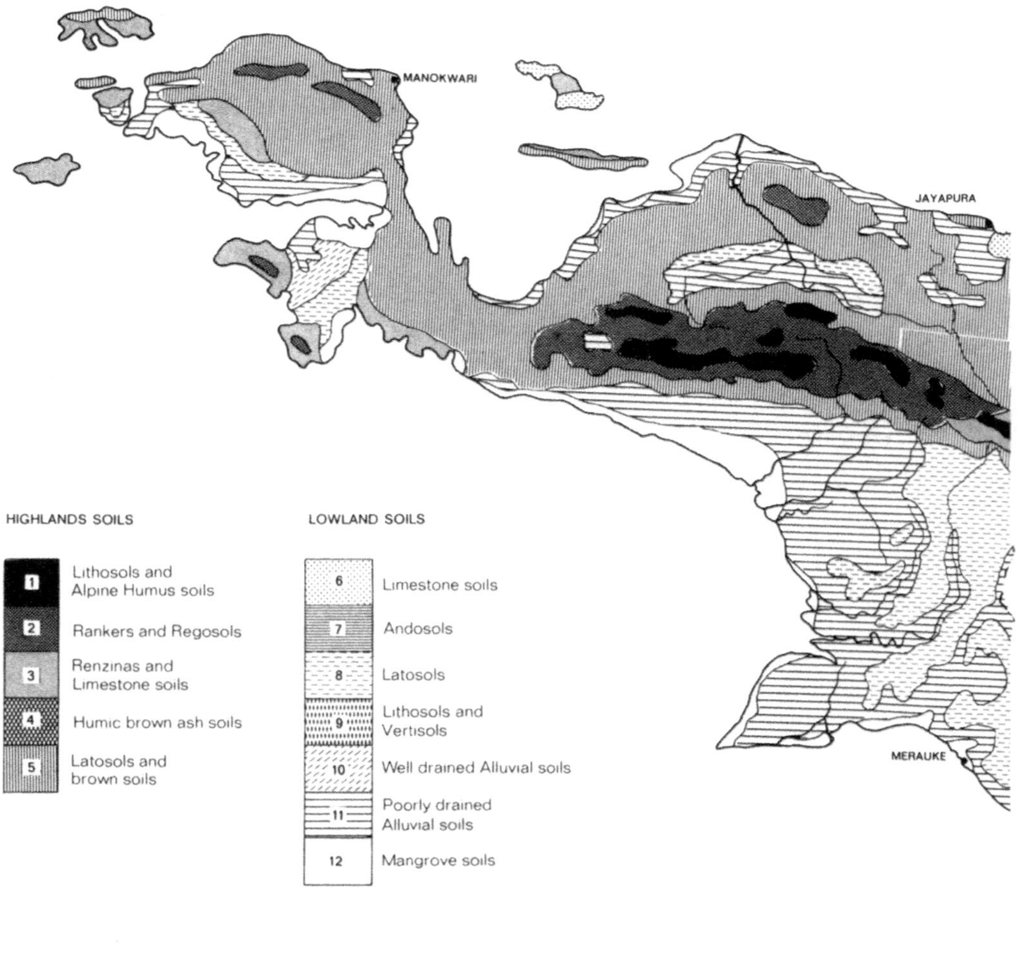

Fig. 1. New Guinea soils.

These soils have been described by Rutherford (1962) as occurring on a wide range of parent materials including volcanic ash, sandstone, greywacke, shale, siltstone and limestone. A distinction is made between soils formed on volcanic ash and those formed on sedimentary rocks, the former having thicker and darker organic topsoils, and the latter having firmer, more massive and more plastic subsoil horizons. Haantjens and Rutherford (1964) have termed these soils zonal soils and suggest that they have developed largely as a response to the cool and wet climatic conditions. Pain and Wood (1976) have questioned this zonal theory and consider that these morphologically distinctive soils are only found in areas where the sedimentary formations have been mantled by thick deposits of volcanic ash (Fig. 2).

4.1.5 Latosols and brown soils. This association includes a wide variety of soil types occurring on hilly and mountainous country, 0–2500 m. The soils are of

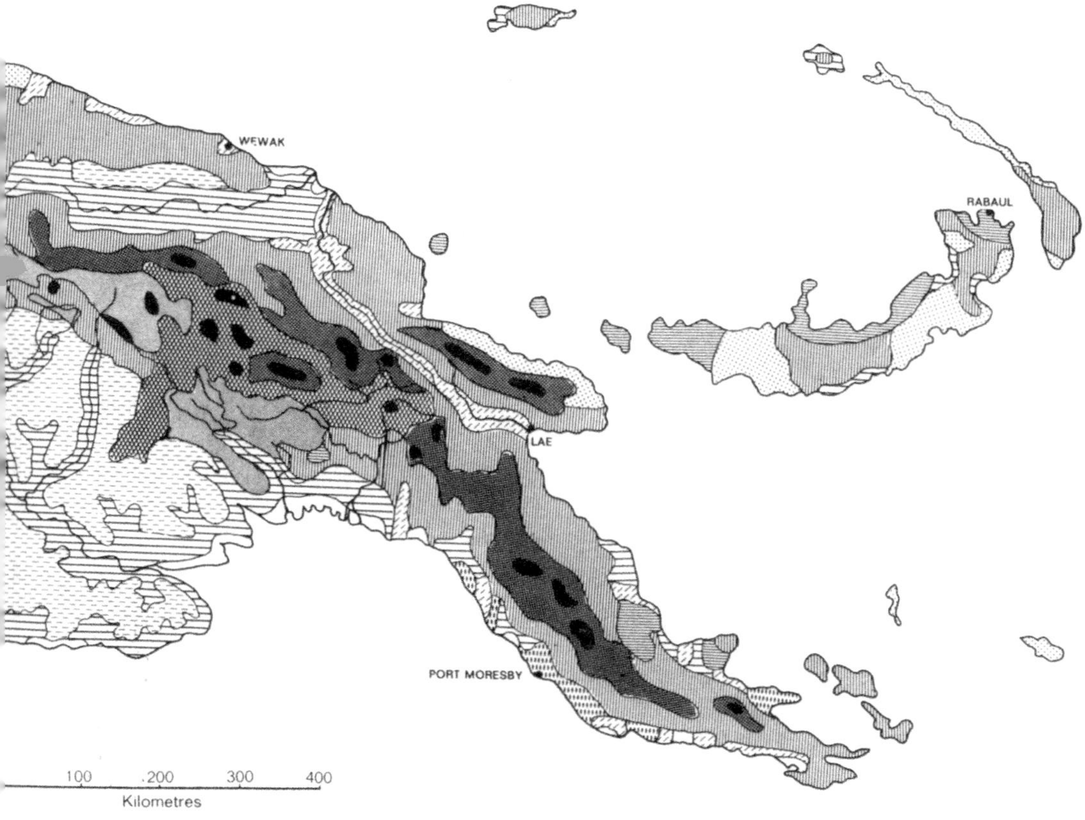

variable depth depending on slope steepness and stability with Lithosols and Regosols on the steepest slopes, Regosolic Brown Soils on unstable slopes, and Brown Forest soils on more stable slopes. Brown Podzolic Soils occur in parts of this map unit, particularly in the western part of New Guinea on more acidic rocks, and where the climate is wet. At lower altitudes Red and Yellow Latosols occur on stable crests and lower slopes.

This is the largest of the map units shown on Fig. 1 and stretches from the eastern tip of the Papua New Guinea mainland to the western tip of Irian Jaya. It contains a wide variety of soil types developed on many different parent materials. It is a unit that has been created for mapping convenience, and would be split up into a number of smaller units on a larger scale map.

4.2 Lowland soil associations

4.2.1 Limestone soils. This association occurs along coastal areas on uplifted coral platforms and limestone hills below 1000 m. Shallow Rendzinas occur on young recently uplifted coral reefs and on steep slopes, whereas deep, red Terra Rossas predominate on older, higher surfaces. Lithosols are found on very steep slopes.

The Rendzinas are characterised by dark brown humic surface horizons overlying dark greyish brown sandy material that merges into weathered coral. The Terra Rossas have dark reddish brown to dark brown surface horizons overlying reddish brown clay, and can exceed 2 m in depth

4.2.2 Andosols. This unit comprises soils that have developed on volcanic ash, lava and agglomerate around active and dormant volcanoes, sea level to 2000 m. Andosols have typically formed on airborne volcanic ash deposits, and can differ greatly in texture, degree of profile development and chemical properties according to the age and nature of the ash.

Hosking (1938) has described some soils from Northern New Britain near Rabaul, developed on volcanic parent materials. The soils are formed in layered pumiceous material, and the profiles have black to dark greyish brown surface horizons overlying yellowish brown to grey subsoils. Textures can vary from fine sand on recent ash showers to loam or clay loam on older, more weathered deposits. On lava and agglomerate flows, Lithosols are common on the youngest flows, and Brown soils and Latosols occur on older flows.

4.2.3 Latosols. This association occurs as an extensive area in the southern part of the island and consists of flat to undulating plains formed on weathered Pleistocene sediments, 10–100 m. The soils are predominantly freely drained Red and Yellow Latosols, which have been classified by Bleeker (1971) as Ultisols according to the American soil classification system. Bleeker describes the soils as being mainly acid with loamy topsoils overlying clayey subsoils. The formation of these soils is thought to be influenced by the alternating wet and dry seasons which characterise this area. The contrast in texture between topsoils and subsoils could be the result of clay movement down the profile caused by seasonal changes in soil moisture (Bleeker 1971).

Within this unit are a series of broad drainage depressions which are inundated during the wet season. Gleyed soils with bleached surface horizons and mottled subsoils commonly occur in these areas.

4.2.4 Lithosols and vertisols This association occurs in drier areas with a distinct dry season. It is typically developed in the Port Moresby area, where the land forms consist of alternating hills and valleys, with a relief of less than 600 m. Parent materials are mainly calcareous with limestones, calcareous sandstones and mudstones. Colluvial deposits occur in hill footslopes and in valleys. Shallow Lithosols occur on hill crests and side slopes; deep stony colluvial soils on footslopes, and vertisols in the valleys. The vertisols are deep, black clay soils with expanding lattice clay minerals which expand when wet and contract when dry. They will only form in calcareous environments and where there are alternating wet and dry seasons (Figs. 4 and 5).

4.2.5 Well drained alluvial soils. This association occurs on gently undulating plains, stable flood plains and alluvial fans. Parent materials consist entirely of Recent alluvial deposits. Soils are predominantly coarse textured and freely drained (Fig. 6).

4.2.6 Poorly drained alluvial soils. This association occurs on poorly drained level floodplains, and swamps of alluvium and peat, 5–150 m. Soils are commonly fine textured and strongly gleyed due to the poor drainage, and during the wet season are commonly subject to inundation. These soils contain expanding lattice clays (van Es and van Schuylenborgh 1967) which cause the soils to crack when they dry out. Soils tend to be strongly alkaline near the coast and near tidal rivers, and increase in acidity inland (Bleeker 1971). Strongly gleyed swampy soils and deep organic peats occur in permanent swamps.

4.2.7 Mangrove soils. This association consists of tidal mangrove plains at sea level consisting of marine sands, muds and peat. Soils are fine textured, poorly drained and are everywhere saline, due to the marine influence.

5. Soil formation

Haantjens (1970b) has summarised the main factors influencing soil type in New Guinea as being weathering, climate, vegetation and drainage, acting on parent material. New Guinea soils are characterised by a scarcity of highly weathered profiles, possibly due to erosion, youth and instability of most landforms (Haantjens and Bleeker 1970). In view of this many soils in New Guinea are formed in only slightly weathered parent materials, and have a broad similarity to the soils of humid temperate areas. Where there are stable landforms, such as the low plains of the southern part of the island (map unit 8), strong weathering has occurred and more typical "tropical" soils have formed.

Fig. 2. Humic brown ash soil near Tari in the Papua New Guinea highlands. Note the abrupt boundary between the organic topsoil and the yellowish brown subsoil.

Fig. 3. Deep peat soil overlying greyish clay, formed under swamp conditions in the Papua New Guinea highlands.

Fig. 4. Soil formed on stony colluvium near Port Moresby, Papua New Guinea.

Fig. 5. Deep, black Vertisol exposed in the bank of a small stream near Port Moresby.

Fig. 6. Stony, well drained alluvial soil on the Gogol river floodplain near Madang, Papua New Guinea.

The effect of parent material on soils is greatest in shallow, young soils (map units 1, 2, 3, and 6). The calcareous parent materials and the drier climatic conditions have given rise to a distinctive group of soils in map unit 9. The Andosols (map unit 7) owe their specific properties to the nature of the coarse grained volcanic ash, and the more highly weathered ash soils (map unit 4) have developed as a response to the parent material and the cool wet climate.

The main effect of poor soil drainage on profiles is to produce gleying, which shows up as mottling or grey or blue colors. Only the alluvial soil units (map units 10 and 11) have been distinguished on the basis of drainage.

Climate is another important soil forming factor. Altitude has an effect on the climatic controls of soil formation, and for this reason the soil mapping units have been divided up into highland soil associations and lowland associations. In the highlands above 1500 m, temperatures are approximately 10°C lower than in the coastal lowlands, and annual rainfall totals are higher. The cool wet conditions which prevail in the Highlands influence the rate of organic matter accumulation. Highland soils generally have thicker, darker organic topsoils, and these are particularly well developed in the Humic Brown Ash soils of the main Highlands valleys of Papua New Guinea. The amount of organic matter in these soils increases with increasing altitude.

6. Agricultural potential

The rugged nature of the topography over much of the New Guinea Highlands and the steep unstable slopes mean that much of this area has a low potential for agriculture. The highland valleys with Humic Brown clay soils in Papua New Guinea, and with poorly drained peaty soils in Irian Jaya support high populations and have a high agricultural potential.

In the lowlands, most soils have a low natural fertility apart from the Andosols (map unit 7). The well drained Alluvial soils of the Markham and Ramu valleys have a high potential, although the dry climate limits the range of crops that can be grown. Limited areas of alluvial soils in coastal areas, particularly those with well drained soils, have a moderate to high potential for agriculture.

References

Bleeker, P. 1971. Soils of the Morehead — Kiunga Area, CSIRO Land Res. Ser. 29: 69–87.

—— 1974. Soils. In: E. Ford (ed.), Papua New Guinea Resource Atlas. Jacaranda Press, pp. 10–11.

FAO-UNESCO. 1974. Soil Map of the World, Volume 1 Legend, 59 pp., UNESCO Paris.

—— Soil Map of the World, IX South East Asia, 1 map sheet, UNESCO Paris.

—— Soil Map of the World, X Australasia, 2 map sheets, UNESCO Paris.

Haantjens, H. A. 1965. Practical aspects of Land System Surveys in New Guinea, J. Trop. Geography 21: 12–20.

—— 1967a. Pedology of the Safia — Pongani area, CSIRO Land Res. Ser. 17: 98–141.

—— 1967b. Use of the Seventh Approximation in the soil classification in the Safia — Pongani area, CSIRO Land Res. Ser. 17: 194–197.

—— 1970a. Soils. In: Ward R. G. and D. A. M. Lea (Eds). An Atlas of Papua New Guinea, University of Papua New Guinea, pp. 40–41.

—— 1970b. New Guinea Soils: their Formation, Nature, and Distribution, Search 1(5): 233–238.
Haantjens, H. A. and P. Bleeker. 1970. Tropical Weathering in the Territory of Papua and New Guinea, Aust. J. Soil Res., 8: 157–177.
—— 1975. Procedures for computer storage of soil and landscape data from Papua New Guinea, II Input of soil capability and soil classification data, Geoderma, 13: 115–128.
Haantjens, H. A., P. Bleeker and L. G. N. Cook. 1975. Procedures for computer storage of soil and landscape data from Papua New Guinea, I General introduction and input of descriptive soil observation data, Geoderma, 13: 105–113.
Haantjens, H. A., J. J. Reynders, W. L. P. J. Mouthaan and F. A. van Baren. 1967. Major soil groups of New Guinea and their distribution. Communication No. 55, Department of Agricultural Research, Royal Tropical Institute, Amsterdam, 87 pp.
Haantjens, H. A. and G. K. Rutherford. 1964. Soil zonality and parent rock in a very wet tropical mountain region, Trans 8th Int. Cong. Soil Sci., Bucharest 5: 493–500.
Hartley, A. C., F. P. Aland and P. G. E. Searle. 1967. Soil Survey of West New Britain, The Balima — Tiauru Area, Soil Survey Report No. 1, D.A.S.F., Port Moresby.
Hope, G. S., J. A. Peterson, I. Allision and U. Radok. 1976. The Equatorial Glaciers of New Guinea, Balkema, Rotterdam, 244 pp.
Hosking, J. S. 1938. Some recent volcanic deposits and volcanic soils from the islands of New Britain in the Territory of New Guinea, Trans. Roy. Soc. S Aust. 62: 366–377.
Knight, M. J. (Ed.) 1973. Land Resources and agricultural potential of the Markham Valley, Research Bulletin No. 14, Land Utilization Section, D.A.S.F., Port Moresby, 10 volumes.
Northcote, K. H. 1965. A factual key for the recognition of Australian soils, CSIRO Div. of Soils Report 2/65.
Pain, C. F. and A. W. Wood. 1976. Tephra Beds and Soils in the Nondugl — Chuave area, Western Highlands and Chimbu Provinces, Papua New Guinea, Science in New Guinea 4(3): 153–164.
Reynders, J. J. 1961. Soil Survey in Netherlands New Guinea, Boor en Spade 11: 78–83.
—— 1964. A Pedo — ecological survey of Soil Genesis in the Tropics from sea level to Eternal Snow, Star Mountains, Central New Guinea. D. J. Drill, Leiden, 159 pp.
Rutherford, G. K. 1962. The yellow brown soils of the highlands of New Guinea, Trans. Int. Soc. Soil Sci., New Zealand, pp. 434–39.
—— 1964. The tropical alpine soils of Mt Giluwe, Australian New Guinea, Canad. Geogr. 8: 27–33.
—— 1968. Observations on a succession of soils on Mt Giluwe, Ann. Ass. Amer. Geogr., 58: 304–12.
Rutherford, G. K. and H. A. Haantjens. 1965. Soils of the Wabag — Tari Area, CSIRO Land Res. Ser. 15: 85–99.
Schroo, H. 1963a. An inventory of soils and soil suitabilities in West Irian, I, Netherlands J. agric. Sci. 11: 308–333.
—— 1963b. An inventory of soils and soil suitabilities in West Irian, 2A, Netherlands J. agric. Sci. 11: 387–417.
—— 1964. An inventory of soils and soil suitabilities in West Irian, 2B, Netherlands J. agric. Sci. 12: 1–19.
U.S.D.A. 1960. Soil classification: a comprehensive system, 7th Approximation, U.S. Govt. Printer, Washington.
—— 1975. Soil Taxonomy: A basic system of soil classification for making and interpreting soil surveys, Agriculture Handbook No. 436, U.S. Govt. Printer, Washington.
van Es, F. W. J. and J. van Schuylenborgh. 1967. Contribution to the knowledge of a solonetzic, magnesium rich alluvial silty clay in the Maro — Keombe plain, Netherlands J. agric. Sci. 15: 11–20.
Wood, A. W. 1976. Soil classification in Papua New Guinea. A review of systems used. Unpub. working paper No. 15, S.P.C. Tech. Mtg., Suva, Fiji, July 1976.
Zijsvelt, M. F. W. 1977. Soil survey and Land Use Potential of the Kapiura — Dagi Area, West New Britain. Research Bull. No. 19, D.P.I., Port Moresby.

PART TWO

Man and his impact on the environment

1. Biological exploration of New Guinea

D. G. Frodin & J. L. Gressitt

1. Introduction

The biological exploration of New Guinea has a relatively long history but for long it was casual, often centered around "curiosa," and for a number of reasons serious work did not begin until the mid-19th century, with penetration of the interior only after 1870.

Many of the early voyages of exploration up to 1840 touched mainland New Guinea or some offshore islands, but only some carried naturalists who took back specimens for study — notably those of Bougainville, Cook, d'Entrecasteaux, Duperrey, Dumont d'Urville and Belcher. Although many important collections were made on these voyages — notably by French naturalists — the first sizeable zoological collection was made by Wallace in 1858. The first major one in botany was in 1872 by Beccari with whom the zoologist-adventurer d'Albertis made the first real ascent into the interior, to only 1100 m, in the Arfak Mts. Beccari was to return to the Vogelkop in 1875, the same year that collecting in SE New Guinea, stimulated by the *Challenger* rivalries and Australian "manifest destiny," began in earnest, leading finally to Forbes' large effort in 1885–86. In the Bismarck Archipelago, 1875 was also the year of the visit of the German circumnavigation in the *Gazelle* as well as of Turner, a naturalist collecting for von Hügel, then working in the S Pacific. NE New Guinea opened up more slowly: although Miklucho-Maclay had resided on Astrolabe Bay in 1871–72 and 1876–77, and Finsch had collected along the coast in 1880–82 and 1884, serious work only began in 1886 with the first large official exploring expedition from Germany.

Aside from limited mountain ascents (notably by Hunstein behind Port Moresby), much of the early work was along coasts, main rivers and on offshore islands (e.g. Duke of York Is., Japen, Saibai, Waigeo). It was MacGregor — one of the few responsible administrators with a marked interest in natural history — and his officers who first greatly expanded knowledge of the mountainous interior. They made several major ascents, notably Victoria in the SE, 1889–98, although Cuthbertson and Sayer reached Mt Obree in 1887 — making a record. Not until the early 20th century was there serious mountain collecting elsewhere: the main cordillera in the west, where the Oranje and Nassau ranges were ascended by Lorentz's (1907–13) and Wollaston's

© *1982, Dr W. Junk Publishers, The Hague.*

(1909–13) expeditions; also the Saruwaged Range on the Huon Peninsula by Keysser in 1912–13. Schlechter unsuccessfully attempted Mt Otto in 1908. Since then, activity has continued in many areas at a higher level, although in Irian Jaya rather less since 1962. By 1939 the major features were more or less known, but gap-filling still continues.

2. Early history

Though New Guinea was visited earlier by Malay (?and Chinese) traders from Dobo and elsewhere, its earliest recorded explorers were Portuguese, in part in search of the great "southern land" then thought to be necessary to balance the large land masses of Eurasia and N America. Not until the 17th century — and this became generally known only much later — did this part of the "great land" prove to be an island. The first to come was Jorge de Meneses in 1527, but he had no idea of its extent. He was followed by Alvaro de Saavedra in 1528 and Ortes de Retes in 1545. The latter named the island New Guinea because the people looked like Africans. All of them landed on the NW coasts and offshore islands, some near Manus. Spanish entry only came later, and then from the Americas: in 1567 Alvaro de Mendana discovered the later–lost Solomon Is., eventually returning with Pedro Fernandez de Quiros, who in 1595 discovered the New Hebrides. In the next decade came the important expedition of Luis Vaiz de Torres who sailed along the south coast in 1606 and traversed the strait later named for him.

The first Dutch voyage to the East Indies was in 1597 and contacts by Dutchmen with New Guinea began in earnest early in the 17th century. Willem Jansz on the *Duyfken* in 1606 may have been the first, but more important was the voyage of Willem Schouten and Jacob Le Maire in 1616, viewing New Ireland, Manus and much of the north coast. Both voyages suffered massacres. In 1623 Jan Carstensz saw the SW coast and was the first to see, in disbelief, the ice-capped highest peaks of the mainland which for long have borne his name (now Gunung Jaya). Two decades later, in 1642, Abel Tasman touched New Ireland and (for the first time) the shores of New Britain, but thought both were continuous with New Guinea and Australia.

In 1700 was the epoch-making visit by the Englishman William Dampier. In his ancient ship *Roebuck* he visited the north coast and discovered Mussau and Amira north of New Ireland, sailed along the north coast of the latter and then past his "St George's Bay" along the south coast of New Britain, discovering the strait between it and New Guinea. Dampier bestowed the name Nova Britannia by which it has since been known (except as Neu Pommern during German rule). Dampier, the ex-buccaneer and explorer, was also a natural historian, and brought back "curiosa" for study: the earliest scientific specimens from the region. The Dutch continued to make discoveries and contacts in the western areas (including from the *Geelvink* in 1707), and in 1714 New Guinea, with the S Moluccas, was ceded to the Dutch by the Sultan of Tidore.

However it was the English and French who were to set an entirely new trend, taking a cue from Dampier. From the 1750's most voyages involved serious scientific work as well as exploration and contact, and carried naturalists. In 1767 Philip Carteret in the *Swallow* visited parts of the SW Pacific, the NW Solomons and "New Britain" — and found that "St George's Bay" was a channel. He thus gave the northern island its name of Nova Hibernia (New Ireland). He discovered a safe anchorage at its SW end, Port Praslin, which was to be used by many expeditions and later became the scene of the tragic "Nouvelle France" settlement scheme. Carteret was followed by Louis Antoine de Bougainville who, in 1768, with Commerson as naturalist, visited among other places Bougainville and Buka islands, the SE coast and the Louisiade Archipelago. In 1770 James Cook, with Banks and Solander, sailed through Torres Strait, verifying New Guinea's distinctness. They only landed on the SW coast for one day, where Banks made some 30 plant collections under protection of the ship's guns and marines. In 1781 Francisco Antonio Mourelle (Spain) discovered more isles of the Bismarcks.

A second French expedition, under Antoine d'Entrecasteaux who was charged with searching for the lost *La Perouse*, was in New Guinea waters in 1792 and 1793 with *Recherche* and *Esperance*. It called at several points, including for the first time Huon Gulf; other important work was done in the Milne Bay islands (many given names), and in New Ireland and on Waigeo. D'Entrecasteaux' naturalists were J. J. de La Billiardière, L. Ventenat and L. A. Deschamps. The expedition broke up in confusion over the consequences of the French Revolution and La Billiardière's collections were taken for a time by the British; the plants are now in Florence (d'Auribean, OIC, was a royalist; La Billiardière a republican). D'Entrecasteaux himself died on Manus. The wars and disruptions of the French Republican and Napoleonic eras were to restrict exploration for the next 20 years or so, but a new flowering took place after 1815, with the growth of mercantile trade and the development of a related activity, detailed charting, first carried out on a large scale by Flinders in the *Investigator*. In New Guinea this was dominated by several great French voyages, all carrying naturalists, which collectively added greatly to collections (at Paris) and to natural history knowledge. First were J. R. C. Quoy and G. J. P. Gaimard with C. Gaudichaud in 1818–19 on the *Uranie* (later lost in the Falklands though many collections survived) and *Physicienne* under L. Freycinet. They collected only at Rawak on Waigeu. Then came Dumont d'Urville, P. Garnot and R. P. Lesson in the *Coquille* under Duperrey in 1822–24, collecting insects, other animals and plants in the Solomons, Port Praslin on New Ireland and Doré Bay (Manokwari, Vogelkop), the last very important for plants. Later Dumont d'Urville led two more expeditions through the region: the first in *Astrolabe* (ex Coquille), 1826–29 with Quoy, Gaimard and A. Lesson as naturalists, calling at Doré Bay, Waigeo and New Ireland; and then in 1837–40 on his *Voyage au Pole Sud* in *Astrolabe* and *Zelée* with naturalists J. B. Hombron and M. Jacquinot. The latter visited at various points: Triton Bay, then recently abandoned by the Dutch, the Louisiades, to complete d'Entrecasteaux' surveys, and along the SE coast. In 1840 the English returned, with a voyage in *Sulphur* under Belcher and carrying R. B. Hinds and A. G. Barclay, as naturalist and gardener-botanist respectively, past the Solomons, Port Praslin, the north coast (Kairuku) and Japen I., collecting animals and plants (London: BMNH; Kew). In 1828 J. Modera led the Dutch *Triton* expedition along the SW coast.

These were the last of the exploring voyages, as discovery gave way to detailed, tedious coastal survey in which the English became preeminent in this region in the mid-19th century. These did, however offer good collecting opportunities, and on the 2 which visited New Guinea (*Fly* in 1839–42 and *Rattlesnake* in 1849–50), John MacGillivray and J. Beete Jukes were naturalists, collecting animals and plants. On the latter, under Capt. Owen Stanley, MacGillivray was joined by T. H. Huxley who collected mollusks and coelenterates (all at BMNH).

Settlement, however, was rare in this early period. An English party had attempted it at Doré Bay in 1795 but the most serious effort was in Triton Bay near Lobo on the SW coast by the Dutch in 1828 soon after formal recognition of Dutch rule in 1824. From a fortified base (Fort DuBus) A. C. Macklot, Salomon Müller and Z. Zipelius (Zippel), all members of the *Natuurkundige Commissie* of the Netherlands East Indies, made important collections of animals and plants (Leiden). However, this failed in the 1830's through illness, which early cut off Zipelius' life. The earliest mission settlement in the eastern regions was also scene of collecting — by Père J. Xavier H. Montrouzier, one of a long line of famous French missionary-naturalists (one of the last and most famous being Teilhard de Chardin). Montrouzier, who collected insects, mollusks and fish at Woodlark, 1847–52 (Paris, partly lost), later worked in New Caledonia.

Thus to the threshold of a new era: that of the independent naturalist, often under private sponsorship, taking advantage not of exploring or surveying expeditions but of the developing steamship network, then hardly reaching New Guinea except in the east through China Strait and St George's Channel. The first was none other than the most famous: Alfred Russel Wallace who in 1858 spent a few weeks collecting primarily birds and insects at Doré Bay, near Manokwari. Ill and coast-bound, his collectors went into the Arfak Mts, achieving new finds. Wallace also collected (and observed) on Waigeo, Salawati and the Aru group, while his assistant C. Allen reached Misool and the Sorong area of NW Vogelkop (collections mainly in BMNH).

Other naturalists, some rather colorful, sought out the NW. Heinrich Bernstein collected animals at Sorong, Waigeo and Salawati in 1864; the sword-carrying German Baron C. B. H. von Rosenberg collected birds (Leiden, BMNH) in 1858–70, meeting Wallace at Doré Bay. And the peripatetic *hortulanus* J. E. Teysmann from Bogor made substantial plant collections in that region in 1869, also for a short time reaching Humboldt Bay (collections Bogor, Melbourne).

At this period a mission station had already been established at Doré Bay, and in 1871 came the first European resident in the east, the now almost legendary Russian ethnologist-naturalist N. N. de Miklucho Maclay. He was landed near Bongu on Astrolabe Bay by the *Vitiaz* — a name now given to the strait between New Guinea and New Britain — and remained there for over a year. He returned in 1876–77, 1878 and 1883 and at other times visited Triton Bay (old Fort DuBus), and briefly at Samarai, 1880, by then something of a trading post. He collected some animals, a few plants, and much ethnographic data, and introduced some fruits and other plants, including *Carica papaya*. Many of his specimens and data were lost, though some insects were described by Wm Macleay in Sydney. His translated diaries (Miklucho Maclay/Sentinella 1975) give an interesting picture of the untouched north coast of a century ago.

The fame of Wallace — his *Malay Archipelago* appeared in 1869 — and even more the fabled birds of paradise, now attracted visitors and the greater ease of travel, with better steamships and the opening of Suez in 1869, brought more naturalists: A. A. Bruijn came in 1871–79 (birds, partly for plume trade — see Gilliard 1969 (collections to Tring: BMNH; AMNH). A. B. Meyer came in 1875, as well as one of the most notorious of adventurer-naturalists ever to visit New Guinea, L. M. d'Albertis, a Garibaldi veteran who had trained at the Museo Civico di Storia Naturale in Genova under G. Doria. He made two trips to New Guinea and his full story is now told (Goode 1977). His first was with the botanist Odoardo Beccari, later a famous palm specialist, to the Vogelkop in 1872–73. They were the first to penetrate far into the Arfak Mts, reaching the Hatam district; and Ramoi (S of Sorong), Mt Epa, Doré Bay and Andai were also visited. Beccari alone collected insects, birds and many plants in 1872–73 and again in 1875 in Vogelkop area, reaching Batanta I., off Salawati, and Kofiau. D'Albertis, however, travelled to the SE for his next three interior trips: in 1875 with the Congregational missionary S. Macfarlane (see below) some way up the Fly River, and in 1876–77 his two famous trips in the steam launch *Neva* far up the Fly to the foothills of the main range. Collections of many groups of animals and also plants were taken back (Genova), with many new species named after him (see d'Albertis 1880; Goode 1977).

The last of the major British surveying voyages was that of Capt J. Moresby in 1873, who that year in his steam-sailship *Basilisk* penetrated the reef and discovered the future site of the capital of Papua New Guinea — Port Moresby. His naturalist, D. R. H. Comrie, collected insects, mammals and birds along the SE coast and up towards Huon Gulf in 1873–74, but very few plants.

3. Southeastern New Guinea (British New Guinea: Papua), 1875–1914

The discoveries and surveys of John Moresby were published in 1876 and reported on in the Australian press; also the voyage of the *Challenger* (1872–76) focused a good deal of attention on natural history. This coincided with the first wave of nationalism in Australia, by then relatively prosperous from mining and wool. Settlement was already looking beyond Queensland, there was fear of other powers invading the region with vested interests in trading and labor recruiting: the great houses of Towns (Townsville named for him), Burns Philp (founded 1875 at Townsville) and others were coming into existence; Miklucho-Maclay's presence in Sydney in the 1870's and his writings also attracted attention to New Guinea in Australia. All this was embodied in Sir Wm Macleay's natural history and marine biological voyage in 1875 in the old sailing ship *Chevert*, whose departure from Sydney was a great social occasion. Natural history was at the height of its social popularity and Macleay was not only a scientist but a wealthy socialite and parliamentarian. Regrettably, however, due to conflicts aboard ship, weather, etc. geographical discoveries were few although the marine zoological collections were considerable (Macleay). At Somerset on Cape York Pen. came the meeting with S. Macfarlane — also an amateur naturalist-explorer, the first of a long line in future British New Guinea. In botany pressure was partly exerted by Ferdinand von Mueller; and d'Albertis' presence at Yule I. had some effect (the rivalries of this period are well told in Goode 1977). Steamship services advanced rapidly and a service through the treacherous Torres Strait permitted establishment of the Thursday I. station in 1877. Cooktown also became a favored transfer point, especially for Samarai and the northern areas. Thus SE New Guinea became relatively accessible and as frequented as the Vogelkop, albeit more by British and Australians. First were Kendall Broadbent (1873–79; coll's to Pittsburgh) and Andrew Goldie (1876–82 mainly). The Sogeri Plateau was reached first by O. C. Stone in 1875 and then by Goldie in 1877. Goldie was joined by Morton from the Australian Museum in 1875–77 (to Louisiades late 1877) and in 1878 by the well-known collector Carl Hunstein who discovered several new birds of paradise in the ranges NE of Port Moresby in the subsequent decades. Goldie also collected animals (Sydney) and plants (Melbourne). With a temporary waning of outside interest (due to disappointing settlement prospects), collecting continued, by Goldie and also by the missionaries J. B. Chalmers, W. G. Lawes and their assistants, particularly Jakoba. Long tours were made by all of them, 1879–83. Goldie went again to d'Entrecasteaux Is in 1882.

The attempted annexation by Queensland in 1883 (through Premier McIlwraith) created new publicity. W. Armit, making a trip for a Melbourne newspaper, came with the three Dentons — father Wm (who died in the field) and sons S. W. and S. F. — that same year, travelling across the Astrolabe Range through Sogeri to the Moroka (Meroka) district (NE of Iawarere, upper Musgrave R.). In 1884 Armit collected plants NW of Port Moresby and Edelfelt in the Astrolabe Range and environs.

Toward the end of 1884 came the annexation as a British Protectorate, and the naturalist H. O. Forbes was employed as geographer-explorer. Though his attempt on the Owen Stanley Range was unsuccessful, he spent a wet season, 1885–86 in E Sogeri (15–20 km E of present Sogeri), collecting many animals and plants (BMNH; Melbourne; etc.). Twice he was joined by Chalmers. Unfortunately due to an argument many of Forbes' plants remained unstudied until the early 1920's though monocots were worked up by Ridley in 1886.

The same period saw the rise of regional geographical societies in Australia and elsewhere (i.e. Nat'l. Geogr. Soc, USA, 1888). Some of the former supported Forbes in 1885, causing the Sydney branch of the Geographical Society of Australasia to organize its own expedition to the Fly and Strickland rivers, also in 1885. Its steam launch *Bonito* carried 12 scientists led by the zoologist Haacke and including W. W. Froggatt collecting insects (AM), and W. Bauerlen collecting plants (Melbourne). In 1887 the Melbourne branch sponsored Cuthbertson and Sayer in their successful ascent of Mt Obree (plants Melbourne). The last of these efforts was in 1890 with the ascent of Mt Yule in conjunction with MacGregor's staff, led by George Belford (plants, Melbourne).

In 1888 came the declaration of British New Guinea as a Crown Colony and the appointment of Wm MacGregor as Administrator (later Lt. Gov.). He vigorously fostered collecting, but made geographical exploration more a government monopoly. In this period most of the major mountains in the SE peninsula were climbed and deep interior penetration made there, in the Fly, and along the hitherto poorly-known Gulf of Papua and present Northern Province coasts. Party leaders included MacGregor himself (Mt Victoria 1889), G. Belford (Mt Yule 1890), R. E. Guise (grandfather of Sir John Guise; Goropu Mts; Mt Suckling 1891), W. E. Armit (Mt Dayman 1894) and A. C. English and A. Giulianetti (Mts Albert Edward & Scratchley 1896–97). Animals went with official collections to Brisbane (C. W. de Vis studied birds), plants to Melbourne until von Mueller's death in 1896, then to Kew for study by W. B. Hemsley (some mosses to V. F. Brotherus in Helsinki; later to H. N. Dixon; liverworts to F. Stephani).

Through these efforts the mountain flora and fauna became much better known, and it was first demonstrated that an alpine flora existed. Earlier efforts in the Vogelkop and the dry SE led to hypotheses over Malesian affinities (Beccari) and Australian affinities (von Mueller) in the flora. In 1889 von Mueller argued for a connection with temperate Australasia for montane plants and to N Hemisphere for high montane plants. O. Warburg in 1891 presented a synthesis for New Guinea phytogeography. But most pronouncements on Papuan biogeography have been on basis of the mammal (and especially marsupial) situation.

Italian naturalists, encouraged by G. Doria at Genova, continued the tradition begun by d'Albertis and Beccari, and in 1889 Lamberto Loria began the first of two expeditions to the Port Moresby region and other parts of Papua. He was, after d'Albertis, the main early entomological collector in this region. He spent long periods in the Sogeri, Meroka (1300 m), Rigo and Mekeo districts, as well as Gulf coast and Milne Bay islands, assisted by Amadeo Giulianetti. (Insects, Crustaceans, etc to Genova; some plants to Florence). After Loria's final departure, Giulianetti worked for MacGregor as Government Agent for collection of natural history specimens, including birds (to Brisbane; plants to Kew). In 1894 W. E. Armit also joined government service. S. Giuseppe in 1892 collected birds in the Gulf (Genova).

The British were not to be outdone by the Italians: in 1888 Basil Thompson had collected birds and mammals in Milne Bay islands (partly York Mus.), followed in 1890 by James Cockerell with the Rev. George Brown. Orchid hunters were also active: in 1887–88 David Burke, for Veitches, collected in Port Moresby region; and in 1894, and again in 1898, Wilhelm Micholitz collected for Sander, at South Cape and in the Milne Bay islands (*Dendrobium atroviolaceum*), also gathering insects when in the Louisiades. C. A. W. Monckton, of MacGregor's staff, collected at Kokoda, etc up to 1907, climbing to top of Mt Albert Edward in 1906 (BMNH). Rev. R. J. Andrew collected insects on Misima I. (BMNH).

But it was bird collecting which took first place in Papua from the mid-1890's until World War I. After MacGregor's departure in 1898 the Government was much less involved and retired altogether after 1906 as an active sponsor, although both Lt. Gov. G. R. LeHunte and his successor F. R. Barton made miscellaneous collections (to Brisbane) of animals and plants. Barton also found fossil dugong, turtle and gavial on Woodlark I., 1904–05. The dominant collector, however, was Albert S. Meek, who under sponsorship of Lord Walter Rothschild and his Tring Museum (UK), collected birds and Lepidoptera from 1894 to 1916, spending 1895–1901 in Milne Bay islands and E Papua. Many assistants worked for Meek: his younger brother W. G. Meek, Albert and George Eichhorn, brothers-in-law W. B. Barnard and Harry Barnard, and Mr Gullivers. After Meek's retirement, the Eichhorns continued work until 1923 in this region (including Owgarra, Bragi, Mambare R.) and in New Britain, 1925, also collecting miscellaneous insects. (The Tring collections are now part of BMNH, although Lord Rothschild's bird collections were sold in 1932 to AMNH.) A. S. Anthony was also active in the region for the British Museum. In the first decade of the 1900s the Pratts — father and 2 sons — visited Mt Albert Edward and some of the Goilala area, following the gradually developing Sacred Heart Mission interior road to the upper Vanapa Valley at Ononge, also concentrating on birds and Lepidoptera (Tring: AMNH & BMNH).

From Germany also came a sponsored collector, Emil Weiske, collecting beetles, Lepidoptera and other fauna, including birds and mammals, in 1897–98 (Dresden and other museums). He worked in Astrolabe Range, Sogeri, Meroka, Elema, Kabadi, Vaitala, Kerema, Brown R., Kemp Welsh R., Paimomo R., Aroa R., Vanapa R., Rigo/Hula and other districts in the Port Moresby region, much as did Loria a few years before. A few botanical contributions were made, notably by Copland King, pioneer Anglican Missionary in NE Papua, collecting ferns described by Copeland at Manila (also Sydney), and Mary Schlendler at the London Missionary Society station of Boku in upper Rigo district described by Bailey (Brisbane). M. S. C. Smith, director of Mines & Agriculture, collected ferns and mosses on Mt Murray, but the specimens were mostly lost in the Kikori River.

With the onset of severe depression in 1893–94, especially in Australia, many financial sources dried up and collectors such as Micholitz (in later years), Meek and others were basically subsidized by wealthy individuals or museums. Specialization was becoming more widespread; and in Papua without government support general botanical collecting declined strongly. Curiosity had been satisfied and changing tastes affected leisure activity. Collecting also focused on more spectacular or saleable items. Thus birds and larger insects were comparatively well known in SE Papua by 1914 but this was not true for interior Gulf and Western provinces, and botanical knowledge was very patchy and disorganized.

4. Northeastern (German) New Guinea, 1875–1914

The opening of large-scale exploration in NE New Guinea (Kaiser Wilhelmsland) developed later than in British New Guinea or in the Vogelkop. Apart from Miklucho Maclay in the 1870s, the only pioneer before 1885 was F. H.

Otto Finsch, who in 1880–82 made extensive coastal surveys in many parts of the area, and in 1884–85 more thoroughly explored the NE coasts with the *Samoa*. Finsch made many natural history collections, particularly in 1881 around Astrolabe Bay, and also worked in the Milne Bay region. In 1885 he sailed 50 km up the Sepik River. Later he turned more to ethnography, a trend discernible in his book *Samoafahrten* (collections Bremen, Brunswick, Leiden; partly destroyed in World War II). Finschhafen was named for Finsch.

Carl Hunstein left British New Guinea in 1884 with Finsch and until 1888 collected in various parts of NE New Guinea, partly with the *Neu-Guinea Kompagnie* expedition of 1886–87 (see below) and also in the Rawlinson Range, where he discovered *Araucaria hunsteinii*. In 1888 he was working in the Kilenge district of W New Britain when he was drowned in the tidal wave of 13 March. As before, he specialized in birds, many of which were bought by Finsch.

Apart from major scientific expeditions, discussed below, other collectors — some of them employees of the *Neu-Guinea Kompagnie* — soon became active with the establishment of "government." However, due to a predisposition to caution, interior penetration by individuals was limited and the only mountain areas reached until the end of the century were the Sattelberg on the Huon (Kai) Peninsula and the Finisterre foothills and ranges westward, south of Astrolabe Bay. Until 1891 Finschhafen was the main government base, after which for health reasons it was shifted to Astrolabe Bay (Stephansort and Friederich-Wilhelmshafen).

J. S. Kubary, a naturalist of Polish origin, collected animals and a few plants around Astrolabe Bay from a base at Constantinhafen (Melanu; first government station in that area), 1887–88, for the Godefroy Museum in Hamburg (partly now in the Senckenberg, Frankfurt). Kubary also made land acquisitions in the area which later led to much local strife, with no "settlement" until 1932. In 1888 R. Rohde collected animals at Stephansort and Finschhafen. From 1886 the pioneer Lutheran missionary J. Flierl collected animals on the Huon Peninsula, as did also his son and grandson. In botany Flierl was complemented by G. Bamler who collected on the Tami Is. and on Huon Peninsula (Berlin). In 1891 H. Fruhstorfer collected Lepidoptera at Finschhafen and elsewhere.

Botanists were also soon active. Apart from Max Hollrung (see below), important collections were made by F. E. Hellwig (1888–89) and C. Weinland (1889–91), both officials of the *Kompagnie* based at Finschhafen, and by the traveller O. Warburg in 1889 (Berlin; largely destroyed, but duplicates in Breslau = Wroclaw, Kew and elsewhere). Hellwig, with whom Warburg made some trips, was the first to reach mountain localities at Sattelberg and (with newspaperman-writer Zoller from Köln = Cologne) the western Finisterres whence the Bismarck range and its Mts Otto, Wilhelm, Herbert and Marien were seen (and named) and the first rhododendrons collected. At Sattelberg, later mission "health station" (900 m), oaks were collected. In 1890–91 Carl A. Lauterbach, a botanist-geographer of private means, made the first of 3 trips to New Guinea. In addition to the Gazelle Peninsula, he collected along the Huon Gulf from Finschhafen toward the Bukaua district and "Burgberg" (Lo Wamung at Lae), on the Sattelberg and pioneered along the Gogol River south of Friederich-Wilhelmshafen (Madang). Lauterbach's collections ranged across the whole plant kingdom (Berlin; private herbarium now in Wroclaw; duplicates elsewhere).

Not only Lauterbach collected non-vascular (as well as vascular) plants. L. Kärnbach, *Kompagnie* horticulturist, obtained many marine algae as well as information on economic plants (1886 onwards), and in the 1890s lichens were collected by L. Biro and E. Nyman (see below) and mosses by the bryologist M. Fleischer.

Warburg's visit, at the end of a 4-year odyssey in Malesia, Asia and Australia, resulted in a paper (1891) in which he argued that while there were several distinct elements in the flora of New Guinea, the predominant relationship was with western Malesia and that in effect no "line" could be recognized as a major floral discontinuity in the island region west of New Guinea comparable with the many lines of fauna (mainly mammal) discontinuity proposed by Wallace, Weber, Lydekker, etc. Later Diels was to fix a major floral line at Torres Strait. Warburg also proposed the term "Papuasia" to refer to the whole region from New Guinea to the Solomon Is. — known to zoogeographers as the Papuan Subregion.

A change in shipping connections in 1889 to Surabaya in Java, followed by transfer of the seat of government in 1891, changed the main focus of activity to Astrolabe Bay where tobacco production was being developed. It also brought many collectors who were also active elsewhere in Malesia. One of the first was Bernard Hagen, brought from Sumatra in 1894 to work in the tobacco plantations. Apart from ethnographic studies, Hagen (1899) collected insects and birds (Berlin) near Stephansort, the government seat, as well as at Berlinhafen and *Astrolabe-Kompagnie* (tobacco) headquarters, around 1894.

From 1893 began the long involvement of the Hungarian collectors, Samuel Fenichel (1893–95) and Lajos Biro (1895–1901). Their primary emphasis was on insects but many other animals were collected (Budapest; birds, reptiles, amphibians, mollusks and Diptera, excluding Pupipara, were destroyed during the 1956 uprisings; part of Fenichel collection to College of Nagyenyed, Transsylvania, Roumania). They also collected plants, including many lichens by Biro, the latter written up by O. Szatela in 1956 (others plants at Berlin). They worked extensively around Astrolabe Bay, at Berlinhafen (Aitape) and the islands offshore (Seleo, Tamara) and in eastern Huon Peninsula (Simbang, Bingala, Sattelberg) as well as on the Gazelle Peninsula, but largely near sea level. Biro was the first real scientist-zoologist to spend much time in New Guinea. He later became something of a hero in Hungary, writing two books of travels popular with young people (Szent-Ivany, pers. comm.). Fenichel died, aged twenty five, at Stephansort of cerebral malaria. There is a memorial to both Biro and Fenichel at the University of Papua New Guinea.

Toward the end of the century came the beginning of Carl Wahnes' long sojourn (1897–1909) as a collector for the British Museum and other institutions (e.g. Odonata at Univ. of Michigan). He collected insects, birds and other animals on the Huon Peninsula (Simbang, Sattelberg, Rawlinson Range), at Bongu on Astrolabe Bay (Miklucho Maclay's old base), and on some offshore islands (Tarawai, Isle Deslacs) and in New Britain. In 1898–99 the Finisterre

Range and the Sattelberg were visited by the Swede E. O. A. Nyman, who collected birds and plants, among the latter many curious lichens (Uppsala, Berlin; epiphyllic lichens described by R. Santesson in 1940s). Robert Koch collected plants and animals in 1900.

Some of the most important contributions came from the large official (or semi-official) expeditions organized by the *Kompagnie* in 1886–87 and in 1896–99. These represented the most successful early efforts deep in the interior, although most collections were of lowland plants, animals and insects. The first (1886–87) was the primary exploring expedition of the *Kompagnie*, led by von Schleinitz. Scientific staff included M. Hollrung (vascular plants), C. Schräder (geography and bryophytes) and F. Grabowsky and C. Hunstein (birds and insects). In over a year's time they worked around Finschhafen, Samoahafen (Salamaua), the coast north of Friederich-Wilhelmshafen (Madang), Astrolabe Bay around Constantinhafen (Melanu), Hatzfeldthafen, and far up the Sepik River (Kaiserin-Augusta Fluss) — the first great voyage thereon — with camps at Zenap (Tsenap) near May River (I. Augusta-Station) and Malu near present-day Ambunti (II. Augusta-Station). The large number of collections went to Berlin (plants partly destroyed, but duplicates at Kew and Melbourne).

The next large expedition was the Ramu Expedition, organized by the *Kompagnie* with Imperial Government aid and led by Lauterbach with Kersting as police-master and E. Tappenbeck as quartermaster. This traversed the dividing range SW of the Gogol River and entered the then-unknown Ramu (Jagei) valley (NW of present-day Usino); from there they travelled by canoe to near the present site of Annanberg and returned by same route. Side trips were taken to Mt Oertzen and the dry Bismarck Range foothills. The expedition was an overall success, with good returns in geographical knowledge and collections of plants, birds and animals (Berlin; plant duplicates in Wroclaw — Lauterbach's private herbarium — and elsewhere). The plants, with those collected by himself in 1890–91 and by others, laid the basis for his consolidated *Flora der deutschen Schutzgebiete in der Südsee*, written with the Berlin botanist K. Schumann (1900–1905).

The Ramu expeditions were continued by Tappenbeck with two associates, H. Rodatz and H. Klink (after whom Klinkii Pine, *Araucari hunsteinii* = *A. klinkii*, is named). In 1898 the inland Ramu was linked to its mouth (Ottilie-Fluss), and Rodatz and Klink lived during winter low water in a temporary camp on the lower river for two months. In 1899, after a change of government, traces of gold on the upper river, near Usino, attracted the *Kompagnie*'s interest. A stern-wheel riverboat was obtained and further trips made. Rodatz and Klink worked during the dry season from a base at Arumene, near Aiome, reaching the foothills of present-day Mt Aiome in the western Bismarcks, and in November Lauterbach returned for a month's trip along the river, leaving early in 1900 (plants, Berlin, with duplicates at Wroclaw). Rodatz and Klink remained in the area at a base camp near Usino, later visited by Schlechter (late 1901), and subsequently they gave many years of service as district officers, Klink eventually at Morobe, whence he explored the mid-Waria Valley (around present-day Garaina), finding stands of *Araucaria*.

The developments in the Ramu Valley were to a large extent promoted by

R. von Bennigsen, first Imperial Governor of German New Guinea. While scientific results under *Kompagnie* rule were impressive, economic results were not, particularly in the wake of the world slump of the mid-1890s, and much capital subsidy was involved. After long negotiations, the *Kompagnie* was relieved of its administrative responsibilities as of 1 April 1899 and von Bennigsen assumed control. He was also a beetle collector of note in New Britain, Duke of York Is. (Neu Lauenberg), and in the mountains behind Stephansort, especially the Kani Range (upper Mindjin River) (collections mostly Dresden). Coleoptera were popular with Germans (S. von Kotze refers to this in *Aus Papuas Kulturmorgen*, 1905), as butterflies were for the British; and von Bennigsen had set the fashion here, for many amateurs also collected beetles which too found their way to Dresden Museum and its noted coleopterist, K. M. Heller.

Von Bennigsen was succeeded as Governor by Albert Hahl, who concentrated more on development in the Bismarck Archipelago, but vigorously promoted mainland exploration and scientific work as circumstances permitted, much as MacGregor had in British New Guinea, although less so in person. Hahl could also rely on goodly numbers of officers and others interested in natural history to make contributions, though the important expeditions, both on the mainland and in the Bismarcks, usually involved visiting scientists.

The first of these was Rudolf Schlechter, sent out by *Kolonial Wirtschaftliche Komitee*, Berlin to find latex-producing trees and vines for gutta-percha and rubber. The first of his two expeditions was exploratory, the second developmental. In 1901–02 he visited the Bismarck Archipelago and, on the mainland, the Torricelli Range behind Berlinhafen, the Finisterre Range, and crossed the Gogol-Ramu dividing range along Bennigsen's "road" to the Ramu gold camp and into the Bismarck Range as far as the present Bundi district. Apart from his gutta-percha and rubber plant discoveries, considered moderately successful, he made large plant collections, including many orchids (Berlin; partly destroyed, with some duplicates elsewhere), as well as some animals. His recommendation to Hahl for a botanical garden was followed up in 1906 at Rabaul, and in 1907–09 he returned to New Guinea, establishing a base camp at Bulu near Bongu on Astrolabe Bay. From here he did intensive field work on the inland divide (especially Kani and Ibo ranges) and in the western Finisterres, and made long trips to the Bismarck Range (with Hahl for an attempt on Mt Saugueti = Otto, in 1908, and to the lower Waria in 1908 and 1909; and to the Torricellis in 1909 with Otto Schlaginhaufen who collected insects and other animals at Paup, east of Aitape and elsewhere (Dresden Mus.). Once again a large collection of orchids and other plants was made (Berlin), but sadly Schlechter's main work areas have never again been re-studied in view of the destruction of so many of his collections during World War II.

With gradually improving economic conditions and more opportunities came other collectors. H. Schoede collected animals in 1906 at Berlinhafen (Aitape) and Simpsonhafen, 1909, E. F. Dahl, Preuss, O. Heinroth and C. Ribbe (also Duke of York Is.) collected on Gazelle Pen., New Britain; P. Nagel in Finisterres. E. Werner collected insects and plants in 1907 south of Astrolabe Bay, partly with Schlechter (plants, Wrocław). In 1908–09 Prof. R. Neuhauss carried out the field research for his encyclopedic *Deutsch Neu-Guinea* (1911), much of it in the Huon (Kai) Peninsula (Sissana, Kelar, Bukaua, Sattelberg, Kai-Imland) but visits were made also to Samoahafen and Dallmannhafen (Wewak), the latter with J. Walis. Insects, animals and a few plants were collected (Berlin) as well as ethnographic material. In the Huon Peninsula Neuhauss travelled with the great missionary and traveller Christian Keysser, who in 1911–12 reached the top of the Saruwaged (Salawaket) Range, Mt Bangeta ("Bolan"), nearly 4000 m; he was to repeat this in 1916 with H. Detzner during the latter's fantastic 4-year odyssey "on the run" from the Australians. The plants found (Berlin) were ultimately summarized by Diels (1929). Detzner became notorious for his semi-fictitious *Vier Jahre unter Kannibalen* (1921).

Other collectors included F. Kunzmann (fish, reptiles, butterflies), and Lothar von Wiedenfeld in 1909–10 (birds and insects at Berlinhafen, Sattelberg and adjacent coast at Simbang and Heldsbach). W. Müller collected insects (Dresden) in 1909, L. Maschmeyer animals, and K. Mailänder collected plants near Morobe (Berlin). P. Nagel collected insects at Komba in the Finisterre Mts. Additional missionaries collecting were Hoffmann, Bergmann, Kunze and Vetter. Ehlers and Piering were lost with collections and notes in a river. The English sea captains Cotton and H. Cayley-Webster (book, 1898) collected birds and butterflies in 1894 (Tring: BMNH).

The latter part of Hahl's reign also was witness to many marine and land surveys and expeditions. Those most concerned with the mainland before 1912 were (1) the British-German border survey of 1909, for which the German team was led by Förster; it was accompanied by Schlechter (see above) on the lower Waria as far as the dividing Maboro ridge (slipper orchid *Paphiopedilum violaceus* found); (2) the German-Dutch border survey of 1909–10, of which the German team was led by the geographer L. Schultze-Jena and which worked along the border (141°E) as far as the Bewani range and then along the Sepik River as far as the Peripatus range (?Sepikberg 1570 m), foothills of the Star Mts, with scattered animal and plant collections (Berlin; plants largely destroyed). Then the Hamburg-Südsee Expedition (after visiting Bismarck Archipelago and Micronesia) in 1909 on the Sepik River, on which G. Duncker collected insects, birds and some plants (Hamburg); a call was also made at Langemak Bay near Simbang.

All of these, however, were outdone by the great Kaiserin-Augustafluss Expedition of 1912–13, the largest and longest of all mounted by Germany and in retrospect a fitting climax to its years of rule. It was led by Capt. A. Stolle, mineral surveyor and the 6-man scientific team was headed by the geographer Dr Walter Behrmann. In the year-and-a-half period, from March 1912 to September 1913 they surveyed most of the Sepik basin by ship and launch. They ascended the foothills of the "Central Range" in several places, as well as most major tributary rivers, also entering the Seerosensee (Chambri Lakes). Their "Hauptlager Malu," the base camp above Malu village, is today occupied by the district town of Ambunti (Ambunti is roughly a translation of "Hauptlager Malu" in local dialect as heard by G. W. L. Townsend in the 1920s: *District Officer* 1968). Localities included Pionierlager, Seerosensee, Peilungsberg,

Hauptlager Malu, Zuckerhut, Maänderberg, Hugellager am Sepik, Sepikberg (1570 m, see above), all on or near the Sepik; Etappenberg (850 m), Lordberg (=Durchblick, 1000 m) and into the Zentralkette (1800 m) as well as Lager 18, all of the Aprilfluss; Felsspitze in the Westkette (1400–1500 m); Friedafluss and Leonhard-Schultze-Fluss; Hunsteinspitze (1500 m); Lehmfluss, Topenfluss, Regensberg (550 m) and Schräderberg (2070 m). Animals were largely collected by Joseph Bürgers, W. Behrmann, L. Schultze and C. Ledermann, the last of whom also secured all the plants (6600 numbers). Collections went to Berlin and Dresden (animals) and Berlin-Dahlem (plants); most animals survived World War II but the plants suffered severe losses; only some duplicates have survived (mainly Wroclaw, Leiden, Edinburgh, Singapore) and little "topotype" collecting has been done. Behrmann did excellent mapping which served for half a century, as well as geomorphological work. But it was also realized that much further progress would have to be aided by aerial survey, for which airships or Zeppelins were first proposed (Neuhauss advocated this; even "stamps" were printed in 1913; he was not confident in airplanes), but the use of aircraft belongs to the next era, starting with Stirling.

Publication of the results of this and other expeditions was deemed to merit special consideration, although much of this was in existing journals. Finance for this came mainly from the *H. & E. Heckmann-Wentzel-Stiftung*. Botanical results appeared within the *Beiträge zur Flora Papuasiens* (1912–42), edited by C. Lauterbach (later by L. Diels) and still the best consolidated work on the plant taxonomy of New Guinea. Some numbers also carried a floristic/vegetational survey by Lauterbach as his "swan song" in 1928–30. Zoological results appeared in *Mitt. Zool. Mus. Berlin*, those on birds by E. Stresemann in 1923.

In summary, while its beginnings were slow, natural history investigation in NE New Guinea rapidly gathered speed under German rule and by 1914 had produced a great mass of material and information, which took years even to partially digest. The botanical losses of World War II represented a severe blow, widely appreciated by the profession, but not sufficiently by agencies in Papua and New Guinea in the decades after the war. Certainly the "data base" had pulled well ahead of Papua, in which collecting, except in some groups, had slackened considerably after 1900 and later after the coming of Australian rule. To the present, the difference, at least on botany, has never been eliminated and, when compared with W New Guinea, former Kaiser-Wilhelmsland retains its "lead," aided by all the postwar biological work done in the Central (Eastern and Western) Highlands and in the present Morobe Province.

5. Netherlands New Guinea, 1875–1915

Beccari's successful explorations in the Vogelkop and penetration of the Arfak Mts to the Hatam district, and the increasing importance of the feather trade, led to a great increase of biological collectors in Netherlands New Guinea from the mid-1870s. However, official interest in the area remained limited; most collectors had outside sponsorship; several were Rothschild collectors. The main attractions, in the tradition of Wallace, were birds and butterflies, especially items of commercial interest. Plants and other animals, except perhaps beetles and orchids, were of secondary importance.

Bird collectors during this period included Leon Laglaize, primarily for birds of paradise, but also other birds, plus butterflies, largely for A. A. Bruijn of the feather trade, in westernmost New Guinea and nearby islands, 1876–84 (birds, Paris; insects, Brussels); Heinrich Kuhn, birds and other groups, on the Onin Peninsula, including Sekar (Kokas) in the McCluer Gulf, 1884; Carl Platen, birds, on Waigeo (Berlin); R. Powell, birds, on Waigeo, Salawati and Vogelkop (Tring: AMNH), both in 1883. F. H. H. Guillemard, on the *Marchesa*, collected mainly birds, plus beetles and some butterflies and shells, on Waigeo, Batanta, Misool, Vogelkop and Japen, 1883–84 (BMNH). W. Doherty collected birds and insects, in many areas on Vogelkop and around Geelvink Bay, including Jobi = Japen I. and Humboldt Bay in 1892–93 (Tring: birds AMNH; insects BMNH); C. W. R. Renesse van Duivenbode collected various animals, in Charles-Louis Range at W end of central cordillera, apparently one of earliest such visits, 1898 (BMNH). J. M. Dumas collected birds, Vogelkop (Mt Moari), 1899 (Tring: AMNH). Johannes Waterstradt collected birds, Waigeo, 1902–03 (Copenhagen; Tring: AMNH). The Pratts (A. E. and son Harry) collected Lepidoptera, at Merauke, 1902, soon after establishment of that post, but hostility curtailed collecting (see below). H. Fruhstorfer, working on Vogelkop (Takar, etc) and elsewhere in 1891, also collected Lepidoptera (Berlin).

Plant collectors were also few. O. Warburg (general) visited Sekar (Kokas) in McCluer Gulf in 1889 on his general tour through Papuasia and E Australia (Berlin, largely destroyed; a few duplicates at Harvard and New York). W. Micholitz, for Sander & Sons, collected orchids in Batanta and elsewhere around 1890 (Kew). Aru and Kei were visited by *Challenger* in 1875 and H. N. Moseley collected plants and animals (Kew; BMNH).

But the western part of New Guinea remained largely neglected by the Dutch until increasing foreign interest influenced them to assert their sovereignty more forcefully in the late 1890s. The accession of the young Queen Wilhelmina in 1898 was in a way a major stimulus, as was the establishment of Merauke post in 1902 to control Marind-Anim (Tugeri) raiders and headhunters; and soon other posts, including Hollandia, Fak Fak and Serui. At the same time came the start of large-scale wholly or partly government-sponsored expeditions similar to those in German New Guinea, but on even larger scale — a tradition which continued until 1939 and was resumed again in the 1950s. Results of the earlier expeditions were largely reported in the special serial *Nova Guinea* (Leiden, 1909–66), sponsored by the Indies Committee for Scientific Research and other bodies.

The first of these major expeditions was the Dutch North New Guinea Expedition of 1903, led by C. F. A. Wichmann. Much of the animal material was collected by H. A. Lorentz and L. F. de Beaufort; no botanist was along, but plants, both living and preserved, were collected by the Indonesian *mantris* Atasrip and Jaheri from Bogor. Most of the collecting was on the Vogelkop and Geelvink Bay islands or coasts: Waigeo, Manokwari (Doré), Karoon, Kwawi, Andai, Wendesi, Tawarin, Bawe, Sageisara, Moso, Oinake, Napan, Angadi,

Jende on Ron I., Timena, Orum, Mios Korwar, Supiori I., Japen I. (Ansus), Moso, Metu Debi, Tjintjan Bay, Matterer Bay, Pokembo, Wakobi, Mansinam, Siari, Kwatisore, Wa Udu, Angadi. Also north coast (Moaif) and Hollandia-Sentani area including Jotefa Bay, Cyclops Mts. Goreda (L. Jamur); and west of Sentani: Timena R., Ibaiso, Jaga, Tanah-Merah Bay (Maffin Bay); and south of Hollandia: Tama R., Sekanto R. (coll's: Amsterdam).

This expedition was followed by the Dutch Southwestern New Guinea Expedition of 1904–05, led by R. Posthumus Meyes and E. J. de Rochemont. Animals and plants were collected by Dr J. W. R. Koch and assistants from Bogor. Much work was done around Merauke but Etna Bay and other places were visited (Amsterdam; plants at Bogor). Plants of this and preceding expedition were reported by Valeton at Bogor (*Plantae papuanae*, 1907).

In 1904 A. van Oosterzee sent living plants to Bogor from Angi Lakes. A. E. Pratt and 2 sons collected insects for months at Angi Lakes in 1907 and 1909 (Tring: BMNH; some plants to Bogor).

From 1907 to 1915 there occurred one of the greatest organized exploring efforts of a territory ever — the Dutch Military Expeditions, which in 8 years obtained an extensive albeit sketchy knowledge of the interior lowlands. A map was published with the final report in 1920. On certain sorties plants and animals were collected, notably by K. Gjellerup in Cyclops Mts and Angi Lakes, as well as inland to the Keerom River during the border survey in 1910–12 (Bogor).

However, the expeditions significant for biology were those with the express intention of reaching the snow, of which the second and third finally ascended Mt Wilhelmina (Trikora). The first two South New Guinea Expeditions were led by H. A. Lorentz and the third by M. A. Franssen Herderschee. All travelled in the first instance up the Noord (Lorentz) River, with a base camp at Alkmaar-Bivouac, from the so-called Asmat (or Casuarina) Coast. The first expedition, 1907–08, worked around Merauke and in the Digul and Noord River basins, and A. C. de Kock and P. Atmodjo reached Mt Goliath (3500 m). Animal collecting was done by H. A. Lorentz and L. F. de Beaufort, van der Ven, van Nouhuys and Lt Habbema; plants were collected by B. Branderhost, G. M. Versteeg and mantris Djibdja and Atmodjo (animals, Leiden; plants, Bogor and Utrecht). The Second (1909–10) and the Third (1912–13) South New Guinea Expeditions both worked all the way from the mouth of the Noord (Lorentz) R. to the snow cap of Mt Wilhelmina, going from Alkmaar and Kloofbivak (Lorentz R.) north along the ridges between Lorentz R. and Noordwest R., through Heuvelbivak, Went Mts, Dromedarisbivak, Hellwig Mts (2000 m), Treub Mts, Wichmann Mts, Hubrecht Mts, Bellevue, Peripatusbivak, Jenjabivak, Kajan Mts, Waterfalbivak, Quarles Meer, Dolomieten, Oranjebivak, Rotsbivak (4300 m) and to the ice-cap (now gone) on Wilhelmina Top (4750 m), Nov. 1909. Most of the animals were collected by Lorentz; some by G. M. Versteeg; plants by L. S. A. M. Roemer, J. W. van Nouhuys and Lt Habbema (under Lorentz) and A. Pulle and Versteeg under Herderschee. Some collecting was also done at Fak Fak and Kaimana (SW coast) 1912–13 (Leiden). (The ice-cap on Wilhelmina, now Trikora, Oranje or Jayawijaya Range, disappeared between 1940 and 1962.)

One of the spurs to the Dutch activity was the series of two expeditions mounted under British auspices: the British Ornithologists' Union Expedition of 1909–10 and the Wollaston expedition of 1912–13, both with a hope of reaching the highest of the "snow mountains", the Carstensz (Jaya) complex in the Nassau (Sudirman) range. The 1909–10 expedition was led by W. Goodfellow, and animals (few plants) were collected by A. F. R. Wollaston, G. C. Shortridge and W. Stalker. Working mostly along the Mimika, Utakwa (Waitakwa) and Setakwa rivers, the expedition met with great difficulties and some loss of life (Stalker drowned) and in addition failed to get high in the central range. However, the challenge was taken up by Wollaston who in 1912–13, assisted by C. Boden Kloss of the Federated Malay States Museums, collected animals and quite a few plants. They finally reached the Carstensz glacier fields but were stopped by ice and rock cliffs, and due to bad weather did not realize they were still some 550 m below the summit. (Carstensz Pyramid, "5030 m," actually 4884 m, which was first climbed by H. Harrer, B. Huizenga and P. Temple in 1962; but see Colijn, below). The animals, by several zoologists, the plants by H. N. Ridley, were published mainly in *Transactions of the Linneaean Society*, and later collectively reprinted in a fine 2-vol. set (Coll's: BMNH).

A large amount of other exploration followed in the wake of these great expeditions until the outbreak of World War I. Some was related to official activities; other explorers worked independently or under outside sponsorship such as by Lord Rothschild. W. Goodfellow visited along the north coast, collecting birds in 1904–06 (BMNH); J. Roux, a Swiss, and H. Merton, a German, collected various animals on Aru and Kei islands, 1907–08 (I. Freiburg expedition?). Thomas Barbour collected amphibians and reptiles on Vogelkop and Waigeo, 1906–7 (MCZ). M. Moszkowski worked along N coast at Teba, Sauwi, Samberi, Assewari, Tama and Mamberamo River in 1910 for insects and plants (Berlin). In the north, J. F. C. ten Klooster, led the Dutch-German Border Expedition and its extension, on which P. N. van Kampen collected animals at Hollandia, Lake Sentani, Zoutbron and Bronbeck, 1910, and Arwo River, 1911, while K. Gjellerup collected plants and animals in the border areas, Cyclops Mts (to 2000 m), Tor River, Nimboran, Sermowai River, Gaudry Mts, Arso R. (SW), Sawia, Angi Lakes and Arfak Mts. J. D. van Gelder collected animals on the lower Mamberamo River in 1910. R. F. Janowsky collected animals at Moesoiro and Legare R., Sawa R., E Geelvink Bay, 1912.

In the south, Hassan collected various animals on the Utumbui R. (SW), Gelib R. and at Akaba, W of Merauke, perhaps with Versteeg, 1909–10 (Leiden; plants Bogor). At the same time (1910) A. S. Meek collected birds and insects at Merauke, along Digul and Eilanden rivers, to Mt Goliath, reaching 2800 m (birds Tring to AMNH; insects BMNH). H. Elgner collected insects at Fak Fak, 1912 (Senckenberg).

In the western islands, Misool was visited by the Second Freiburg Moluccas Expedition with O. D. Tauern who collected animals in 1911 (Leiden). Finally a significant field trip was made in 1913 by Miss Lillian Gibbs to Angi Lakes for botanical and ecological studies, which were among the earliest such in New

Guinea; a report appeared in 1917 (coll's BMNH). This intrepid field botanist had also visited Fiji, Canary Is. and other places of particular plant-geographical interest. Miss Gibbs was followed by W. K. H. Feuilletau de Bruijn in the Meervlakte and Schouten Is. in 1914 and by Palmer van den Broek in the Cyclops Mts and Humboldt Bay in 1915 (Bogor). In 1914 on the 3rd Mamberamo expedition under Capt J. V. L. Opperman, Lt L. A. C. M. Doorman went up the Mamberamo and Rouffaer rivers to summit of Mt Kemboe, 3550 m, later renamed Doorman Top.

Field activity largely came to an end due to World War I which among other effects disrupted world shipping. Though the Netherlands (and the Indies) were neutral, it became infeasible to continue exploratory work after 1915. Large-scale activity resumed, however, in 1920 — far earlier than in the eastern territories which as from 1914 were both under Australian administration. Although there was more continuity in exploration in W New Guinea, the zenith of great expeditions had largely passed and publication of results generally became more modest.

The first great review of New Guinea exploration, by A. Wichmann, was published as the first two volumes of *Nova Guinea*, 1909–12. This covers in considerable detail the period up to 1902, with an exhaustive search of nearly all published material. C. C. F. M. le Roux (in *Nieuw Guinée*, 1, 1935) reviewed exploration, but in less detail, up to 1934, for the whole island. C. Lauterbach was at work on a history and topography of New Guinea up until his death in 1937, but the whereabouts of his manuscripts is unknown.

6. New Guinea: The inter-war period, 1918–1942

An increasing amount of interrelationship characterized biological work in the 1920s and especially in the 1930s. Also, more collecting was done by individuals working the field for long periods rather than by large expeditions. Some individuals encountered difficulties, and some official obstacles, as related by Evelyn Cheesman ("Six-legged snakes in New Guinea") and others. John Todd Zimmer collected animals on Woodlark I. December 1917–July 1918, and worked on coconut pests (BMNH). He also collected birds on Fly R., 1929–31 (AMNH). W. Goodfellow collected vertebrates in S Papua, 1925 — also 1908–9 — L. S. Crandall and H. Hamlin (SE mts) collected birds in 1928 (AMNH).

In Netherlands New Guinea the postwar era was opened by the great New Guinea Expedition of 1920–21 led by Capt. A. J. A. van Overeem. Animals were collected by W. C. van Heurn; plants by H. J. Lam, who wrote up the plants in *Fragmenta Papuana*. Entry was made up the Mamberamo R; stations including Prauwenbivak, Bivak Batoe, Pionierbivak, Doormanpad-bivak, Mamonbivak, Kikkerbivak, going up the Idenburg, Rouffaer and Doorman rivers. The Swart Valley was reached, and Mt Doorman climbed (coll's Bogor, duplicates Leiden & Utrecht). However, a major aim was to make a link with the terminus (at Mt Wilhelmina) of the Herderschee expedition of 1912–13. So, with 800 men (!), Capt. J. H. G. Kremer returned in 1922, retraced much of the former route, and from the Swart Valley crossed into the western

Baliem system (missing, however, the Grand Valley), and finally reached Lake Habbema (first seen in 1909) and Mt Wilhelmina (Trikora). On this trip was Paul Hubrecht who collected some plants (Leiden), and who had been there also on the 1912–13 expedition.

Although accomplished in the early 1920s, these really represented the last expeditions in the pre-World War I form, style and tradition. Future major expeditions would rely partly on aircraft and radio. Never again would a battalion-sized party take to the field as with Kremer; for its large size, scientific results were not commensurate.

The "Netherlands-America Expedition" of 1926 under M. Stirling pioneered the serious use of aircraft in New Guinea as a transport aid, although the real air age began in 1927 when commercial services started between Lae and Wau (see J. Sinclair; *Wings of Gold*, 1978). A route similar to that of van Overeem and Kremer was followed: up the Mamberamo through the van Rees Mts and then along the Rouffaer up to near the central ranges, reaching 2400 m. Zoological and botanical collections were made by Doctors van Leeuwen, but the plane had to be abandoned and some field data were lost (Bogor, Leiden). Stirling worked on anthropology. Other collectors in W New Guinea in the 1920s included T. Jackson around Merauke (birds, MCZ) in 1920–24 and P. Wirz in the same area (animals, Leiden) before 1923, as well as to Doorman Top with Kremer in 1922. Wirz died in the Maprik area of NE New Guinea in 1955. P. T. Putnam collected amphibians and reptiles in the Merauke area, 1927 (MCZ).

After the Australian mandate in NE New Guinea was granted in 1921, some small-scale exploration resumed in that region. A geological-geographical survey was made on the Ramu R. about 1922, and from 1922 to 1924 came the first forest survey in the territories, by C. E.Lane-Poole who collected many tree specimens in various areas (Brisbane). His report of 1925 remains a classic. Key areas were the Ramu, Huon Pen., Central, Northern and Gulf provinces. L. J. Brass collected plants in the then Central (to Owen Stanley range divide), Eastern and Gulf divisions of Papua in 1925–26 for the Arnold Arboretum (Harvard). He was preceded in the Central Division by C. T. White, Queensland Government Botanist, in 1918, collecting during a vacation from Brisbane. White later studied Brass' early collections. In 1928 A. Jeswiet collected plants around Lake Daviumbu and on the Fly, Strickland and Oriomo rivers and around Port Moresby (Utrecht) on a search for sugar cane, with C. E. Pemberton of Honolulu, who collected insects (HSPA: Bishop).

These field trips marked a significant resumption of botanical work in the east after a relative lull of nearly 20 years, but this and later work was largely externally sponsored. Later the size and extent of the Archbold Expeditions were to cause some ill-feeling. The two 1926 expeditions — Brass and Stirling — also marked the entry of American interests which came to play a leading role in 1929–39. A second projected Arnold Arboretum expedition for 1929 came to naught, but under joint American, British and German sponsorship Prof. Ernst Mayr, while still a graduate student under Prof. E. Stresemann (author of report on birds of the German Sepik expedition of 1912–13), collected many birds and mammals, and some insects and plants, in 1928–29 in Vogelkop (Siwi, Momu), Cyclops Mts (to summit), Huon Pen. including Saruwaged Range (to summit, following in Keysser's and Lane-Poole's tracks) and in the Herzog Mts south of Lae (incl. upper Snake Valley) (birds: Tring: AMNH; insects and plants, Berlin, plants largely lost).

In 1928–29 the Whitney South Sea Expedition, ongoing and of wide scope, collected birds on Milne Bay islands and E Papua (AMNH). In 1929 the Crane Expedition on the *Illyria*, headed by Shurcliff, collected marine specimens, many fresh-water fish, other animals, and a few plants at Rabaul, in Huon Gulf, Sepik R., Manokwari and Waigeo; A. W. C. T. Herre, ichthyologist, was chief biologist (Field Mus.). From the U.K., Rothschild sponsorship was continuing on a smaller scale; from 1928 Fred Shaw Meyer spent nearly 40 years on intermittent or long stays in New Guinea. Early trips were for the Tring Museum, collecting birds and mammals (Vogelkop 1928; Weyland Mts 1930; Huon Pen. 1931; SE Papua isles 1935). Later he made the first major collections from the Central Highlands as these became opened up, an activity continuing after World War II, 1946–53 (BMNH). Herbert Stevens collected, 1932–33, in the Wau-Bulolo area, the first to do so in this famous gold-mining region (birds, insects, herps, a few mammals; MCZ). He visited Wau, Mt Missim, Bulolo, Bulowat and Watut Valley. W. J. Eyerdam collected shells and corals (AMNH).

Other zoological collections in eastern New Guinea in this period included C. T. MacNamara in 1928–30 on Mt Lamington, SE of Popondetta, notorious for a Peleean eruption in 1951 which killed 3000 people and destroyed vegetation over a considerable area (beetles; Sydney, Adelaide). The Rev. L. Wagner collected at Lutheran stations at Finschhafen, Wareo and Komba and in the Cromwell Range in 1929 (beetles; Adelaide). In Netherlands New Guinea in 1929 a major Belgian operation was the expedition led by Crown Prince Leopold, accompanied by the Princess and Victor van Straelen. Insects and other animals were collected in far west (Brussels); and a stately series of reports resembling *Nova Guinea* resulted. W. G. N. van der Sleen collected insects in upper Digul River in 1929 (Amsterdam); W. J. C. Frost obtained birds on Vogelkop, Batanta, Waigeo and Salawati in 1930 (BMNH); and Dr and Mrs G. Stein collected birds and some plants in Vogelkop, Weyland Mts and Japen I. in 1931 (Berlin; Bogor). S. L. Brug worked on mosquitos on the SW coast and Aru Is. in 1932 (Amsterdam; BMNH).

The great depression starting in 1929 caused a decrease in participation notably by Australia and the Netherlands Indies, both basically primary producers which suffered considerably. However, much work in the 1930s was accomplished by the Archbold Expeditions and by several private individuals, notably two indomitable women: Evelyn Cheesman (UK) and Mary Clemens (USA). Miss Cheesman told her adventures in a number of entertaining books. Some work on Mary Clemens is being undertaken by P. Holzknecht at Lae, but regrettably some materials were lost (see below). In the years before World War II some Japanese expeditions made brief or extended visits, and Dutch work resumed. The valuable general survey by W. C. Klein, *Nieuw-Guinée*, appeared during this period (1934–38). Australian activities revived somewhat, aided by gold revenues from Bulolo and Wau.

The 1930s were dominated in biological exploration by the three pre-war Archbold Expeditions, led by Richard Archbold. They represent the most ambitious and well-organized general biological undertakings in New Guinea, and were based on a special unit "Archbold Expeditions" in the American Museum in New York. Archbold was an oil millionaire, an amateur pilot and a mammalogist/associate at that museum. The expeditions grew in size and the third was largest.

The First Archbold Expedition, 1933–34, worked a complete altitudinal transect from the south coast to the top of Mt Albert Edward in the Owen Stanleys. Access was facilitated by the Sacred Heart Mission track to Ononge (S of modern Woitape), thence overland to Murray Pass. Stations included Kubuna, Dieni, Mafulu, Mt Tafa and Ononge. Toward the end the expedition worked on the Oriomo Plateau NW of Daru. A. L. Rand collected birds and other animals (AMNH); L. J. Brass, plants (N.Y. Botanical Garden). The next two expeditions were notable for their extensive use of aircraft, radio, aerial survey and photography (see Sinclair: *Wings of gold*).

The Second Archbold Expedition to New Guinea, 1936–37, was along the Fly River and parts of the Palmer and Strickland tributaries, as well as some of the lakes, such as L. Daviumbu and L. Murray, as well as the Oriomo R. One objective was the eastern Hindenburg (Dap) limestone range. Besides Archbold, Rand and Brass, G. H. H. Tate was mammalogist. This expedition had an amphibious airplane, commuting from Daru I., and also a ketch, but the plane was lost at Port Moresby, terminating that phase, before the mountains far up-river were penetrated. However, important field work was accomplished in the still little-known Oriomo Plateau and other areas south of the Fly (AMNH; Arnold Arboretum).

The Third Archbold Expedition (Netherlands Indian–American Expedition 1938–39), the most ambitious, was done jointly with the Dutch. It traversed the great Mamberamo River from the sea to its highest headwaters, partly by boat and mainly by amphibious airplane, the "Guba," commuting from Hollandia. Field work was mainly from the middle Idenburg R. region to the summit of Mt Wilhelmina. Collecting stations included Bernhard Camp, 50 m, on Idenburg R., and Araucaria 800 m, Rattan 1200 m, Sigi 1500 m, Mist 1800 m, Top 2150 m, Baliem 1600 m, Bele 2200 m, Moss Forest 2800 m, L. Habbema 3225 m, Letterbox 3560 m and Scree Valley 3800 m. Parts of the large Baliem (Grand) Valley were explored, and very extensive collections amassed. The expedition first recognized *Nothofagus* in New Guinea, although specimens had been collected earlier and not correctly identified. Zoologists with Archbold included A. L. Rand, W. B. Richardson, and entomologists L. J. Toxopeus and J. Olthof; botanists were L. J. Brass, E. Meijer Drees and C. Versteegh. (Vertebrates AMNH, insects Leiden after processing at Bogor; plants Arnold Arboretum, duplicates elsewhere). Brass played a considerable role in preparing general reports of the first three expeditions.

Evelyn Cheesman of England did pioneer collecting of insects in several parts of New Guinea. In 1932 she did extensive work in the Cyclops Mts. On trying to enter NE New Guinea from Hollandia by small boat, she was sent back over the border to Dutch New Guinea by the first Australian officer because she did not have an entry permit. In 1933–34 she collected insects at Kokoda, Orrori and Oquali, north side of Own Stanley Mts, and at Mafulu 1200 m, Mondo 1500 m,

Diene 600 m, Mt Tafa 2550 m, Ishurava 900 m on S slopes of Owen Stanleys. During part of that latter period she joined with the First Archbold Expedition. In 1936 she collected briefly at Kavieng, New Ireland, and again at Hollandia. In 1938–39 she worked on Waigeo I. and on Japen I., then at Hollandia, Humboldt Bay, then at Aitape (having largely walked from Hollandia!), the Torricelli Mts, and hills between Vanimo and Hollandia. For quarantine and bond reasons, in spite of advance notice and application, she was again forced to go back to Dutch New Guinea, largely by foot, after encountering the first Australian official. Many species have been based on Miss Cheesman's collections (insects BMNH; NE NG 1939 collections at Adelaide; plants BMNH, Kew).

Mary Clemens first came to Finschhafen with her husband, a retired U.S. Army chaplain, in 1935, after work in China and Borneo. From then on she worked extensively in Morobe Prov. until the onslaught of war in 1942, collecting perhaps 14,000 numbers of plants. Stations included Malolo (nr Salamaua), Wau, Lae, Kaiapit, Wantoat, Boana, Matap, Samanzing, Amieng, Mt Salawaket, Sambanga (above present Kabwum), Ogeramnang (also visited by E. Mayr 1929), Yunzaing, Sattelberg, Wareo and Quembung, many of them Lutheran mission stations. Collections went to Berlin (much destroyed but duplicates elsewhere) until 1939, and until 1941 to Michigan University Herbarium. All transport was cut off during the war and some collections had to be abandoned at Boana or Finschhafen when all women and children were hurriedly evacuated in Jan. 1942. The plants were said to have been destroyed but in the 1960s bundles of specimens were uncovered at the two major herbaria in Tokyo (Tokyo Univ.; National Science Museum). Clemens' collections were largely reported on by Diels and others at Berlin and from 1939 in Archbold reports by Merrill, Perry and others.

In 1935–36 C. E. Carr, sponsored in part by the British Museum (Natural History), collected plants extensively in the Port Moresby region and around Mt Victoria: Locales included Kanosia, Koitaki, Boridi (Central Prov.) and The Gap, Alola, Lala R., Usuarava, Yodda Creek and Kokoda (Northern Prov.). Over 5500 non-orchid and some 1000 orchid collections were obtained (BMNH; Singapore; many duplicates elsewhere, especially Canberra, Leiden). Collections were reported in same way as for Mary Clemens, but less thoroughly.

In the Bismarck Archipelago, Father G. Peekel had collected plants around Lemahot and Ugana mission stations in New Ireland, and partly ascended the Lelet Plateau, 1000 m (not, however, collected for plants until "Noona Dan" expedition 1962 and Frodin 1973). Until World War II Peekel was engaged with compiling his valuable, never-to-be-published *Illustrierte Flora des Bismarck-archipels für Naturfreunde*, which he saved from destruction under his robes, and terminated in 1947. It has now been microfilmed, with a microfiche version commercially available. It is the most valuable single pre-war contribution on New Ireland flora.

G. Bateson ("Naven"), a former husband of the late Margaret Mead, collected some plants in the Gazelle Pen. while studying Baining people in 1931; and Beatrice Blackwood collected a few plants near Arawe and Kandrian and also in the upper Watut Val., in 1937. But outside the Gazelle Pen., New Britain remained floristically virtually unknown until the late 1950s. Manus, too, was virtually ignored until World War II.

On the mainland and closer islands, other collectors in the 1930s included E. Jacobson on Waigeo in 1936 (birds, insects; Bogor); Lt J. M. van Ravenswaay Claasen on Berau Pen., Vogelkop in 1937, and Mappi, Digul R., Merauke and Ajamaru in 1938 (insects; Leiden); W. Stüber in the Hollandia-Sawia area and hills to S and E: Sentani, Kressi Rd, Korime, Mamda, Ajiop, Tarafia 600 m, Koffo = Komfe Hills, Vokwar, Njau, Bougainville Mts 400 m, Bewani Mts, Nonno = Japoe, Cyclops Mts and Pim R. (insects, Leiden; also orchids for commercial purposes — he discovered the "Sepik Blue" Dendrobium). S. Dillon Ripley (on the Denison–Crockett expedition) with his wife in 1937–38 collected birds in the Arfak Mts and on Salawati, Batanta and Misool islands (Philadelphia). R. G. Wind collected butterflies and other insects along south coast, at Fak Fak, Merauke, etc. in 1939 (sold to various museums).

In the east, Ivan Champion and C. T. J. Adamson, inspired by the Archbold Expeditions proceeding concurrently (and by whom some aerial support was provided), collected animals on their Bamu-Purari patrol through the Southern Highlands, in 1936–37, including Lake Kutubu, first discovered in 1936. In 1938 there was a small Papuan-Australian Expedition (BMNH).

Dutch government activity picked up in the years before 1941, and apart from expeditions, forest surveys were carried out, especially by Z. Salverda in 1936 in the McCluer Gulf, Bomberai Pen. and along SW coast (Bogor). In 1936 A. H. Colijn, with J. J. Dozy and pilot F. J. Wissel, made a largely successful general survey of the Carstensz complex, reaching the top of Ngga Pulu (dense fog obscuring Carstensz Pyramid), although much remained for Harrer and Temple in 1961–62. A valuable small plant collection was secured by Wissel (Leiden, Bogor). In 1938 Wissel made another discovery: the three Wissel Lakes; soon a government station, Enarotali, was opened; this inspired two more large expeditions after the close of the Third Archbold in 1939.

The Royal Netherlands Geographical Society's C. C. F. M. le Roux Expedition of 1939, with zoologist Prof. H. Boschma, collected insects and other animals in the Nassau Mts, Wissel Lakes and Etna Bay (Leiden). Explorer and *controleur* J. P. K. van Eechoud — one of two "Bapa Papua" (papa Papua) collected birds, insects and other animals during 1939–40 in Wissel Lakes area as well as in Mamberamo basin and van Rees Mts (Leiden). In the mountains, both expeditions were accompanied by P. J. Eyma as botanist (Leiden, Bogor); part of the field data, however, were lost.

The Negumy Expedition in 1941 included insect collecting by the forester E. Lundquist on Berau (Vogelkop) Pen., McCluer Gulf, Agonda, Bomberai (Onin) Pen., Etna Bay (SW), Oeta and Najeju (S coast); plants were collected by the *mantri* Aët from Bogor (insects Leiden; plants Bogor, Leiden). Anta collected on the Digul in 1941. J. J. van der Starre collected insects at Kaimana (SW) in 1941 (Leiden). This and the preceding expedition did much to improve knowledge of the poorly-known Bomberai Pen.

J. B. McAdam, Terr. N.G. forestry officer, collected plants, 1938–41, in New Britain and around Wau (Canberra, Brisbane). J. L. Froggatt, B. A. O'Connor and Gordon Dun were entomologists based at Keravat, New Britain before the war (coll's mostly destroyed). G. F. Hill collected mosquitos and other insects in New Britain (Macleay; CSIRO).

Two expeditions were made by the Japanese botanists R. Kanehira and S. Hatusima. In 1937 they paid a hurried visit, based on a tour ship, to Kavieng, Rabaul, Lae and Wau; but in 1940 they mounted a significant expedition to Angi Lakes and to Dalman (S of Nabire, S Geelvink Bay), the latter the site of a damar (*Agathis*)-gum enterprise (coll's Fukuoka). The plant collections survived the war, both at Fukuoka and their duplicates at Bogor (where Hatusima was botanist 1942–45).

Europe-based activity largely ceased with outbreak of war in 1939, but as noted above some work continued until 1941–42. However, its cessation was abrupt following Japanese declaration of war in December 1941, and established collections at Rabaul and Keravat were abandoned (and subsequently largely destroyed). Mary Clemens' collections are mentioned above. Australian activity remained low until the 1950s. However, as with geographical exploration — declared "closed" by Sir Hubert Murray in 1939, at least in E New Guinea — a sketchy but useful knowledge had been obtained of higher plant flora and vegetation and the mammal, bird and butterfly fauna. The appreciation of these, however, was still largely static and it remained so for another generation before the biota could begin to yield more secrets in relation to dynamic concepts of biology, ecology and biogeography. For such work, the level of knowledge in almost any group was still much too sketchy.

7. War–time collecting

During World War II a number of entomologists and others with the armed forces of USA, Australia, New Zealand and Japan worked in New Guinea. Some of the collections did not survive the war, but many important lots are preserved. P. J. Darlington, Jr. collected many carabid beetles, and ants, frogs, etc, primarily at Dobodura near Buna, NE Papua, and (on leave after being mauled by a crocodile) on slopes of Mt Wilhelm, and to its summit, but also at Milne Bay, Aitape and Sansapor (Vogelkop), 1943–44 (MCZ). K. V. Krombein collected Hymenoptera and other insects, primarily at Nadzab, Markham Valley, 1944 (USNM). E. S. Ross and S. G. Jewett collected all groups of insects at Finschhafen, Hollandia, Maffin Bay or elsewhere, mostly in 1944 (CAS). J. Frank Cassel collected birds and amphibians and reptiles (herps) at Finschhafen, 1944 (Cornell). L. W. Jarcho, C. W. Moren, W. M. Beek, A. M. Keefe and W. H. Stickel collected herps along N coast, mostly in 1944 (MCZ, USNM); and by Melvin Kurz (AMNH). W. H. Wagner, Jr. collected ferns (UC), and with D. F. Grether collected butterflies, especially on Manus I. (USNM), reported by them and A. H. Clark. Lee Burcham (USNM) and John Reeder (Arnold) collected grasses. Borys Malkin collected mostly beetles, several areas (USNM). Harry Hoogstraal collected generally, and for mosquitoes and ectoparasites, in Cyclops Mts (to summit) and elsewhere, 1944–45

(Field, CAS, USNM), but much of his collection was lost. Willard V. King, W. E. Brewer, H. W. Cook, J. Forbes, W. R. Fullen, D. P. Furman, Donald R. Johnson, W. T. Nailon, George H. Penn (also herps), L. W. Saylor, C. J. Steinhauer and J. P. Toffaleti collected many mosquitoes along north coast, 1944 (USNM and CSIRO). Carl Mohr and W. D. Fitzwater collected mites, ticks and other medical arthropods in the Buna-Gona area, Papua and at Owi I. and Sansapor in 1944–45 (USNM). Glen Kohls and Cornelius B. Philip collected chiggers and other arthropods, etc, at Dobodura, Purdy Is or elsewhere, 1943–44 (USNM). Kenneth L. Knight and Lloyd E. Rozeboom collected mosquitos in several N coastal areas and islands, 1944–45 (USNM).

Mosquitoes, chiggers and other medical arthropods were collected in NE Papua and elsewhere by Frank H. Taylor (and Wewak), Anthony R. Woodhill, Carl Gunther, Ian M. Mackerras, D. O. Atherton, D. A. C. Cameron, F. Chippendale, D. H. Colless, H. A. Grandall, R. N. McCullock, M. H. Wallace and R. Harry Wharton, 1943–45 (Brisbane, Macleay, CSIRO). Carl Gunther also collected mites, etc at Bulolo both before and after the war (Brisbane).

L. J. Dumbleton collected mosquitoes etc on Nissan I. (CSIRO and Nelson). Marshall Laird studied mosquitos and parasites at Jacquinot Bay, New Britain in 1945 (Nelson). George E. Bohart, J. M. Fritts, Ashley B. Gurney, Paul Hurd, L. A. Posekany, Barnard V. Travis and George W. Wharton (USNM) and E. Eldon Beck, E. Reimschüssel, Harry P. Chandler and Dorald Taylor (BYU) and others collected in the Solomon Is, some of them, including Gurney, on Bougainville, and Reimschüssel also in Admiralty Is.

Several of the above entomologists published on some of their discoveries, and J. N. Belkin, R. M. Bohart, Joanna Bonne-Wepster, Robert Domrow, D. S. Farner, Gressitt, D. J. Lee, Elizabeth Marks, C. B. Philip, Alan Stone, F. H. Taylor, Herbert Womersley and others also published early on the collections, and many others later. (See taxonomic index in Gressitt and Szent-Ivany, 1968.) Most of those mentioned above (except Bonne-Wepster, Dumbleton, Laird) are American or Australian.

Some insect and plant collections made by Japanese just before the war, or transported to Japan during Japanese occupation, are mentioned in the preceding section. During the Japanese occupation, Dr S. Issiki collected insects, also in the "neck" area: Windesi, Majosi, Wasior (Wandammen Pen.), Rumberpon (Amberpon) I. (coll's Taipei; partly reported by Gressitt). Yoko-oji collected birds at Manokwari, 1942–44 (Tokyo; all but 7 destroyed). M. Satake studied general Nat. Hist., Wandammen "neck" area, 1942–43 (1963 book in Japanese, Tokyo); and Prof. Toyohi Okada (chapter IV:10) was at Aitape as a soldier.

In 1942–43 Carl de Zeeuw collected plants in various parts of Papua (MEL). N. A. Wakefield collected plants, principally ferns, in various parts of E New Guinea and the Solomons, 1943–45 (BMNH, MEL). H. S. McKee collected plants in 1944–45 in various parts of NE New Guinea. In 1944 the NGF-series of forest tree collections was begun at Lae; until 1945 this was under the Australian Forces forestry unit directed by J. B. McAdam. Over 2000 numbers of collections were made before cessation of operations. In 1946 the series was resumed by the Dept of Forests, TPNG (see below).

8. The post-war period

8.1 Flora, West New Guinea (Dutch N.G., 1946–63; Irian Jaya, Prov. of Indonesia, 1963—)

In 1948 A. J. G. H. Kostermans collected at Angi Lakes, Arfak Mts; in 1964 he visited Baliem valley (Bogor). In 1954 P. van Royen with the 1st Rijksherbarium expedition (partly with C. Versteegh) explored many areas of Vogelkop and Batanta I.; also in south from Merauke to Fly R., and then the Cyclops Mts in north. In 1955 he worked Waigeo and then to TPNG (Leiden). In 1955, etc, Gressitt collected a few plants (Bishop). In 1957 C. W. Grassl collected grasses in lowlands and at Angi and Wissel Lakes (Leiden).

A few plants were collected on the New Zealand expedition to the Carstensz, 1961, by D. E. Cooper and Philip Temple; and in 1962 by Temple and Heinrich Harrer (Auckland).

The Dutch developed an agricultural station at Kota Nica near Lake Sentani. The research sections were later moved from there and Hollandia to Amban near Manokwari as a research station was developed there over the period 1953–62. Botanists were C. Kalkman and later W. Vink, assisted by Peter and Gerrit Iwanggin and Cris Koster. Foresters J. F. U. Zieck and botanist C. Versteegh worked at Angi Lakes, 1954. In 1961 P. van Royen with H. O. Sleumer, on the 2nd Rijksherbarium expedition worked on Vogelkop: Kebar Val., Tamrau Mts, Neltoti range, etc (Leiden). The Dutch left in 1962 and early 1963 and West New Guinea became the Irian Barat, later Irian Jaya, Province of Indonesia on 1.V.1963.

In 1963 Indonesia established Cenderawasih (Tjenderawasi) University at Abepura near Jayapura (former Hollandia). Its agriculture and forestry college was set up in the Manokwari area, at Amban near the above experiment station, in 1964.

In 1968–70 a Japanese expedition, under Y. Kobayashi, investigated lower plants in the Wamena area of the Baliem valley and elsewhere (Tokyo). During 1971–73 the Australian Universities expeditions to Irian Jaya, organized at Univ. of Melbourne, made glaciological and biological investigations on and around the Gunung Jaya (Carstensz) glaciers in 2 operations. Biologists were Geoffrey S. Hope and Judy A. Peterson (Canberra). In 1972 ex-King Leopold III made his 2nd expedition to W New Guinea (1st, 1929, above), with J. Raynal (into 1973) collecting near the Mt Jaya area, etc (Brussels, Bogor, Paris, Leiden). In 1976 P. Hiepko, W. Schultze-Motel and W. Schiefenhövel (latter medicinal plants), worked with German doctors and anthropologists in the Eipomek Val. and elsewhere, C Highlands (Berlin-Dahlem; Leiden).

8.2 Fauna, West New Guinea (Dutch N.G., 1946–63; Irian Jaya, Prov. of Indonesia, 1963—)

On Sten Bergman's Swedish-Netherlands Expedition, ornithologist Bergman studied and collected birds (Stockholm); and Dr M. A. Lieftinck collected insects on Misol, Salawati and at Sorong, 1948–49 (Leiden). Bergman again worked in W New Guinea in 1952–53 and 1958.

In 1952 L. D. Brongersma, with W. J. Roosdorp, collected animals in the north, on Vogelkop and at Wissel Lakes and Merauke (Leiden). In 1954 L. van der Hammen collected Acarina etc on Vogelkop, Geelvink Bay, L. Sentani, Hollandia area and Wissel Lakes (Leiden). In 1954–55 Brongersma collected animals again, with M. Boeseman and L. B. Holthuis, at many localities on Vogelkop, Biak, Japen, Digul River and elsewhere (Leiden).

In 1955 J. L. Gressitt collected insects for a month, mainly in the Wissel Lakes area, including Kamo Valley, and also at Hollandia, Sentani, Cyclops Mts and Biak I. largely in company with R. T. Simon Thomas, newly arrived government agricultural entomologist. In 1957 he revisited the Cyclops and Biak; in 1958 he spent three weeks in the Swart Valley and area to its west. In 1959 he collected with T. C. Maa in the Cyclops, on Biak and at Fak Fak (incl. caves) and inland on the Bomberai (Onin) Peninsula. In 1962 he collected in the Cyclops, Biak, Nabire and Wissel Lakes, partly in company with J. Sedlacek, N. Wilson and H. C. Clissold (Bishop). In 1977 and 1979 he made brief visits to Jayapura, Biak, Manokwari, Arfak Mts and Sorong, mainly to make contacts at the Cenderawasih University branches.

Also on the Bishop Museum entomological survey, D. E. Hardy collected at Manokwari, Angi Lakes and elsewhere on the Vogelkop, and Sentani and Hollandia, 1957; T. C. Maa collected at Waris, Sarmi, Holmafin, Baliem Valley and Merauke, besides the above-mentioned, 1959; L. W. and Stella Quate collected at Bokondini, L. Archbold and Sibil, Star Mts 1961. In 1962 and early 1963 besides Sedlacek, Wilson and Clissold, collecting was done by L. P. Richards, Max C. Thompson, Philip Temple, Heinrich Holtmann and Ray Straatman (Bishop).

During these last few years of Dutch rule in W New Guinea, besides R. T. Simon Thomas as agricultural entomologist (coll. Manokwari), Rudolph Sloof, Dirk Metzelaar, Johannes van den Assem and Willem J. O. M. van Dijk were medical entomologists, mainly concerned with mosquitos and malaria (Leiden, Amsterdam).

The Star Mts expedition of 1959, led by Dr L. Brongersma and G. F. Venema worked from Sibil Valley E of the Baliem, and scaled several of the peaks west of the Dutch-Australian border, including the ice-capped Juliana Top, 4700 m. W. Verwoort also collected animals (Leiden). Plants were collected by C. Kalkman and B. O. van Zanten, up to 3380 m on Mt Antares (Leiden; mosses to Groningen). Bär and Danselaar made a complete S coast to N coast crossing of the island in this area. At almost the same time a French movie group, with Gaisseau ('The sky above, the mud below'), made a similar S-N crossing, just slightly to the west. In 1960–61 S. Dillon Ripley and wife collected birds on N slopes of Snow Mts (Yale).

Little animal collecting has been done in the west since the transfer to Indonesia. In 1963–64 a Tsendrawasih expedition with S. Soemadikarta and Mr Boeadi collected some animals, E of Wissel Lakes (Bogor). In 1964 there was a joint Cenderawasih University and Explorers' Club of Nanzan Univ., Japan (Collembola: Yosii) climbing trip to Ngga Pulu (Carstenz, Gunung Jaya).

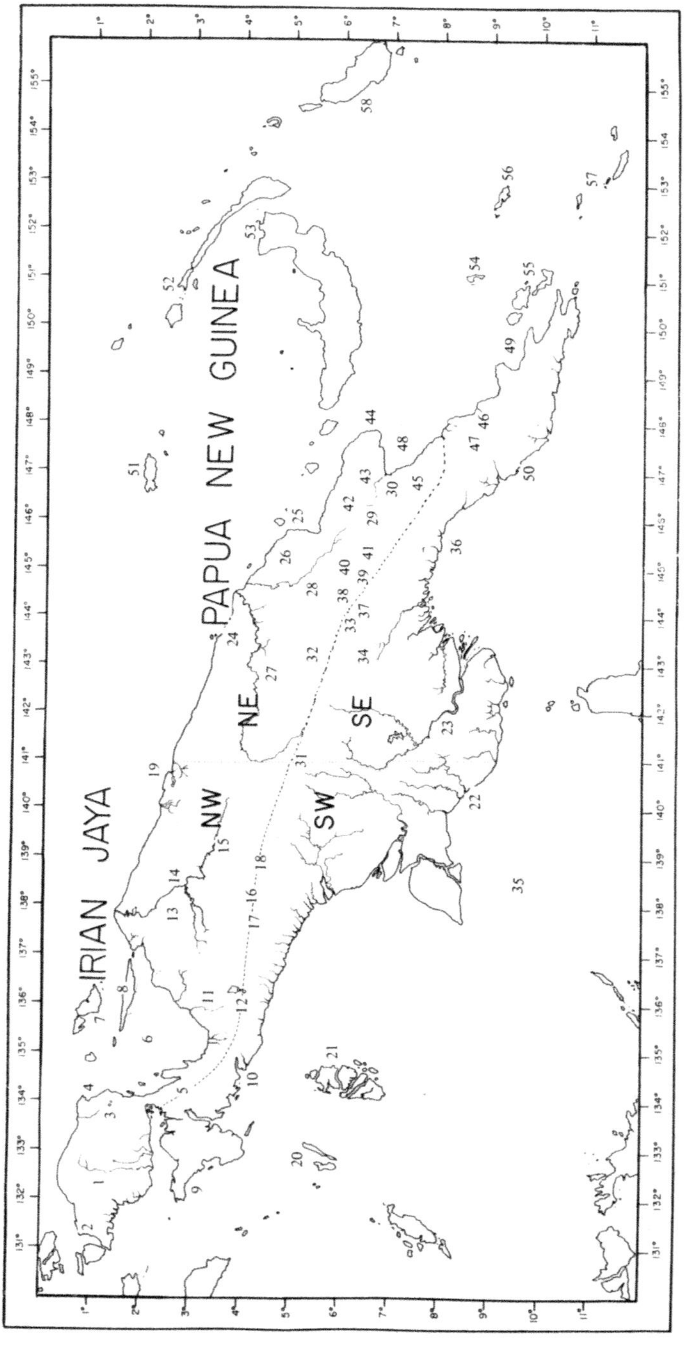
IRIAN JAYA
PAPUA NEW GUINEA
NW
NE
SW
SE

Fig. 1. Key to localities. This map is one used in many publications on New Guinea insects to help locate places and refer to them, in 4 quadrants of the island: NW, SW (Irian Jaya); NE, SE (Papua New Guinea). The latter 2 are the former German and British (Papua) territories.

1. Vogelkop Peninsula
2. Sorong
3. Arfak Mts
4. Manokwari (Doré Bay)
5. Wandamen Mts
6. Geelvink Bay
7. Biak I.
8. Japen I. (Jobi)
9. Fak Fak
10. Etna Bay
11. Weyland Mts
12. Wissel Lakes
13. Van Rees Mts
14. Mamberamo R.
15. Idenburg R.
16. Central ranges
17. Gunung Jaya (Carstenz)
18. Mt Wilhelmina (Trikora)
19. Jayapura (Hollandia) and Humboldt Bay; Cyclops Mts
20. Kė (Kei, Key) Is
21. Aru Is
22. Merauke
23. Fly River
24. Wewak
25. Madang
26. Adelbert Mts
27. Sepik R.
28. Ramu R.
29. Markham R.
30. Lae
31. Star Mts
32. Enga Prov.
33. Mt Giluwe
34. Lake Kutubu
35. Arafura Sea
36. Gulf of Papua (and Kerema)
37. S Highlands
38. W Highlands
39. Chimbu (Simbu) Prov.
40. Mt Wilhelm
41. E Highlands
42. Finisterre Mts
43. Saruwaged (Salawaket) Mts
44. Finschhafen
45. Wau (and Mt Kaindi, Bulolo, Mt Missim, Bulldog Rd)
46. Popondetta (nr Buna, Gona, Mt Lamington, Kokoda)
47. Owen Stanley Mts
48. Huon Gulf
49. Collingwood Bay
50. Port Moresby
51. Manus I. (Admiralty Is)
52. Kavieng (New Ireland)
53. Rabaul (New Britain)
54. Trobriand Is
55. d'Entrecasteaux Is
56. Woodlark I.
57. Louisiade Archipelago
58. Bougainville I. (N Solomons)

a
b
c

A Kyoto University science expedition, 1963–64, worked from Wissel Lakes to Ngga Pulu; Y. Yasue chief biologist. In 1971 a museum was established beside the campus of Cenderawasih University at Abepura. In 1979 Prof. Jared Diamond studied birds along the Mamberamo River.

8.3 Flora, East New Guinea (Territory of P. and N.G.=TPNG, 1946–1971; Papua New Guinea, 1971-; independent from 1975)

The leading contributions have been made through the Division of Botany, Department of Forests, established in 1946. Until 1975 it was under the direction of John S. Womersley, who developed it extensively starting with the 1944–45 collections assembled largely by L. S. Smith and C. T. White of the Queensland Herbarium. The Lae Herbarium now houses a very large collection of PNG plants and has jointly sponsored many botanical expeditions from abroad. Among staff of the herbarium have been or are currently Michael Galore, from 1953 (now chief); A. G. Floyd (1953–55); E. E. Henty, from 1957 (now senior botanist), Andrée Millar (1954–71, since then National Capital Botanic Garden); Pieter van Royen (1962–65, now Bishop Mus.); J. Buderus (1964–66); A. Gillison (1964–67, later Forestry College, Bulolo to 1972, now CSIRO); D. G. Frodin (1965–66, since 1971 Univ. PNG); C. E. Ridsdale (1966–68, now Leiden); A. W. Dockrill (1966–70, now CSIRO); M. J. E. Coode (1966–72, now Kew); J. Vandenberg (1967–70); D. B. Foreman (1968–74, now Armidale); R. J. Johns (1968–71, later Bulolo, since 1979 PNG Univ. Tech.; see chapter III: 3); P. F. Stevens (1970–73, now Arnold Arb.; see chapter III: 4); G. Leach (1971–73, from 1979 Univ. PNG); W. R. Barker (1974–76, now Adelaide); N. Clunie (1974–78); and currently J. Croft (since 1973), Paul Katik and Yakas Lelean. Many collections were also made by forestry officers, chiefly before 1960.

Among visitors who have worked at the Lae Herbarium are R. F. Thorne (various, RSA); M. M. J. van Balgooy (Leiden); H. St. John and B. C. Stone (see chapter III: 7), *Pandanus* (Bishop, Kuala Lumpur); M. J. S. Sands (Kew, Begoniaceae); W. R. Philipson (Christchurch), rhododendrons, Araliaceae; P. Kores (WEI), rhododendrons; B. Barlow (Adelaide), Loranthaceae; H. U. Stauffer (Zürich), Santalaceae; B. Verdcourt (Kew), Leguminosae; Fred Essig (Tampa) palms; Alistair Hay (WEI, Bulolo) palms; D. E. Symon (Adelaide) Solanaceae; S. Sohmer, also collecting in various areas in 1979 (USNM, Bishop), *Psychotria*: and A. Touw (Leiden), bryophytes.

Other herbaria have also been developed in PNG to meet particular needs. In 1957 the forerunner of the Forestry College at Bulolo was established and gradually an herbarium developed (J. J. Havel, H. Streimann, A. Kairo,

---

Fig. 2. a) Some of higher summits of Arfak Mts, Vogelkop, N of Angi Lakes, 2500 m; *Nothofagus* on summit ridge and mixed oak forests on parts of slopes; bracken on fired area in foreground; June 1979, Gressitt;
b) Some of higher summits of Arfak Mts, 2400 m, N of Angi Lakes; some fire effects, and emergent podocarps;
c) Same area, 2350 m; some fire effects, with *Gleichenia*, other ferns, *Styphelia*, *Rhododendron*, etc; June 1979, Gressitt.

A. Gillison, R. J. Johns, B. Conn, A. Hay). Active collecting has been carried out in the Wau-Bulolo region and elsewhere. At the adjacent Forest Research Station work in forest pathology commenced about 1969 (currently J. Simpson); the fungus herbarium there currently has the best local collection of macrofungi (F. Arantz).

At Konedobu the Plant Pathology herbarium of the Dept Primary Industry (DPI) was begun about 1955 and built up largely by D. E. Shaw (see chapter III: 10); now G. R. Kula. Its contents comprise largely pathogenic microfungi, but small collections of other thallomorphic plants are also held.

At the University of Papua New Guinea (UPNG), the herbarium was started in 1968 by R. Robbins and M. Pulsford but from 1971 was greatly enlarged by D. G. Frodin (see chapter III: 12) with contributions from others (incl. A. Millar, J. Powell, C. Huxley, J. Dodd, I. Johnstone, A. Gebo). In 1978 the herbarium was badly damaged by an accidental fire, but most collections were saved. Frodin studies NG flora, especially *Schefflera*, and has collected in many areas. Currently the distinctive seasonal flora of Port Moresby region is under study. In 1974–77 Camilla Huxley studied ecology and biology of ant-plants (*Hydnophytum* and *Myrmecodia*) and concluded that the ants serve to disperse material which may serve in nutrition of the plants (see fig. 4). In 1974–79 Ian Johnstone studied sea grasses and mangrove ecology (see chapters III: 11, 12).

At Wau Ecology Institute (WEI) (see Fauna 8.4.) field studies and propagation of rhododendrons were carried out by Paul Kores (1974–78), supported by the Stanley Smith Horticultural Trust. Kores visited many mountain areas of PNG, among them Owen Stanleys, the Ekuti, Bismarck and Saruwaged ranges, Mt Hagen, Mt Giluwe and the Bürgers (with Veldkamp) and Victor Emmanuel

Fig. 3. Moss forest, or stunted subalpine forest, near summit of Ekuti range, 3200 m, near Wau; 1969, Gressitt.

Fig. 4. a, Ant plants, *Myrmecodia* sp., Arfak Mts, N of Angi Lakes, 2400 m; June 1979, Gressitt; b, bisected *Myremecodia* plant, Mt Kaindi, 2200 m; 1976, Gressitt.

Ranges. The greatest species concentration was found on the upper part of the Bulldog Road on the Ekuti range S of WEI. A large arboretum of native trees and other plants was gradually developed at WEI through the efforts of Gressitt, Sedlacek, A. Allison, B. Gagné, A. Hay and others. Hay worked on palms and B. Gagné on agro-silviculture with W. C. Gagné. In 1979 D. I. Axelrod visited the Wau area to study vegetation history (chapter VI:2). Many plants were also collected by T. Pratt and A. Allison in the course of zoological studies, 1974–80. A small station herbarium has gradually been developed.

Numerous botanical expeditions have been conducted since 1946. Among them are those which the Div. of Botany organized, or was involved with, as well as the later Archbold Expeditions, those of CSIRO for land-survey work and those of the Rijksherbarium, British Museum, Kew, Copenhagen and others. These have more or less covered most parts of PNG as of 1980, although gaps remain and other areas remain undercollected, including many worked by the Germans before 1914. Original sets or duplicates of nearly all collections have been deposited at Lae.

The Fourth Archbold New Guinea Expedition, 1953 (following on the Archbold Cape York Peninsula Expedition of 1948) worked in E Papua, between Collingwood Bay and Mt Dayman, as well as on Goodenough I. L. J. Brass was leader and collected insects as well as plants. H. M. Van Deusen was mammalogist and G. M. Tate was general collector. Latter was evacuated fatally ill after 6 months. (Animals AMNH; plants Harvard.)

The Fifth Archbold New Guinea Expedition, 1956, worked on Normanby and Fergusson islands (D'Entrecasteaux) and Misima, Sudest and Rossel (Louisiades), Woodlark I, the Trobriands and at Milne Bay on the mainland. Brass was again leader and collected plants and some insects. R. F. Peterson was mammalogist. (Animals AMNH; plants Leiden).

The Sixth Archbold New Guinea Expedition, 1959, worked in the E Highlands including Mt Wilhelm, Mt Otto, Mt Michael, Mt Elandora, and at Okapa, Kassam Pass, and from Lae to Mt Kaindi and Edie Creek, as well as Markham valley. Brass was again leader, with Van Deusen as mammalogist. J. Womersley and J. D. Collins joined part of the time. T. C. Maa of Bishop Museum joined for a month on the collection of vertebrate ectoparasites, during the Okapa and Kassam Pass periods (Vertebrates AMNH: ectoparasites Bishop; plants USNM). (Chapter IV: 12.)

The Seventh Archbold New Guinea Expedition, 1964, was led by Van Deusen and worked on the Huon Peninsula. S. A. Grierson was general zoologist and photographer; R. G. Zweifel (see chapter V: 1) was herpetologist and R. D. Hoogland was botanist. Work was done in the Rawlinson range, Cromwell Mts and near Finschhafen (Hoogland also in Saruwageds). (Animals AMNH; plants CSIRO.) Many reports, primarily botanical and in vertebrate zoology, have resulted from the Archbold expeditions, but many insects were collected by the 1939 Third Archbold (Neth. Indian-American) Expedition in the west.

The CSIRO Land Research Surveys, 1953–70, covered many aspects, including vegetation studies, particularly by R. D. Hoogland assisted for various periods by A. Kanis, P. Darbyshire, J. Flenley, L. Craven, R. Pullen,

R. Schodde, T. G. Hartley (medicinal plants, 1962–65) and P. C. Heyligers. These teams worked in many areas, but especially in highlands, N Sepik basin, S coast, etc (CSIRO).

C. R. Stonor collected plants on Mt Hagen and Mt Wilhelm, 1948–49. In 1954 H. S. McKee collected in the Wahgi valley (Brisbane; Lae). In 1955 P. van Royen collected at Bulolo, Bulldog track, Lae and Port Moresby (Leiden). Dr D. C. Gajdusek, U.S. National Institutes of Health, collected plant and animal material as samples of food and drugs in E Highlands and many other areas, 1958–1979, while studying Kuru disease and related studies, for which he became a Nobel Laureate.

In 1956 on a CSIRO-Lae operation, Hoogland, R. Pullen, Womersley and Galore collected on Mt Wilhelm. In 1957 R. C. Robbins worked on vegetation on Mts Wilhelm, Hagen and Giluwe; and J. C. Saunders, also 1957, in Kubor Mts. In 1960 Hoogland, Schodde and Robbins worked on Mt Sugarloaf near Hagen, and J. A. R. Anderson at Wilhelm. In 1961 Hoogland and Darbyshire worked at Aitape and Torricelli Mts, and F. R. Fosberg collected in Goroka, Lae and Rabaul areas (USNM). The Danish Noona Dan Expedition, 1962, collected plants in Bismarcks (see Fauna, 3.4.). In 1961 P. van Royen collected with M. O. Sleumer on Mt Wilhelm and in New Britain. In 1962 van Royen collected on Mt Wilhelm with F. W. Went, and in 1963 with S. Carlquist (RSA) on Huon Peninsula. In 1963 W. Vink of Leiden and R. Pullen of CSIRO collected on Kubor range, and E. E. Henty went with Carlquist to Mt Piora. F. Kleckham of DASF (DPI) collected plants on Mt Strong in 1963; and G. Rosenberg collected plants and insects on Mt Amungwiwa. In 1964 Craven and Croft worked on Mt Victoria and in 1965 Gillison, van Royen and J. Buderus worked on Mt Biota, W of Albert Edward, and B. Craig in the Star Mts. In 1963 Roger Heim collected fungi in highlands (Paris).

In 1964–65 A. Clive Jermy led the BMNH — Newcastle-upon-Tyne Univ. expedition, with Trevor G. Walker, Alan Eddy and M. E. Bacchus, joined by C. D. Sayers of Lae and R. Pullen of CSIRO, stressing ferns and bryophytes, in Finisterres and hills S of Lae and on Mt Wilhelm.

The vegetation history and pollen analysis studies carried out from 1960 under the direction of Prof. Donald Walker (ANU) also included substantial field work. Early studies were largely carried out in W Highlands and Enga provinces. In 1965 Walker established the Mt Wilhelm research laboratory at Pindaunde, 3300 m, with help from Bishop Museum and the Territory administration. Much research was done here on vegetation, palynology and climate. The station was ceded to the National Parks Board in 1974. Some specific studies include: 1965, Prof. Walker and J. R. Flenley on pollen deposits in Enga Prov. (see chapter III: 1); 1966, Jocelyn Wheeler Powell on deposits in the E Highlands (see chapter II: 4); 1966–67, intensive studies on Wilhelm by D. McVean and L. K. Wade; 1968, Walker, B. O. van Zanten, W. A. Weber, R. J. Johns and McVean on vegetation studies including the cryptogamic flora at Wilhelm; 1968–69 G. S. and Jeannette Hope on vegetation history and animal ecology at Wilhelm, G. S. Hope also working at Mt Giluwe and later in the Owen Stanleys; and 1972, J. A. Peterson on glacial history and J. M. B. Smith on alpine plant ecology at Wilhelm and on Kubor range (see chapter III: 2). In

1976 S. Garrett-Jones worked on palynology, L. Wanum. A 1977 field trip to Lake Trist E of Wau was aborted, but some work was tried at L. Kutubu and a small lake near Mt Ialibu.

Other vegetation studies include those by R. Hynes, J. Ash (see chapter III: 5) and N. Clunie on *Nothofagus* (Clunie also on New Britain); N. Enright on *Araucaria* (see chapter III: 6), P. C. Heyligers in the Port Moresby region, 1963–67 (see chapter III: 8); K. Paijmans on Mt Albert Edward in 1970 and in Galley Reach in 1975; and P. Edwards and P. J. Grubb near Mt Kerigomna above Marafunga in 1970–71.

Among special plant groups, the collecting of orchids has been accomplished by A. Millar, N. Howcroft (silviculturist at Bulolo Forestry Station), T. Reeve (Laiagam) and J. Dodd among others. J. Croft and R. J. Johns have specialized in ferns, for which R. E. Holttum and B. J. Croxall have also paid visits. Holttum also studied bamboos. B. Verdcourt studied legumes in 1976 and 1978.

In the 1960s and 1970s a number of the above botanists were assisted in the field by Aubita Kairo, botanical assistant at Bulolo Forestry College, who recognizes most genera of trees from sterile material, and who has identified many in the WEI arboretum. Paul Katik, Artis Vinas (until 1979) and Yakas Lelean, and in earlier years Michael Galore and John Koibua have also assisted many in the field, including DGF. Paul Katik has acquired the ability to recognize almost all genera and many species of seed plants at sight, thus ranking with *mantri* Nedi at Bogor.

In 1965 M. J. van Balgooy visited TPNG and worked on Mt Wilhelm. In 1966 on a joint Leiden-Lae project, C. Kalkman, W. Vink, A. Gillison and D. Frodin worked at the Doma Peaks (Leiden; Lae). In 1968 C. Ridsdale and P. Woods collected on Mt Albert Edward (Leiden; Lae). In 1969 M. J. S. Sands and M. Coode from Kew worked on New Ireland (Kew; Lae); and Coode also collected on Mt Giluwe. (Sands again in N. Ireland and E NG with G. A. Pattison and J. J. Wood in 1975; Kew). In 1970 H. F. Winters and J. J. Higgins of USDA searched for Ericaceae etc in several mountain areas (U.S. Nat. Arboretum). In 1971 F. R. Mitchell, American, came from Christchurch and collected plants in several areas (Chch; Lae). In 1971 M. Coode and P. Stevens worked on the summits of Mt Strong and Mt Scratchley (Lae; Kew). In 1970–71 Y. Kobayashi collected lower plants on Mt Wilhelm, Mt Hagen and at Oksapmin. In 1972 there was a large Leiden-Lae expedition to Mt Suckling with W. Vink, J. F. Veldkamp, P. Stevens, D. Frodin, N. Cruttwell, R. Pullen, F. Essig and entomologists T. L. Fenner and Gressitt (Leiden; Lae; DASF =DPI; Bishop; etc). In 1973 M. Jacobs (Leiden) and K. Paijmans (CSIRO) collected on Mt Bosavi. In 1975 another Leiden-Lae expedition worked in the E Star Mts, PNG, with Veldkamp, A. Touw, W. R. Barker, B. Conn, J. Croft (Leiden; Lae). In 1977 Veldkamp worked in W Highlands (Leiden). In 1975 a group of Japanese botanists led by S. Kurokawa and M. Inoue collected mainly cryptogams on Mt Albert Edward, Wau area and Saruwaged Mts. G. Argent (Edinburgh) worked on *Musa* in early 1970s.

In 1976 a Bishop — WEI expedition, primarily botanical, with P. van Royen as leader, worked on Mt Victoria (v. Royen, P. Kores, Frodin, R. T. Corlett); Mt Giluwe (Royen, Kores, B. Gagné); Victor Emanuel Mts (Royen, Kores, B.

and W. Gagné, Gressitt); Mt Amungwiwa (Royen, B. Gagné); and Finisterres (Royen, Kores, B. Gagné, Gressitt), mainly at higher altitudes (Bishop; WEI). B. and W. Gagné collected plants and insects in the Louisiades (1979) and elsewhere, 1976–80 (Bishop; WEI). Corlett (ANU) also collected on Mt Giluwe, Amungwiwa, etc, 1975.

Canon Norman E. G. Cruttwell has been an Anglican missionary in E NG from 1946 into 1980, long stationed at Daga in NE Papua and since 1976 at Goroka. He knows many of the rhododendrons and orchids and visited Mt Simpson in 1947 and 1968, Mt Dayman 1951, 1974, etc, Mt Aniata 1959, Mt Suckling 1972, etc, and is establishing a high altitude botanic garden above Goroka. Among other missionaries doing botanical work have been the Rev. H. A. (Bert) Brown, based at Elema but also travelling to the Kunimaipa (nr Mt Strong); and currently Br Borrel at Kairiru off Wewak. Fr E. Borgmann studied chromosome numbers of Mt Wilhelm plants in 1960. Many collections have also been made by anthropologists in connection with field studies; some, like that of N. Bowers, have been substantial and of good quality. Medicinal plants have been studied by D. K. Holdsworth at UPNG, and collections deposited in the UPNG herbarium. P. Eddowes in 1960s made considerable collections in connection with wood-anatomical work (Lae). Barbara S. Parris collected ferns in several areas, 1971–72 and 1977 (Cambridge; Lae). In 1978–79 K. Arisumi and staff studied rhododendrons in the NE (Kagoshima).

8.4 Fauna, East New Guinea (Territory of P. and N.G. = TPNG, 1946–1975; Papua New Guinea, 1971—; independent from 1975)

N. B. Blood collected birds in the highlands 1945–47 and later (AM). Fred Shaw Meyer collected birds in Hagen area 1946–47, Mt Wilhelm 1949. Blood started the bird-of-paradise (plus other wild life) sanctuary partly funded by Sir Edward Hallstrom, at Nondugl in the Wahgi Valley, starting in 1948. Shaw Meyer took over managing it in 1954. It was turned over to the government and moved to Baiyer River in the early 1960s. It is now a National Park. It has been operated by Graeme George and now by Roy Mackay.

The Gyldenstolpe expedition, 1951, in the Hagen-Wahgi area, concentrated on birds, but also collected other animals and plants. It consisted of Nils Gyldenstolpe and wife (Stockholm: Ark. Zool. 8, 1, 1955).

In the 1950s Stan Christian, Harry A. Standfast, Wallace Peters and others worked on mosquitoes, malaria transmission, taught mosquito control and built up reference collections (now UPNG). Gordon Dun was agricultural entomologist pre-war and post-war to about 1960, partly overlapping with Dr Szent-Ivany.

Wm W. Brandt, earlier from Europe, was engaged by Sir Edward Hallstrom to make a large collection of Lepidoptera, primarily butterflies, in TPNG. Brandt worked on this project, at many localities, from 1949 to 1955, and then Hallstrom terminated the operation and gave the collection for the future PNG national collections, while depositing it for safe-keeping with CSIRO at Canberra. Brandt then collected half-time (miscellaneous insects) for Bishop

Museum from 1956 to 1960 (Bishop), with the other half of his time on Lepidoptera for TPNG (collections to add to the above collection, CSIRO). Following this he was employed by CSIRO to curate the collection and to do some collecting in TPNG (CSIRO).

E. Thomas Gilliard made several trips, 1948–59, to collect birds: Astrolabe Ra. 1948, Wilhelm, Giluwe, 1950, Kubor, 1952, Victor Emanuel and Hindenburg R., 1954, Hagen, Finisterres 1956, Adelbert 1959, New Britain 1958–59 (AMNH). He also photographed for the National Geographic Magazine, and collected butterflies, obtaining new species of *Delias* (AMNH) plus a few plants (Arnold).

For Archbold Expeditions, 1953–64, see section 8.3. In 1954 Ellis Troughton assisted by Norman Camps and helped some by N. Blood, collected mammals in W Highlands (AM; Au. Mus. Mag. 11: 246). Blood and Camps also assisted Gilliard and Gressitt.

In 1954 Dr Joseph J. H. Szent-Ivany, a refugee from Hungary to Australia (formerly curator of Lepidoptera at Budapest and university lecturer in zoogeography) joined the Dept of Agriculture (DASF, now DPI) in 1954. On field trips and vacations he collected extensively in various areas, greatly adding to the national insect collection at Konedobu (part of Port Moresby). Dr Szent-Ivany remained until 1966 (later at WEI). Other government entomologists who helped add to these collections included J. H. Barrett, T. L. Fenner, J. Healy, L. Smee, A. Catley, R. M. Stevens, T. V. Bourke, G. Baker, E. Hassan, Stuart Smith, J. N. L. Stibick, Jan Greve. Sir Alan Mann, Chief Justice in 1950s, collected many beetles and others, adding them to the collection.

In 1955 Edward O. Wilson, then graduate fellow at Harvard University, collected and observed ants in various areas, mainly lowlands on N and S sides and to top of Saruwaged range (MCZ). In 1955 Rev. O. Shelly collected some frogs in the Wahgi Valley (AMNH).

In 1955 J. L. Gressitt commenced his 26 years of annual visits or extended stays collecting insects: 1955: E and W Highlands, Mt Otto, Wilhelm, Jimi valley, N coast, Manus, New Ireland and New Britain. 1956: Kokoda, New Britain (coconut beetle study for DASF), Solomons; 1957: Solomons, N coast, Biak; 1958: S Highlands, Baiyer R.; 1959: N coast, Cyclops, Fak Fak; 1960: Oriomo, C. Rodney; 1961: with wife (M.) to many highland areas, Karimui, Mt Giluwe; also Maprik, Dreikikir, Lae, Popondetta; also establishing Bishop Museum Field Station at Wau (10 years later to become Wau Ecology Institute), placing Joseph Sedlacek, his wife Marie and son J. H. Sedlacek, in residence. From then, for over 10 years, there was continuous insect collecting for Bishop Museum, also including collection of vertebrate hosts with their parasites and plants for identification of insect hosts. Besides the Sedlaceks' 10 years work following on that of Brandt, Ray Straatman collected 1961–66, Abid Beg Mirza 1967–74 and part-time 1975–80 while manager of WEI. For shorter periods many others participated, including T. C. Maa, Wallace A. Steffan, G. A. Samuelson, L. W. and S. Quate, Y. M. Huang, S. Sirivanakarn, Nixon Wilson, Frank J. Radovsky, and A. C. Ziegler of Bishop Museum research staff, as well as Dr and Mrs Szent-Ivany for periods after retirement from DASF in 1966; also E. J. Ford, Jr., Peter Shanahan, Geoff Monteith (see chapter IV: 8), P. H. Colman, N. L. H. Krauss, H. Clissold, local assistants Tawi Bukam, Rennie, Spanis, Wita, Gewise, G. Nalu and others. Cooperating researchers from other institutions largely funded by or through Bishop Museum included Drs C. D. Michener, E. N. Marks, John Smart, Peter Mattingly, D. Elmo Hardy, Robert Traub, M. Nadchatram, J. Balogh (1969 trip only) and Y. Hirashima. Emphasis in 1963 and 1966–67 was in Enga, Tari, Kubor and Mt Michael areas; in 1965 with T. C. Maa to Mt Ialibu; 1966 with E, C. and R. Gressitt to Wilhelm, Kubor and Schrader ranges; in 1969 at Bulldog Road, Garaina, with M. and E. Gressitt, Mt Wilhelm and Angoram with Balogh and Hirashima; and many other areas in other years (Bishop).

Much collecting of vertebrates, especially mammals, with ectoparasites, was done by M. C. Thompson and P. Temple; R. Traub (USNM) with Abid Beg Mirza, M. Nadchatram, E. Mann; Wilson, Ziegler, R. Greene (Bishop; WEI), D. Schlitter and S. Williams (also Carnegie).

Starting in 1970 work at Wau gradually shifted to local ecological research, then education, conservation and preparation of handbooks. Wau Ecology Institute was established in late 1971. Participants in these phases included T. Tigner, B. Cheary, A. Allison, M. Simon, Paul Kores, Wayne and Betsy Gagné, Bruce Beehler, Nalini Nadkarni, Alistair Hay (Bishop; WEI). Also, many scientists from other institutions visited for shorter or longer studies, including Dr Michael and Barbara Robinson and Dr Yael Lubin from Smithsonian Tropical Research Institute, Thane Pratt (Smithsonian fellow on 2nd stay), Drs van der Vecht, S. Asahina, R. Kano, H. Kurahashi (chapter IV: 12), S. Shinonaga, H. Shima, Leonard Thien (chapter III: 13), Hampton Carson and T. Okada (chapter IV: 10), K. Kaneshiro, Paul Ehrlich, Paul Hebert, Rudolph Harmsen, M. B. Fenton, D. P. Furman, E. I. Schlinger, Herbert Levi, A. R. and Barbara Y. Main, Wm Peters besides many people mentioned elsewhere in this chapter. F. R. Allison collected Protozoa (NZ). P. Deharveng collected Collembola, 1978, both there and elsewhere (Toulouse). D. F. McMichael collected land and freshwater mollusks at many areas, both E and W on 2 trips, 1955, 1956 as a Bishop Museum fellow (AM). Y. Kondo collected some land mollusks, 1965 (Bishop).

In 1957 Drs Eugene Munroe and George Holland from Ottawa made extensive collections of moths, mostly at highland localities (CNC). M. E. Bacchus collected beetles, especially water beetles near Wau, Huon Peninsula, etc on the BMNH — Univ. Newcastle-upon-Tyne exped. in 1964 (BMNH). In 1959 a Reimer collected some animals in Kokori area, Papua Gulf (München; Frankfurt). In 1966 Prof. J. Illies searched for Trichoptera in the E Highlands (Schlitz). In 1964 and 1970 Prof. T. C. Schultze-Westrum studied mammals and conservation, Mt Bosavi, etc, for the IUCN and PNG.

Dr R. W. Hornabrook, government doctor studying Kuru disease at Okapa and first director, Institute of Medical Research (initially Inst. of Human Biol.), Goroka, made a large beetle collection, mostly in highlands and Huon Pen., 1962–74 (Wellington; some types in BMNH and Bishop). Henry Ohlmus, amateur beetle collector, assembled much material from highlands and elsewhere, 1961–1980 (mostly Canberra and Brisbane; some types in Bishop). R. Parrott, Canadian amateur beetle collector, made a large collection in the

NE, 1970. Angus Hutton collected animals at Garaina. J. Allen, J. M. Carlisle, D. Hutton, I. Johnson, D. Price, M. Stevens, M. Erben and G. Rosenberg collected insects on various mountain tops in 1962–63 (DASF).

Janos Balogh, partly with I. Loksa, made many collections, mostly of soil arthropods, on six visits, 1956–80 (Budapest). Harold Cogger collected reptiles and amphibians in several areas, especially in Papua, and D. K. McAlpine (see chapter IV: 9) collected Diptera, in 1963–64 (AM). Elizabeth Marks made several visits collecting mosquitos, mostly in Papua between 1957 and 1979 (QIMR). G. Monteith collected insects, especially Hemiptera on several visits, 1960's (Bishop, Brisbane). R. Schodde collected many birds, and some insects, on various trips in several areas, with S. Schodde at L. Kutubu and Mt Giluwe 1961, Central Prov., 1962. W. B. Hitchcock collected birds, with botanists W. Vink and R. Pullen (CSIRO), and Pullen also collected some insects (CSIRO). I. W. B. Thornton (LaTrobe) and C. N. Smithers (AM) collected Psocoptera in several areas of mainland, 1970, and Bismarcks, 1974 (see chapter IV: 6).

Animal fossils were studied in the Bulolo-Watut area by Michael Plane (CSIRO). Ethnological and archeological animal material was studied in highlands, Jimi V, 1955, etc, by Ralph and Susan Bulmer (AM; Auckland, see chapter II: 3); and archeological animals by Jeanette Hope, 1969 (ANU) and James Menzies, 1970's (UPNG).

Michael J. Tyler, relieving Shaw Meyer at Nondugl in 1960, collected many frogs and reptiles in Wahgi valley and on Wahgi-Sepik divide; and in 1967 made a frog survey of New Britain (Adelaide, AM, BMNH and AMNH). Richard Zweifel collected many frogs in several areas, 1964, 1968 and 1969 in highlands, Wau area and N coast (AMNH, see chapter V: 1). Fred Parker made large herpetological collections in many parts of E NG, 1960–78 (MCZ, AM, etc). Alden H. Miller collected birds, and W. Z. Lidicker Jr mammals, in the Wau-Bulolo area in 1962 (MVZ).

Barry Gray and Ross Wylie, and later Peter Shanahan, John Dobunaba and Dr H. Roberts, built up a large collection and much data for forest insects at Forestry Research Station, Bulolo, from 1962 (Bulolo). Schedl described many scolytid beetles from the Bulolo collection (Schedl coll'n). Donald Sands (DASF; CSIRO) collected butterflies in the early and mid-1970's. The CSIRO set up a screw-worm fly laboratory in Port Moresby about 1964. P. Spradbery, Donald Sands and R. Tozer have been entomologists there.

The Danish Noona Dan Expedition, 1962, visited many islands of the Bismarck Archipelago, and animals, especially birds, insects and fresh water animals, were collected by Leif Linneborg, Torben Wolff, Finn Salomonsen and Wm Buch (København).

Prof. Jared Diamond, starting in 1964, made detailed surveys of birds of certain mountains, such as Karimui and Torricellis, on several visits, pioneering certain aspects of ecology and biogeography, and making recommendations for conservation, in New Guinea (coll's mainly AMNH). Pierre Jolivet studied ecological aspects of several arthropod groups, including parasites, in E Highlands and elsewhere, 1968–69. H. Heatwole studied reptiles on some smaller islands in E Papua, 1969, on the Fairbridge New Guinea coral reef expedition.

The R/V Alpha Helix New Guinea Expedition, May-Nov. 1969, sponsored by the U.S. National Science Foundation, involved scientists from universities and museums in several countries, especially USA and Australia. Much work was done on physiology and ecology, primarily of vertebrates (C. Sibley, R. Zweifel), as well as bioluminescence in fireflies etc (J. P. Buck, J. E. Lloyd), marine life and fungi, mostly in NE NG and the Bismarcks. Australian collaborators included John Calaby, H. Cogger and R. Schodde (USNM; Yale; CSIRO; AMNH etc). George Zug collected herps at Mt Kaindi, etc. in 1971 (USNM). In 1975 there was a large British speleological expedition with P. Chapman and P. Beron (see chapter V: 6).

9. Collections

9.1 Major collections of New Guinea biota

Vascular plants: Leiden, Lae, Arnold, Kew, BMNH, CSIRO, New York, Sydney, Bogor, Bishop, USNM.

Non-vascular plants: Fungi (PNG: Konedobu, Bulolo); bryophytes (BMNH, Helsinki); lichens (Geneva, Budapest); general (Tokyo, Farlow, Nichinan, Berlin).

Vertebrates: Mammals (AMNH, Bishop, BMNH); birds (AMNH, BMNH, CSIRO, Bishop); reptiles (MCZ, Leiden, AMNH, AM, Bishop); amphibians (AMNH, Adelaide, AM, Leiden, Bishop); Fresh-water fish (Leiden, BMNH, USNM, København, AM).

Insects and other arthropods: In general (Bishop, BMNH, Leiden, CSIRO, AM, AMNH, CAS, København, Konedobu); Odonata (Leiden, Bishop, BMNH); Lepidoptera (BMNH, CSIRO, Bishop, CNC, Leiden); Diptera, Coleoptera, etc (Bishop, BMNH, Leiden, Dresden, AM); Muscoid Diptera (TMDU); arachnids and parasites (Bishop, BMNH, Leiden, USNM, Budapest, Amsterdam); soil arthropods (Budapest, Bishop).

Mollusca: (terrestrial): AM, Leiden, BMNH.

9.2 Principal repositories

Adelaide	South Australian Museum, Adelaide
AM	Australian Museum, Sydney (now including Macleay Museum of Sydney Univ.)
AMNH	American Museum of Natural History, New York
Amsterdam	Natura Artis Magistra (Zool. Mus.), Amsterdam
ANU	Australian National University, Canberra ACT
Arnold	Arnold Arboretum, Harvard University, Cambridge, Mass. (A)

Bishop	Bishop Museum, Honolulu
Berlin	Botanisches Museum, Dahlem (B); Humboldt University (Zool. Mus.), E Berlin
BMNH	British Museum (Natural History), London
Bogor	Bogor, Java, Indonesia: Herbarium Bogoriense (BO); Museum Zoologicum Bogoriense
Brisbane	Brisbane, Queensland, Australia: Queensland Herbarium (BRI); Queensland Institute of Medical Research; Queensland Museum; University of Queensland
Brussels	Institute Royale d'Histoire Naturelle de Belgique, Bruxelles
Budapest	Termeszettudomanyi Museum (Hungarian National Museum), Budapest
Bulolo	Forestry College herbarium; Forestry Research Station (insects, fungi)
BYU	Brigham Young University, Provo, Utah
Canberra	(see CSIRO; ANU)
CAS	California Academy of Sciences, San Francisco
CNC	Canadian National Collection, Ottawa: Insects, at Agriculture; others, National Museums of Canada
CSIRO	Commonwealth Scientific and Industrial Research Organization, Camberra ACT: Herbarium Australiense (CANB) and Entomology Div.
DASF	Department of Agriculture, Stock and Fisheries (now DPI), Konedobu
DPI	Department of Primary Industries, Konedobu, Port Moresby, PNG (insects, fungi) (see also Lae)
Dresden	Staatliche Museum für Tierkunde, Dresden
DSIR	Department of Scientific and Industrial Research, New Zealand
Farlow	Farlow Herbarium, Harvard University, Cambridge, Mass.
Field	Field Museum of Natural History, Chicago
Florence	Italian State Herbarium, Firenze (FI): Palmae, Pandanaceae, Beccari coll's
Frankfurt	Senckenbergisches Institut und Muzeum, Frankfurt
Geneva	Museum d'Histoire Naturelle de Geneve
Helsinki	Finnish State Herbarium, Helsinki
HSPA	Hawaiian Sugar Planters Association Experiment Station, Honolulu (insects now in Bishop)
Kew	Royal Botanic Gardens, Kew, England (K)
København	University of Copenhagen (Zool. Mus.), København
Konedobu	DPI (DASF) Port Moresby, PNG (insects, fungi)
Lae	Herbarium, Div. Botany, Dept Forests (DPI), Lae, PNG (LAE)
Leiden	Rijksherbarium, Leiden (L); Rijksmuseum van Natuurlijke Historie, Leiden
Macleay	See AM
Manokwari	Forestry and Agriculture Experiment Station; Agriculture College of Cenderawasih University, Amban, Manokwari, Irian Jaya, Indonesia
MCZ	Museum of Comparative Zoology, Harvard University, Cambridge, Mass.
Melbourne	National Herbarium of Victoria (MEL); National Museum of Victoria, Melbourne
München	Zoologische Staatsammlung der Bayerische Staat, München
MVZ	Museum of Vertebrate Zoology, University of California, Berkeley
Nelson	Cawthron Institute, Nelson, New Zealand (now part of DSIR, insects to Auckland)
New York	New York Botanical Garden
Nichinan	Hattori Botanical Laboratory, Nichinan, Japan
Paris	Museum Nationale d'Histoire Naturelle, Paris
PNG	Papua New Guinea National Museum and Art Gallery, Waigani (vertebrates)
QIMR	Queensland Institute of Medical Research, Herston, Brisbane
RSA	Rancho Santa Ana Herbarium, Claremont, California
Sydney	National Botanic Garden and Herbarium of New South Wales; University of Sydney; also see AM
Tokyo	National Science Museum, Shinjuku, Tokyo
TMDU	Tokyo Medical and Dental University, Bunkyo-Ku, Tokyo (Muscoid Diptera)
Toulouse	Universite Paul Sabatier, Toulouse, France (Collembola)
Tring	The Rothschild Museum at Tring, England. Bird collections sold to AMNH, 1932. Insects now BMNH
UC	University of California Herbarium, California
UniTech	University of Technology, Lae, PNG
UPNG	University of Papua New Guinea, Waigani (Port Moresby), PNG
USNM	National Museum of Natural History, Smithsonian Institution, Washington DC
WEI	Wau Ecology Institute, Wau, PNG
Wellington	National (formerly Dominion) Museum, Wellington, New Zealand

Acknowledgement

We are indebted for assistance to Drs F. R. Fosberg, L. B. Holthuis, Peter Raven, P. van Royen, J. J. H. Szent-Ivany and to curators and librarians of Bishop Museum, California Academy of Sciences, American Museum of Natural History, Museum of Comparative Zoology, and others.

Selected references

Because a bibliography of all historial sources of information plus results of collecting mentioned in this review would fill a large volume, most works mentioned by author and/or title in the text are not repeated here, and only a few more general sources or items are listed below. Some of the more important general sources are Nova Guinea (Leiden); Archbold Expedition Reports (American Museum of Natural History; Arnold Arboretum; New York Botanical Garden); Flora Malesiana; Bishop Museum publications (Pacific Insects; Pacif. Ins. Monographs; J. of Medical Entomology; Insects of Micronesia); Papua New Guinea Scientific Society; PNG Agricultural J. and PNG Medical J.; PNG Wildlife Div. reports; Encyclopaedia of Papua New Guinea; CSIRO Land Research Reports; Wau Ecology Institute handbooks; and publications of Queensland Museum, Queensland Herbarium, University of Queensland, Australian Museum, South Australian Museum, University of Sydney, Linnean Soc. of New South Wales. For plants principal reference is Flora Malesiana; for insect references, Gressitt and Szent-Ivany, which has a taxonomic/subject index; and for vertebrate animals American Museum of Natural History.

d'Albertis, Luigi Maria. 1880. New Guinea. Sampson and Low, London.

American Museum of Natural History: Bulletin; Am. Mus. Novitates (Results of the Archbold Expeditions).

Archbold, R. and A. L. Rand. 1935. Summary of the 1933–34 Papuan Expedition. Bull. Am. Mus. N. H. 68: 527–579 (see also Am. Mus. N. H.).

Beccari, O. 1924. Nuova Guinea. Firenze, Soc. Ed. 'La Voce'.

Bishop Museum (J. Med. Ent.; Pacif. Ins.; Pacif. Ins. Monogr.; Insects of Micronesia; Gressitt and Szent-Ivany).

Brass, L. J. et al. 1935–1964. (Summaries of Archbold Expeditions). J. Arnold Arboretum, Harvard. Bull. Amer. Mus. Nat. Hist.

CSIRO. Commonwealth Scientific and Industrial Research Organization, Canberra. Land Research Series; and other reports.

Diamond, J. M. 1972. Avifauna of the eastern highlands of New Guinea. Nuttall Ornith. Club, Cambridge, Mass. 438 p.

Encyclopaedia of Papua and New Guinea. 1972. Univ. Melbourne Press and Univ. PNG, 3 vols.

Flora Malesiana. (van Steenis, ed.) Series I. 1950— Djakarta and Leiden.

Gilliard, T. 1969. Birds of paradise and bower birds. American Museum, NY. 485 p.

Goode, John. 1977. Rape of the Fly. Thomas Nelson, W Melbourne, 272 p.

Gressitt, J. L. 1956. Entomological investigations in New Guinea mountains. Proc. Hawaiian Entomol. Soc. 16: 47–69.

Gressitt, J. L. and J. J. H. Szent-Ivany. 1968. Bibliography of New Guinea entomology. Pacif. Ins. Monogr. 18: 1–674. Bishop Museum, Honolulu.

Hagen, B. 1899. Unter den Papua's. Beobachtungen und Studien über Land und Leute, Thier- und Pflanzenwelt in Kaiser-Wilhelmsland. Kreidel, Wiesbaden, 327 p.

Huxley, Camilla. 1978. The ant-plants Myrmecodia and Hydnophytum (Rubiaceae), and the

relationships between their morphology, ant occupants, physiology and ecology. New Phytol. 80: 231–268.
Jolivet, P. 1971. La Nouvelle-Guinée Australienne. Introduction ecologique et entomologique. Cahiers du Pacifique, Paris 15: 41–70.
Journal of Medical Entomology. 1964— Bishop Museum, Honolulu. Many articles on New Guinea arthropods.
Krieger, Maximilian. 1899. Neu-Guinea. (parts by von Danckelman, von Luschan, Matschie and Warburg). Bibliothek der Länderkunde. Alfred Schall, Berlin, 535 p.
Lorentz, H. A. 1913. Zwarte menschen — witte bergen. Brill, Leiden. (see also Nova Guinea, v. 8).
Montrouzier, J. X. H. 1855. Essai sur la faune de l'ile de Woodlark ou Moiou. Soc. Imp. Agr. Hist. Nat. Lyon, Ann. ser. 2, 7: 1–114.
Noona Dan Papers. Reports from the Noona Dan Expedition, 1961–62. See No. 24, 1966, by B. Petersen, Insects and other land arthropods (Ent. Meddel., 34).
Nova Guinea. Resultats de l'expedition ..., vol. 1–19, 1909–1936. Zoology vol. 5, 9, 13, 15, 17. Botany v. 8, 12, 14, 18. History of discoveries, v. 1, 2, 4, 8, etc. New series, A journal of botany, zoology ..., 1937–1959, vol. 1–10, no. 2. New series, Botany, 1–24, 1960–1966. New series, Zoology, 1–38, 1960–1967. Leiden, Brill.
Pacific Insects. 1959- Bishop Museum, Honolulu. Many articles on New Guinea insects.
Pacific Insects Monographs. 1961- Bishop Museum, Honolulu. No. 2, 4, 5, 6, 10, 13, 15, 16, 17, 18, 19, 20, 26, 28, 29, 30, 33 and 36 concern New Guinea.
Papua New Guinea. Agriculture Journal. Medical Journal. Publications of Wildlife Div., DASF, DPI, etc. Div. Botany, Dept Forests. Botany Bulletins, 1969–; Handbooks of the flora of Papua New Guinea, 1978–.
Papua New Guinea Scientific Society. Ann. Reports and Proc. 1952–; Trans. 1960–.
Rand, A. L. and L. J. Brass. 1940. Results of the Archbold Expeditions. No. 29. Summary of the 1936–37 New Guinea expedition. Bull. Amer. Mus. Nat. Hist. 77: 341–390.
Royen, P. van. 1980- The alpine flora of New Guinea. J. Cramer, Braunschweig, 4 vols.
Science in New Guinea. 1971- Journal published by University of Papua New Guinea.
Steenis, C. G. G. J. van. (Ed.) see Flora Malesiana.
Szent-Ivany, J. J. H. In press. The economic insects of Papua New Guinea. Pacific Insect Monograph, Bishop Museum, Honolulu.
Thomson, J. P. 1892. British New Guinea. B. Philip and Son, London, 336 p.
University of Papua New Guinea Library. 1970. Brief notes on scientific exploration in New Guinea. 15 p, 37 ref.
Wallace, A. R. 1869. The Malay Archipelago. MacMillan, London.
Wau Ecology Institute, Wau, Papua New Guinea. 1972— Handbooks (1, Frogs; 2, Beetles; 3, Pidgin glossary; 4, Birds; 5, Mt Kaindi; 6, Rodents; 7, Reptiles). Leaflets, 1–2; Pamphlets, 1–2; Biennial reports, 1–4.
Wollaston, A. F. R. 1912. Pygmies and Papuans. Smith and Elder, London. (see also Linnean Soc. Trans., Botany and Zoology, 1916, etc).

2. Physical anthropology, linguistics, and ethnology

Ann Chowning

Introduction

With a population of about three million, New Guinea has long been famous for its linguistic diversity — the greatest for its area in any part of the world — and for an almost equal diversity of cultures. On the whole, however, the physical diversity of the inhabitants has tended to be much underrated, and only in the past few years has this point begun to be appreciated. Much is known about the cultures of New Guinea, though field work has tended to be concentrated in coastal regions and the Highlands, with Irian Jaya the most neglected area. The classification of languages has received intensive attention only in the past few decades, and the picture is constantly changing. In all of these fields new data are constantly brought forward, and attempts to correlate them and draw deductions from them are likely to be at best soon out-of-date, and at worst quite erroneous. What follows is accordingly only a tentative presentation of what is known and what is assumed about the distributions and relations of peoples, languages and a few cultural items.

Especially as regards culture, New Guinea is part of the larger area usually labelled Melanesia (see discussion in Chowning 1977). When "Melanesia" is used in this text, it will include New Guinea, the Bismarck Archipelago, the Torres Straits Islands, the Solomons, New Hebrides and New Caledonia, with the status of Fiji left indeterminate. "Indonesia" is used in the cultural rather than political sense, to include Borneo and the Philippines. Unless the island of New Guinea is specified, "New Guinea" includes the Bismarck Archipelago. "Highlands" refers to the intermontane valleys that extend from Papua New Guinea to West Irian. When necessary the provinces of Papua New Guinea will be designated by their names (See Map 1).

Languages will be discussed in more detail below; here only the principal terms used will be mentioned. The languages of New Guinea have traditionally been divided into two categories, Austronesian (AN), otherwise called Malayo-Polynesian or Melanesian, and non-Austronesian (NAN), otherwise called Papuan. The Austronesian languages of New Guinea are related to others in what is usually called a family, extending from Madagascar to Easter Island and including all of the languages of Polynesia and Micronesia, many of Melanesia, most of those of Indonesia, the indigenous languages of Taiwan, and some of mainland southeast Asia, including Malay.

Monographiae Biologicae, Vol. 42, ed. by J. L. Gressitt.

© *1982, Dr W. Junk Publishers, The Hague.*

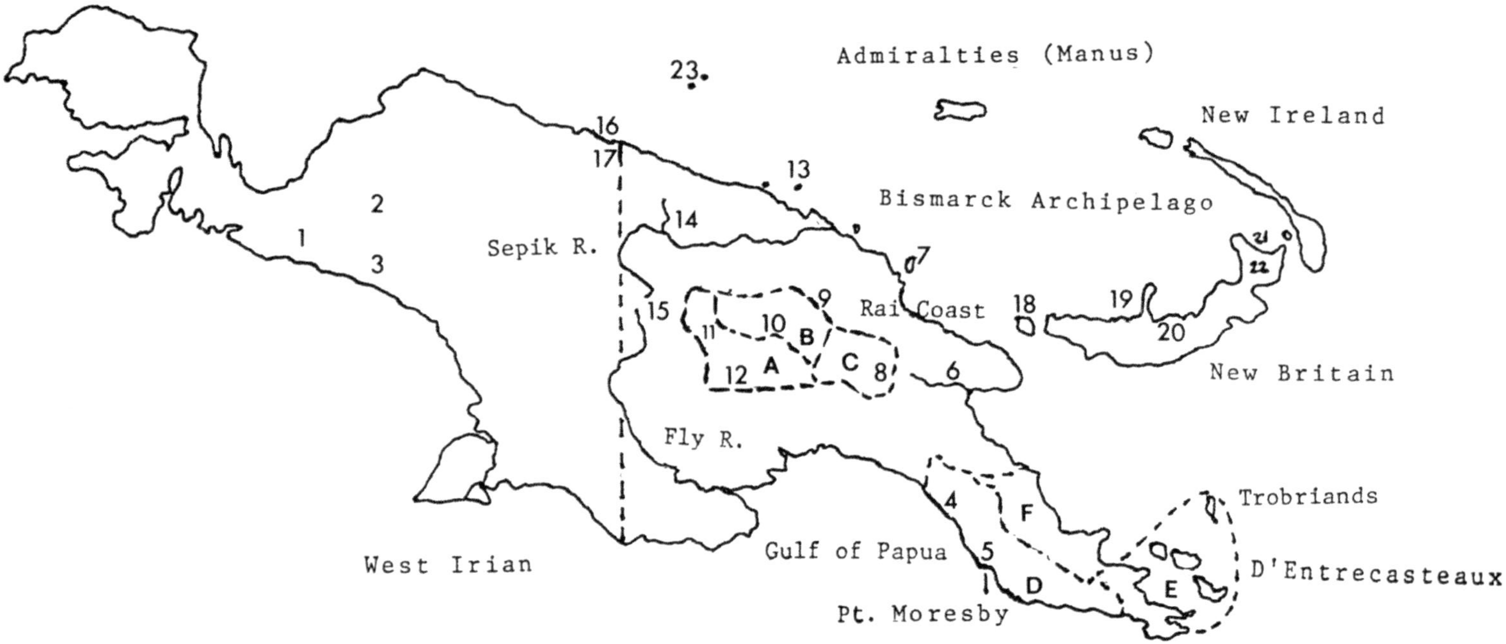

Map 1. Locations of places and people mentioned in the text.
—Former Papua New Guinea administrative districts:
A = Southern Highlands; **B** = Western Highlands; **C** = Eastern Highlands; **D** = Central District; **E** = Milne Bay; **F** = Northern District.
—People and places mentioned in the text:
1. Tapiro; **2.** Dani; **3.** Asmat; **4.** Yule Island; **5.** Motu; **6.** Markham River; **7.** Karkar; **8.** Awa; **9.** Aiome; **10.** Mt Hagen; **11.** Hewa; **12.** Lake Kutubu; **13.** Wogeo; **14.** Yellow River; **15.** Telefomin; **16.** Humboldt Bay; **17.** Sentani; **18.** Umboi; **19.** Kove; **20.** Lakalai; **21.** Tolai; **22.** Baining; **23.** Aua and Wuvulu.

By contrast, NAN is a catch-all for non-Austronesian languages spoken in Melanesia and in a few nearby islands of Indonesia. Even today, all of these languages have not been demonstrated to be related and probably never will be, although the amount of diversity has been greatly reduced by Wurm and his colleagues in recent years (Wurm 1975a). (Bellwood 1978:119 is mistaken on this point.) Wurm has classified languages in terms of closeness of relationship, with the highest-order group called a phylum, followed by "stock" and "family". In these terms, all AN languages in the world fall into a single phylum, while the island of New Guinea contains ten separate NAN phyla (see Map 2). However, one of these, the Trans-New Guinea phylum, covers most of the island, so that the large majority of NAN languages spoken within the island are now thought to be related to each other. Much disagreement still exists about the classification of individual languages, for reasons which may become apparent in the discussion that follows consideration of the physical anthropology.

Physical anthropology

Attempts to deal with the physical anthropology of New Guinea have been of four sorts. One could be called impressionistic: an observer describes the appearance of the inhabitants, but often so vaguely that a second observer is not sure whether his impressions tally with those of the first one, and is still less able to compare a different group with the one described. Such descriptions are, however, usually supplemented with photographs. While often misleading as to skin color and height, these are invaluable comparatively except where physical features are obscured by face paint, nose-piercing and other cosmetic treatments. Another type of description depends upon anthropometric measurements, particularly of height, weight and the dimensions of skull and face. These are not always so easy to replicate as is commonly supposed, and in most cases deal with samples so small that it is dangerous to assume that they represent a particular population (see Swindler 1962:17), but for comparative purposes they are certainly superior to verbal description. In a very few cases careful measurements have been made of hair form (van der Sande 1907:331–336) and skin color, the latter usually by matching unexposed areas to standard color charts (Littlewood 1972:102). Finally, most recent investigators have tended to concentrate on traits controlled by single genes, especially those exhibited in the blood, and deal with other factors little if at all (e.g., for a region adjacent to New Guinea, Friedlaender 1975 includes photographs but barely mentions surface features).

Often the same observer uses all four methods, although of course special training is needed for the last two. Professional physical anthropologists seem just as likely as the general public to inform us simply that hair color is dark brown and hair form "medium helical" (Littlewood 1972:73), if indeed they do not characterize a population as looking like "small ogres" (Howells 1973:163). It is consequently difficult to document not only the variation between New Guinea populations but the degree to which they resemble and differ from

Fig. 1 (Above). A woman uses a shell knife to clean yams that she is harvesting — MOLIMA (AN language) D'Entrecasteaux. She wears a skirt of shredded fiber and a shell armband; a spare skirt is covering the yams in front of her, and a section of thatch, made of sago leaves, is behind her. Her physical type is characteristic of this region.

Fig. 2 (Below). A woman from southwest New Britain (AN-speaking) holds her baby, whose head is bound with barkcloth to produce the artificial deformation typical of this region (sometimes called ARAWE). Head deformation is also practiced in the New Hebrides. The women's heavy browridges (bone beneath the eyebrows) are characteristic of the north coast of New Guinea and west New Britain, regardless of language group, but are more conspicuous in males.

Fig. 3. A KOVE man (AN language, north coast of New Britain uses a decorated adze to carve an ornamental walking stick. (The decoration of utilitarian objects is typical of much of lowland New Guinea.) Above his head are a bamboo mast and shell net weights. His earlobes have been split and stretched. His features are typical of the north coast of New Guinea and west New Britain, including heavy bone structure and musculature, a long face, and relatively thin lips.

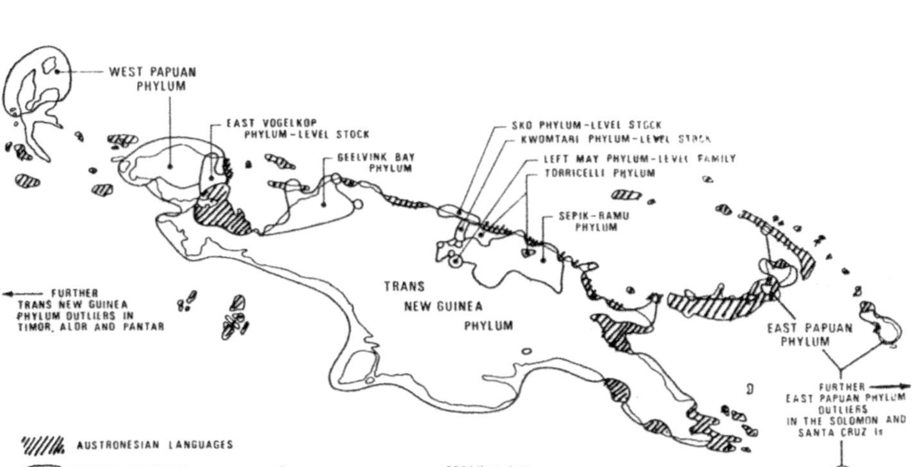

Map 2. The distribution of the Papuan languages. From Bellwood 1978: 120.

Fig. 4 (Left). A woman wearing a shell necklace and fiber skirt, and armband, with her loaded *bilum* in front of her. Her nasal septum is pierced and has a wooden ornament through it. HEWA, Southern Highlands (NAN-speaking).

Fig. 5 (Below). A man in the typical dress of a limited part of the Southern Highlands, including a cap of barkcloth. Women in the background. His nasal septum is pierced and has a wooden ornament through it. HEWA (NAN-speaking).

Fig. 6 (Right). Man beating a *kundu*, Southern Highlands. He wears a cap of bark-cloth and a shell necklace, the shells having travelled from the coast by a traditional trade network. His nose is pierced and has an ornament through the septum.

Fig. 7 (Below). Women and children, Southern Highlands. The women have loaded *bilums* beside them. One is giving her daughter a drink from a gourd container. The fence in the background keeps pigs from the garden.

Fig. 8 (Above). Middle-aged man sewing up a sail, north coast of New Britain (AN-speaking). His slit ear lobe is looped over the top of the ear. He is greying and beginning to bald; rates and incidence of both are typically later and lesser than among Europeans. The long face and muscular body build are typical of this region.

Fig. 9 (Below). Woman chewing betelnut, north coast of New Britain. Her face, arms, and hands are tattooed in traditional designs. The house wall is made of sago leaves, sewn with rattan, the baskets of coconut leaves, and the broom of coconut-leaf midribs. Her ornaments, in addition to beads of European origin, include bracelets of shell, tortoiseshell, and fiber (AN-speaking).

Fig. 10 (Above). Masked figures accompanying the celebrants of a girls' puberty ceremony, northwest coast of New Britain. (The girls being honored are partly obscured, wearing feather headdresses.) Such ceremonies are typical of New Guinea and masks are found throughout much of the lowland region, though differing in design from one area to another. The men are beating *kundus*. Many of the participants have powdered lime on their foreheads.

Fig. 11 (Below). A father (left), his son, and grandsons, LAKALAI, New Britain (AN-speaking). The older men, whose features are typical of the region, and whose earlobes are artificially stretched, is displaying paddles he has painted in a traditional style. His wife is seated in the background. The son's lips (a shape inherited by his children) are fuller than is usual in this area.

Fig. 12 (Above, left). An old man holds a miniature *kundu* (still lacking a head) and a cup he has carved for his grandson. His nostrils are unusual; otherwise his face is typical of this region (west New Britain — north coast of New Guinea), AN-speaking, north coast of New Britain. (A cigarette is in his mouth.)

Fig. 13 (Above, right). Old KOVE (AN-speaking) woman, north coast of New Britain. Note long face, thin lips, and texture of hair. She wears tortoiseshell earrings in her stretched lobes, and tortoiseshell bracelets, and holds a coconut-leaf basket.

Fig. 14 (Below). LAKALAI women at a funeral. Note the prominent bone structure, typical of this region. Shiny marks on the upper arms are ornamental scars. Note the heavy and prominent bone structure. The woman on the left is making a fiber bag; the one in the center is chewing betelnut. (Her roughened skin results from a form of ringworm.)

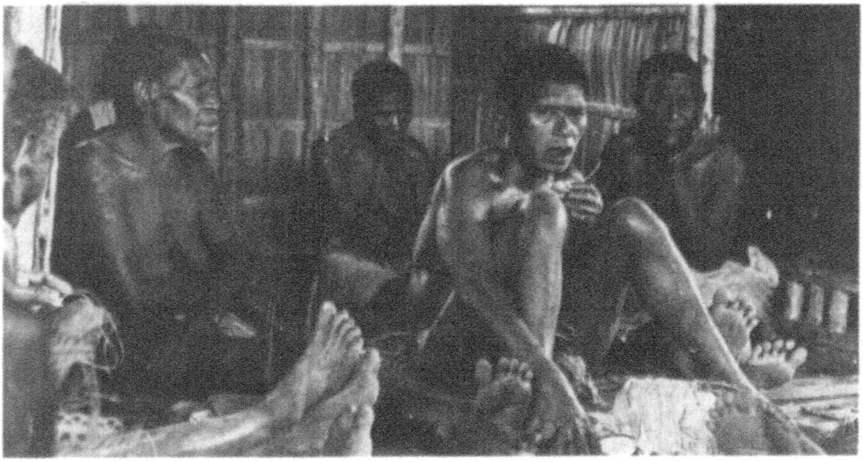

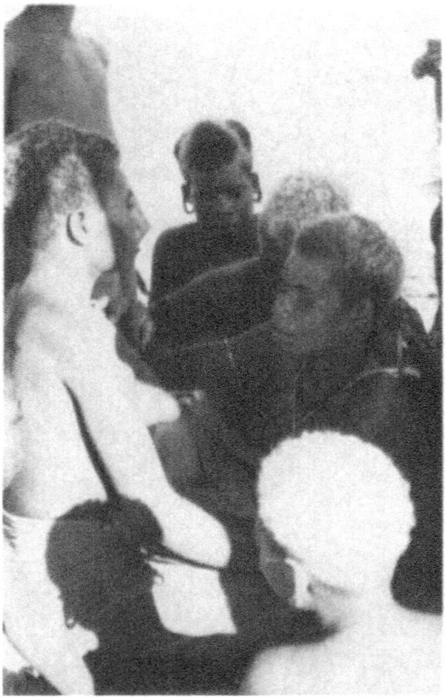

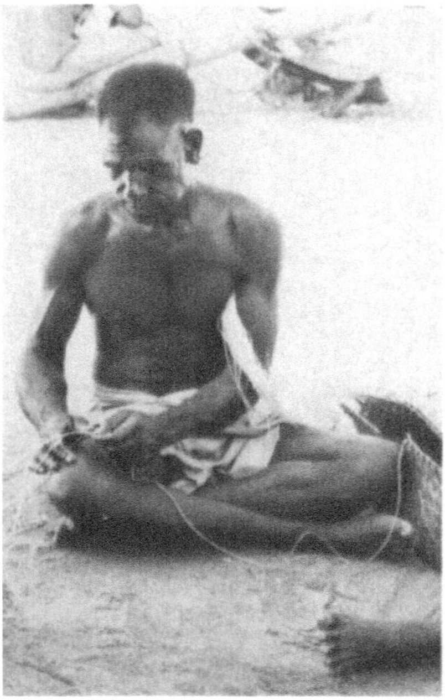

Fig. 15 (Above, left). KOVE (AN-speaking) girls tattooing each other. Tattooing is associated with relatively light skin, and is as common in certain parts of the Highlands as in some coastal AN-speaking regions, as here (KOVE, northwest New Britain). The girls' hair is artificially bleached.

Fig. 16 (Above, right). Balding middle-aged man making thread by rolling fibers on his thigh. His physical type is very typical of this area (KOVE, AN-speaking), north coast of New Britain.

Fig. 17 (Below). KOVE men on the porch of a men's house which is tabu to women. Such structures are very widespread in New Guinea, though the art is peculiar to this region. *Kundus* are hanging up under the eaves.

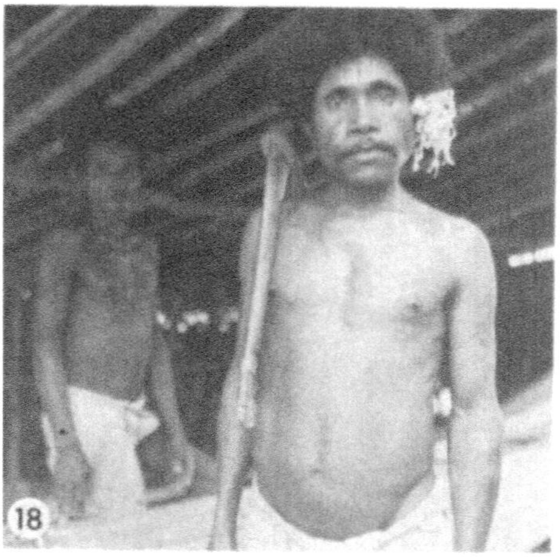
18

19

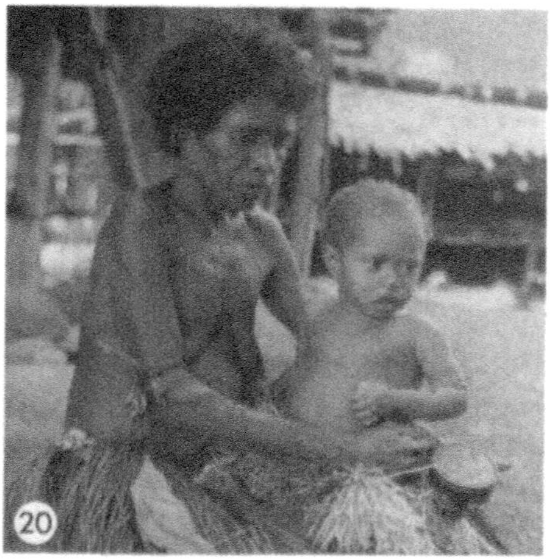
20

21

22

23

Fig. 18. Canoe-carver from D'Entrecasteaux, Milne Bay (AN-speaking), with his adze over his shoulder. The light body build and bone structure is typical of this region and of SE New Guinea as a whole.

Fig. 19. Women sitting beside a cooking pot (acquired in interisland trade). D'Entrecasteaux, Milne Bay region. The woman in the foreground has natural red hair, as do two of her sisters.

Fig. 20. Old woman with her grandchild, D'Entrecasteaux, Milne Bay. Note her hair form, very common in this region, and delicate bone structure. The child is very light-skinned, though not albinotic, and remained so into adulthood, though her hair darkened.

Fig. 21. Visiting traders, D'Entrecasteaux, wearing the shell armbands typical of this region. Note the relatively light and slender body build.

Fig. 22. Women with bananas on their heads, and man and boy carrying taro, D'Entrecasteaux. The woman at the left is plumper than the average. Somewhat slanting eyes like hers are also found on the south coast of Papua. Note the wavy hair of the boy in the background and the woman on the right.

Fig. 23. D'Entrecasteaux man carrying a club of the kind formerly used for killing victims in cannibalistic raids. The decoration is typical of the Milne Bay region. He wears tortoiseshell earrings in his slit ears. Note his hair form and that of the boy in the background.

Fig. 24 (Above, left). Trobriand woman carrying her daughter on her hip. The daughter wears tortoiseshell earrings in her slit ears. The mother, in accord with Trobriand custom, has shaved her eyebrows. A modern innovation is the name tattooed on her chest.

Fig. 25 (Above, right). A Big Man from the Western Highlands, wearing on his chest the rods that symbolize his success in competitive pig exchanges. His hair is covered with barkcloth and netting. Beards are symbolically important and worn by most traditional Highlands men. Most lowland New Guineans cannot raise such heavy beards.

Fig. 26 (Below, left). Western Highlands women, the one in the foreground holding a *bilum* with her baby in it. Note the round faces, and the traditional tattooing which shows up against the relatively light skin.

Fig. 27 (Below, right). Boy having his foreskin slit (superincision) in an operation meant to mark his separation from women and entry into adulthood. Similar operations are widespread. Note hair form, less frizzy on the children in the foreground (hair artificially bleached) than on the adults performing the operation (hair white with age).

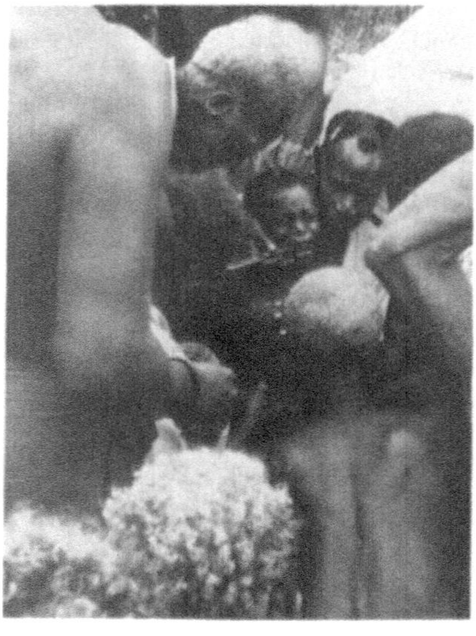

Fig. 28 (Right). Back view of a highlander carrying a *bilum* containing a pandanus rain hood and sweet potatoes. Photo: J. L. Gressitt.

Fig. 29 (Below). Young married woman (a teenager) and a child with the artificially deformed skulls characteristic of southwest New Britain (AN-speaking). These people lack pottery and cook most food (here, taro) directly on the fire, using bamboo tongs. The child's hair is artificially lightened.

peoples in other parts of the world. Furthermore, for many areas, especially in West Irian, no useful descriptions exist. What follows is only a preliminary synthesis of available data.

On the basis of their physical appearance, the people of New Guinea have often been classed with those of some neighboring islands as Oceanic Negroids, a term that suggests a connection with African Negroes. The features noted are hair form and skin color, which also inspired the other terms by which they have been designated: Papuans (from a Malay word meaning "frizzy hair"), or Melanesians (from a Greek word for "black"). The term Papuan has also been applied, especially by the Dutch, to the dark-skinned frizzy-haired inhabitants of a few of the eastern Indonesian islands, such as Timor and Alor, but has rarely been used for those of the islands east of New Guinea. By contrast, Melanesian has been used to designate the inhabitants of the main islands lying between New Guinea and Fiji: the Bismarck Archipelago, the Solomons, the New Hebrides and New Caledonia. Sometimes New Guinea, Fiji and the Torres Straits Islands are included, sometimes not (for a full discussion, see Chowning 1977:1). When New Guinea is excluded from Melanesia, the reasons have nothing to do with the physical appearance of its inhabitants, who are acknowledged to have their closest ties with the people of the islands to the east of them. Here the inhabitants of New Guinea will, as far as possible, be described separately, before their external relations are considered.

Beginning with Deniker in 1900, a number of physical anthropologists divided New Guineans into two physical types, confusingly labeled Melanesian and Papuan, the latter characterized particularly by height, long and rugged faces, consistently frizzy hair, and hooked noses (see Coon 1966:176; Cranstone 1961:13). But these distinctions were set up and popularized largely by anthropologists who had never visited New Guinea and depended on photographs and a few skulls in museum collections. From very early, those familiar with even a small part of the area complained that the scheme was too simplistic (Seligmann 1909), but it still appears in the literature. Recently anthropologists working in the Highlands have reinterpreted the term Papuan (originally applied to coastal peoples) to designate the distinctive physical type of the Highlanders of New Guinea, lumping almost all the lowland coastal people together as Melanesians (Brown 1978:28; Littlewood 1972:73). This practice equally oversimplifies the enormous physical diversity of non-Highlanders (see Howlett 1973:1).

The degree of this diversity has only occasionally been noted in print, but those who do mention it also point out how much physical diversity exists within a single society (Forge 1971:60). A sample of different members of a single society, the West Nakanai or Lakalai of New Britain, is divided by Howells into one subgrouping of peoples of New Ireland and Espiritu Santo in the New Hebrides, and another (presented as very different from the former) of some Australian aborigines and people of the Loyalty Islands (Howells 1973:40). This grouping was based on a computer analysis of stature, head length and breadth, face length and width, and nose height and breadth. When such characters as skin color, slope of forehead, development of brow ridges and projection of nose are added the complexities increase.

Despite the internal diversity, a number of regional types have been distinguished, and it is often possible for an experienced observer to identify an individual's approximate place of origin with considerable accuracy, though not absolute certainty. Littlewood notes that it is very rare to find someone typical of the coast in a Highlands population, though he states (contrary to the impression of some observers) that typical Highlanders "can be found in many areas on the coast" (1972:9). It would probably be impossible to mistake a Lakalai for a Trobriand Islander, but not difficult to confuse a Lakalai with an inhabitant of the Rai Coast of New Guinea, or a Trobriander with someone from the Port Moresby region. Examples of some regional types follow. Highlanders "tend to shorter stature, longer trunks, lower and larger vaults, heavier browridges, a high frequency of convex noses (the so-called "semitic" nose), and larger dentition, which gives these relatively small people a prognathous appearance" (Littlewood 1972:9). (Here the comparison is with the lowland populations considered together.) To these features might be added relatively light skin (apart from the so-called "redskins", to be discussed below) and uniformly frizzy hair. The untrained observer is struck by the relatively large heads, compared with body size, and round faces of many Highlanders, especially women, even when they do not show the other signs of dwarfing (see below). According to Howells (1973:35), "in the south and west of New Guinea men are long-headed, with a relatively delicate build and skull ... with its sloping forehead and flat temples ... These same natives often have what has been called a 'Semitic nose' because of its relative prominence and narrowness." Obviously their nose form does not distinguish them from Highlanders, but along with skull shape, their dark skins do. The Lakalai of New Britain also have dark brown skins, but lack arched noses as well as prognathism. Skulls are characterized by height above the ears and narrow foreheads (Swindler 1962:28–29). However, though Swindler does not mention the fact, they are also rugged, often having heavy browridges like the Highlanders, but with a very different long face. By contrast, the people of the Milne Bay region of east New Guinea and its offshore islands are lightskinned with a high frequency of hair that is merely curly or wavy (Howells 1973:163); furthermore, their skulls are much less rugged than those of Highlanders or the Lakalai, and they also lack arched noses.

Clearly the few features just mentioned barely hint at the actual and often conspicuous phenotypical differences between these peoples. As more measurements were taken and pooled, it became possible for anthropologists discussing very different populations to say that they fell well within the "Melanesian" range (here a term designating all the inhabitants of New Guinea and the islands, rather than just certain coastal peoples). Swindler says this of the Lakalai as regards "most characters" (1962:29), and Littlewood, referring to Swindler's data, says the same of his Eastern Highlands samples as regards "many measurements and indices" (1972:9). But the range keeps expanding, and before considering the genetic data we need to have some idea of the boundaries of the features just mentioned. These have often been expressed in terms of comparisons with neighboring populations, especially Polynesians and Australian aborigines, and with Africans. The only reasonably up-to-date summary and assessment is to be found in Howells 1973 (but see also Bellwood 1978). Except where indicated, the descriptive material which follows is taken from Howells.

The New Guinea population as a whole is distinguished from those of Polynesia, Indonesia and Micronesia by generally darker skin. What this means is that many New Guineans are much darker than these others, but many others are just as light. The darkest-skinned Melanesians are concentrated in Bougainville and other northern Solomons. Of the groups tested within New Guinea, none were as dark as Australian aborigines (Munro 1966:129); possibly this finding might not hold if more populations were tested. New Guineans were also considerably lighter than the Africans and American Negroes in the sample compared with them.

Eye color is brown, but different observers describe it as medium or dark (see Swindler 1962:23). Head hair is normally dark brown, though likely to bleach somewhat from the effects of sun and salt water. Many New Guineans color their hair artificially, but apart from this, at least two separate mutations seem to have produced lighter hair in certain regions. Particularly in the Milne Bay area, natural red hair is common and can readily be distinguished from dyed hair by the accompanying light eyelashes. Red-haired people are somewhat lighter-skinned than the average, with pinkish lips and nipples; sometimes the skin freckles. The hair color may fade or darken somewhat with age, but remains distinctive. Such red hair appears sporadically in a number of coastal populations. In East New Britain and New Ireland, light brown or blondish hair is seen along with red; it is not clear whether this represents a different mutation. But certainly different are the "redskins" of interior New Guinea; not only is their skin reddish, but the hair is often a bright blond. According to Walsh (1972a:542), "this condition is characterized by deficient production of melanin but is easily recognized because of the presence of a reddish-brown pigment which is not normally present." He notes that it is most common in the Western Highlands and the Maprik region of the Sepik. Finally, pure albinos, albinos with pigmented patches on the skin, and partial albinos with notably light but not totally depigmented skin, hair and eyes have been noted throughout New Guinea. All of these conditions seem to be inherited as autosomal recessives; but although they all indicate deficient melanin production, they are not identical. The point is worth making because Howells, after noting the occurrence of several of these conditions, ends by lumping them with the "juvenile blondism" of Australian aborigines and, he reports, some Malaitans (Solomon Islands) (Howells 1973:33, 169–170). The actual situation is considerably more complex than Howells believes and does not afford evidence for a link between Australian and New Guinea populations.

Hair form is predominantly frizzy, but often curly or wavy in the southeast — Milne Bay and its offshore islands, and the south coast as far as Yule Island. Less frizzy hair appears sporadically in other coastal populations, particularly in children and women. The prevailing frizziness distinguishes New Guineans and other Melanesians from the inhabitants of the surrounding islands (apart from the "Papuan" populations of eastern Indonesia) and from Australian aborigines, though not from the now extinct inhabitants of Tasmania. But hair curl is not so spiral as that of African Negroes (Howells 1973:33; Swindler 1962:24–25; van der Sande 1907:331–336).

The stature of New Guineans ranges from medium to extremely short. The tallest populations include the Lakalai, averaging about 164 cm. for adult males, and range down to the Aiome of the Highlands at 148.8 cm. (Swindler 1962:60; Walsh 1972b:899). Some groups, such as those along the south coast of New Britain, are said to be taller; Chinnery found the men of one island to be almost 170 cm. However, Swindler, who cites this figure, notes that the samples are extremely small, and warns of the danger of relying too much on them (1962:17). If we temporarily except the "pygmoid" groups such as the Aiome, New Guineans do not differ greatly in height from Indonesians and Micronesians, but are shorter than Australian aborigines and much shorter than Polynesians (who average 170 cm.) (Howells 1973:161). As Howells also notes, they further contrast with Polynesians in their lack of body fat, a characteristic accentuated by the tendency of both sexes to lose weight with age and, in the case of women, childbearing (Sinnett 1977:80), unlike Polynesians as well as Europeans. New Guineans can and do become fat on a European diet, but very rarely on a traditional one, even in regions where it hardly differs from a traditional Polynesian diet.

In several parts of the Highlands, average male height is so short as to fall within the pygmy range — that is, below 151 cm. The groups so described include the Aiome, the Awa of the Eastern Highlands (Littlewood 1972:36), and the Tapiro of West Irian. They have aroused considerable interest because other peoples with very short stature, frizzy hair and dark skin, usually called Negritos, are found not only in Africa (the true Pygmies) but in intervening areas, notably the Andaman Islands in the Indian Ocean, the Malay Peninsula and the Philippines, where full-sized Negroid peoples do not occur. It has been suggested that the Negritos linked Africa and Melanesia, and that in New Guinea they were the source of dark skin and frizzy hair (Haddon 1912; Howells 1973:171–177). Very short peoples also exist in the interior of some of the Solomon Islands and New Hebrides but, except in myths (as the Menehune of Hawaii), no farther east. As more thorough investigations were carried out, the theory of a single Negrito population distributed from Africa to Melanesia has been discarded by most anthropologists, with Birdsell an outstanding exception (see below). The "pygmies" of New Guinea may look dwarfish in head and body proportions (see plates in Littlewood 1972), but otherwise their facial features are like those of their full-sized neighbors (Howells 1973:173–4). Genetic studies confirm their connections with other Highlanders and their differences from Africans. Furthermore, an altered diet produces greater adult height. It now seems to be generally accepted that short stature in the Highlands represents adaptive response to certain local environmental conditions, including diet, though these remain to be identified with certainty. The pygmoid peoples are said to grow slowly and mature late (Gajdusek, reported in Howells), but observers agree that the adults are notably strong and hardy. They appear to be extreme cases of a general Highlands pattern of slow maturation and relative shortness (Malcolm 1970, 1979). The question of whether they are related to the pygmoid peoples in the islands to the west will be considered below.

Apart from the "pygmy question", the problem of how closely New Guineans and other Melanesians resemble Africans remains. Recent writers agree that the resemblance is much less than popularly supposed. The shape of the skull is very different, as are most features; almost never does an experienced observer mistake an African man for a Melanesian, even in a photograph. (Because some of the differences depend on bony development, they are less conspicuous in women.) Some distinctive, though not universal, African features, such as small squarish ears and a "lip seam" encircling everted lips, do not appear in Melanesia. Howells sums up the evidence from measurements as follows:

The Pacific group is ...

> distinguished from the African by heaviness of the brow ridges over the eyes, and general prominence of the middle of the face, especially below the nose. Thus, although Melanesians appear to be "negroid", like the Africans, in some external features such as darkness of skin, bushiness of hair and breadth of nose, they (and Australians) are decidedly different in the skull. This emphatic difference ... makes improbable any basic connection between Africans and Melanesians, as is also hinted by subtle differences in skin pigment and hair ... In fact, while of course there are resemblances, such features as lip form, hairiness and overhanging brows in the Pacific peoples do constitute differences from Africans; and the nose, between the eyes, is not at all flat as in Africans (1973:45–46).

The subtle difference in skin color mentioned above lies in the distribution of melanin below the skin — dispersed in Africans and Australian aborigines, but grouped in clumps in other peoples of the world, including very dark-skinned Bougainvilleans (reported in Howells 1973:32–33). "Hairiness" refers to the heavier body and beard hair of Melanesians.

Finally as regards external appearance, it is difficult to draw a clear boundary between Highlanders and other New Guineans or even to treat all Highlanders as being of a single distinctive physical type, as is sometimes done (Brown 1978:28). Still less is it possible to distinguish between New Guineans and the inhabitants of other Melanesian islands to the east. On the other hand, the physical diversity within New Guinea and within Melanesia as a whole makes it dangerous to assume that a single population can be considered representative, as when Howells (1973:45) uses the Tolai of New Britain as his sole example of Melanesians. It is even more dangerous to assume that an individual represents a region or population (see Munro 1966).

The diversity of New Guineans has become particularly obvious in genetic studies, which have focussed on blood antigens and proteins, though various other features have been investigated, such as finger and palm prints, color blindness and dental anomalies (see Littlewood 1972, Swindler 1962, Walsh 1972b). Unfortunately only occasionally has the same series of tests been carried out on different populations, and the problem of comparison is increased by the constant discovery of new genetic markers and improved means of detecting them. In several recent cases, conclusions have been proposed about the possible historical significance of the known distribution of certain genes without an adequate knowledge of their actual distribution (see, for example, Booth and Oraka 1968).

Genetic investigations have had several related aims: determination of the genetic makeup of New Guinea populations, tracing of connections with outside groups, and reconstruction of the history of migration and settlement, with particular attention to correlation between linguistic and physical groupings. On the first point, a frequent discovery has been the amount of variation between villages of the same language and culture. This variation, usually attributed to drift, has been documented for gene frequencies in three of four blood groups tested (ABO, MNS, and P) in villages two miles apart by Giles and his colleagues. They conclude that: "This survey of the Markham valley area should act as a caveat to the implication often found that blood group gene frequencies based on samples from small semi-isolated primitive villages represent with any accuracy anything of any wider area or cultural significance than those particular villages" (Giles *et al.* 1966:152). Littlewood, who found that his figures for specific linguistic groups and for particular villages within them diverged greatly from published figures for the same peoples, warns of the danger of "relying on a single village or two to represent the whole language area" (1972:85). Unfortunately much of the published material does indeed rely on small samples from single villages, and so must be treated with caution.

With these warnings in mind, we can examine some of the available data (from Gajdusek 1978; Howells 1973:50–79; Walsh 1972a,b). To start with blood groups, all three genes of the ABO group are found throughout New Guinea, though in varying frequencies. Except in one family, only the A_1 allele has been identified. The significance here is that B is absent among Australian aborigines, though high in southeast Asia. In the MNS series, high M is found in eastern New Guinea, New Britain, and New Ireland, but low M in the Highlands on the Papua New Guinea side except at Telefomin, near the border. However, Gajdusek found high M (near 20%) in the Highlands of West Irian. Consequently Walsh was probably wrong to attribute the distribution of M in the east to Polynesian admixture along the coast (high M being characteristic of Polynesia). According to Bellwood (1978:32), "the gene *n* in the MNSs blood-group system has very high frequencies in Australia and southern New Guinea." S is enormously variable even within the Highlands but is present in most populations, though absent in Australia.

The most interesting data concerning serum proteins and red cell enzymes are those which seem to distinguish or link specific groups. Although Giles had thought that gamma globulin antigens differentiated AN and NAN speakers in the Markham Valley (Giles *et al.* 1966), later investigations indicated that the really significant difference in these antigens separated people of the mainland of New Guinea from those of New Britain and the Solomons, regardless of their languages (Howells 1973:66–68). Gajdusek *et al.* suggest that some rare red cell enzymes are "Papuan markers" and that migration between Australia and New Guinea is indicated by the presence of a different rare allele found in northern Australia and among the Asmat of the south coast of New Guinea (but also in Malaysia) (Gajdusek *et al.* 1978:57). Howells (1973:63,69) repeatedly warns that it is too early for firm conclusions to be drawn, if indeed it is ever possible, given the possibilities of genetic drift in small populations. With any characters controlled by single genes, we must also consider the possibility of differential selective pressures. The distribution of some of these alleles, if not this one in particular, is often assumed to represent adaptation to the presence of endemic malaria in lowland regions (see Walsh 1972a:542).

Nevertheless, a few cases of suggestive coincidence between the physical and linguistic evidence have emerged. Giles' villages that differed in the distribution of gamma globulin alleles between AN and NAN speakers also differed significantly in the proportions of wet and dry ear wax, wet normally near 100% in Highlanders but lower in Mongoloids and some AN-speaking Pacific populations. (In neither case did the distinction appear in Solomons populations speaking different languages — Howells 1973:74.) A particularly striking case of coincidence between language and genetics is one unsuspected by those who reported it. Testing ABO and MSS blood groups in New Britain, Booth and his colleagues note among other surprises that "the Kaulong group appears much more closely to resemble the Mangsing rather than the Arawe Pilelo series reported by Simmons *et al.* (1961), notwithstanding possible linguistic affiliations with the Arawe people" (1969:115). In fact, the Kaulong language belongs to the same linguistic subgroup as the geographically distant Mangsing and is not closely related to the contiguous language of the culturally identical Arawe (see Chowning 1976). Other findings of that survey are also intelligible in the light of better knowledge of New Britain languages.

On the other hand, a recent survey of the Madang region by Serjeantson (1979) showed very little relation between genetic and linguistic distance; with a few odd exceptions, not explicable in simple terms, adjacent groups were genetically most alike regardless of their linguistic affiliations. Similarly, Gajdusek and his colleagues report their surprise at the genetic closeness between the Asmat of coastal West Irian and the Dani of the Highlands, saying that no contact between them seems to have occurred in the recent past (1978:61). But the linguistic map of West Irian shows the Asmat-Kamoro family contiguous with the Dani family, so the entire groups are perhaps not so separate as some of their members.

At present the most surprising genetic information concerns the distribution of a single gene controlling a blood group called Gerbich. Gerbich negative (Ge (a–)) is to date unique to New Guinea and highly localised there. Extensive testing has turned up in the Highlands one case who might be an immigrant, and several in Manus. Otherwise it is confined to an extensive stretch of the lowlands of northeast New Guinea, between the western tip and Finschhafen; it does not appear on the southern side of the island, nor elsewhere in the Highlands. But where it is present, it tends to be very common, with the highest incidence (64%) among the Namie, where the Yellow River joins the Sepik, and another focus at the head of the Markham Valley. It does not reach the coast in the Sepik region, but does elsewhere. Northeastern New Guinea is an area of great linguistic complexity — indeed, as regards NAN languages, the greatest — and Gerbich negative extends across lines dividing a number of NAN phyla from each other and from AN languages, which themselves are very diverse. It seems likely that this gene was introduced by one migrant group and, through time, spread by intermarriage over a very wide area, but the identity of the migrants is far less certain. At one point Wurm and his colleagues seem to suggest the gene came in with Austronesian speakers, but later tie it firmly to speakers of Torricelli Phylum languages, so the AN-speakers with this gene are called "now Austronesianised original speakers" of Torricelli languages (Wurm *et al.* 1975:939, 952). But the spread of the gene is much wider than this theory implies; high incidences occur along the Waria River, far to the east of any existing Torricelli languages. Booth and his colleagues offer the "speculative hypothesis" that "in the distant past the coast from the Huon Gulf to possibly the eastern tip of Papua was settled by a people or peoples with high *Ge* frequency, the ancestors of the current non-Austronesian speakers", but these were later displaced and especially driven inland by later arrivals (1972b:264). The only argument against this sensible suggestion is that there is no known relationship between the NAN languages currently spoken by populations high in Gerbich negative; just because the incidence is highest there, Booth *et al.* assume that the language of the carriers was Namie, of the Sepik-Ramu Phylum, an interpretation opposed to that of Wurm *et al.* This is the most conspicuous case of genetic evidence differing from that derived from other sources; other rare single genes either occur in no discernible pattern, or are thought to influence adaptation to particular environments, particularly malarial ones.

Where a combination of genes is examined, as is usually done, it obviously becomes less likely that long-distance ties will be readily discernible; often the combined genetic evidence seems to be pointing in several directions simultaneously (see below). Occasionally particular groups living far apart are said to be genetically similar, as when Booth points to resemblances between the people around Port Moresby and the Tolai of New Britain (Booth and Vines 1968:72; Booth *et al.* 1972:270). This finding, if correct, runs counter both to the phenotypic and the linguistic evidence; not only do the people look very different, but their languages, though both AN, serve as the examples of the two main types into which Capell (1969) divides the AN languages of New Guinea. Superficially at least the Motu look like Milne Bay people, and the languages seem to be closely connected as well (Pawley 1975:92).

In some other cases, genetics indicate the biological separateness of people who live side by side, whether they look similar, as with the people of Karkar Island (Serjeantson 1979) or do not, as with the Tolai and Baining of New Britain (Kariks and Walsh 1968). Occasionally genetics confirms the distinctiveness of a population which is also difficult to classify linguistically, like the people around Lake Kutubu (Booth and Hornabrook 1972:231; Wurm 1975b:491).

On the whole, however, the data from linguistics and genetics coincide to a very limited extent. Genetic drift, intermarriage and the adoption of new languages are prominent among the factors that have confused the picture. Even if we had all the genetic data concerning the distribution of various genetic systems throughout New Guinea and beyond it, we would need to exercise great caution in assuming that particular living populations represent any original migrant groups. These warnings apply even more strongly to basing conclusions on one or two visible traits such as skin color and hair form. When conclusions about racial affiliation are based on fragmentary skulls, the difficulties are multiplied. Nevertheless, several physical anthropologists and prehistorians show little hesitation in reconstructing human history on decidedly flimsy bases.

The origins of New Guinea populations

The main lines of evidence derive from physical anthropology and linguistics, with contributions from archaeology, the geological history of the region and, to a minor extent, the distribution of culture traits. The various theories are discussed more fully in the chapter on Prehistory, so the following account will be brief.

The archaeological evidence indicates that man reached New Guinea before plants are known to have been domesticated anywhere and before watercraft suitable for long-distance travel had been invented. People are consequently thought to have come via the land bridges that once connected New Guinea with Australia and with eastern Indonesia. Coon, Howells and Bellwood all assume that this whole region, extending from the present mainland of Asia, was once occupied by a type of Homo sapiens descended from fossil forms found in Java, and labelled Australoid by Coon. Because of the present distribution of such peoples in fringe and refuge areas, their apparent contribution to the populations of regions such as Borneo, and references in early literature, it is assumed that this population had frizzy hair as well as dark skin, and so resembled the Tasmanians rather than the majority of inhabitants of the mainland of Australia. Accordingly Howells had labelled the whole region "Old Melanesia", and thinks that 25 000 years ago "Melanesian" peoples extended from the Asiatic mainland as far as New Ireland. (He expresses some doubts about how early the islands off New Guinea were inhabited because he underestimates the variety of wild food, from shellfish to starch vegetables, available on these for hunter-gatherers — Howells 1973:177–98.)

Measuring only the Tolai, AN-speakers of northern New Britain, Howells finds them anthropometrically closer to the Tasmanians and the Australian aborigines, in that order, than to any other populations in the world (1973:45–6). Coon, who calls the Australoids a subspecies, considers the New Guineans one of its races (1962:427). That there is a shared component has been widely accepted, given the presence in many New Guineans of such "Australoid" features as heavy browridges and deep-set nasal roots. Once this resemblance was said not to appear in the blood groups (which of course are not available for the extinct Tasmanians). Macintosh summed up the evidence available to him by saying: "In brief, there is a characteristic pattern for Australia, another for Melanesia, and they differ radically from each other" (1965:61). The differences applied to Highlanders, who, for example, are very high in S, which is absent in Australia. More recently, however, Kirk is reported as saying that on genetic grounds, "Australian Aboriginals and New Guineans, particularly the New Guinean Highlanders, are closely related to each other" (ANU Reporter 27 May 1977). To date, however, archaeological evidence linking Australia with New Guinea is lacking. Greenberg's suggestion about a connection between Tasmanian and New Guinea languages is dismissed by Wurm (1975:929).

These contradictions can, of course, be explained on the assumptions that New Guinea received new genes that never reached Australia, or that the now-extinct Tasmanians resembled New Guineans in ways not apparent in mainland Australians. It can be said that a physical connection between New Guineans and Australians seems highly likely, even apart from Kirk's evidence, but these peoples have diverged so through time that few anthropologists would like to follow Coon and Bellwood in simply labelling New Guineans "Australoids". All other objections aside, doing so hearkens back to a generally outdated theory that all present populations are made up of various mixtures of a few original pure races. (For discussion, see Howells 1977.)

The same difficulty applies in a slightly different form to the question of whether these "Old Melanesians", or even present New Guineans, are related to Africans. Here the Andamanese are a critical case. Unlike other Asiatic "Negritos", including those of the Philippines, the Andamanese look like African pygmies, with bulging foreheads, very spiral hair, and heavy fat deposits on the buttocks (steatopygeia). Anthropometrics confirm the apparent resemblance. Howells, baffled, simply suggests that the Andamanese "should be set aside as another problem" (1973:176). By contrast, Coon sees no difficulty in assimilating them to his Negrito variety of Australoids, but does not consider them related to Africans (1962:418, 427). On the other hand, Birdsell, also stressing the outward resemblance of Andamanese to African Pygmies, labels the Andamans "Oceanic" — an unusual extension of the term — and then says that the sharing of these Pygmy traits by Oceanic Negritos (with no other mentioned) "weighs heavily in favour of the migratory hypothesis": the derivation of New Guineans, other Melanesians and a component of the Australian population from Africa (1975:516–17). Specifically, he insists that the presence of spiral hair in New Guinea, other parts of Melanesia, eastern Indonesia and parts of Australia, particularly Tasmania, is to be attributed to a Pleistocene migration from the west into New Guinea and Australia of Oceanic Negritos. These Negritos were the source not only of the Melanesian hair form but of the comparatively short stature of many Melanesians (Birdsell 1977:153–57). Skin color is not mentioned, and Birdsell notes that the facial features of New Guineans, particularly adult males, tend not to be "Negritoid" but "Australomorphic", reflecting a component (his "Murrayians") which he assumes derives from a later wave of migration represented by the majority of present-day Australian aborigines. He admits being baffled by the question of how the Murrayians reached Australia without passing through New Guinea and greatly affecting the appearance of its population (1977:156, 160–64).

Slight support for Birdsell's hypothesis was originally suggested by Greenberg, who saw relations between Andamanese and some New Guinea languages. Wurm thinks such a relationship might exist, especially as regards a possible "archaic substratum" in what now seem to be completely different languages (1975c:929–30). On the other hand, the early reports indicate that serologically the Andamanese do not group with short-statured New Guineans (Simmons 1960), but it must be remembered that a number of genetic markers have been discovered and traced since that study was made.

If we do accept the serological evidence that New Guineans are not related to

Africans (and therefore should not be called Oceanic Negroes) (Macintosh 1965:59), it is then necessary to explain the association of dark skin with frizzy hair in these parts of the world (see Coon 1962:426). The alternative is to assume that hair form as well as skin color is adaptive to tropical conditions, which is exactly what Brace and Montagu suggest (1965:295), or that hair form is correlated with other traits which are selected for in these environments. Even with skin color, usually assumed to be adaptive, many problems are raised by the actual distribution, and even greater problems arise when we try to explain why so many dark-skinned peoples are not frizzy-haired, or why frizzy hair is as common outside the tropics (southern Africa, Tasmania) as inside them. Migrations and the dominance of straighter hair form introduced by intermarriage with other populations may be invoked as explanations, but proof is lacking at every point. It is not surprising that Birdsell feels justified in assuming that New Guineans do share a common ancestry with Africans.

To turn to the populations to the east, the differences in body build between New Guineans and Polynesians have been mentioned, but there are differences also in facial features. Although people from eastern New Guinea are sometimes said to look Polynesian, the resemblance is one of skin color and hair form rather than faces, which would fit more easily into a Micronesian or Indonesian population than a Polynesian one. Nor are there notable anthropometric resemblances between New Guineans and Polynesians, though some New Guineans do resemble some Micronesians (Howells 1973:40–41). Blood groups equally separate New Guinea from Polynesia (*ibid.*:60). Finally, because the facts are so frequently misrepresented in the literature, it should be noted that AN-speakers of the north coast of New Guinea, New Britain and New Ireland look like other New Guineans and not like non-Melanesians; the light skin and less frizzy hair of some AN-speakers in the Central and Milne Bay Province have led to unwarranted generalizations about AN-speakers elsewhere in New Guinea.

It is generally taken for granted that AN languages were spread by people of what has been called a generalized Mongoloid type, a category broad and vague enough to include many Indonesians and all Polynesians. Later migrations from the mainland of Asia would have affected the people now living in Indonesia, as did admixture with the previous inhabitants of the islands; at the same time, some Micronesians show the effects of admixture with Melanesians. But if AN languages were brought to New Guinea by people who were not "Australoid" or "Negroid" physically, the present distribution of physical traits along the north coast of New Guinea and in New Britain and New Ireland cannot be explained on the assumption that the AN-speakers simply intermarried with previous residents similar to present-day Highlanders, who are, says Bellwood, "probably the most direct descendants of the original Australoid settlers of western Melanesia" (1978:27). (By contrast, Birdsell would assume that Highlanders represent a mixture of at least two groups of settlers.) For one thing, such a mixture would not be very dark-skinned; the darkest skin in New Guinea is found in the lowland regions, many of them, as in New Britain and New Ireland, inhabited by AN-speakers. It is probably safe to assume that New Guinea was well-populated before AN-speakers arrived, and that intermarriage of the newcomers with the numerically dominant previous inhabitants obscured their foreign traits. If NAN-speakers had been as plentiful in New Britain and New Ireland as in New Guinea, it is not clear why most of them abandoned their languages while through intermarriage so affecting the physical type of the invaders. The possibility that the AN-speakers brought new culture traits that gave them and their languages high prestige is not to be ruled out, however. It may be that at least many of the present inhabitants of the Bismarck Archipelago represent peoples who intermarried with NAN-speakers elsewhere and then entered the islands. To the extent that the admixture took place locally, it was presumably with people more like the Bainings of New Britain than like New Guinea Highlanders (see Howells 1973:40). Furthermore, wherever the mixture took place, the NAN languages spoken by the previous inhabitants presumably did not share those grammatical features that are assumed to have so greatly affected the AN languages of New Guinea itself, since the AN languages of New Britain and New Ireland show little evidence of NAN influence (see below).

Language and migrations

As has been noted, several attempts have been made to correlate physical type, language and the history of migration and settlement. Several assumptions are usual. First, the earliest colonizers are likely to be represented by peoples living in the remote interiors of islands, or in less desirable fringe or "refuge" areas. Second, not only will these people probably speak NAN languages, but these languages are likely to be connected only remotely with any other NAN languages; languages which are obviously closely related would have diversified recently and so are likely to have been spoken by later arrivals. (Linguists assume that languages which share a common ancestor tend on separation to diverge from each other at fairly equal and constant rates; therefore the smaller the resemblances between two related languages, the longer they have been separated.) Third, even if the earlier settlers adopted new languages, they probably altered them in ways that reflect their original speech (the substratum theory). Finally, ethnographic evidence may point to the presence of what are believed on comparative or historical grounds to be archaic culture traits, recent inventions, or foreign importations. All this information may be used to reconstruct the history of migrations and contact between peoples. The ethnographic evidence will be discussed more fully below, but remember that one clue to the antiquity of a trait comes from comparative linguistics; if the names for it in many languages are cognate and do not represent recent borrowings, the trait is likely to be an old one among speakers of these languages.

In a recent (1975) survey, Wurm, Laycock, Voorhoeve and Dutton discuss the linguistic evidence for the settlement of New Guinea and population movements within it, especially as concerns the NAN languages. They suggest that the only trace of the first languages spoken there may be a few pronominal forms found widely scattered throughout the region, representing a substratum in different phyla. Possibly some of the small phyla such as Torricelli are descendants of the earliest languages, but if more than one such phylum dates back to a single original language, the connection between the descendants is no longer evident. Concerning the phyla now present in New Guinea, it is assumed that those now confined to the far eastern and western extremities of the island represent earlier groups which were driven apart by the expansion of later comers, particularly speakers of the now numerically dominant Trans-New Guinea Phylum languages. (The reasons for the expansion are unknown.) This expansion would also have separated related languages, such as those of the Asmat and Sentani regions of West Irian. It is further assumed that often the expanding languages were adopted by people whose original language was different. The history of migrations is postulated very tentatively, the authors pointing out that at one point they may have reversed the actual order of events. Ideas about the distribution of New Guinea languages are changing so rapidly that little is to be gained by summarizing what is already likely to be an out-of-date synthesis.

Only with the arrival of the Austronesian languages at a postulated date of 5000 years before present (BP) do the suggestions become firmer. AN-speakers entering the Markham Valley are thought to have influenced Trans-New Guinea Phylum languages there, so that as the latter expanded they took some words of AN origin through the island. On the basis of their distributions, it is assumed that two NAN phyla, Sepik-Ramu and Sko, entered New Guinea first on the central part of the north coast as did the AN languages, but the writers barely discuss the sequence of events and the origin place of these NAN languages. (In contrast, the earlier NAN languages are assumed to have come in from the west.) After the initial move into the Markham Valley, it is plausibly suggested that later AN-speaking migrants settled in many coastal areas and on all the offshore islands except Rossel Island in Milne Bay. It is frequently asserted that the AN languages of the mainland of New Guinea, of the offshore islands (including those of Milne Bay) and of some parts of New Britain show effects in pronunciation, grammar and vocabulary of contact with NAN languages, and Capell in particular has advocated the theory that these particular AN languages were all adopted and heavily modified by speakers of NAN languages. He consequently thinks these languages are not fully AN but represent a sort of "pidgin" amalgamation of AN and NAN languages, contrasting with most AN languages of the Bismarck Archipelago, which have not been so influenced and so are fully AN (Capell 1969). Few linguists accept the denial of full AN status to the languages of the mainland, but all concede that contact with NAN languages has affected them, while disagreeing about the precise nature of the effects (see Grace 1976).

Questions still abound concerning the AN settlement of New Guinea: when it took place, how many separate settlements were involved, and the precise origin place of the settlers. It may be that the first arrival long preceded 5000 BP. Almost certainly several different "migrations" produced the original settlements of AN-speakers in New Guinea: and as might be expected from seafaring people, many subsequent shifts occurred. Sometimes, as with the NAN languages, distribution and diversification of AN languages makes it possible to reconstruct a possible sequence of settlement, as has been done for New Britain (Chowning 1976:379–80). It has often been suggested that some of the AN languages of New Guinea, notably some of those of the south coast of New Guinea and north coast of New Britain, represent back-migrations from regions much farther east (Goodenough 1961; Pawley 1972). Certainly such back-migrations did occur, as the Polynesian "Outlier" islands scattered through Melanesia indicate, but the languages of the AN-speakers most often said to show Polynesian admixture, the Trobrianders and the Motu, are very unlike Polynesian. A thorough investigation remains to be carried out on the resemblances between Micronesian languages and those of various parts of New Guinea, especially the Trobriands and the northern islands of the Bismarck Archipelago. Smythe (1970) spoke of Micronesian influences on the Admiralty languages, but on the basis of incomplete and outmoded knowledge of the characteristics of Micronesian and Melanesian languages.

It is probable that most of the diversification of Melanesian languages took place within the New Guinea area, with the greatest diversity found along the central north coast of New Guinea and in the islands just to the north, including New Britain. Dyen proposed that this area was the origin place for all AN languages (1965:44), but this theory has not been generally accepted. Instead, it is taken for granted that the point of origin and dispersal lay to the west. Even today the AN languages of eastern Indonesia show closer connections with those in the New Guinea region than with Indonesian languages spoken farther west (Blust 1978). As with the NAN languages, but with more promise of certainty because the languages involved are all related, better answers will depend on the results of much ongoing research.

Cultural items

The linguistic diversity of New Guinea sometimes seems to be matched by a cultural diversity that makes generalization very difficult. But in fact the biological analogy is more accurate. All of the languages of New Guinea are not related, but all the people are; ultimately interbreeding has produced what can be called a single population. Admittedly some inhabitants of the more remote islands of the Bismarck Archipelago have had less contact with the rest of New Guinea, but trade networks have reduced their degree of isolation.

The unity of New Guinea is shown in the technological base and in certain aspects of social organization. Although frequently said to exist, no pure hunter-gatherers have been reliably reported; all New Guineans seem to make some use of cultivated foods, though their main source of calories may be wild sago. A few coastal groups did not garden but obtained cultivated vegetables by trade (Mead 1930; Held 1957). Everywhere, within limits imposed by the climate, particularly that of the Highlands, the traditional vegetable foods are essentially the same: root and tree crops, supplemented by greens. The staples, with the exception of the sweet potato, are indigenous to the Indo-Pacific region, though several crops of American origin have become increasingly popular in recent years (see Powell 1976). Techniques of cultivation vary (see next chapter) but within a range which almost always includes a period of fallow, and gardening tools are few, with the stone axe/adze and simple wooden digging stick the most common, and spades and hoes confined to a few areas. Horticultural techniques are similar to those of the other Pacific islands, but a difference from some parts of Polynesia, which may have had social consequences, is the failure of New Guineans to develop or utilize techniques for the long-term storage of foods which otherwise spoil rapidly. More food is usually grown than is needed, but only in a few yam growing areas, notably the Trobriands, are surplus crops harvested and stored for lengthy periods. On the other hand, the relative abundance of wild foods reduces, though it does not eliminate, the possibility of famines caused by natural disasters that affect the crops.

The rest of the technology is also similar to that of the other Pacific islands. It is what Linton (1955) calls the Southeast Asiatic Neolithic, though his particular reconstruction of it is out-of-date. Essentially the technology includes the traits that accompanied reliance on cultivated plants and domestic animals, particularly the pig, but preceded later inventions like metallurgy, loom weaving, the cultivation of grains (particularly rice) and the domestication of other animals such as the water buffalo. The distinguishing traits of the so-called Neolithic include tools of polished stone, shell, wood and bamboo; shifting horticulture based on taro, yams, bananas and sago; domestic animals confined to the dog, pig and chicken; cooking in clay pots, or with heated stones, often in earth ovens; clothing of barkcloth and shredded fiber; sea transport by the single outrigger canoe; fishing techniques which included nets, traps and poison. All of these traits are associated with speakers of AN languages, but need not have been introduced to New Guinea by them. The distribution of pottery-making in regions close to the coast, even though many of the present potters speak NAN languages, has led to the assumption that this trait was an AN introduction, but the case is not proved for all areas. Almost certainly the outrigger canoe did arrive with AN speakers, given not only its distribution but the AN names for its many parts. But archaeological discoveries make it seem increasingly likely that pigs and some crops, especially taro, reached New Guinea well in advance of AN speakers, and there is no reason to assume that such traits as the manufacture of barkcloth and fish-traps are not equally old.

It is consequently dangerous to correlate cultural items with language groupings, as is sometimes done when particular traits are described as "Papuan" (i.e., NAN) and others as Austronesian. For example, in New Guinea most NAN-speakers used bows in hunting and warfare and most AN-speakers did not; but this correlation does not hold in the Solomons, and a widespread AN word for "bow" exists; obviously AN-speakers were familiar with the bow before they reached New Guinea. With any cultural item, questions about its origin and associations depend to some extent on whether it is likely to have been invented just once and spread from a single center, or to have been re-invented separately in different regions. Such factors as the distribution of the trait in other parts of the world, and its complexity (which presumably reduces the likelihood of re-invention in the same form), must be considered; so too must the actual likelihood of particular migrations which are invoked to account for the distribution of certain traits. Where decisions can be made, they often depend on subjective judgments; and consequently anthropologists and others interested in culture history often disagree on the probable significance of the data. The most suspect judgments are those that presuppose, often covertly, that certain peoples are less inventive or less intelligent than others. It is often taken for granted, without examining the reasons, that traits shared between New Guineans and outside groups (other than aborigines) must have spread into New Guinea, and not vice-versa. When the reasons are examined, as when art historians sometimes suppose that New Guinea art styles originated in a foreign complex civilization, racial prejudice may be involved.

There are a few cases in which foreign influence is readily detectable. The physical type of the people, and their technology, makes clear the heavy Micronesian influence on the people of Aua and Wuvulu, west of the Admiralties. Loom-weaving, found in some of the islands north of New Ireland, almost surely derives from Micronesia as well. In West Irian, a long history of trade and political contacts involving not only Indonesia but possibly China has introduced foreign artifacts, such as glass beads and Chinese porcelain, and also in one or two places the technology necessary to produce not only woven cloth but metal tools (van der Sande 1907:236–7; Bellwood 1978:266). The effects of these contacts on other parts of the island, including the interior of West Irian, seem to have been minimal; it is still possible to differentiate New Guinea from Indonesia, even from those islands where the people look like New Guineans and speak NAN languages, on technological grounds. Equally, but on a different basis, it is possible to draw a line between New Guinea (south of the Torres Straits Islands) and the mainland of Australia, which culturally is very different. But it is impossible on any cultural grounds, or indeed as regards physical anthropology and linguistics, to draw a firm line between New Guinea and the other islands of Melanesia, particularly the Solomons and New Hebrides; the only distinguishing traits are those that characterize the special adaptations to the Highlands. In recent years, anthropologists who have worked in the Highlands have sometimes written as if the coastal regions simply do not exist or are not really part of New Guinea because they "have been affected by outside influences over a long period of time and are diverse" (Brown 1978:264). But the work of linguists, prehistorians and geographers indicates that people in the interior have shifted around a great deal and have been affected by events occurring well outside their own regions (see Hughes 1977:202). Highlands cultures have their peculiarities, but these are not so great as is sometimes imagined; even those traits which are not found in other parts of New Guinea (such as large-scale pig feasts) sometimes occur elsewhere in Melanesia.

Accordingly, it is difficult to discover culture traits that characterize New Guinea to the exclusion of Indonesia on the one hand and the rest of the Pacific on the other. We can only say that some traits are more common in New Guinea. (The single exception is the type of intensive culture of sweet potatoes found in some parts of the Highlands; where intensive agriculture exists elsewhere in Melanesia, other crops are involved.) Judging from surveys of material culture (see Cranstone 1972 for a recent example, though it contains gaps), almost the only wide-spread artifact which is peculiar to New Guinea is the women's net carrying bag (called *bilum* in Pidgin English) worn suspended from the forehead and often constituting a permanent part of female dress. Particularly common in the Highlands, it is also found in many coastal areas, but not on the offshore islands, such as the D'Entrecasteaux and New Britain. (Although net carrying bags exist there, women usually balance loads on top of their heads, and do not carry their babies in bags.) By contrast, another typical New Guinea artifact, the hour-glass drum (Pidgin *kundu*) is used on the offshore islands as well as being found throughout New Guinea. In the past, a characteristic New Guinea (and New Britain) weapon was the stone-headed club; these are comparatively rare elsewhere in Melanesia. They did not however, have such a wide extension as the *bilum* and the *kundu*.

As regards social organization, New Guinea is noted for its small-scale egalitarian societies in which leadership tends to be achieved by the exceptionally ambitious and able man who can attract a following (the "Big Man" system). The political unit is usually much smaller than the population that shares a language and culture, so that this population is split into rivalrous and often hostile groups. This is the usual pattern; but in a number of New Guinea societies, mostly coastal, leadership is hereditary, and a number of others, mostly located between the coast and the Highlands, are reported to lack recognizable leaders of any sort (see Chowning 1979). Throughout traditional New Guinea, warfare or the threat of it was incessant, and inhibited the development of larger political units even where people were numerous and settlements dense. The causes of warfare differed from one area to another; sometimes it was so regulated that relatively few fatalities resulted, though the system of repaying one death with another tended to sustain hostile relations indefinitely. Fights over land were particularly common in the Highlands, where it was often a scarce resource. Headhunting was largely confined to the Sepik and Fly River regions and the coast to the west of the Fly. Cannibalism, as major motive for warfare rather than a way of disposing of the dead, occurred in a few widely scattered areas, the best known being the New Ireland — Duke of Yorks — Tolai region, and the D'Entrecasteaux Islands.

A feature of social organization peculiar to the island of New Guinea is a strong tendency towards patrilineal descent, affiliation with one's father's kinship group (clan or lineage). Most societies on the island of New Guinea have patrilineal descent groups within which marriage is forbidden; and although most permit resident outsiders to join and eventually be adopted, the majority of men who live, work and fight together tend to be related through their fathers. The prominence of this pattern suggests that it has a long history in New Guinea. Patrilineality is not characteristic of Indonesia, and although it occurs in Australia, there it is part of a peculiar and elaborate system of marriage classes that is not duplicated in New Guinea. By contrast, most of the islands in Milne Bay, eastern New Britain and New Ireland are characterized by matrilineal descent, in which a person belongs to his mother's clan or lineage, and a man shares certain property rights with his mother's brother. (Contrary to popular belief, women do not necessarily have high status or special property rights in matrilineal societies.) Matrilineality continues through Bougainville, and also appears in an attenuated form at a few points on the north coast of New Guinea, as near Lae. It is the characteristic method of reckoning descent in Micronesia, and the distribution of this trait in New Guinea and its environs makes it tempting to suggest that Micronesian influence was involved, particularly since the distribution of these descent systems is in no way correlated with linguistic groupings. Matrilineality has not, however, been reported from the one part of the Bismarck Archipelago in which, as was noted above, Micronesian influence has certainly been heavy, the islands of Aua and Wuvulu. Unilinear descent systems, whether patrilineal or matrilineal, are rare in southeast Asia and Indonesia, and take a different form where they occur in Polynesia; consequently they are usually not thought to be part of the AN heritage. They may represent local adaptations, possibly to pressures on land (see Goodenough 1955:81, Alkire 1977:87), or, as is suggested here regarding patrilineality, may have been taken over from speakers of NAN languages.

Another widespread characteristic of New Guinea societies is the extreme separation between men and women, correlated with fears of menstrual blood and of contamination caused by sexual intercourse. Frequently associated institutions include men's houses which are forbidden to women; ceremonies, especially those concerned with the initiation of boys into manhood, which attempt to separate them from women and neutralize the effects of contamination during childhood; often, low status for women; and, extending across the central part of the island, institutionalized male homosexuality. These patterns of avoidance do not permeate all of New Guinea, being particularly underdeveloped at the eastern end; but they are well-developed in New Britain and appear through much of the Bismarck Archipelago. Extreme patterns of avoidance are found among AN-speakers, as in Wogeo and southeast New Britain, as well as NAN-speakers; furthermore, similar patterns recur in Malaita in the Solomons, which is wholly AN-speaking (Keesing 1976:398–401). Nevertheless, the extreme focus on sexual fears and avoidance in much of New Guinea suggests that at least for this region the trait is an old one which antedates the coming of AN-speakers and may have been acquired by them from the earlier residents. It is, incidentally, one of the few traits that link aboriginal Australia with New Guinea; extreme forms of avoidance are not found in the other Pacific islands or in Indonesia.

As in most small-scale societies, marriages are normally arranged, and payments are normally made from the man's kin to the woman's, or exchanged between the two groups, though especially in the interior lowland regions, sister-exchange, in which two men marry each other's sisters, may be practiced without payments being involved. There is nothing peculiar to New Guinea in these practices, nor is the use of marriage payments, so-called "brideprice", correlated with the status of women.

Some of the ceremonial practices and religious beliefs found in New Guinea are more distinctive, and sometimes show clear evidence of diffusion from a single center. The best documented of these is the use in male ceremonies of a pair of sacred flutes, the "birds", to represent the voice of the ancestors. Ceremonies focussing on these flutes extend from the Eastern Highlands to the north coast of New Guinea and along it in both directions, from the Northern Province to west of Humboldt Bay (see Allen 1967; van der Sande 1907:294). A supposed history of successive migrations by peoples whose cultures contain distinctive mythologies and religious cults has been postulated by Schmitz to account not only for the use of paired flutes but for the distribution of such traits as myths about twins and cannibalistic ogres, the depiction of snakes in art, and a variety of other motifs (see Schmitz 1963:57–66; Buehler 1962:72–77). These specific theories continue a 19th Century tradition, still maintained among some German-speakers, which assumes the firm association of certain artifacts, ideas and customs with particular migrant groups, so that one can talk about the "round-adze peoples" (Schmitz), the "megalith-builders" (Riesenfeld 1950), or the "betel people" (Rivers 1914), and deduce their influence by the presence of a supposedly associated trait, such as a burial practice. Such theories remain popular among art historians concerned with New Guinea, but most anthropologists consider them too simplistic. Real problems remain unexplained, such as the very scattered distribution of kava drinking in New Guinea (portions of the north and south coasts, plus Manus), or the break between the Sepik and Gulf of Papua regions in the manufacture of elaborate ceremonial masks (Laycock 1975). It is still possible to assume that migrations accounted for these distributions, as Haddon suggested long ago (1920). On the other hand, many of the supposedly significant traits, including myths about ogres and twins and special attitudes towards snakes, are so wide-spread among human beings that they are better explained in terms of the universal properties of the human mind. That may also be the explanation for the wide but discontinuous distribution of chopping off a finger in mourning, which is practiced among the Dani of West Irian and in the D'Entrecasteaux. Apparently more trivial beliefs and customs may be complex enough so that diffusion is almost surely involved: the wide-spread belief that sneezing indicates that someone is talking about you and that the speaker can be located by extending the arm in various directions till the joints crack; the assumption, found from Milne Bay to New Ireland, that lines in the palms of the hand indicate which birds are tabu for the owner's matrilineal kin to eat. Items as small as these indicate how New Guineans who live far apart are connected by a network of cultural relationships.

As has been noted before, however, these ties do not end with the region we have been considering, but extend through the rest of Melanesia, aided by the elaborate trade systems for which the area is so famous. Any cultural boundary between New Britain, New Ireland, and Bougainville is as artificial as one between Bougainville and the rest of the Solomons. The question remains of whether the New Guinea region differs notably from the rest of Melanesia. Perhaps the most conspicuous difference is the absence of obvious resemblances to Polynesia, which become increasingly common in eastern Melanesia. If we do not assume, as has often been done, that hereditary chieftainship in some New Guinea societies, and the use of kava in others, resulted from Polynesian settlement, we are left without any of the detailed resemblances in such fields as mythology and religion which are so conspicuous farther east. On the other hand, the evidence for contact between Micronesia and the islands north of

New Guinea is clear in a few cases, and confirmed by oral tradition in some (see Riesenberg 1965–66). Manus religion, with its emphasis on the ghosts of ancestors and its moral quality, sounds distinctly Micronesian (see Fortune 1935); although mortuary observances loom large throughout New Guinea, ancestors are not usually the focus of religious activity, which is characteristically concerned with spirits or gods of non-human origin, or with the manipulation of supernatural forces by magicians hoping to control the weather, harm their enemies or attract a woman's interest. (Ancestors are, however, important in the religions of some regions, including the Sepik, parts of the Highlands, and the Gulf of Papua; see Lawrence and Meggitt 1965.) If the culture of the Trobriands, which is the one in New Guinea most often said to be of foreign origin, actually owes anything to outsiders, geography, physical anthropology and possibly linguistics make Micronesia a more likely source than Polynesia.

Conclusions

Physically, the inhabitants of New Guinea exhibit a diversity that undoubtedly reflects a lengthy and complex history of settlement from outside the area, internal migration and intermarriage. Genetic drift presumably further affected the composition of many groups, and it is plausibly argued that some also show the effects of adaptation to local conditions, such as endemic malaria in the lowlands and poor protein supplies in some parts of the Highlands (Howells 1977:176–7).

Both the location of New Guinea, and the physical resemblances between its inhabitants and the Australian aborigines on the one hand, and some dark-skinned, frizzy-haired Indonesians on the other, indicate that New Guineans shared some of their ancestry with peoples now living to the south and west of them. Their ultimate origins, are, however, hotly disputed, and it is possible that the dispute may never be settled. The presence of AN languages in coastal New Guinea points to an additional physical component, presumably introduced by people who, like AN-speakers elsewhere, were relatively light-skinned and straight-haired. The fact that present-day AN-speaking New Guineans rarely manifest these traits, however, but instead look like their NAN-speaking neighbors, therefore reflects a long period of intermarriage between these later migrants and those already resident in the islands. By the time the former arrived, the population of New Guinea was probably already very diverse. Certainly at present, even in regions where there is no reason to suspect AN-admixture, there is no single "Papuan" (NAN) physical type that would include the inhabitants of the Baining Mountains in New Britain, the Sepik River, the Highlands of New Guinea and the south coast of West Irian. The variation between regions, and the differential degree of admixture with AN-speakers in some, make it unlikely that any gene would actually be so distributed as to serve as a "Papuan marker" (cf. Gajdusek *et al.* 1978:57). Finally, AN-speakers probably did not reach New Guinea and the Bismarck Archipelago in a single migration, and it is quite probable that when they arrived, the AN-settlers differed from each other even before they mixed with NAN-speakers.

New Guinea was probably never densely settled, and scattered populations were able to develop regional cultures linked by distinctive traits and attitudes as well as by trade and shared ceremonies. It is possible to list a number of such traits that tie together the people of the Milne Bay (otherwise called Massim) region (such as matrilineality, fear of female witches, relatively high status for women, a distinctive art style in wood carvings, and a ceremonial trade system using particular shell valuables) or the people of the Highlands (such as patrilineality, ritualized antagonism between the sexes, virtual absence of art except in self-decoration, and ceremonial exchange centering on pigs). Similar lists, which might include types of housing, hunting and fishing equipment, preferred forms of marriage, and ceremonies, could be used to characterize other regions. Some components of these distinctive complexes do reflect the local ecology (intensive cultivation of sweet potatoes in the Highlands, harvest festivals centering on yams in Milne Bay), but most do not. Furthermore, the boundaries are never clearcut, and rarely coincide with geographical ones other than major mountain ranges or bodies of water. Furthermore, these so-called culture areas unify people who often speak very diverse languages, even AN and NAN speakers sharing virtually identical cultures in some regions (southeast New Guinea, Karkar Island, Umboi Island).

An outsider entering a previously undescribed New Guinea society could expect to encounter a number of widespread traits, such as clans; economic exchanges at marriage; formal separation of the sexes; rituals associated particularly with puberty and death; magic to promote the fertility of gardens, attract lovers, control weather and harm one's enemies; wealth in the form of shell ornaments and pigs; and chronically hostile relations with outsiders. But any or all of these might well be absent, and when they are present, they do not distinguish New Guinea societies from others in Melanesia (see Chowning 1977). Despite their common economic base, the combination of local geography and a long and complex history has made diversity of physical type, language and culture as characteristic of New Guinea as any widely shared traits might be considered to be.

References

Alkire, W. 1977. An Introduction to the Peoples and Cultures of Micronesia (second ed.). Menlo Park, California, Cummings.

Allen, M. 1967. Male Cults and Secret Initiations in Melanesia. Melbourne Univ. Press.

Anon. 1977. New light on origins of Pacific populations. ANU Reporter, 27 May, Canberra, Australian National University.

Bellwood, P. 1978. Man's Conquest of the Pacific: the Prehistory of Southeast Asia and Oceania. Auckland, Collins.

Birdsell, J. B. 1975. Human Evolution: an Introduction to the New Physical Anthropology (second ed.). Chicago, Rand McNally.

—— 1977. The recalibration of a paradigm for the first peopling of Greater Australia, in Sunda and Sahul: Prehistoric Studies in Southeast Asia, Melanesia, and Australia. J. Allen, J. Golson and R. Jones (eds.). New York, Academic Press.

Blust, R. A. 1978. Eastern Malayo-Polynesian: a subgrouping argument, in Second International

Conference on Austronesian Linguistics: Proceedings, S. A. Wurm and L. Carrington (eds.), Pacific Linguistics, Series C: 61, Canberra, 181–234.
Booth, P. B. and R. eka Oraka. 1968. Blood group frequencies along the south coast of Papua, Arch. and Phys. Anth. in Oceania 3: 146–155.
Booth, P. B. and A. P. Vines. 1968. Blood groups and other genetic data from the Bismarck Archipelago, New Guinea, Arch and Phys. Anth. in Oceania 3: 64–76.
Booth, P. B., A. P. Vines and J. J. Saave. 1969. Further blood group genetic data from New Britain, Arch. and Phys. Anth. in Oceania 4: 113–119.
Booth, P. B. and R. W. Hornabrook. 1972. Blood group genetic data from the Lake Kutubu region, Southern Highlands, Human Biology in Oceania I (3): 229–333.
Booth, P. B., K. McLoughlin, R. W. Hornabrook, A. MacGregor and L. A. Malcolm. 1972. The Gerbich Blood Group System in New Guinea, II. The Morobe District and the North Coast, Human Biol. in Oceania I (4): 259–66.
Booth, P. B., K. McLoughlin, R. W. Hornabrook, A. MacGregor. 1972. The Gerbich Blood Group System in New Guinea, III. The Madang District, the Highlands, the New Guinea Islands and the South Papuan Coast, Human Biol. in Oceania I (4): 267–272.
Brace, C. L. and A. Montagu. 1965. Man's Evolution: an Introduction to Physical Anthropology, New York, MacMillan.
Brown, Paula. 1978. New Guinea: ecology, society, and culture. Annual Review of Anthropology 7: 263–91.
Buehler, A. 1962. Oceania, in The Art of the South Sea Islands, by A. Buehler, T. Barrow and C. P. Mountford. New York, Crown.
Capell, A. 1969. A Survey of New Guinea Languages. Sydney Univ. Press.
Chowning, Ann. 1976. Austronesian Languages: New Britain, in New Guinea Area Languages and Language Study, S. A. Wurm (ed.), Pacific Linguistics C: 39 (2).
—— 1977. An Introduction to the Peoples and Cultures of Melanesia (second ed.). Menlo Park, California, Cummings.
—— 1979. Leadership in Melanesia. J. Pacific Hist. 14 (1–2): 66–84.
Coon, C. S. 1962. The Origin of Races. New York, Knopf.
—— 1966. The Living Races of Man. London, Jonathan Cape.
Cranstone, B. A. L. 1961. Melanesia: a Short Ethnography. London, British Museum.
—— 1972. Material Culture, in Encyclopaedia of Papua and New Guinea, Melbourne Univ. Press.
Dyen, I. 1965. A Lexicostatistical Classification of the Austronesian Languages. International Jour. of American Linguistics Memoir 19.
Forge, A. 1971. The people and their culture, in Papua-New Guinea: Prospero's Other Island, P. Hastings (ed.). Sydney, Angus and Robertson.
Fortune, R. 1935. Manus Religion. Memoirs of the American Philosophical Soc. III.
Friedlaender, J. 1975. Patterns of Human Variation: the Demography, Genetics, and Phenetics of Bougainville Islanders. Cambridge, Mass., Harvard Univ. Press.
Gajdusek, D. C., W. C. Leyshon, R. L. Kirk, N. M. Blake, B. Keats and E. M. McDermid. 1978. Genetic differentiation among populations in Western New Guinea. Amer. J. of Physical Anthropology 48 (1): 47–63.
Giles, E., E. Ogan, R. J. Walsh and M. A. Bradey 1966. Blood group genetics of natives of the Morobe District and Bougainville, Territory of New Guinea, Arch. and Physical Anth. in Oceania I: 135–54.
Goodenough, W. H. 1955. A problem in Malayo-Polynesian social organization, American Anthropologist 57 (1): 47–63.
—— 1961. Migrations implied by relationships of New Britain dialects to Central Pacific languages, J. Polynesian Soc. 70: 112–36.
Grace, G. W. 1976. History of Research in Austronesian Languages of the New Guinea Area, in New Guinea Area Languages and Language Study, S. A. Wurm (ed.), Pacific Linguistics C: 39 (2), Canberra.
Haddon, A. C. 1912. The pygmy question, in Pygmies and Papuans, A. F. R. Wollaston. London, Smith, Elder and Co.
—— 1920. Migrations of Cultures in British New Guinea. Royal Anth. Inst. of Great Britain and Ireland Jour. 50: 237–80.
Held, G. J. 1957. The Papuas of Waropen. The Hague: Martinus Nijhoff.
Howells, W. W. 1973. The Pacific Islanders. New York, Scribner's Sons.
—— 1977. The sources of human variation in Melanesia and Australia, in Sunda and Sahul, J. Allen, J. Golson and R. Jones (eds.). New York, Academic Press.
Howlett, Diana. 1973. Papua New Guinea: Geography and Change. Melbourne, Thomas Nelson.

Hughes, I. 1977. New Guinea Stone Age Trade, Terra Australis 3, Dept. of Prehistory, R.S.P.S., Australian National University, Canberra.
Kariks, J. and R. J. Walsh. 1968. Some physical measurements and blood groups of the Bainings in New Britain, Arch. and Phys. Anth. In Oceania 3: 129–142.
Keesing, R. M. 1976. Cultural Anthropology: a Contemporary Perspective. New York, Holt, Rinehart and Winston.
Lawrence, P. and M. J. Meggitt. 1965. Gods, Ghosts and Men in Melanesia. Melbourne, Oxford Univ. Press.
Linton, R. 1955. The Tree of Culture. New York, Knopf.
Littlewood, R. A. 1972. Physical Anthropology of the Eastern Highlands of New Guinea. Seattle, Univ. of Wash. Press.
Macintosh, N. W. G. 1965. The physical aspect of man in Australia, in Aboriginal Man in Australia, R. M. and C. H. Berndt (eds.). Sydney, Angus and Robertson.
Malcolm, L. A. 1971. Growth and Development in New Guinea. Madang, Institute of Human Biology.
Mead, Margaret. 1930. Growing Up in New Guinea. New York, Wm Morrow.
Munro, R. R. 1966. Histological aspects of skin pigmentation in indigenes of the Territory of Papua-New Guinea, Arch. and Phys. Anth. in Oceania 1: 119–34.
Pawley, A. 1972. On the internal relationships of Eastern Oceanic languages, in Studies in Oceanic Culture History, R. C. Green and M. Kelly (eds.), Vol. 3. Honolulu, B. P. Bishop Museum Dept. of Anth. Special Publication.
—— 1975. The relationship of the Austronesian languages of Central Papua, in Studies in Languages of Central and Southeast Papua, T. Dutton (ed.). Canberra, Pacific Linguistics C: 29.
Powell, Jocelyn. 1976. Ethnobotany, in New Guinea Vegetation. K. Paijmans (ed.). Canberra, Australian National Univ. Press.
Riesenberg, S. 1965–66. Table of voyages affecting Micronesian Islands. Oceania 36: 155–60.
Riesenfeld, A. 1950. The Megalithic Culture of Melanesia. Leiden, E. J. Brill.
Rivers, W. H. R. 1914. The History of Melanesian Society. Cambridge Univ. Press.
Schmitz, C. 1963. Wantoat: Art and Religion of the Northeast New Guinea Papuans. The Hague, Mouton.
Seligmann, C. G. 1909. A classification of the natives of British New Guinea. Royal Anth. Inst. of Great Britain and Ireland, Journal, 39: 246–74, 314–33.
Serjeanson, Sue. 1979. Genetics, linguistics and distance in Papua New Guinea. Paper presented to the 49th Congress of ANZAAS, Auckland.
Simmons, R. T. 1973. Blood group genetic patterns and heterogeneity in New Guinea, Human Biology in Oceania II (1): 63–71.
Sinnett, P. F. 1977. Nutritional adaptation among the Enga, in Subsistence and Survival: Rural Ecology in the Pacific, T. Bayliss-Smith and R. Feachem (eds.). New York, Academic Press.
Swindler, D. R. 1962. A Racial Study of the West Nakanai. Philadelphia, Univ. Museum.
van der Sande, G. A. J. 1907. Nova Guinea III: Ethnography and Anthropology. Leiden, E. J. Brill.
Walsh, R. J. 1972a. Human Genetics, in Encylopaedia of Papua and New Guinea. Melbourne Univ. Press.
—— 1972b. Physical Anthropology, in Encylopaedia of Papua and New Guinea. Melbourne Univ. Press.
Wurm, S. A. 1975a. Language distribution in the New Guinea area, in Papuan Languages and the New Guinea Linguistic Scene, S. A. Wurm (ed.). Canberra, Pacific Linguistics 38.
—— 1975b. East Central Trans-New Guinea Phylum languages, in Papuan Languages and the New Guinea Linguistic Scene, S. A. Wurm (ed.). Canberra, Pacific Linguistics 38.
—— 1975c. Possible wider connections of Papuan languages: Papuan and Australian: Greenberg's Indo-Pacific Hypothesis, in Papuan Languages and the New Guinea Linguistic Scene, S. A. Wurm (ed.). Canberra, Pacific Linguistics 38.
Wurm, S. A., D. C. Laycock, C. L. Voorhoeve and T. E. Dutton. 1975. Papuan linguistic prehistory, and past language migrations in the New Guinea area, in Papuan Languages and the New Guinea Linguistic Scene, S. A. Wurm (ed.). Canberra, Pacific Linguistics 38.

3. Human ecology and cultural variation in prehistoric New Guinea

Susan Bulmer

1. Introduction

Humans have been present in New Guinea since late Quaternary times, possibly up to 50 000 years. It is apparent from the complexity of cultures and of ecological adaptation in 19th century New Guinea, the end of prehistory, that the country has had a complex prehistory. This is of interest in its own right, in the adaptation of tropical hunter-collectors to a new and varied environment. Humans were not passive in their relationship with the environment, in the past any more than today (Winslow 1977) and the understanding of processes of change in landforms, vegetation and animal life must take into account the effects of humans, however imperfectly these are known at present.

New discoveries in both archaeological and palaeoenvironmental studies in recent years have expanded our knowledge of the prehistoric past: it is the task of this paper to discuss the archaeological evidence now available in terms of 3 major ecological themes which have come to be central to New Guinea prehistory:

(a) Environmental variation and change and their relationship to human settlement and technology;
(b) The extension of indigenous plant and animal resources through the cultivation and domestication of local and imported plants and animals, and other cultural reflection of this process;
(c) The extension of access to industrial resources through trade, and other related cultural exchange.

2. Non-archaeological background

2.1 Origins and migrations and prehistoric populations

The ultimate or immediate origins of the earliest settlers of New Guinea are not known, but it is assumed that they came from somewhere in the islands to the west or the SE Asian mainland. The earliest colonists may have come as early as 50 000 BP. This estimate is based on the assumption that the earliest dates so far

Monographiae Biologicae, Vol. 42, ed. by J. L. Gressitt.

© *1982, Dr W. Junk Publishers, The Hague.*

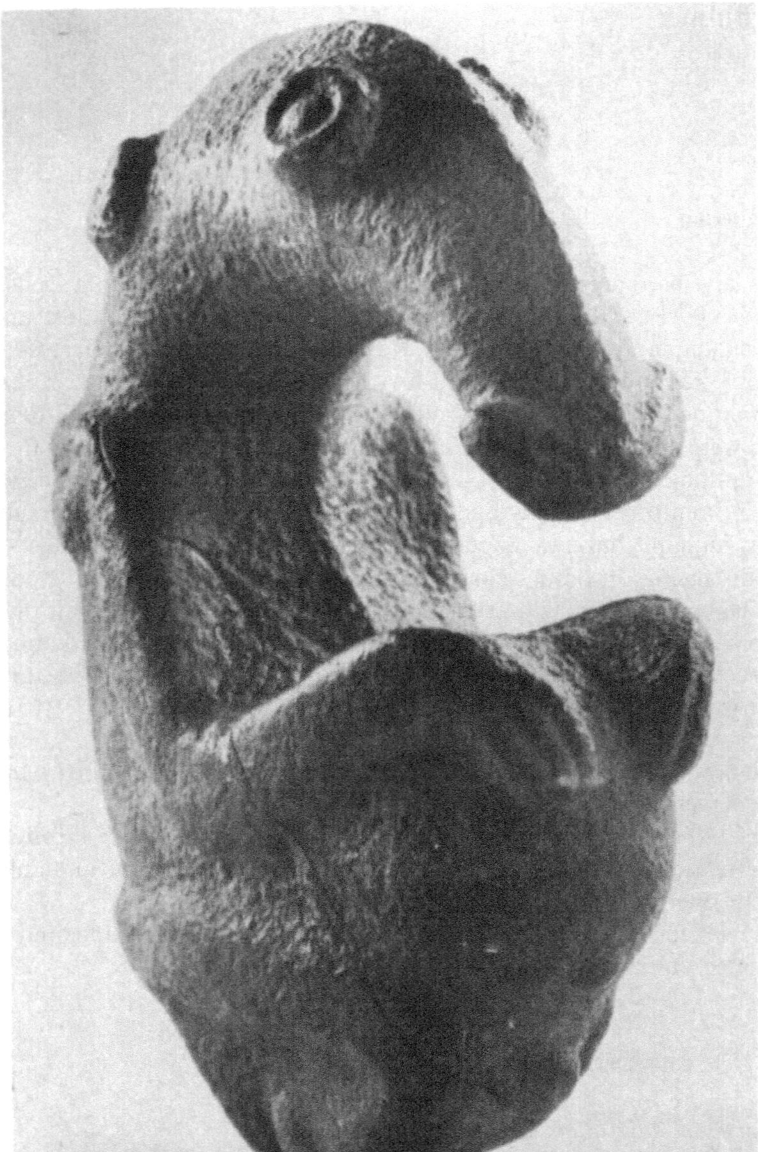

Fig. 1. Zoomorphic stone figure from the Western Highlands Province. Photograph originally published in J. Polynesian Soc. 74(1). (Probably young of *Zaglossus bruijni*, Monotremata.)

in Australia, 40–30 000 BP are not likely to be the earliest found, and that New Guinea would have been settled as early as Australia. The earliest date so far in New Guinea, 26 000 BP, is not likely to be the oldest.

The earliest colonists must have come by boat or raft, for there has never during the period in question been less than 100 km of open water between Sahulland and Sundaland (Calaby 1976:23). Presumably the earliest colonists lived on the coast. Rising sea levels at the end of the Pleistocene might have submerged most of their occupation sites, but areas where the land has risen may have former coastal sites some distance inland. The Aitape site (Hossfeld 1965) on an old beach line now 50 m a.s.l., is dated 5000 BP, but at Collingwood Bay, archaeological deposits of 1000 BP are 30–40 cm below sea level (Egloff 1971:25). Many archaeological sites are obscured by volcanic ash, tropical vegetation or dramatic changes in landform, with landslips and erosion on a major scale. However, further early human habitation remains may be expected to be found.

Whether the earliest immigrants to New Guinea were merely a handful of people or a number of groups may be difficult to establish. Present language distributions (see chapter II:2) indicate a common origin for people now inhabiting most of New Guinea, and it is apparent that there was a very long period of occupation by non-Austronesian language speakers prior to the arrival of the Austronesian speakers perhaps 6000–5000 BP.

While the Austronesian languages are thought to have spread from Indonesia and the Philippines throughout the entire Pacific, the non-Austronesian languages may belong uniquely to New Guinea. Only a few are found in the islands to the west, and Bougainville is the only island to the east with a large number of non-Austronesian languages, although they are found as far east as the SE Solomons.

Other contacts of cultural significance have also occurred. These include accidental as well as deliberate contacts such as trade, and artifacts of unusual character, such as the bronze weapons found at Lake Sentani, and other signs of outside influence such as style changes in pottery, may be attributed to such contacts.

So far there is little direct evidence of the prehistoric people of New Guinea, and little investigation has been focused on prehistoric human osteology as such. Only 2 large archaeological burial groups have been excavated to date, both from sites in the vicinity of Port Moresby and relatively recent, about 700–250 BP (S. Bulmer 1978).

Other than these studies, only the Aitape remains (Hossfeld 1949) and human bone fragments from 4 Highlands sites (White 1972) have so far been described.

The island of New Guinea and adjacent islands supported, at the end of prehistory, about 2 500 000 people, the majority of the population of the Pacific islands. About one third of the people were living in the intermontane valleys of the central mountain chain, and there was a sparsely populated belt between about 750 and 1000 m altitude, which on the basis of oral histories had become depopulated only in the relatively recent past. The remainder of the people occupied coastal, riverine and mountain areas in highly varying densities, and it is clear from oral history that the distribution of the population was changing, and was related not only to natural resources but to cultural factors such as warfare and disease. This change is important in the interpretation of archaeological sequences, as it is apparent that the distribution and densities of the population of the immediate past could be very different from those of earlier times.

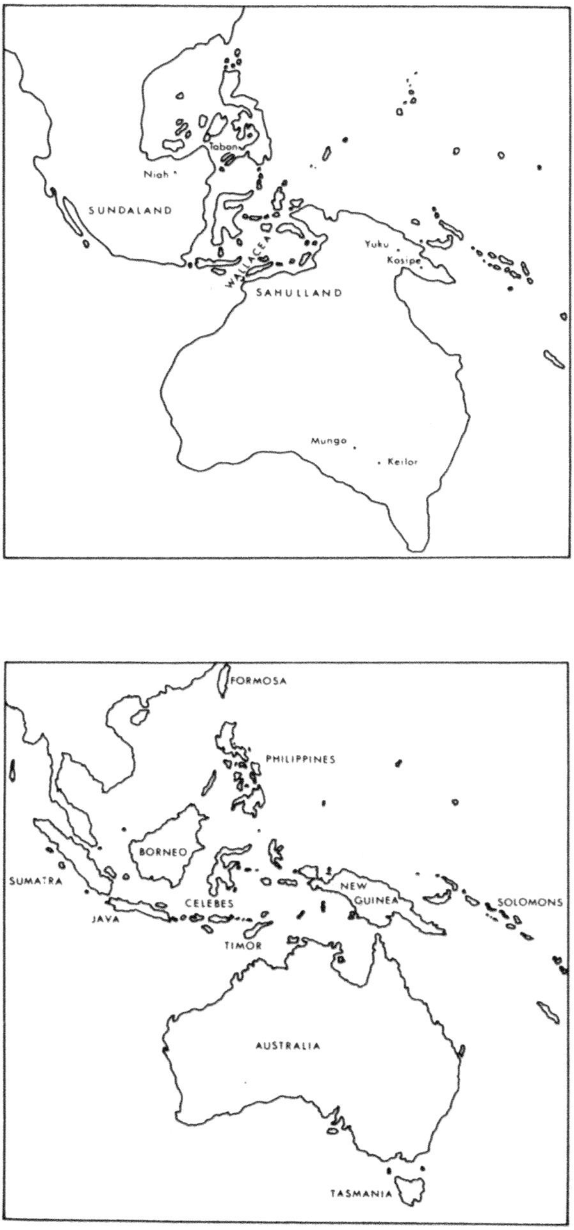

Fig. 2. Sundaland and Sahulland. 50 000 years BP, above and modern sea level, below.

2.2 Environmental variation and change

Variation in past climate in New Guinea is discussed in chapters I:3, III:1 and is only briefly summarised here as framework for interpreting archaeological data. Climate reconstructions have been based on patterns of glacial advance and retreat, changes in altitude of forest vegetation, and variations in composition of vegetation (Hope and Peterson 1975, Galloway *et al.* 1972). The data relate mainly to the Highlands, but some extrapolation based on studies of past climate in Australia and detailed studies of sea level changes assist in reconstructing past environment in the lowlands.

Climate change in New Guinea can be subdivided into 4 major periods relating to the period of human occupation:

2.2.1 The final glacial period From about 50 000 BP, when human populations probably first settled New Guinea, until 15–14 000 BP when the final major glacial retreat began. During the period of greatest glaciation, from about 17–15 000 BP, 2000 km^2 of glaciers covered the highest mountain peaks of New Guinea, with a much larger surrounding area of alpine and subalpine grass and shrub vegetation. At lower altitudes forests were compressed downhill in response to the generally colder and drier climate conditions, and large areas of coastal plains were covered in grasslands and savanna, with the sea level 150 m below its present level at the maximum low. During this and the following period of glacial melting New Guinea was part of a single landmass referred to

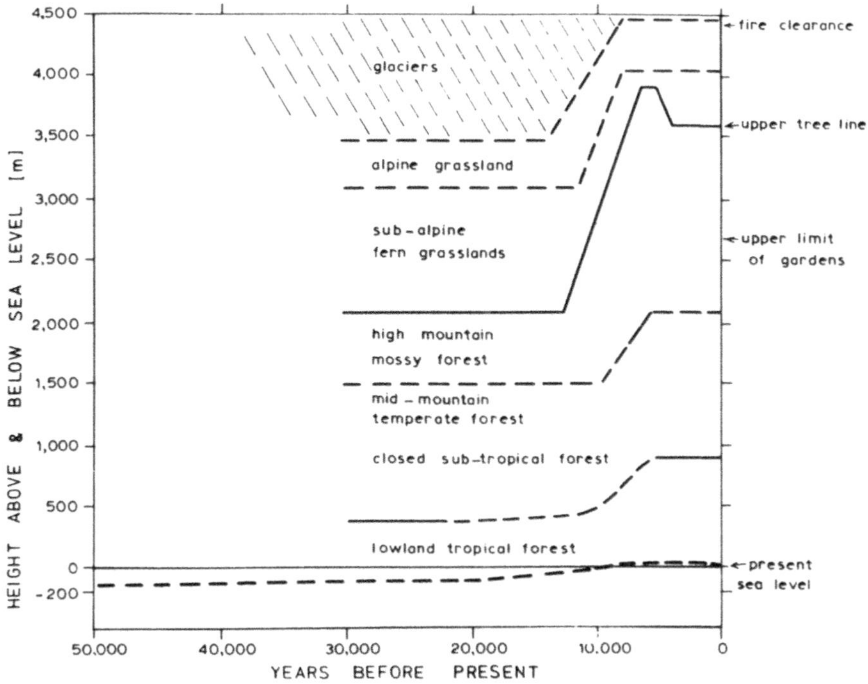

Fig. 3. Chronology of altitude changes in vegetation zones and limits of cultivation.

as "Sahulland," in contrast to the SE Asian "Sundaland" from which it was separated by at least 50–100 km of sea.

2.2.2 The period of glacial melting From 15–14 000 BP, when melting began, until about 9–8000 BP when glaciers had disappeared from the eastern mountains of New Guinea and retreated to above 4300 m in the higher western New Guinea mountains. At about 11 000 BP there was a period of marked increase in rainfall (Galloway 1971:21; S. Bulmer 1975:11), and at about the same period the upper level of the mountain forest began to rise, until it reached about 200 m above its present level at about 4100 m. The mountain forest zones expanded to cover former subalpine areas; and 3 enlarged major mountain human habitats were thus created: the high mountain temperate forest, the mid-mountain subtropical forest and the low mountain tropical forest.

2.2.3 The early Holocene period of warmer, wetter conditions From 9–8000 BP until 5000 BP. The higher upper treeline during this period, about 200 m higher than at present, is thought to indicate warmer wetter climate, a condition that has been found in certain areas of Australia during the same period (Nix and Kalma 1972). The land bridge to Australia was flooded some time between 8000 and 5500 BP and is thought to have affected prevailing wind and rainfall patterns.

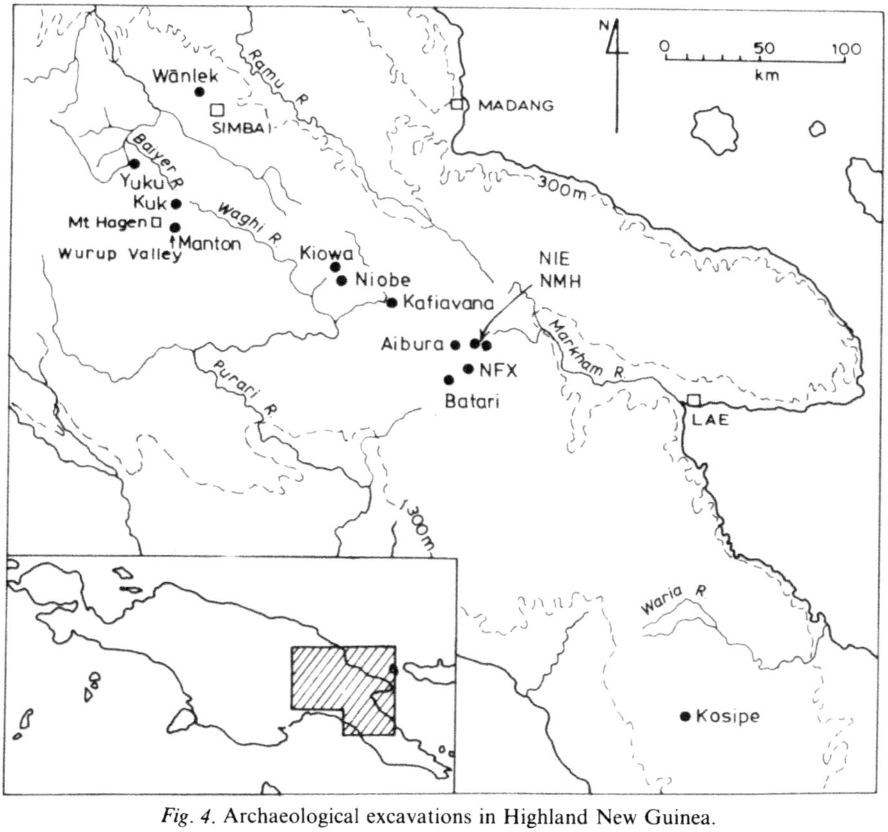

Fig. 4. Archaeological excavations in Highland New Guinea.

2.2.4 The period of modern climate With generally cooler and more fluctuating conditions, began about 5000 BP. Ice readvances in the western mountains, a lowering of the upper tree line, and changes from grassland to tundra conditions on the summit of Mt Wilhelm, are thought to be evidence of a slightly deteriorating climate period. A minor environmental change of major importance to human life was a slightly higher sea level, which seems to have occurred generally in the Pacific, although its interpretation in the New Guinea area is complicated by land movements. The raising of the level of reefs and coral islands related to this sea level greatly expanded the maritime niche in some areas, such as SE Papua.

New Guinea is environmentally complex, in contrast to most islands of the Pacific, and this has contributed to the development of cultural regionalism. Thus different regions must be expected to have different prehistoric sequences, and detailed archaeological data is so far only available for three major areas:

(a) *the northern islands* of New Britain, New Ireland and the Admiralties, together with the mainland adjacent to the Vitiaz Straits;

(b) *the Highlands*, the intermontane valleys of the central cordillera which runs the length of the island, and

(c) *the coastal plains and offshore islands of SE Papua.*

Major areas about which we at present know little archaeologically include all of Irian Jaya, the northern mountains from the west Sepik to the Huon peninsula; the major lowland river basins of the Ramu, Sepik, Markham; and the Papuan gulf and rivers.

3. The archaeological background: prehistoric settlements

3.1 Mountain settlements

Archaeological research in the highlands since 1959 has provided information about a variety of prehistoric occupation sites, including caves, rockshelters, open settlements and gardens, where 27 sites have so far been excavated archaeologically. The sites contain evidence spanning at least 26 000 years, but the data still constitute a patchy prehistoric sequence. Nonetheless they illuminate some topics, in particular the history of agriculture, which are of international importance, as well as bearing testimony to more mundane matters of everyday prehistoric life.

Varied sequences of habitation have been found at the various sites, none including the majority of the total sequence. The two mountain caves so far excavated had relatively brief occupation. Batari cave (White 1972a) had an early date of about 17 000 BP but this was not thought to definitely relate to human occupation. Batari was later occupied between 8200 and 900 BP. Aibura cave (White 1972) was used from about 3800 BP. Both caves contained stone, bone and shell artefacts, and faunal remains that indicated they were hunting camps or occupation sites where wild game was cooked.

Eight rockshelter sites have been excavated and these cover interleaved periods over the past 15 000 years. Their main occupation as hunting camp sites seems to have been in the second and third climate periods, but they have been occupied for various uses since. Yuku (S. Bulmer 1975:29) for example, was first a hunting camp, then a burial site, and finally used for pit cooking. Most caves and rockshelters continue to be used in the highlands, as hunting camps, ritual cooking sites, burial ossuaries, and travellers shelters.

The caves and shelters are spread over some 225 km of the highlands and do not supply coherent regional prehistoric sequences. The only group of rock-shelter excavations in one particular region were investigated by Christensen in 1973 in the Wurup valley, a side valley of the upper Wahgi. There 4 shelters, in a sequence of altitudes from 1770 to 2450 m, supplied evidence of the occupation of the upper limits of cultivation and the domestication of the mountain nut pandanus. Unfortunately, due to Christensen's untimely death, only a preliminary account of this work is available (Christensen 1975).

In spite of their geographical spread and the fact that they were selected for their promising depth of deposit rather than specific archaeological questions, the cave and shelter sites provide a technological and faunal sequence for the central highlands, and some general cultural patterns are now apparent.

The Pleistocene stone industry consisted of a variety of large/heavy tools and of smaller flake tools. Some previous summaries of highlands archaeology have emphasised the pebble choppers and unretouched flake tools, but as the data accumulate the highlands Pleistocene industry looks more and more like the Australian core and flake industry of the same period. In addition to the variety of pebble tools, i.e. waterworn pebbles with unifacial or bifacial edges trimmed on one or more margins, there are large flake tools, "hand axes" (bifacially trimmed tools with one pointed end-edge and usually a convex edge at the other end), axe-adzes of both unwaisted and waisted varieties, and "horsehoof" cores. Undated surface collections also include "keeled" and "disc" cores, in addition to the morphologically undifferentiated stone industry characteristic of E

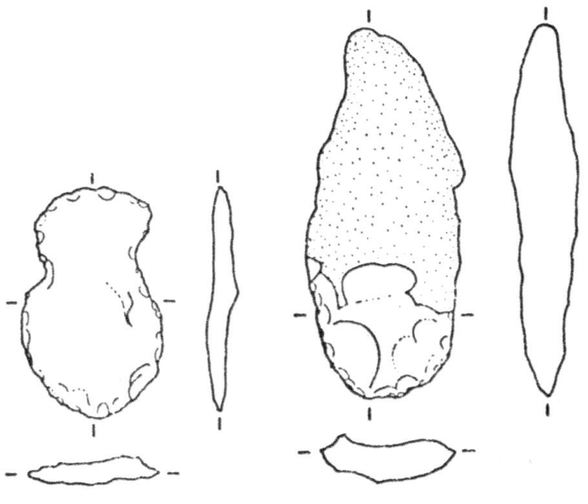

Fig. 5. Pleistocene stone tools from Kosipe. Right. Pebble axe (E5248, Papua New Guinea National Museum and Art Gallery). Left. Waisted axe-adze (E5215, Papua New Guinea National Museum and Art Gallery).

Highlands assemblages (White 1972; Watson and Cole 1977). The affinity of Australian and New Guinea technology in the period when the 2 areas were joined by land and therefore likely to have had to some extent interacting populations is of interest. The "waisted" tools are axe-adzes, that is heavy end-edge implements that were probably hafted in cane hafts. These tools are thought to have been implements used in arboreal hunting (S. Bulmer 1977a), and in their final form they were edge-ground. These tools have been found in 4 of the 5 shelter sites of early date in the highlands and at Kosipe. The edge-ground axe-adze is a tool that may be as old in New Guinea as it is in Australia where it has been found in deposits dated to 22 000 BP (C. White 1967).

Open settlements and garden sites have provided a different range of data than shelters and caves, and in some cases can be directly associated with paleoenvironmental data (see next chapter). The earliest open site, Kosipe, is located at 2000 m in the Papuan mountains, occupied between 26 000 and 9000 BP (White *et al.* 1970). At the edge of the mountain forest of that period and adjacent to a large swamp with *Pandanus* stands, this site is thought to have been a seasonal camp occupied during the collecting of mountain *Pandanus* nuts. Kosipe fits technologically with the earliest shelter sites, containing waisted and unwaisted axe-adzes and some other large tool types, but lacking in smaller flake tools or other evidence of habitation except scattered charcoal.

Two other early open settlements have been excavated, Site NFX near Kainantu, occupied from about 13 000 to 11 500 BP (Watson and Cole 1977), and Wañlek, at which the first period of occupation is dated to between 15 000 and 12 000 BP (S. Bulmer 1977b). The remains at both sites possibly reflect seasonal mountain settlements of some sort.

A second occupation at the Wañlek site, and 3 other open settlements

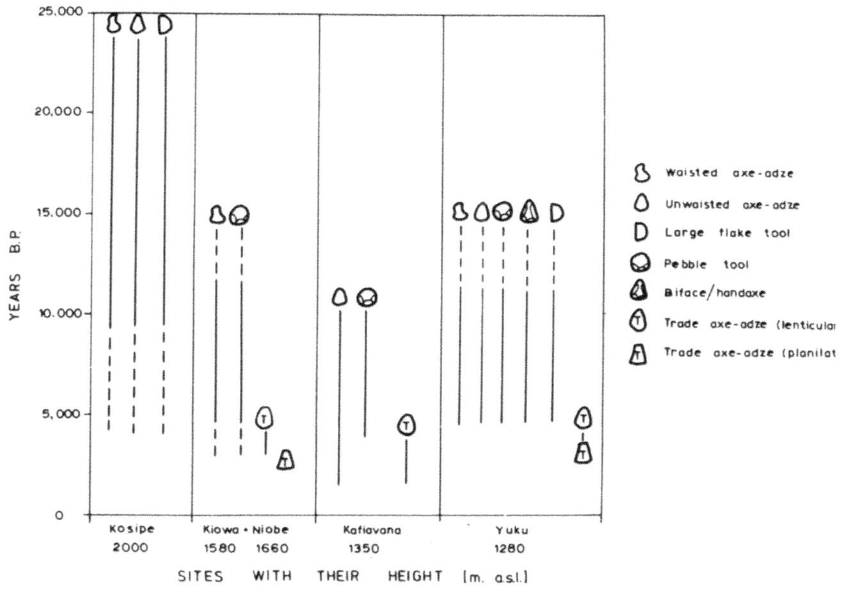

Fig. 6. Estimated chronology of Pleistocene stone tool assemblages and trade axe-adzes.

(Watson and Cole 1977) show new settlement in the highlands during the period 4–3000 BP, and possibly earlier. The Wañlek site was occupied by communities of axe-adze makers and gardeners, possibly also making salt for trade at a nearby salt spring. Containing sherds of trade pottery from the lowlands and tanged slate tools thought to be cultivating or weeding implements, this is obviously a settlement engaged in agriculture and probably also in trade with both highlands and lowlands.

Archaeological excavations have been carried out in 4 cultivation sites in swamps in the upper Wahgi valley (Manton, Kuk, Minjigina, Mugumamp). These show evidence of complex drainage and water control, with the first use of the rich soils of the valley floor for cultivation sometime before 9000 BP, and a series of ditch complexes dating from before 5000 BP until less than 400 BP. These sites are discussed further in the next chapter. Figure 7 shows the plan and posthole pattern of a house of the upper Wahgi area; similar house sites have been excavated at the Kuk site in association with the most recent phase of agricultural use.

It seems that fluctuations may provide an explanation for the apparently

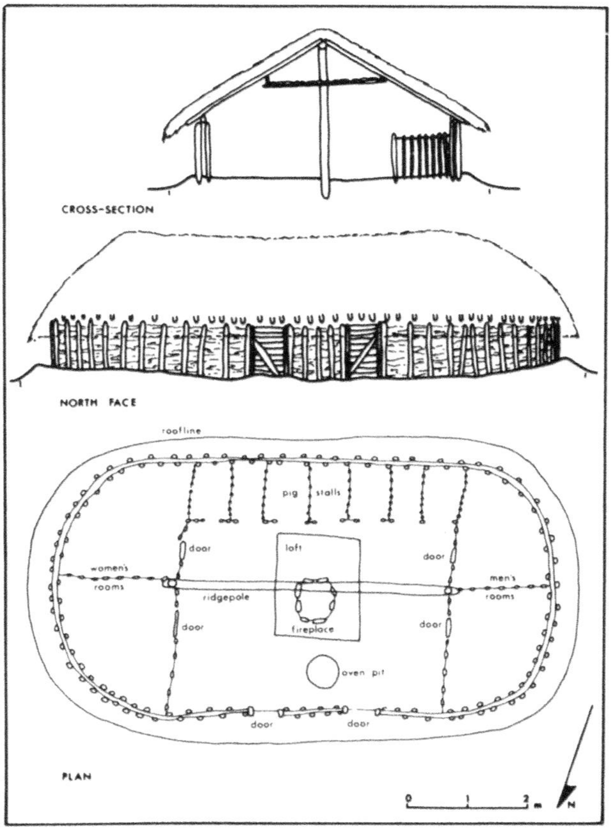

Fig. 7. Plans and posthole patterns of a Western Highlands house (after Axel Stensberg, Oldtidslevn i New Guineas Hverdag, Forum, Kobenhavn, 1973, Figure 28).

interrupted archaeological sequence in the highlands. The reconstruction of temperature changes based on pollen analysis is compared to the dates from the highlands sites in Fig. 8. The highest site, Kosipe, near the upper edge of the forest, was occupied at the same time as a period of higher temperature in the Glacial period, 27–26 000 BP. The early open settlements at Wañlek, NFX and Yuku were first settled when the temperature returned to within 2° of its modern state. Interestingly, the early date at Batari, questioned by the excavator, if the radiocarbon dating is taken at the earlier end of the 2 standard deviation age-range, falls into the only intervening episode of temperature rise, and on this evidence may possibly be related to human occupation.

The coincidence of fluctuations of higher temperature and settlement in the mountains seems likely to be significant. The colder conditions in the final glacial period would certainly have discouraged settlement in mountain areas that were peripheral to the main concentration of occupation even in post-glacial conditions, such as Kosipe. Regular year-round settlement in the highlands, where recent population concentrations were found, may only have begun in climate conditions of the period of glacial melting. J. Hope (Hope and Hope 1976:236) reported investigations at 2 high altitude rock shelters (3996 m, 3450 m) and a pollen site at 3630 m in Irian Jaya which indicated human activity in the area from at least 5500 BP.

Although these general patterns of technological and settlement changes are

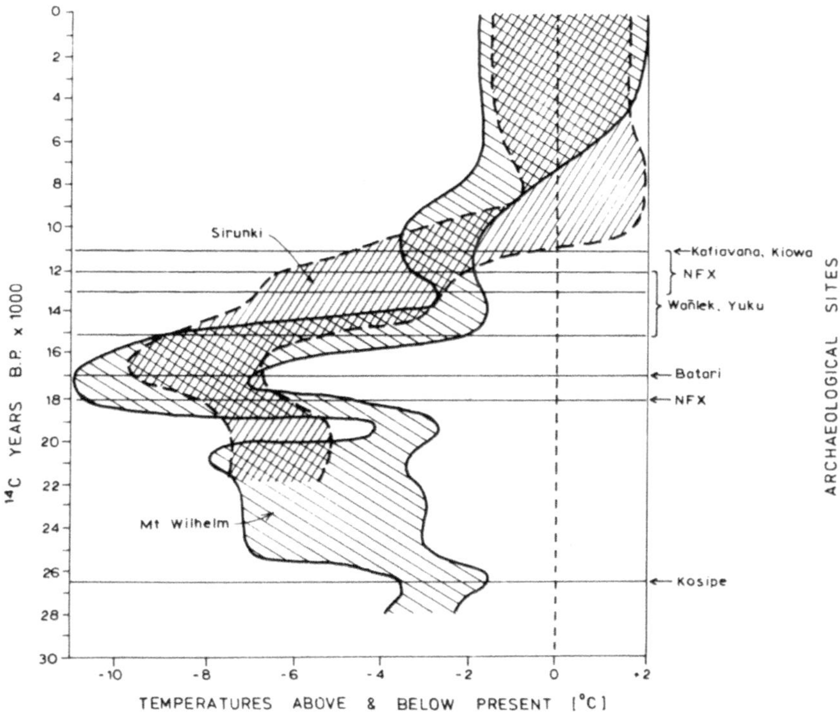

Fig. 8. Temperature variation and radiocarbon dates of early Highlands settlements (after Bowler et al., 1976: Fig. 2).

now becoming apparent there are still many basic questions left unanswered. We know little of the anthropomorphic and zoomorphic stone carvings (Fig. 1), for example. A mortar has been dated by radiocarbon to between 3500 and 3000 BP (Watson and Cole 1977:193), but archaeology has so far added nothing to the speculations about use and significance (e.g. R. Bulmer 1964). The distribution of some of these objects indicates some sort of close connections between the highlands and New Britain and New Ireland (see Fig. 9).

3.2 *Lowlands settlements*

It is more difficult to generalize about lowlands prehistory at this stage of knowledge, largely because archaeological research has been more scattered geographically. Even in the areas where the most concentrated work has been done, i.e. the northern islands and Papua, we are far from having established coherent local sequences. However, an attempt will be made to make some

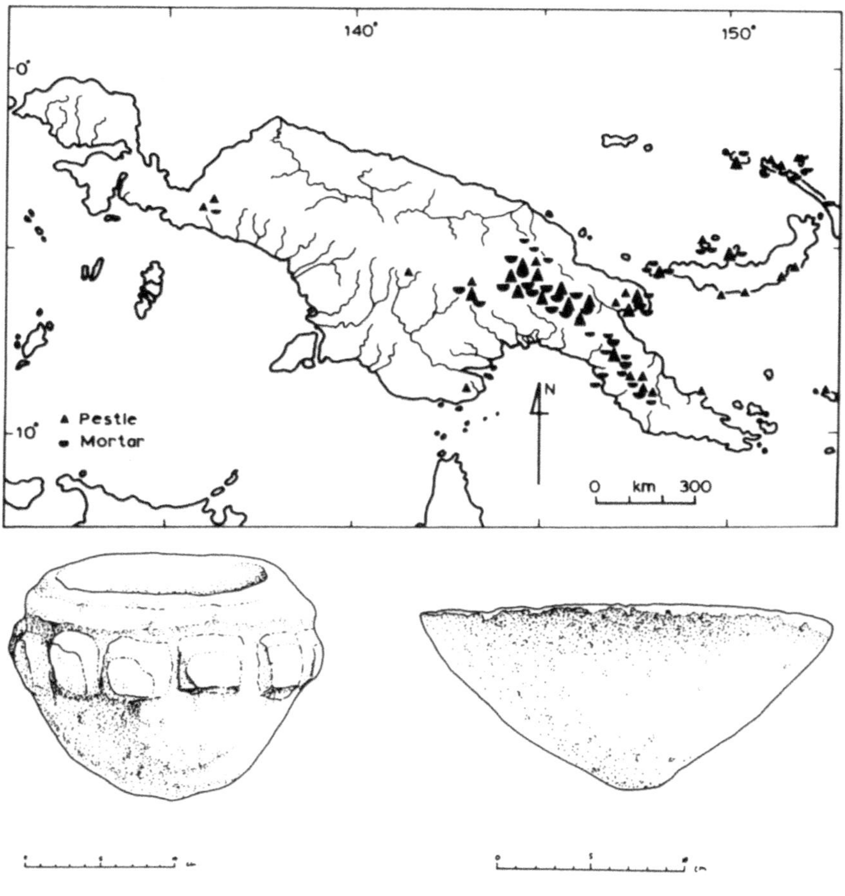

Fig. 9. Map showing the distribution of pestles and mortars in New Guinea, and two stone mortars.

general points. Comprehensive excavation reports are only available so far for 1 site on New Ireland (Downie and White 1978) and 10 sites in coastal Papua (Egloff 1971; Vanderwal 1973; Irwin 1978; S. Bulmer 1978). Preliminary reports have been published for 1 site in E New Britain (Specht 1968), 2 in New Ireland (Clay 1974; White and Specht 1971), 2 on Elouae I. (Egloff 1975), 10 on the Madang coast and the islands of the Vitiaz Straits (Egloff 1975), and 1 near Port Moresby (Allen 1976, 1978), with reports of a variety of investigations of lesser proportions (e.g. S. Bulmer 1971b; Abramson 1969). Most of these sites are open settlements, mainly middens, with only 2 rockshelters (Eriama, near Port Moresby, and Balof, on New Ireland) and 1 cave (Kukuba, near Yule I.). Nearly all pertain to the last 2500 years, with Kukuba cave having a preceramic deposit dating to about 4000 BP and Balof shelter having the oldest lowland archeological deposits so far, a pre-ceramic deposit dating to over 7000 BP. The only possible evidence so far of Pleistocene age occupation has been found in the Passismanua area of SW New Britain, where surface collections over a large area of low altitude limestone plateau are closely similar to those of the highlands Pleistocene industry (S. Bulmer 1977a).

The analysis of pottery sites is proving useful in recognizing and investigating prehistoric lowland settlements. On the mainland of New Guinea there is a widespread style of coil-made pottery, from the Sepik mountains in the west to the Northern Province, which on technological and distributional grounds may indicate relatively early lowland settlement. (S. Bulmer 1971a). This pottery is comparable technologically and stylistically to early pottery of Japan and E Asia, the Japanese pottery being contemporary with the second climate period proposed for this discussion of New Guinea, i.e. between 15 000–14 000 and 9000–8000 BP (Bleed 1978). On grounds of its widespread distribution among only non-Austronesian speaking groups in New Guinea, it is thought to have been present earlier than the arrival of the Austronesian speaking people, 6000–5000 BP. Sherds of this coil-made pottery tradition have been excavated from layers at Wañlek, in the highlands, dated 4000–3000 BP, but the tradition may eventually prove to be even earlier. Whether this tradition reflects a "migration" of new people to New Guinea or the introduction of new technology to existing groups is not yet clear.

One well-established general pattern in the lowlands settlement is the presence of Lapita pottery using groups at scattered localities in New Ireland, New Hanover and New Britain. On grounds of close similarity in pottery style and other cultural items, and the presence of New Guinea obsidian in comparable sites in E Melanesia as far away as New Caledonia, and evidence of trade in a variety of other commodities, it seems that the New Guinea Lapita sites, at Elouae I., Watom, Talasea and Ambitle, are settlements of a widespread culture group of maritime people. Green (1978) is of the opinion that the Lapita tradition may have had its origins in the New Britain area, and could therefore have begun even earlier than the earliest dates for the E Melanesian and Polynesian Lapita sites, dated to around 3500 BP. None of the excavations at Lapita sites in the New Guinea area have yet been reported in detail, but all have been discussed in preliminary reports (Specht 1968b; White and Specht 1971; Egloff 1975).

One of the elusive aspects of lowlands prehistory is the presence of a variety of stone structures, some of which are known to have been used in the immediate past, but others of which appear to be ancient. These have led some to speculate about a "megalithic culture," but it appears rather that many groups built a variety of stone structures and that these will have to be understood in terms of local sequences rather than as a general "culture." Stone structures of known use include stone "tables" and carved boulders said in oral history to have been used in ritual cannibalism (Fig. 11). Rectangular enclosures of upright stones and "pillars" were associated with burials in a number of areas. Upright stones were also arranged in circles as "seats" or backrests for chiefs. Stone carvings and motifs are found on some of the stone structures, and rock engravings and paintings are found at many other sites.

4. Prehistoric subsistence and economy

4.1 Hunting, animal collecting and domestic animals

4.1.1 The highlands

Fossil evidence of late Pleistocene fauna in New Guinea (Plane 1966, 1972) comes mainly from archaeological collections, providing a faunal sequence beginning about 15 000 BP, at the start of the period of glacial melting

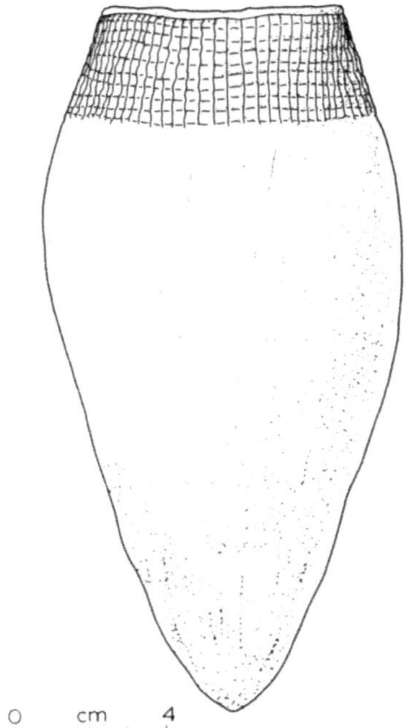

Fig. 10. Pot of the non-Austronesian-speaking potters of the Ramu valley.

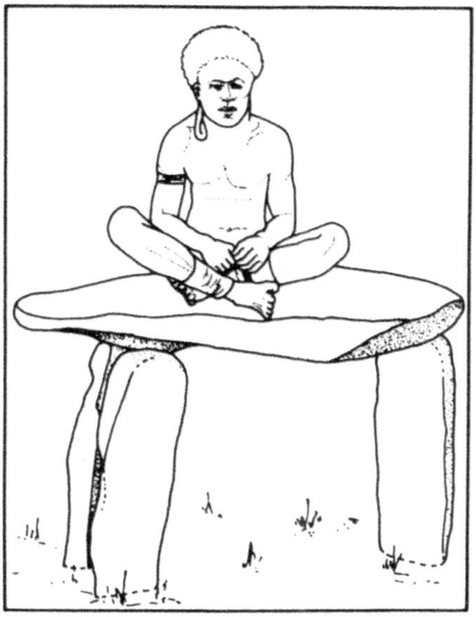

Fig. 11. Stone structure on Unea I., after Parkinson, R. 1907, Dreissig Jahre in der Südsee. Strecker and Schröder, Stuttgart, Tafel 14.

Fig. 12. Designs pecked in stone, Milne Bay Province. (After illustrations in J. Royal Anthropol. Inst. Vol. 61, F. E. Williams, Papuan Petrographs.) Upper left, Plate XV, Fig. 2, Upper right, Plate XV, Fig. 3, Lower left, Pl. XVI, Fig. 2, Lower right, Pl. XVI, Fig. 3.

(S. Bulmer 1979). An earlier deposit of vertebrate fossils, dated to about 33 000 BP, has been found in the Pureni swamp near Tari in the S Highlands (Williams *et al.* 1972) These fossils included a diprotodon (a giant wombat), as well as phalangerids and rodents, which, on the assumption people were already living in New Guinea, were contemporary with humans, although not in this case the remains of hunting.

Unlike SE Asia where there appears to have been no substantial faunal change over the past 40 000 years, New Guinea, like Australia, has experienced considerable change in distribution of species and in the extinction of Pleistocene species. This is not well known as yet, but 2 patterns of change seem to have occurred: the extinction of larger species and the reduction in the number of extant species. Whether or not humans ever hunted the "megafauna" in New Guinea (Hope 1971), it seems likely that the earliest hunters in New Guinea were general small game hunters, such as is the case with their contemporaries in Australia (Bowdler 1977).

It cannot, of course, be assumed that the animal remains at archaeological sites represent the available game in the immediate environs of the site, or even the full range of game actually taken by the residents (R. Bulmer 1976). Hunters travel long distances to obtain game and transport their prey long distances. Wild animals, alive or as meat, are traded from group to group. As well, certain kinds of animals may not be consumed at the normal residential site, or their remains may be disposed of elsewhere. Thus the faunal assemblage at a site must be seen as the end result of a variety of selective processes, based on hunting or trading, with different consumption, use, and disposal of the bone and other animal remains.

There are 6 faunal collections from highlands archaeological sites from which information is now available, but it has to be emphasized that none of these has yet been adequately studied. White (1972), working in the mid-1960s before adequate reference collections of most animal groups were available in

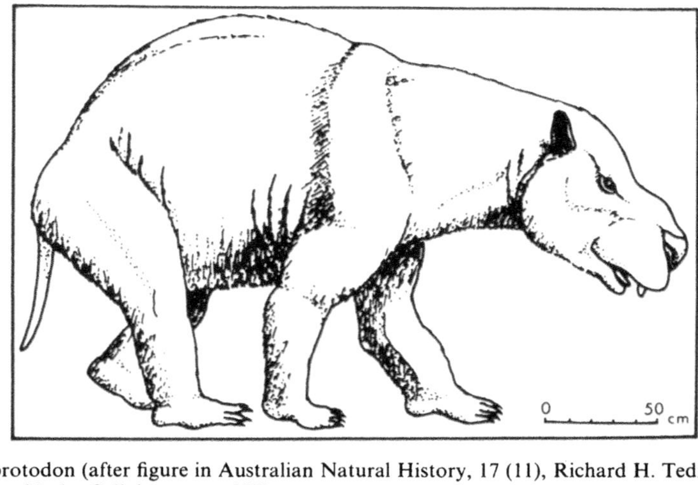

Fig. 13. Diprotodon (after figure in Australian Natural History, 17 (11), Richard H. Tedford, The diprotodons of Lake Callabonna, p. 350).

Australasia, studied only the cranial bones from his sites and was not able to identify the majority of these bones to the species level. A much larger proportion of the cranial bones from Yuku and Kiowa have been identified to species level by G. Fenner and J. I. Menzies, using reference collections at the University of Papua New Guinea Biology Department Museum. However, the post-cranial bones from all these sites remain to be studied, and until this is done a detailed understanding of prehistoric hunting is impossible.

None of the archaeological collections of animal bone can yet be attributed to the glacial period prior to about 15 000 BP. The Hopes (1976) have argued that with more extensive subalpine and alpine zones during the Pleistocene, hunting conditions and animal resources may have been different and more favorable, but there is no evidence yet that Pleistocene people hunted in the subalpine zone or for that matter what animals were present there. The Hopes (1976) described faunal specimens obtained on an expedition to high mountain areas in Irian Jaya. These included some species not previously reported at such high altitudes, including a new species of the wallaby *Thylogale*. However, the presence of these animals at archaeological sites does not demonstrate that they lived at this altitude (3996 m) as they may have been transported by man from lower altitudes. All sites so far indicate that people, during the period of glacial melting lived in the mid-mountain zone and exploited the game of a number of vegetation zones below the treeline.

The period of glacial melting from about 15–14 000 BP to about 9–8000 BP, is reflected in the faunal collections from at least 3 highlands sites, Yuku, Kiowa, Kafiavana and probably Niobe, although this site is as yet undated. The animal bones show that during this climate period considerable faunal change took place. This does not appear as a loss of "giant" fauna, although 2 species extinguished, the fruit bat (*Aproteles bulmerae*) and the wallaby (*Protemnodon* sp.), were rather larger than their modern equivalents, but rather a displacement of species, and possibly a reduction in the number of extant species. (But *A. bulmerae* not extinct; see below.)

The faunal collections indicate that the hunters were probably taking nearly all mammals available, which is to be expected on basis of modern hunting. At Kiowa 43 mammal species, and at Yuku (a much smaller site), 29 mammals were identified (S. Bulmer 1979) which can be compared to the 43 mammal species killed and eaten by the modern people of the Kaironk valley (R. Bulmer and Menzies 1972:488) where all species were taken and all but 2 eaten. At all 6 sites, cuscuses (*Phalanger* spp.) were among the commonest animals identified comprising half or more of all animals other than fruit bats at Yuku, Kiowa and Niobe, and a substantial proportion at the other 3 sites.

Three species of *Phalanger* are so far identified at Yuku and Kiowa: *P. carmelitae* (the commonest throughout); *P. orientalis*, present throughout the sequence at Yuku at 1280 m, but apparently at Kiowa at 1530 m (with less direct access to lower altitude hunting) only from about 9000 BP on; and *P. gymnotis*, which varied considerably in relative proportion to the others. Another as yet tentatively recognized cuscus species is present in the earliest layers at Yuku only, prior to 12 000 BP, and is not comparable to any living species to which it has been so far compared. Because of their importance for

Table 1.

Animal group	Batari n=183*	Aibura n=334	Kafiavana n=67	Niobe n=923	Yuku n=1688	Kiowa n=4639
A. Percentages of animal groups identified on the basis of cranial bone, at 6 highlands sites (White 1972; S. Bulmer 1979)						
Macropod (3 spp)	18	27	27	15	21	7
Cuscus (4 spp)	14	20	22	37	67	21
Ringtail possum (3 spp)	17	5	10	9	3	4
Other phalangers (4 spp)	3	4	0	1	0	1
Bandicoot (5 spp)	6	5	6	4	2	3
Dasyurid (2 spp)	3	1	5	1	1	1
Large fruit bat (2 spp)	2	1	12	32	3	56
Small bat (3 spp)	1	0	0	0	0	2
Giant rat (4 spp)	9	4	8	1	3	4
Other rat (8 spp)	28	34	10	0	1	1
B. Presence or minimum numbers of other animal groups so far identified at 6 highlands sites X=present; 0=not yet identified; Numbers=minimum numbers						
Echidna (1 sp)	0	0	1	0	0	X
Snake, lizard, eel (4 spp)	5	X	5	2	X	X
Brush turkey (egg)	?	0	0	?	X	X
Cassowary (egg or bone)	X	X	X	X	X	X
Pig	2	2	6	2	X	X
Dog	0	1	0	0	X	1
Chicken	?	X	0	0	0	0
Fish	0	0	X	X	0	0
Bird	X	X	X	X	0	1
Tasmanian wolf	0	0	0	X	0	1

*White established minimum numbers of animals by dividing cranial bones into left and right mandibles and maxillae, taking the commonest category total as minimum number for the relevant level or horizon. The total numbers for White's 4 sites shown here are much smaller than the minimum numbers of animals in the excavated collections because, lacking adequate reference collections, White was not able to identify the bones to genus or species level.

The Yuku and Kiowa percentages are calculated here on the basis of total numbers of cranial bone identifications. The vast majority of these are mandibles. Minimum numbers have not yet been calculated as the study of these collections is still in progress. However, the relative proportions of different animal groups indicates the wide range of animals taken and the lack of specialized hunting, except for the large proportions of large fruit bats taken at caves and the higher proportion of rats in the E Highlands sites.

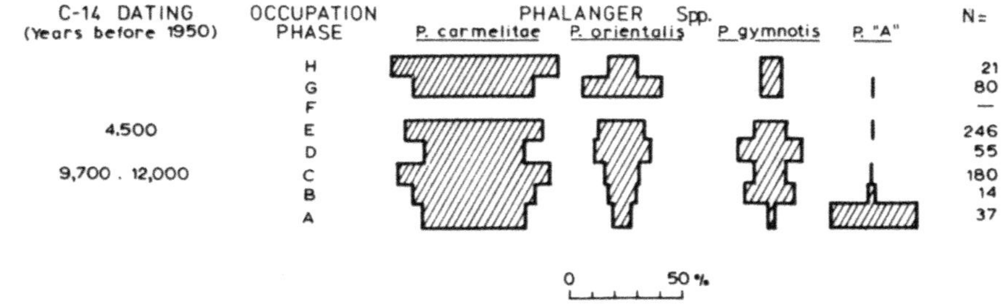

Fig. 14. Percentages of cranial bone identifications of *Phalanger* spp. at Yuku rockshelter, by occupation phase (n=633).

fur, as well as being the largest common game, and the ease with which they can be taken asleep in their nests, it is not surprising that phalangers were always important in prehistoric highlands hunting.

Other animals hunted included the full range of modern highlands species of bandicoot, striped possums, ringtail possums, wallabies, and rats, as well as *Zaglossus*, echidna, cassowary, brush turkey, eels and lizards. Interestingly no small bird remains or frogs were found, perhaps indicating not that they were not eaten but that they were either totally consumed, or eaten elsewhere.

Large fruit bats were the commonest game at Kiowa, where they comprised the majority of remains in all but the earliest layer, the bones in which were attributed to the regurgitation of an owl. In the earliest layers at Kiowa a previously unknown species, *Aproteles bulmerae* (Menzies 1977), was dominant, although the modern *Dobsonia moluccensis* comprised about $\frac{1}{4}$ the large bats. In the layers dated to about 9000 BP the *Aproteles* had been replaced by the

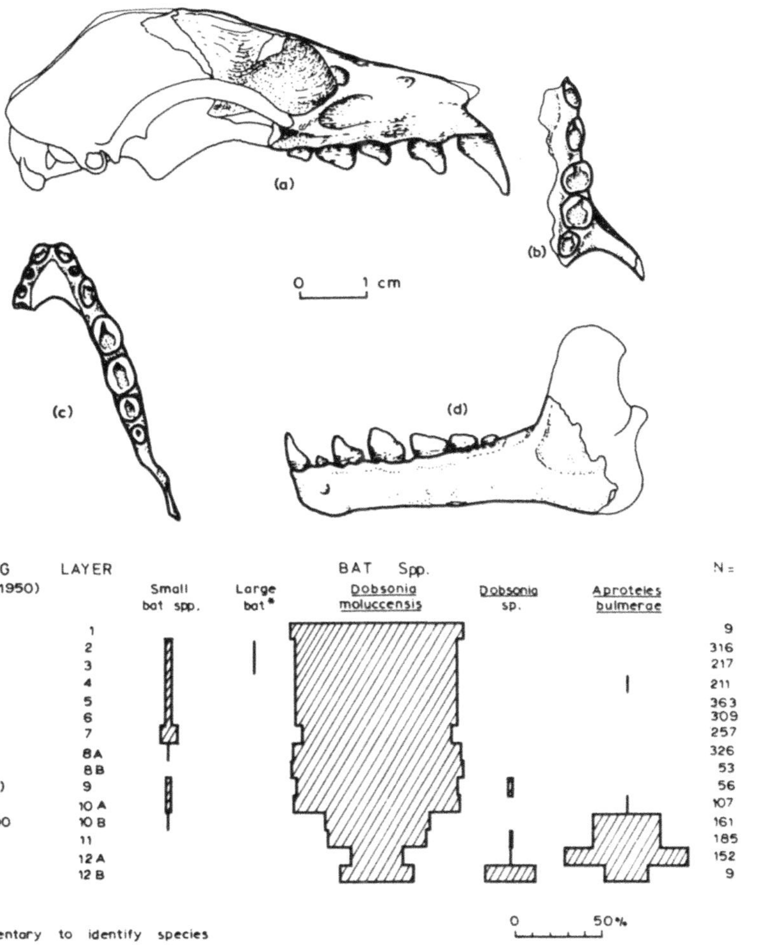

Fig. 15. Aproteles bulmerae. a. Reconstruction after Menzies, 1977. b. Percentage distribution of bat spp. at Kiowa by layer.

Dobsonia. Note: The *Aproteles* has recently been found to be still living in the Star Mountains (Hyndman and Menzies 1980).

Other animals apparently extinguished in the same period include the Tasmanian Wolf, *Thylacinus*, only one mandible of which has so far been described (Van Deusen 1963), but which is also present at the Niobe site (J. Hope, pers. comm.), and the wallaby *Protemnodon* also at Niobe (J. Hope, pers. comm.) but not yet recognized at other sites, other than in a pre-human layer at Kafiavana (Plane 1966).

Other evidence of faunal change comes from the rodents at Kiowa, the only archaeological rat collection so far studied in detail. This is essentially modern in composition, except in the deepest layer (J. I. Menzies, pers. comm. 1976), dated to about 11 000 BP. There is a small rat, probably an arboreal *Melomys* and a large rat, neither of which is an extant species, but which are too fragmentary to describe in publication. There is also a surprisingly large number of the large rat *Macruromys* remains, an animal which is now very rare in the area.

Yuku and Kiowa contain evidence of the early advent of pig in New Guinea. A single incisor has been found at each site, one in Layer 5A at Yuku (dated c. 10 000–12 000 BP, and one in Layer 12A at Kiowa (dated to 10 350 BP). Until post-cranial bones from these and other sites are studied, the significance of these finds is not clear. If the pig was self-introduced to New Guinea (it is a strong swimmer, S. Bulmer 1975:18), or brought as a wild animal by human agency, there is little objection to its reaching the highlands by this period. If they are the only pig remains at these sites, the teeth could have found their way to the highlands as ornaments, and therefore need not indicate the presence of pig in the highlands.

On the other hand pig could have accompanied agriculture into the Highlands, either as a domestic animal or as a wild bush-fallow volunteer. A close study of the archaeological bone collections may result in clues in respect to these questions. At present pig thrives in a feral state in the upper mountain forest (J. I. Menzies, pers. comm. 1976, reported abundant evidence of wild pig on Mt Albert Edward up to 3600 m). There is no reason for pig, once having arrived in New Guinea, not to have found its own way into the mid-mountain forests. The general presence of pig remains in highland sites at a later period

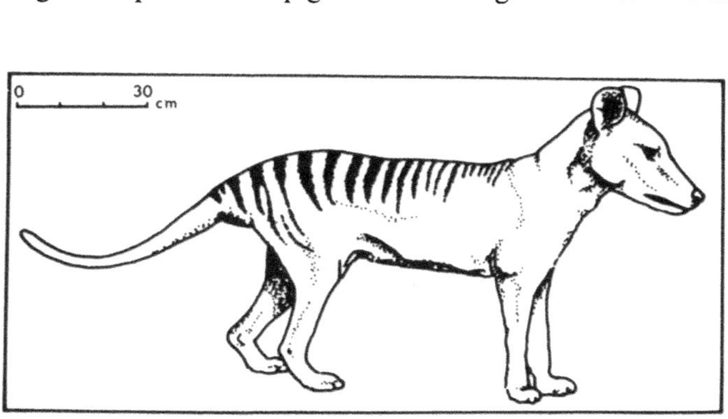

Fig. 16. Thylacine ("marsupial wolf" or "tiger").

(S. Bulmer 1966, White 1972), c. 6000–4500 BP, and more recently could indicate a change in site use. In any case, the quantity of pig bone in the site is small, and therefore its absence from other deposits inconclusive.

There is no indication as yet of dog or chicken, the other 2 introduced domestic animals, in highlands sites. It has been argued that both may have arrived in New Guinea relatively recently. However, considering the antiquity of domestic dog elsewhere in the world (over 9000 BP in America, Lawrence 1967), and the lack of detailed study of the highlands post-cranial bone collections, such speculations seem premature. The same can be said for lack of evidence of chicken, in the absence of detailed study of bird bone in highlands archaeological collections.

4.1.2 The lowlands: Northern islands and mainland

Balof shelter, 1.5 km inland, 60–75 m, contained the earliest evidence of lowland prehistoric hunting and fishing. In all levels shells from marine species comprised the bulk of faunal remains at this site. Bones from humans, pigs, and a variety of wild animals were present in the occupation layers. Pig is present in Horizons I–IV only, but other animal remains were spread throughout the layers, in contrast to pig; crabs, sea urchins, *Thylogale* wallabies, the cuscus *Phalanger orientalis*, skinks, and other reptiles, and large fruit bats were in all layers. There is only a single individual of *Macropus* (large grassland wallaby) or *Dendrolagus* (tree kangaroo) in levels II–III, in contrast to *Thylogale*, the little forest wallaby, in all layers. However, this is of interest in that neither of the large macropods now occurs anywhere but on mainland New Guinea. Land snail shells throughout the deposit are interpreted to be a sign of intermittent occupation.

Two middens, at Lesu and Pinikindu, the former dated to 2500–1500 BP, and the latter thought to be of similar age on grounds of pottery styles, contrast with the Balof shelter animal remains. The Lesu site consisted of shell middens 20–80 m from the present beach and resting on former beach level. This material has not yet been reported in detail, but a preliminary account reported a large number of shells, a few fish bones and some pig bones. Pinikindu test excavation was also of a beachside midden. It contained pigbone, and a few fragments of human skull, taken to be from a burial as the site is in a men's house enclosure. Land snails were found at this site and as at Balof taken to indicate discontinuity of settlement.

Shellfish at Pinikindu were all common reef species, which are not now eaten due to the small amount of meat, but were traditionally eaten. *Trochus niloticus* and *Tridacna* were species thought to be present not only as food but also as industrially important in the manufacture of tools and ornaments.

Test excavations at a mainland midden near Madang (Egloff 1975:34) showed an occupation associated with paddle-decorated pottery ancestral to the modern Bilbil and Yabob ware. The middens contain marine and lacustrine shells and sparse bone from food refuse, being mostly pig and wallaby (*Thylogale bruijni*).

A site on Eloạue I., to the north of New Ireland, is an extensive shell and bone midden 250 m from the beach, which on grounds of its Lapita pottery probably dates to the period 3500–2500 BP (Egloff 1975). The bone identified included human, wallaby, turtle, fish, pig, large bird, crocodile, dugong, dog, cuscus. This is consistent with animal remains at other Lapita sites, indicating a heavy reliance on sea animals. The presence of pig is consistent with the finding of domestic pig at Lapita sites in the SE Solomons (Green 1976, 1978), but may, in the Eloaue case, indicate either domestic pigs or hunting of feral pigs. Although it has been argued that the Lapita people did not keep pig, this is not supported by the archaeological evidence.

The wallaby identified as *Thylogale* is interesting in light of the fact that this animal is not now present in the St Matthias group, and may have been brought by people from New Ireland, where *Thylogale* bones were found in all levels of Balof shelter, dating from over 7000 to less than 2500 BP.

The other Lapita site so far tested in the NE islands is the Watom site, but the faunal collections have not been reported in detail. Marine shell was present, and the commonest animal was pig, with fish next common. Two bird bones were identified, including a chicken bone from an early level, presumably associated with the Lapita pottery, and a swamp hen (*Porphyrio*) bone from a late level.

4.1.3 The lowlands: SE Papua

The more extensive research and detailed reports available for sites in coastal SE Papua indicate the general use of a wide range of animal resources, but also indicate some change and variation in their use.

The Rainu site, a midden dated to c. 1050–500 BP, provides the most detailed midden study so far from New Guinea. Three different midden mounds were excavated, the fauna in one of which was mainly pig, but another contained a larger proportion of shell. Egloff (1971) suggests that this may have to do with the social context of the mounds and disposal of refuse, possibly indicating variations between groups in hunting and collecting practices.

Among the animal remains at Rainu, the large grassland wallaby (*Macropus agilis*) and pig were commonest. This is consistent with descriptions of traditional hunting practices of the Wanigela people who had large communal hunts using nets and spears to despatch large numbers of pigs and wallabies. Fire is frequently used and occasionally bandicoots are flushed from the burning grass. Small mammal and bird and fish remains were rare in the site, which reflects the paucity of mammal species on the Collingwood Bay coast, but could also reflect scavenging dogs, as the few coprolites found at the Rainu site contained many small fishbones. However, the historic Wanigela people did not gain very much of their sustenance from the sea, although the presence of shell net weights indicates there was fishing.

The absence of cassowary and crocodile bones from the Rainu site was commented on, as they are important game in historic times. Bats are also common and frequently eaten, but not present in excavations. Dog was represented in all levels of occupation, with both bone and bored canine ornaments.

The relative proportion of different species of shellfish in the Rainu mounds shows an early heavy dependence on marine and mangrove species. Later there was an increase in the proportion of riverine species used, and finally a decrease in the riverine shellfish and an increase in mangrove species. Egloff suggests this may be related to local environmental change, particularly the gradual subsidence of the land that has occurred, and changes in the river course.

The two excavations at Mailu on the S coast of Papua, in contrast, showed no radical change in the exploitation of shellfish over the course of the past 2000 years (Irwin 1977). The mainland site (Selai) had much larger proportions of mangrove swamp and intertidal sandy and muddy shore shellfish than the site on the island. The only possible change registered is a decline in some of the heavily exploited reef species and a broadening of the range of ecological zones used for shellfish collecting.

The vertebrate remains indicate a very wide variety of hunting and the keeping of pigs and dogs throughout the sequence. Wallaby (both grassland *Macropus* and forest *Dorcopsis*) are only present in the 1st millenium A.D. but are still hunted, so must be absent due to sampling error. This is thought to be the case with the remaining list of vertebrates too; for example, only one bat bone (*Pteropus*) was found and one eagle claw. Snake and goanna were present, as well as *Rattus ruber* (on the island), and the giant rat *Uromys anak* at the Selai site.

Notable is the evidence for dependence on sea animals; turtle, dugong, fish (7 genera), sea urchins, and crabs are present throughout the sequence. There were no bandicoots, arboreal marsupials or crocodiles, but otherwise the fauna is consistent with the hunting practices of the historic period, and there is no positive sign of change in animal exploitation. Shell net sinkers throughout the sequence indicate long continuity of historic net fishing.

The Port Moresby area has pottery-using communities that have made use of the grasslands and lowland forests for hunting a wide range of animals, and have kept pigs and dogs, for over 2000 years (S. Bulmer 1978). Two coastal sites show the exploitation of a wide range of shellfish, which is not surprising as the shellfish beds along this coast are very rich. 50 spp. of rocky shore and reef shellfish, but only 4 species of muddy shore origins, were used by the coastal villages. They ate all these species and in addition used at least 8 species for manufacturing ornaments and implements. The 3 inland sites also made use of marine shellfish as food at all 3 sites, and at Nebira 4 (Allen 1972) also probably in the manufacturing of ornaments from *Chama* sp shell. The residents furthest inland, at Eriama, ate a much higher proportion of mangrove and muddy shore species. They may have got these through trade, as in the modern situation, or had their own access to shellfish beds not used by coastal communities.

The inland communities had access to forest hunting territory as well as grassland. At least 13 spp of vertebrate have been identified, including 2 spp. of bandicoot, a cuscus (*Phalanger*), a large rat (probably *Uromys* sp.) a large bat, a small and a medium sized bird, 2 species of wallaby, dugong, turtle, and goat. Goat bones from one individual were used as burial goods at Eriama and a single tooth at Nebira was presumably also there for the same purpose. This goat could have come from Indonesia through trade or from indirect contact with European voyagers, as a burial with which it is associated is dated to probably 1623 A.D. ±70 (S. Bulmer 1978:213).

There is little fishbone in the mainland sites, but the good condition of the tiny scraps in the Taurama site suggest that the bones were being scavenged by dogs or pigs. At the riverside site at Nebira fish was one of the commonest animals present (with pig and *Macropus* wallaby), consisting mainly of catfish, which would have come from the nearby river, and 3 reef species, which could have been taken on trips by boat to the sea, which may have been closer to the site during that period (S. Bulmer 1978:12–15). Net weights at Taurama and Nebira 4 indicate net fishing in prehistoric times, similar to modern fishing. Only 3 net weights were found on other inland sites, which may indicate grave goods rather than fishing nets, or reflect the relatively higher value of this item to inland people.

Only one Port Moresby area site has been found to contain possible evidence of specialization in hunting and fishing, i.e. Motupore, which Allen (1976) suggested was a settlement of maritime trading specialists, obtaining wallabies through trade with mainland groups. The majority of animal bones at this site was fish, almost entirely shallow reef and harbor spp., and aside from one rodent, a lizard, and a relatively small amount of pigbone, the rest was *Macropus*. However, it should be remembered that Egloff found varying proportions of different animals in different middens at the same village site at Rainu. As well, there is little direct evidence of trade, as such, at Motupore, other than stone axe-adzes of inland origin, and it could equally be argued that they had gardens on the mainland and hunted for themselves. The predominance of grassland wallaby may indicate extensive forest clearance on the mainland.

The remaining SE Papuan area for which archaeological faunal data are available is the vicinity of Yule I., where 3 excavations (Vanderwal 1973) have shown variation and change in faunal exploitation. There was no fauna in Kukuba cave, but at Oposisi, a hill top site occupied c. 2000–1000 BP, pig was by far the most important food source, followed by wallaby, turtle and fish. Shellfish was also used in large quantities, and dugong, reptiles and cassowary were also eaten. Dog is known only from bored canine ornaments, so may not have been eaten or even have been kept by the occupants.

The faunal remains on the Abe site 4 km inland on the mainland north of Yule I. show a rather different emphasis of animal exploitation. There is a relatively smaller amount of turtle and fish and a relatively greater amount of wallaby. There is still a considerable use of marine resources, however. This site dates to about 1600 BP.

Another site, Urourina a beach site on Yule Island, shows its occupants about 700 BP were largely dependent upon marine resources, particularly turtle, fish and dugong. They also ate cassowary and dog, the latter represent not only by bored canines, but also charred post cranial bone.

There are a number of general patterns emerging in the faunal record from lowland sites. First is the persistence of the importance of broad spectrum hunting over the entire period. The prehistoric people of lowland New Guinea were at least for 7000 years wide-ranging small game hunters, as were prehistoric Highlanders. They made use of a wide range of shellfish species for food, and the shells of some also served as raw material for manufacture of ornaments and tools. Another general pattern is the relative importance of the pig over the past 4000 years, including the first evidence of the presence of the pig in a site that probably dates to 4000 BP.

There is ample evidence for the continuity of net-fishing, and no sign of fishhooks so far in the archaeological sites, although the bone bi-points at Balof could possibly have been fish gorges (Downie and White 1978:779). Everywhere the *Anadara*, or other equivalent bivalve, net weight was and is used. On the other hand, the ethnographic record of 19th and 20th century New Guinea indicates a variety of fishing gear, much of which would only rarely survive in archaeological context. These include fish spears and arrows, harpoons, basket traps, hand nets with wood or bamboo frame, casting nets, fish poisoning. A variety of fish hooks are known to have been used in the past century, so it is clear the archaeological record is so far inadequate (Cranstone 1961:77–82).

4.2 Gathering of plants

The wide range of plants utilised in prehistoric New Guinea have been discussed by Powell (1976 and next chapter), so need not be reviewed again here. The plant remains so far found in highlands archaeological sites are very restricted, with the exception of the Wahgi agricultural sites and the rockshelters of the Wurup valley (Christensen 1975). A few plant remains have been found at Yuku (*Pandanus* nuts, bamboo and sugar cane), Niobe (probably *Planchonella* seed, and Palmaceae), pollen in Horizon I at Kafiavana (Myrtaceae, *Gamneae*, *Casuarina*, and *Wahlenbergia*), Aibura (bamboo, and a rainforest tree, probably *Cryptocarya*), and *Lagenaria* was reported to have been found at the Manton site. The details of these have already been reported in S. Bulmer 1975. Plant remains in lowlands sites are even more restricted, having been found only in a swamp deposit at Watom (Specht 1968b:126) and a rockshelter (S. Bulmer 1978).

4.2.1 Upper mountain. The nuts of *Pandanus julianetti* and *P. brosimos* (chapter III:7) are still important and highly valued food items, and archaeological evidence suggests that this has been so since the Pleistocene. The one site that is definitely dated to Period I is Kosipe (AER), an open settlement in the mountains of Papua at about 2000 m. This site would have been at the edge of the mountain forest and subalpine shrubland, and so is likely to have been visited either to hunt in the alpine zone or possibly to gather mountain *Pandanus* nuts. Located on a flat-topped ridge extending out into a 28 km^2 swamp which today contains *Pandanus* stands, it has been interpreted as a camp occupied during nut-collecting. The evidence of the settlement there consists only of stone artifacts and charcoal, with no organic or structural remains to indicate site use.

Pandanus nut collecting is also in evidence at the Yuku site, where nuts were present in layers dated to 12 000 BP or earlier (S. Bulmer 1975:31), and at the Manim site (Christensen 1975:31), in layers dated to earlier than 9000 BP. The nut-bearing mountain *Pandanus* species occur naturally in the mountains above about 1500 m, and would have grown lower during the glacial period. This resource could have been a motive for settlement in the intermontane valleys, particularly in Period 2 when the stands would have moved uphill with the general pattern of forest movement.

4.2.2 Mid-mountain. Although early New Guinea hunters and collectors would have exploited the resources of the upper mountain *Nothofagus* and mixed broadleaf forest up to 2000 m and of the alpine grassland between about 2000 and 3500 m, they probably dwelt mainly below this, making periodic visits to the upper mountain zones. The lower altitude zones would have been more favorable for living in terms of general comfort, but it is also possible that the mid-mountain forest contained a wider range and greater density of plant as well as animal resources. Unfortunately so much of the mid- and lower-mountain forest has been ecologically transformed by the activities of more recent agriculturalists that its former condition can only be extrapolated. However, it is apparent that there is a marked contrast between the biogeographic character of the forest between about 1200 and 1700 m and that above; the former would have provided a rich ecological niche for hunter-collectors, while there are few resources in the latter. The mid-mountain forest contains a variety of trees that bear edible nuts or fruits, including *Elaeocarpus*, *Sloanea*, *Finschia* and *Sterculia* species, and there are many other sources of plant food, including vines with edible fruits, wild edible tubers (*Dioscorea* spp.), many trees, shrubs and ferns with edible foliage and many kinds of edible fungi. The acorns of the "true" oak (*Quercus* s.l.) could have been used as a staple food if techniques of processing were known.

4.2.3 Lowlands. There are as yet no early lowlands or coastal sites discovered, but it can be assumed that people did live in these areas as they did in Australia and SE Asia. The extensive savanna woodland and swamp zones brought about on the S side of New Guinea by lower sea levels were probably inhabited by hunters and collectors similar to contemporary people in Australia, but people on the N side would have been more dependent on the tropical forest, with more abrupt coastlines. A wide range of lowland and lower mountain plant resources would have been available, including sago (up to about 300 m at present, but probably strictly coastal during Period 1), and a wide variety of nuts and fruits, including breadfruit and coconut.

It has been suggested (Glover 1972:158) that early hunter gatherers in SE Asia may have been more dependent upon riverine and sea resources than upon the forest itself, but this interpretation is based on the conspicuous shell middens of coastal sites. Gorman (1970) found that 15 of 18 Hoabinhian sites so far investigated were upland in location and non-maritime in orientation. On the other hand, this preponderance of non-maritime sites might be due to the drowning of early coastal sites through rising sea levels. Gorman found that the 3 maritime settlements also exploited forest resources, while in contrast the settlements located in the inland hills did not utilise coastal resources. He concluded that SE Asian hoabhinian hunter-collectors were "broad-spectrum" exploiters of several environmental zones.

The plant remains from Spirits Cave (Yen 1977) show that these SE Asian contemporaries of early New Guinea lowland hunter-collectors used a similar range of nut and fruit bearing treees, bamboo, and other plant products during the period 11 000 to 7500 BP. The firmly identified ones that bear edible seeds, include *Aleurites*, *Areca*, *Canarium*, *Madhuka*, *Terminalia*, *Castanopsis*, *Prunus*, *Celtis*, *Calamus*, and *Mangifera*. Other plants such as *Piper*, with a variety of uses in modern times are also present. This is the appropriate period in New Guinea to recognise signs of domestication of plants, and indeed these plants at Spirits cave are all ones that are later cultivated both there and in New Guinea. However, this is not evidence of cultivation as such as these could equally have been collected from uncultivated trees and plants, but it is a sign of continuity and similarity of plant use in the 2 areas.

4.3 Plant cultivation

The prehistory of New Guinea plant cultivation and domestication is discussed in detail by Powell (next chapter). Aside from the pollen and soils, the most notable highlands evidence of agriculture comes from excavations of the actual cultivations, primarily wet-land drained sites, and the recovery of artifacts directly associated with agriculture.

It has long been pointed out that a large proportion of New Guinea's grasslands are directly related to agriculture and have their origins in forest clearance for cultivation (Robbins 1963; Eden 1974). However, until recent years there has been little idea how old this forest clearance might be. The most recent archaeological information (Golson 1977b) indicates that plant cultivation in the highlands had its origins in the period of glacial melting. By 9000 BP, at the Kuk site in the Wahgi valley, patterns of soil deposition and disturbance indicate that forest clearance and cultivation had already begun (Golson 1977b:612), and by above 5000 BP forest clearance was widespread in the highlands.

It is not the case, however, as alleged by Bellwood (1917:64) that the characteristic implement of the late Pleistocene assemblage, the "waisted blade" is an implement used in forest clearance, and therefore could be an indication of the earliest horticulture in the world. This implement is likely to have been used primarily in arboreal hunting and collecting (S. Bulmer 1977a). The general appearance of edgeground and quarried fine-quality axe-adzes in archaeological deposits dated to 6000–4500 BP in various highlands sites, and the application during the same period of grinding to the waisted blades at Yuku, are more likely to be indications of forest clearing on a large scale. Although the earliest agricultural evidence is much earlier than this, it need not have involved extensive forest clearance and the waisted blade used 26 000 BP at Kosipe takes this tool from the realm of possible agricultural association.

There are 4 archaeological excavations of cultivation sites that have contributed data about the history of highlands agriculture, but much of this so far available only in preliminary summaries (see Powell, next chapter). Manton, near Mt Hagen, was the first open agricultural site investigated. The only possible food plant remains so far found is from this site, *Lagenaria*, not in the edible, but the gourd stage of growth, dated to about 2500 BP. Excavations at Manton in 1966 yielded fence posts, pointed digging sticks and paddle shaped spades, all of which are closely similar to modern or recent traditional highlands cultivating implements (Powell 1974).

Golson (1977a,b) has reported much earlier agricultural evidence from the Kuk site, where 5 distinct phases of swamp management have been found, beginning earlier than 9000 BP and ending before 100 BP. An interesting form of artifact that has been found in the Wahgi valley swamp cultivation is a tanged stone tool with a wide curved or pointed blade. These may have been cultivating or ditch-digging tools (Allen 1970) and similar tools have been found at the Wañlek site, Madang Province, where they have been interpreted as tools associated with cultivating (S. Bulmer 1978). Other stone tools, in use in the early historic period in the Chuave area, Chimbu Province, also show signs of wear from digging in the earth (S. Bulmer 1973), so the use of stone for the manufacture of tools associated with cultivating may be more widespread than previously thought.

4.4 Prehistoric trade

Prehistoric human ecology in New Guinea involved not only the "exploitation" of resources, either natural or man-induced or modified. It also involved the

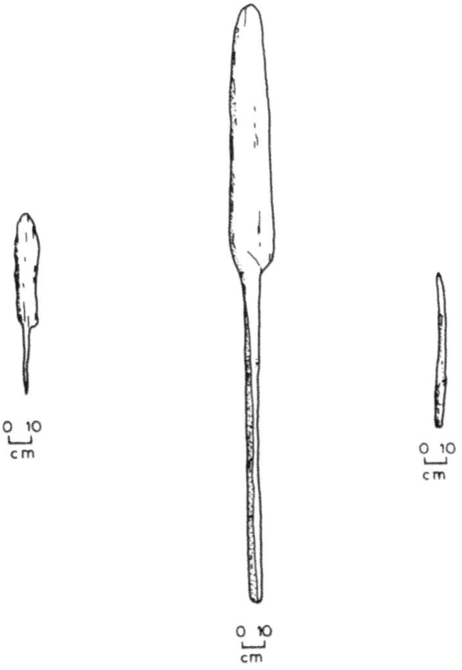

Fig. 17. Wooden gardening implements from the Manton site, Wahgi valley (after Powell 1974, Figs. 2, 3).

moving about of resources through trade, meaning that humans could live in areas where the natural resources could not sustain the population or where humans had so over-exploited the resources that they had deteriorated significantly.

Western scientists, archaeologists amongst them, tend to paint a picture of prehistoric traditional economies based on the scarcity of resources, but this is true only for a few New Guinea societies. Most had diversified economies and mixed domains, so that in case of drought, flood, or volcanic disaster alternate resources or neighbors could be called upon. Most communities were thoroughly familiar with their resources and had the potential for self-sufficiency, but in fact most were not self-sufficient. This is not because of a lack of food or adequate resources, but because of the culturally-created dependance on commodities of restricted supply, such as stone or shell for manufacture, or cosmetic or ornamental supplies. Thus an understanding of the trade patterns of prehistoric times is essential to an understanding of human ecology, and the examination of food resources is only part of the picture.

At its extreme, trading took the form of a kind of commercialism, with certain groups living in places with insufficient food supplies, making their living by manufacturing portable goods for trade and by acting as middlemen in trade between other groups. At the other end of the trade spectrum were groups, particularly those in relatively isolated mountain areas, that produced all or nearly all their own food and industrial materials, and traded in a variety of minor items. In between were the majority of New Guinea societies, complexly interdependent with their neighbors in the exchange of a wide range of goods.

The wide variety of goods produced and exchanged have been ably reviewed by Hughes (1977a,b). These included goods, such as stone axes, obsidian, ochre and salt that were available to only some communities and were highly useful or desirable, if not strictly necessary for survival. However, at the other end of the spectrum there were goods that were not economically important, such as items of clothing and wooden bowls, manufactured by certain communities as a part of a trading complex. Pigs were a most important exchange item, converting

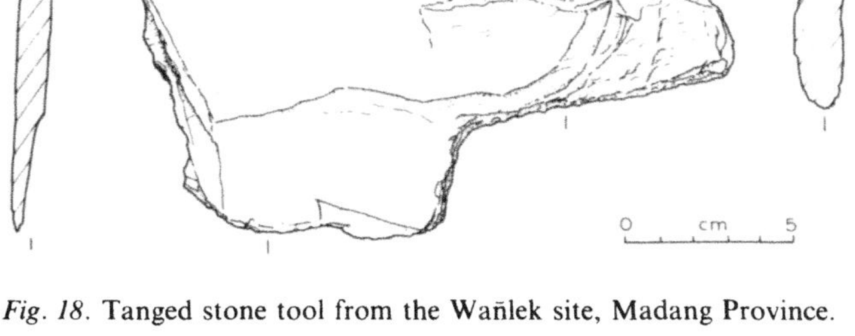

Fig. 18. Tanged stone tool from the Wañlek site, Madang Province.

garden produce into animal protein, particularly in areas lacking means of otherwise storing garden produce.

Nearly all New Guinea archaeological sites contain materials that can be interpreted as trade goods because they are "foreign", i.e. have come from such a distance that it is very unlikely that they were obtained directly by the residents of the site. This interpretation has to be accepted with caution, however, for it is known from the historic period that traders travelled long distances, such as the Chimbu traders who conveyed bird-of-paradise feathers throughout the highlands, and the Mailu who went as far as Rossel I. in the Milne Bay Province to obtain shell for ornament manufacture. The extent to which traders travelled long distances, particularly by sea, in the prehistoric past may be difficult to recognise archaeologically, but must be taken into account as a possible explanation of the presence of "foreign" goods.

Nevertheless, much trade is known to have been short-distance neighborly exchange, although certain groups specialised in trading, some moving hundreds of km by boat to redistribute goods and sell their own wares.

The archaeological evidence for trade, although very restricted so far, indicates a very long history of exchange of useful goods over long distances, as well as changes during different periods of time. Four important trade commodities will be discussed briefly here; stone, salt, pottery and shell ornaments. In all these items but salt, there was prehistoric exchange not only around the coast and between islands, but from lowlands to highlands and vice versa.

Fine quality rock was used for making axes and adzes and other implements and non-utilitarian objects, and was very localised in its natural occurrence in New Guinea. Axe-adzes were economically important in being essential to major forest clearance and were nearly everywhere an important trade item. Archaeological sites in the highlands have contained evidence of the earliest trade in quarried axe-adzes in the period 6500–4500 BP. Trade in a variety of rock materials and artifacts, including axe-adzes, began in New Ireland at least 7000 BP. Another highly valued stone that was traded even greater distances than axe-adzes was flakable quality obsidian (Smith *et al.* 1977; Ambrose 1976). Obsidian trade began in the New Britain-New Ireland area by 7000 BP, with the earliest source so far recorded being one at Talasea, and several Admiralty Is sources also being used in more recent times. Very little obsidian reached the highlands, but the earliest examples excavated so far date to about 4500 BP, at the Kafiavana site in the E Highlands.

Figure 19 shows the approximate location of rock used for axe-adzes and recorded obsidian sources. Cherts and flints were also used widely for manufacturing artifacts and cutting tools, and their natural distribution is even wider than obsidian. The most notable examples of exchange in stone resources are the communities that used "Lapita" pottery, which were responsible for distributing artifacts of New Guinea origins as far away as the SE Solomons and New Caledonia.

The archaeological record is so far mute in respect to prehistoric salt trade in New Guinea, but its relatively early beginnings should be anticipated, as salt-making is at least 5000 years old in Asia (Hughes 1977). Trade in shell ornaments, on the other hand, is on present evidence even earlier than stone trade. The presence of marine shell is considered evidence of trade at sites that are almost certainly out of direct contact with shell sources, such as Kafiavana, where marine shells were found in deposits of 9500 BP. It is more difficult to establish shell trade in coastal and lowland sites, but it is sometimes possible, through the presence of manufacturing debris, to show that certain kinds of shell ornaments have been made at a particular site, such as at Taurama, near Port Moresby (S. Bulmer 1978). However, evidence from the communities to which the ornaments were traded is also needed to demonstrate the practice of trade, and archaeological evidence is not yet adequate to show this.

The product so far most satisfactorily linked with trade is pottery, particularly in Irwin's (1977) study of the progressive specialisation of the Mailu potters of S Papua. Egloff (1971) has also shown trade links with the northern Milne Bay Province and Collingwood Bay through the sourcing of potsherds. Sherds from pottery of lowland origins have been found in the Wañlek site in the Madang Province, showing small quantities of pottery reached the fringes of the highlands at least as early as 3000 BP.

5. Summary and conclusions

The use of climate periods as a framework for interpreting the archaeological evidence emphasizes the association between the processes of changes in climate and vegetation and those of human ecology and technology. Humans chose from their changing environment a variety of alternate strategies of living, they explored new and changing resources, and they in turn themselves caused environmental changes.

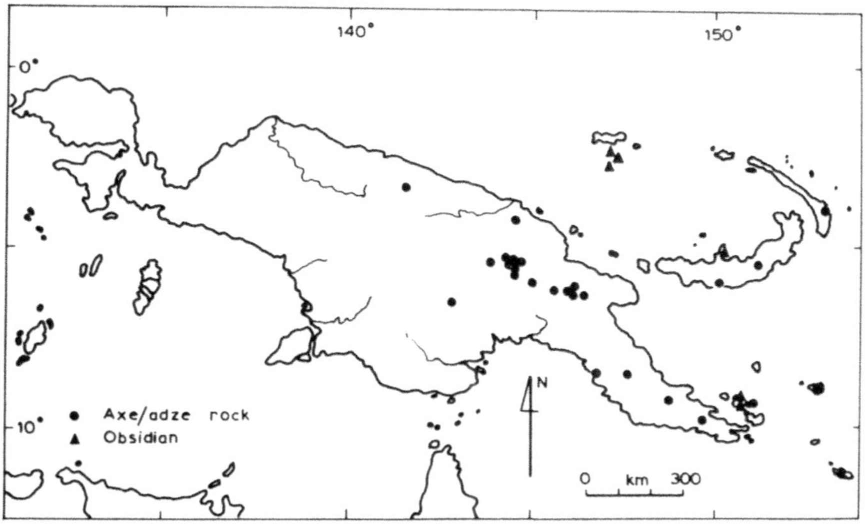

Fig. 19. Obsidian and axe-adze stone sources in New Guinea.

The following patterns can be seen during the 4 major climate periods discussed:

a. *The final glacial period*, from about 50 000 to 15–14 000 BP, is only known through one archaeological site, Kosipe, in the mountains of Papua and one other, NFX, in the E Highlands Province. Kosipe is thought to be a camp site occupied during the seasonal collecting of mountain pandanus nuts. A restricted range of large stone artifacts in the assemblage from this site contrasts to the wide range of artifacts and other materials present at later sites, and supports the interpretation of this as a temporarily occupied site, rather than a general settlement site. Its location at 2000 m, near to the edge of the upper tree line at that time, i.e. 26 000 BP, would have had relatively uncomfortable living conditions in contrast to sites further down the mountain. When they were in residence they may also have hunted in the nearby upper mountain scrubland of the sub-alpine zone, but no faunal collections were present at the site to indicate what animals might have been living there. The faunal remains from the Pureni swamp, dated to an earlier part of the final glacial period, i.e. about 33 000 BP, is located at 1500 m, well below the edge of the mountain forest.

This early settlement at Kosipe has been seen to coincide with a minor improvement of temperature (Fig. 8) and could indicate the beginnings of the use of the higher mountain resources in slightly warmer conditions. Another site, NFX, similarly coincides in its earliest occupation with the next rise in temperature, about 18–17 000 BP. These coincidences could possibly indicate that humans only moved into the highlands when the temperature returned to within a few degrees of modern temperature, but this must remain a tentative hypothesis until further archaeological evidence is found.

On these grounds, and evidence from the same general climate period in Australia, it seems possible that the Pleistocene hunters and collectors were mainly coastal and riverine, moving into the mountains during the warmer seasons. Bowdler (1977) has argued that the earliest people in Australia were small game hunters and fishers and shellfish collectors. Allen (1977) is of the opinion that the modern suite of New Guinea fauna is not a viable basis for man's subsistence, but all the early sites so far show that New Guinea hunters, like their Australian counterparts, were small game broad-spectrum hunters. If humans and the "giant" fauna were contemporaries in New Guinea, it could be that the human populations moved to be near this game, wherever it lived. It has been argued that the subalpine scrubland may have been rich in game, but evidence for this is so far lacking.

b. *The period of glacial melting*, from about 15–14 000 to 9–8000 BP, is the time when some New Guinea groups began to specialise in mountain living, if they hadn't done so before. Some probably moved into the mid-mountain forest and intermontane valleys for the entirety of the year, and perhaps visited the lowlands only seasonally. This is a period of marked change in the size and distribution of vegetation zones, of changes in animal life related to the vegetation changes and probably to human exploitation, and of the relocation of coastlines and river courses in response to rising sea levels.

The archaeological evidence from the highlands points to the beginnings of plant cultivation by the end of this climate period. This probably consisted of the use of indigenous plants and trees. The natural changes during this period in location and size of vegetation zones presumably encouraged the replanting and nurturing of particular useful species in more convenient locations, such as the vicinity of settlements. This would have led to the discovery that the replanting of some species caused an increase in the size and quantity of fruits and nuts. The general motive of extending the range and altitude of indigenous food plants, or at least maintaining the status quo in the face of climate change, may have been the basis of the initiating of cultivations in the Wahgi valley, the immediate origins of which may have been mid-mountain cultivations.

The importance of the environmental changes of this period, probably due more to non-human than human causes, is evident in its effects on fauna. The 4 archaeological collections of animal bones relating to this climate period show the extinction of a number of species, including the Tasmanian wolf, a large fruit bat, a large wallaby, and an as yet unidentified cuscus, as well as the local extinction or reduction in numbers of other species now not or rarely found in the highlands. Also, the earliest examples of pig appeared in the highlands during this period. Whether self-introduced as wild animals, or brought to New Guinea by humans, pigs could have maintained themselves in the lowland and mountain forests or could have been kept as domestic animals on a diet of sago or other indigenous crops.

A variety of new settlements began in this period in the highlands, including Wañlek, possibly Niobe, Kiowa, Yuku, Kafiavana, Batari NFB, and Kuk. There is a variety of technology, but no obvious patterns of technological change recognised so far, but trade with the lowlands commenced before the end of this climate period.

The archaeological evidence for this climate period has so far been found only in the highlands. However, a few non-archaeological clues indicate what may be occurring in the lowlands. On linguistic grounds it has been argued that there may have been further immigrants to New Guinea, prior to the earliest Austronesian-speakers and some culture traits also suggest this, such as the N coastal distribution of the non-Austronesian coil-made pottery industries. The products of these industries seem notably similar to the pottery of Japan during the final glacial period. Although far removed in time and space from New Guinea, it seems plausible that the earliest pottery-using people of New Guinea may have been the ancestors of these potters, and that they may derive from some common general E Asian tradition, but this hypothesis will have to await archaeological evidence.

c. *The early Holocene period of warmer wetter climate*, from about 9–8000 to 5000 BP, was an era of marked change both in the highlands and lowlands. Sea levels returned to about their present level during this period, and the "drowning" of the Torres Strait accompanying this is thought to have been responsible for changes in wind and rain patterns.

Looking at the lowlands first, little archaeological evidence is available. This is, on linguistic evidence, the period of the movement of Austronesian-speaking groups into New Guinea and the rest of Melanesia. Probably arriving first on the N coast of the island, they spread all around the N coasts and N islands in what may have been a complicated series of moves and settlements. They apparently did not make and use pottery, but sought out rock sources and exchanged a variety of useful rocks among themselves. They colonized some islands, such as the Admiralties, which were apparently not settled before by non-Austronesian-speakers, and moved in with, around or absorbed non-Austronesian-speakers on New Britain, New Ireland, and on the coast of the mainland. The Austronesian-speaking groups may not have moved far inland on the mainland, except in the Markham valley, and this latter move is likely to be related to the next climate period, discussed below. In most areas the Austronesian speakers were closely adapted to the coasts and coastal plains. They were cultivators of a repertoire of SE Asian root and fruit crops, and could have been responsible for their introduction to the New Guinea area during this period.

In the highlands, archaeological evidence of this early Holocene period has been found at a variety of sites, including the Wahgi agricultural sites where it appears that the swampy ground in the valley was not much used for cultivation, but the slopes of the hillsides above were, presumably with forest-clearance and cultivating leading to the sedimentation of deep soils on the valley floor below. It seems likely that these cultivations contained not only the local cultigens, but at least some of the tropical crops of SE Asian origins which were introduced by Austronesian immigrants to the lowlands and from there introduced to the highlands. In this slightly warmer climate period, the establishment of tropical crops in the mountains would have been easier than in the earlier or later less tropical periods. In any case, pollen studies show that by 5000 BP widespread forest clearance had taken place in the highlands, so agriculture must also have been widespread.

This period shows technological change in the highlands as well, particularly during c. 6000–4500 BP. Although few open sites can be attributed to this period, a number of rock shelters show the establishment of trade in axe-adzes quarried at certain localities in the Wahgi and Jimi valleys, an increase in the presence of marine shell ornaments and other artifacts from the lowlands, and the presence of pig bone. The latter could have a number of explanations; it could reflect pigs present in a domestic state, or in the advent of cooking pigs in rock shelters, or the hunting of wild pigs.

Another notable change is the replacement of the Pleistocene style axe-adzes by quarried polished trade axe-adzes. The introduction of new axe-adze technology and commencement of their trade within the highlands should certainly be considered a sign of trade contacts with the lowlands. In the absence of lowlands sites of this period to the north, there is no indication of whether highlands axe-adzes were traded to the lowlands, but some products presumably were, in exchange for the marine shell ornaments that are present in highlands sites.

d. *The period of modern climate*, from about 5000 BP until the historic period has to be seen as one of complexity and diversification of human culture, in a period of deteriorating climate and adaptation to new and various ecological change. Two other environmental phenomena were of major importance for human ecology. The first was the slight rise in sea level that is thought to have been felt in a number of regions, although the evidence for this is complicated by local uplift and subsidence in many areas. The other environmental change directly related to man's own interference with the landscape, is the process of erosion and sedimentation of the river valleys and coastal plains. The silting up of valleys and embayments, and the extension of swamps, like the raising of barrier reefs and coral islets all affect transport and communication, marine resources and land use. Although these issues have to be understood in terms of local geography and land movements, their importance in respect to human settlements and their use of resources is considerable.

Agricultural development in the highlands seems to have been affected by the slightly deteriorating climate of this final period. The Wahgi cultivations show a series of adaptive re-designings of drainage and water-control ditching, which appear to be aimed at coping with changes in sedimentation and water level, induced by the cultivation at this site and on hill slopes above. One of the explanations of this process of adaptation is that it is most likely these cultivations were designed for use of the tropical crops of SE Asian origin, which are at the upper limits of their effective cultivation in the intermontane valleys, and which therefore required special cultivation techniques to ensure their maximum productivity.

The most momentous introduction to New Guinea agriculture was the sweet potato, a crop of S American origin. This crop permitted in highlands context both the great increase in yield on ordinary agricultural land, but also the extension of settlement based on effective root crop cultivation to higher altitudes and onto land of more marginal productivity. The high densities of highlands human and pig population are unquestionably related to the acquisition of the sweet potato. This is thought to have occurred during the period of European voyaging in the Pacific, probably less than 400 years ago at most.

The lowland archaeological remains that have attracted the most attention so far are ceramic sites, largely because they are so visible. The research of the last decade has shown the presence of a variety of pottery styles in different parts of Melanesia between 4000 and 3000 BP. The best known pottery using groups of this period used the style of pottery called "Lapita," but other contemporary styles indicate the diversification of cultures in Melanesia. A wide variety of settlement locations and ecological bases were utilised, with trade and exchange making it possible or more profitable to occupy an ever wider variety of ecological niches. At the end of prehistory some groups were even living where it was impossible or impractical to cultivate, obtaining food supplies through trade.

The first 2 decades of archaeological research in New Guinea indicate many millenia of human adaptation and ingenuity in the changing land. In spite of the fact that there is as yet by no means an adequate archaeological coverage of New Guinea to write its prehistory, it is possible to begin to appreciate the wide variety of human ecological adaptation.

Acknowledgement. Caroline Phillips drew, or redrew, all the illustrations.

References

Allen, J. 1970. Prehistoric agricultural systems in the Wahgi valley — a further note. Mankind 7 (3): 177–83.

—— 1972. Nebira 4: an early Austronesian site in Central Papua. Arch. Phys. Anthr. Oceania 7 (2): 92–124.

—— 1976. Fishing for wallabies: trade as a mechanism of social interaction, integration and elaboration on the Papuan south coast. In Rowlands, M. and J. Friedman (Eds) The evolution of social systems, London, Duckworth.

—— 1977. The hunting neolithic: adaptations to the food quest in prehistoric Papua New Guinea, pp. 167–88. In J. V. S. Megaw (ed.) Hunters, gatherers and first farmers beyond Europe, Leicester Univ. Press.

Bellwood, P. 1978. Man's Conquest of the Pacific. Auckland, Collins.

Bowdler, S. 1977. The coastal colonisation of Australia, pp. 205–216. In Allen, J., J. Golson and R. Jones (Eds) Sunda and Sahul, London, Academic Press.

Bulmer, R. N. H. 1964. Edible seeds and prehistoric stone mortars in the Highlands of East New Guinea. Man 64, No. 183: 147–50.

—— 1976. Selectivity in hunting and in disposal of animal bone by the Kalam of the New Guinea Highlands, pp. 201–20. In G. de G. Sieveking, I. H. Longworth and K. E. Wilson Problems in Economic and Social Archaeology, London, Duckworth.

Bulmer, R. N. H. and J. I. Menzies. 1972–3. Karam classification of marsupials and rodents, J. Polyn. Soc. Part 1, 81: 472–99; Part 2, 82: 86–107.

Bulmer, S. 1966. Pigbone from two archaeological sites in the New Guinea Highlands, J. Polyn. Soc. 75 (4): 504–5.

—— 1971a. Pottery variation and change in New Guinea and its implications for the reconstruction of prehistoric settlement. Mimeo 41 pp. Dept Anth. Soc. UPNG.

—— 1971b. Archaeological investigations at the Koigen site, near Wewak, East Sepik District. Rec. Papua New Guinea Museum 1 (2): 16–36.

—— 1973. Notes on 1972 excavations at Wañlek, an open settlement site in the Kaironk valley, Papua New Guinea. Working Paper No. 29, Dept Anth. U. of Auckland.

—— 1975. Settlement and economy in prehistoric Papua New Guinea. A review of the archaeological evidence. J. Soc. Océanistes 31 (46): 7–75.

—— 1977a. Waisted blades and axes. A functional interpretation of some early stone tools from Papua New Guinea. Pp. 40–59 in R. V. S. Wright (ed.) Stone tools as cultural markers. Change, evolution and complexity. AIAS Canberra and Humanities Press, New Jersey. Paper presented to a symposium of the 1974 meeting of the AIAS, Canberra.

—— 1977b. Between the mountain and the plain; prehistoric settlement and environment in the Kaironk valley. Pp. 61–73. In Winston 1977.

—— 1978. Prehistoric Culture Change in the Port Moresby region. Ph.D. thesis, UPNG.

—— 1979. Archaeological evidence of prehistoric faunal change in Highland Papua New Guinea. Paper read to ANZAAS, Auckland.

Calaby, J. H. 1976. Some biogeographical features relevant to the Pleistocene movement of man in Australia. In R. I. Kirk and A. G. Thorne (Eds), The origins of the Australians. Australian Institute of Aboriginal Studies, Canberra.

Chappell, J. M. A. 1966. Stone Axe factories in the Highlands of New Guinea. Proc. Prehistoric Society 32 (5): 96–121.

Christensen, O. A. 1975. Hunters and horticulturalists: a preliminary report of the 1972–4 excavations in the Manim valley, Papua New Guinea. Mankind 10 (1): 24–26.

Clay, R. B. 1974. Archaeological reconnaissance in Central New Ireland. APAO 9 (1): 1–17.

Downie, J. E. and J. P. White. 1978. Balof shelter, New Ireland — report on a small excavation. Rec. Aust. Mus. 31 (19): 762–802.

Eden, M. J. 1974. The origin and status of savanna and grassland in Southern Papua. T. I. Br. Geog. 63: 97–110.

Egloff, B. J. 1971. Collingwood Bay and the Trobriand Islands in recent prehistory. A.N.U. Ph.D. thesis.

—— 1975. Archaeological investigations in the coastal Madang area and on Eloaue Island of the St Matthias Group. Rep. Papua N.G. Mus. 5.

Galloway, R. W., G. S. Hope, E. Löffler and J. A. Peterson, 1972. Late Quarternary glaciation and peri-glacial phenomena in Australia and N.G., Pp. 125–138. In E. M. van Zinderen Bakker (ed.) Palaeoecology of Africa and Antarctica, Capetown, Balkema.

Glover, I. C. 1972. Settlements and mobility among the hunter-gatherers of south-east Asia.

Pp. 157–61, in P. J. Ucko, R. Tringham and G. W. Dimblby (Eds) Man, settlement and urbanism, London, Duckworth.
Golson, J. 1977a. Simple tools and complex technology. Agriculture and agricultural implements in the New Guinea Highlands. Pp. 154–61. In R. V. S. Wright. Stone tools as cultural markers, AIAS, Canberra.
—— 1977b. No room at the top: agricultural intensification in the New Guinea Highlands. Pp. 601–38. In Allen, J., J. Golson and R. Jones (Eds), Sunda and Sahul, London, Academic Press.
Golson, J., R. J. Lampert, J. M. Wheeler and W. R. Ambrose. 1967. A note on carbon dates for horticulture in the New Guinea Highlands. J. Polyn. Soc. 76 (3): 369–71.
Green, R. C. 1976. Lapita sites in the Santa Cruz group. Pp. 245–65. In R. C. Green and M. M. Cresswell (Eds.), Southeast Solomons culture history. Royal Soc. New Zealand, Bull. 11.
—— 1978. New sites with Lapita pottery and their implications for an understanding of the settlement of the western Pacific. Paper pres. at IXe Congrès Union Internationale des Sciences Prehistoriques et Protohistoriques, Nice.
Hope, G. S. and J. A. Peterson. 1975. Glaciation and vegetation in the high New Guinea mountains. Pp. 155–62. In R. P. Suggate and M. M. Cresswell (Eds.), Quaternary Studies. Royal Soc. New Zealand.
Hope, Jeanette H. 1971. The effect of prehistoric man on the fauna of New Guinea. Pp. 21–7. In Winslow 1977.
Hope, J. H. and G. S. 1976. Palaeoenvironments for man in New Guinea. Pp. 29–54. In R. L. Kirk and A. G. Thorne (eds.) The origin of the Australians. Human Biology Series No. 6, AIAS, Canberra.
Hossfeld, P. S. 1949. The stratigraphy of the Aitape skull and its significance. Royal Soc. S Aust., Trans. 72: 201–7.
—— 1965. Radiocarbon dating and palaeoecology of the Aitape fossil human remains. Proc. Royal Soc. Vict. 78 (2): 161–5.
Hughes, I. 1977a. The use of resources in traditional Melanesia. Pp. 28–34. In Winslow 1977.
— 1977b. New Guinea stone age trade. Terra Australis 3. Dept Prehistory, ANU, Canberra.
Hyndman, D. and J. I. Menzies. 1980. Aproteles bulmerae (Chiroptera) of New Guinea is not extinct. J. Mamm. 61 (1): 159–60.
Irwin, G. J. 1977. The emergence of Mailu as a central place in the prehistory of coastal Papua, Ph.D. thesis, ANU.
Lawrence, B. 1967. Early domestic dogs. Zeitschrift fur Saugetierkunde 32 (1): 44–59.
Menzies, J. I. 1977. Fossil and Sub-fossil fruit bats from the mountains of New Guinea. Aust. Jour. Zool. 25: 329–36.
Nix, H. A. and J. D. Kalma. 1972. Climate as a dominant control in the biogeography of northern Australia and New Guinea. Pp. 61–92. In Walker, D. (ed.), Bridge and barrier: the natural and cultural history of Torres Strait. Dept. Biogeog. and Geomorph., Australian National University. Pub. BG/3.
Plane, M. D. 1966. Fauna from the basal clay at Kafiavana. Appendix 5, pp. 168. In White 1972.
—— 1972. Fossil Mammals. Pp. 464–5. In Ryan, P. (ed.) Encyclopedia of Papua and New Guinea, Melbourne University Press.
Powell, J. M. 1970a. The impact of man on the vegetation of the Mt Hagen region, New Guinea. ANU, Ph.D. thesis.
—— 1970b. The history of agriculture in the New Guinea Highlands. Search, 1 (5): 199–200.
—— 1974. A note on wooden gardening implements from the Mt Hagen region, New Guinea. Rep. Papua NG Mus. 4: 21–8.
—— 1976. Ethnobotany. Pp. 106–83. In Paijmans, K. (Ed.) New Guinea vegetation. Canberra CSIRO and ANU Press.
Robbins, R. G. 1963. The anthropogenic grasslands of PNG. In: UNESCO Symposium on the impact of man on the humid tropics vegetation. Goroka 1960 p. 313.
Smith, I. E. M., G. K. Ward and W. R. Ambrose. 1977. Geographic distribution and characterization of volcanic glasses in Oceania. APAO 12 (3): 173–201.
Specht, Jim R. 1968. Preliminary report of excavations on Watom Island. J. Polyn. Soc. 77 (2): 117–34.
Vanderwal, R. L. 1973. Prehistoric studies in central coastal Papua. Unpub. Ph.D. thesis, A.N.U.
Van Deusen, H. M. 1963. First New Guinea record of Thylacinus. J. Mammalogy 44: 279–80.
Watson, V. D. and J. D. Cole. 1977. Prehistory of the Eastern Highlands of New Guinea. Seattle, U. of Wash. Press.
White, C. 1967. Plateau and plain: prehistoric investigations in Arnhem Land, Northern Territory. Ph.D. thesis, Australian National University.

White, J. P. 1972. Ol tumbuna: archaeological excavation in the Eastern Central Highlands, PNG. Terra Australis 2. Canberra, Dept Prehistory, RSPS, A.N.U.
White, J. P., K. A. W. Crook and B. P. Ruxton, 1970. A late pleistocene site in the Papuan Highlands. P. Polyn. Soc. 36: 152–70.
White, J. P., J. E. Downie and W. R. Ambrose. 1978. Mid-recent human occupation and resource exploitation in the Bismarck Archipelago. Science 199: 877–9.
White, J. P. and J. Specht. 1971. Prehistoric pottery from Ambitle Is, Bismarck Archipelago, Asian Perspectives 14: 88–94.
Williams, P. W., I. McDougall and J. M. Powell. 1972. Aspects of the Quaternary geology of the Tavi-Koroba area, Papua. J. Geol. Soc. Aust. 18 (4): 33–47.
Winslow, John H. 1977. The Melanesian environment. Canberra, A.N.U. Press.
Yen, D. E. 1973. The origins of Oceanic agriculture. APAO 8 (1): 68–85.
—— 1977. Hoabinhian horticulture; the evidence and the questions from northwest Thailand. Pp. 565–66. In Allen, J., J. Golson and R. Jones (Eds.), Sunda and Sahul. London, Academic Press.

4. The history of plant use and man's impact on the vegetation

J. M. Powell

1. Introduction

The natural flora of New Guinea comprises 1465 genera (in 246 families) of widespread tropical, Malesian and Australian affinities (van Balgooy 1976). The island is extremely diverse in environments and vegetation, ranging from the lowland coastal and riverine swamps with mangroves, nipa and sago palm forests, through dry plains covered with grassland and open woodland to rich lowland, foothill and high mountain rainforests (Paijmans 1976). Within this landscape some 3.4 million people live. Population densities are generally very low in the coastal and lowland swampy areas and in the grassland and open woodland areas with marked rainfall seasonality (2–4 persons per km^2), somewhat higher in lowland and mountain rainforest areas (8–16 persons per km^2) and very high (locally up to 200 persons per km^2) in the intermontane valleys and basins of the central cordillera.

A few predominantly hunting and gathering groups are known in New Guinea today (Brookfield with Hart 1971) but the majority of people are subsistence agriculturalists, tending and planting indigenous crops such as sago, sugarcane, bananas and breadfruit, and exotic crops such as taro, yams and sweet potatoes, as well as gathering plant products (and hunting small animals) from the forests, open woodlands and grasslands.

While in some areas the natural vegetation may remain in a relatively undisturbed state, in many areas it has been greatly changed or even removed by man's action. Man has modified various communities and has extended the range of distribution of others. Thus many of the shrublands and grasslands of the highland valleys and basins appear to be the result of man's agricultural and hunting practices: forest and forest regrowth have been repeatedly cleared and burned for grasslands and associated plantings, and grasslands burned to drive out small game, to ward off enemies and for various other purposes. The high altitude grasslands of the mountains have been greatly extended by man's use of fire in particular and many of the open woodlands and grasslands of the lowland valleys and plains have been extended greatly by anthropogenic influences.

Following disturbance and removal of vegetation other changes in the environment occur: the microclimate is altered, soil properties change and the

Monographiae Biologicae, Vol. 42, ed. by J. L. Gressitt.
© *1982, Dr W. Junk Publishers, The Hague.*

hydrological cycle is affected (Löffler 1976). In particular, slope wash is accelerated (once ground cover has been destroyed) and soil erosion may be severe, especially on steep slopes. The washed out sediment may accumulate in the valley floor, block the natural drainage and result in the development of swamps.

Direct evidence for such changes must come from palaeobotanical, geomorphological and archaeological studies, and the data interpreted on the basis of knowledge of the present day use of plants, of extant patterns of agriculture and subsistence life-styles. At present the vegetation history and archaeological records for the New Guinea area are fragmentary, geomorphological studies of man's influences have only just begun and the present plant use data are far from complete. In this chapter, discussion will center on: (a) the vegetation types present and the environmental changes which occurred during the probable period of man's occupancy of the island; (b) the plants available for food (and other purposes) during this period; and (c) the evidence for the development of agricultural systems in the highland areas of Papua New Guinea.

2. Man's entry into New Guinea, early environments and vegetation types

Archaeological evidence indicates that man has been in upland New Guinea for at least 26 000 years (Allen 1976; Bulmer 1975 and chapter II:3); he probably reached coastal areas 50 000 BP or earlier. At present nothing is known of the vegetation history of lowland and mid-mountain areas, but the degree of vegetational shift indicated for the highlands during the last glacial period (Walker and Hope chapter III:1) and earlier (Powell unpubl. data) implies that there was at least some modification of lowland and mid-mountain vegetation distribution patterns. The composition of vegetation types may have been different also from that known today but many extant taxa were probably already present, and some, with broad ecological tolerances, may have been more abundant than at present. Chappell and Thom's (1977) discussion of late Quaternary sea level changes in the Sunda-Sahul region, Nix and Kalma's (1972) reconstruction of former climates (and vegetation) in the Torres Strait area and Kershaw's (1976) vegetation history study in N Queensland, together with the New Guinea studies, provide a basis for speculation on early environments in these areas.

During the last 120 000 years there have been a number of periods of low sea level (Chappell and Thom 1977) during which man may have entered New Guinea from NW Malesia. Between 60 000 and 55 000 years BP the sea level dropped to −120 m or more, between 40 000 and 35 000 BP to possibly −80 m and between 25 000 and 15 000 BP to −150 m. During these periods New Guinea was linked to Australia across the Arafura Plain. Vegetation changes for the latter period in highland areas have been discussed by Walker and Hope (chapter III:1). They suggest that by 18 000–15 000 BP snowlines were at least 1000 m and locally more below present snowlines (implying temperatures of 6–8° below present levels) and that forest tree-lines stood at c. 2300 m altitude at that time. Nix and Kalma (1972) consider that by 17 000–14 000 BP rainforest would have been restricted in distribution and open sclerophyllous forest would have occupied the Markham, Ramu and lower Sepik valleys as well as parts of the Arafura Plain. Other areas of the plain would have been covered with open woodland and grassland. Shrub-rich grasslands occupied areas above the mountain forests.

In highland New Guinea deglaciation began c. 15 000 years BP, temperatures ameliorated and by 9 000 BP forest was present at 2500 m altitude (in the Sirunki area) and had reached 3910 m by 8600 BP on Mt Wilhelm (Walker and Hope chapter III:1). In NE Queensland, Kershaw's (1975) pollen diagrams indicate the replacement of sclerophyllous forest by angiospermous rainforest between 9500 and 7000 BP. Rising sealevels submerged the Arafura plain by 6000 BP.

Evidence of earlier glaciation is available (Löffler 1977) which can probably be linked with the earlier sea level changes. Vegetation changes similar to those outlined by Walker and Hope could also be expected to be associated with the earlier glaciations. Pollen analytical data from the Upper Wahgi Valley sites (at 1550–1850 m) indicate that upper mountain forests ("mixed" and "subalpine" of some authors) replaced lower mountain forest (including beech and oak forests) between 50 000 and 40 000 years BP and subalpine shrubs and forbs were present also. By 30 000 BP, lower mountain forests, dominated by beech (*Nothofagus* spp.) were widespread. In NE Queensland subtropical rainforest was present on the Atherton tableland from before 60 000 BP to c. 38 000 BP according to Kershaw (1976) and "it existed under about half the present-day annual rainfall." Sclerophyll vegetation replaced it between 38 000 and 27 000 BP as a "result of a decrease in precipitation, a decrease in temperature or the activities of man."

During the whole of this period, the central areas of the broad Sunda shelf, of low relief, would have been occupied by rapidly migrating shores, according to Chappell and Thom (1977). At times of regression there would have been broad intertidal areas with vegetated flats (occupied by mangroves) and shallow lagoons (which would become hypersaline) behind the migrating dune ridges. During periods of transgression mangroves would occupy shallow lagoons behind the seaward ridge and a richer estuarine biota would have developed as hypersaline conditions would be absent; freshwater would be more readily available behind earlier dune ridges.

If man entered S New Guinea at some time between 40 000 and 35 000 BP or between 60 000 and 55 000 BP then the Arafura shelf may have been partly or more or less fully exposed respectively (about 1200 km across) and the ecosystems most readily available for exploitation would have been coastal, with widespread mangrove vegetation, beach ridge and littoral thicket and some areas of brackish swamp within estuaries, and the adjacent inland grassland, woodland and sclerophyllous forest, with some riparian vegetation and possibly small areas of freshwater swamp. Areas of rainforest were probably very restricted in distribution on the southern lowland plains but more extensive on the northern side of the island and 100 km or so inland from the south coast. Lowland rainforest may have extended up to c. 300 m altitude while above it the lower and upper mountain forests occupied valleys and slopes up to about 2000–2400 m. Shrub-rich grasslands and "alpine" grasslands, similar to those present in more recent times (J. and G. Hope 1976) may have occupied the higher altitude zone above the rainforests.

3. The plant resources and subsistence systems

The plants used for various purposes in New Guinea today are numerous (Powell 1976a,b) and many remain unrecorded. Over 200 food plants are known to be used and of these some 65 are cultivated, 48 are both cultivated and gathered from the wild and 158 are harvested as wild resources. An analysis of the biogeography of the genera involved indicates that the major relationship is to the NW and suggests the very early transfer of knowledge of particular plant use from the Malesian region to New Guinea. Detailed analysis is hampered by the lack of adequate taxonomic treatments of the genera but in one case at least, *Terminalia*, greater use of species is indicated for New Guinea (7 spp.) than elsewhere (2 spp.).

The plant resources of the coastal fringe and of the freshwater swamps and riparian vegetation are considerable (Table 1). Seasonal foods may have included the mangrove, *Bruguiera* (recorded as a staple today in the E Cape York peninsula; Harris 1977), and some of the fruit and nut crops, such as *Canarium*, *Barringtonia*, *Inocarpus*, *Pandanus*, *Pometia* and *Terminalia*. *Cocos*, *Metroxylon* and other palms yielding edible nuts, sago or "heart cabbage," together with various bananas (Australimusa section) and *Artocarpus* spp. may have been important as non-seasonal food sources. Wild sugar canes (*Saccharum* spp.) and a number of ferns and edible forbs may have been available also together with wild gingers and possibly wild yams (Table 2).

The animal resources of these areas would have been considerable also, the maritime, intertidal and estuarine areas yielding fish and shellfish and together with the freshwater swamps, birds and small animals.

The grassland, woodland and sclerophyllous forest areas probably yielded far less plant foods, but they may have been used seasonally and been an important source of animal food. Fires are important as a hunting strategy today in such areas (Bulmer 1968) and may have been used then. If *Cycas* spp. were present then the pith and seeds of various species may have been used for food and the stands managed with fire, as suggested by Harris (1977) for areas of the Cape York Peninsula. Wild yams (*Dioscorea* spp.), some bananas, *Gnetum*, *Citrus* spp., *Mangifera* spp. and possibly *Castanopsis* spp. may have been available, together with ferns such as *Dicranopteris*, *Gleichenia* and *Pteridium*.

The rainforests are extremely rich in plant foods. Many of the fruit and nut species of the littoral thicket grow also in lowland rainforest areas, for example, *Canarium*, *Pangium*, *Pometia*, *Spondias* and *Terminalia*. Some occur over a very great altitudinal range: included here are *Artocarpus*, *Musa*, *Pandanus*, *Castanopsis* and some of the Palmae, while others, such as *Sterculia*, *Syzygium*, *Finschia*, *Elaeocarpus* and *Chisocheton*, are more important in the mountain forests. Shrubs and forbs are abundant in the understorey levels of the rainforests and many of them have edible leaves (Table 2 and Powell 1976b). Ferns are plentiful also, and the aroids, *Alocasia* and *Cyrtosperma*, yielding edible stems and tubers respectively, may have been present in swampy areas.

Animal food resources are considered to be high also in rainforest areas.

It seems most likely that initially a broad spectrum subsistence system evolved. Populations probably centered on the coastal fringe and on the rainforest-woodland or rainforest-grassland ecotones of the mountains, but hunting and gathering occurred over a relatively wide range of ecological zones. Man may have altered the vegetation to some extent: by his use of fire he may have extended the sclerophyllous forests, open woodlands and grasslands and by his utilization of certain species altered the composition of the littoral thickets, the sclerophyllous forests and the mountain rainforests. His effects were relatively minor and short term, however, when considered against the background of climatic change.

When sea levels rose (after c. 55 000 BP, c. 35 000 BP and after 15 000 BP) the brackish and freshwater swamps became more extensive and some areas of woodland became seasonally inundated. Lowland rainforest occupied areas formerly covered by woodland and sclerophyllous forest and the mountain forests replaced subalpine grasslands at the higher altitudes. As well, there was a loss of land: with sea levels rising at the rate of 1.3–1.5 m per 100 years the lateral migration of shorelines would have been several hundred meters per decade on the Arafura shelf (Chappel and Thom 1977).

It might be expected that at such times the plant and animal resources would have been reduced due to the decreased area of land available for hunting and gathering. The much richer food sources available in the lowland rainforest and swampland may have offset this decrease somewhat but if population pressure existed, then a gradual change of emphasis onto certain plants such as sago, coconuts, bananas, breadfruit, other fruit and nut crops together with yams and aroids, may have occurred and led to the development of horticulture. Whether or not plants were domesticated and agricultural systems developed independently of outside influences is unknown but it is obvious that it would have been feasible on the plant base available. The possible importance of the fruit and nut species should not be overlooked in this context. In general, they are nutritionally superior to the edible root crops, to the palm products and the green vegetables (Powell 1976b) and are recognized as such by some coastal and highland people. Yen (1974a) has indicated their importance today in the subsistence of the Santa Cruz area of the E Solomon Islands and has recorded also the use of some of them (*Canarium*, *Terminalia*, *Madhuca*, *Castanopsis* and *Aleurites*, among others) in Thailand in very early times (dating from c. 11 000 to c. 7600 BP; Yen 1977). They may well have formed the staple food of the early inhabitants in New Guinea and been one of the first groups of plants to be tended and transplanted into groves and later, garden sites.

The entry of the SE Asian cultigens, in particular bananas of the Eumusa section, a number of yams, and taro, would have reinforced any early horticultural experiments being undertaken by the New Guinea people. The time of entry of these crops is unknown (Table 3); it is generally considered to be prior to 4500 BP, the date of transfer of rice out of S China, but could have been

Table 1. Plant resources: Indigenous fruit and nut crops and palm products

Taxon	No. of species	Area of natural distribution	Vegetation type	Part used in NG and status	Areas of known similar use
Aglaia	250–300	Asia to SW Pac.	Rf.	Fruit; w	Malay P., Java, Phil., Fiji
Aleurites	2	Asia to SW Pac.	Litt., Rf.	Nut; w, t	Malaya, Java, Sol., Aust. (Qld)
Arenga	11	Malesia to N Aust.	Litt., lowl. and mt., Rf.	Heart "cabbage"; w	Wide
*Artocarpus	47–50	SE Asia to Pac.	Litt., Rf.	Fruit, seeds; c, w	Indochina, Indo-malaya, Micr., Poly., Sol.
Barringtonia	100	E Africa to Pac.	Litt., ripar.	Nut; c, w	Indochina, Malay P., Poly., Sol.
**Bruguiera	6	E Africa to Pac.	Litt.	Fruit; w	Celebes, Mol., Torres St. Is, Aust. (C. York)
Burckella	11	Moluccas, NG to Samoa	Litt., ripar.	Fruit; w	Sol.
**Canarium	100	E Africa to Pac.	Litt., lowl. 1°, 2° Rf. dry f.	Nut; w, t	China, Phil., wide. Malesia, Sol.
Caryota	12	Malesia, NE Aust.	Litt., lowl. Rf.	Pith of young trunk; w	Borneo, Java, Sumatra
Castanopsis	120	Asia, Malesia, SW N Amer.	Rf. (1°, 2°): lowl., mt.	Nut; w, t	Indochina, Malay P., Java
Chisocheton	50	Asia, Malesia, N. Hebr., Aust.	Rf.: lowl. and mt.	Nut; w	Malay P.
Citrus	12	China, India, Malesia, Phil., Aust.	Seas. dry f.	Fruit; w	Wide.
*Cocos	1	Asia to Pac.	Litt.	Nut; cs, c, w	Wide.
**Cycas	200	Madagascar, Asia to Pac.	Dry f., open woodlands	Pith, seeds; w	India, Japan, Phil., Malaysia, Mol., Aust. (C. York, Arnhem)
Diospyros	500	Tropics + warm-temp.	Rf.	Fruit; w	India, Japan, Phil., Java, Celebes
Elaeocarpus	200	Asia to Pac., Aust.	Rf.	Nut; w	Indochina, Malay P., Phil., Borneo, Java, Mol.
Ficus	800	Indo-Malaya to Pac. + warm-temp.	Rf.	Fruit, leaves; c, w	Wide.

Finschia	7	Palau, NG, Sol., N. Hebr.	1°, 2° Rf.: lowl., mt., swamp f.	Seeds; w	N. Hebr. important throughout range
Garcinia	400	Africa, Asia, Malesia	Rf.	Fruit, leaves; bark; w	Wide. Malesia, China
**Gnetum	30	Africa, Asia to Pac., S Amer.	Rf.: lowl. and lower mt., dry f.	Leaves, fruit; c, w	Malay P., Phil., Java, Celebes, Sol., Fiji
Gronophyllum	13	E Malaysia to N Aust.	Rf., mt.	Heart "cabbage"; w	—
**Inocarpus	4	NG, Poly.	Rf.: lowl., litt.	Nut; w	Sol., Poly.
Madhuca	85	Indochina, Malesia, Aust.	Litt., Rf.: lowl.	Fruit; w	India, Phil.
Manilkara	70	tropics	Litt., Rf.: lowl.	Fruit; w	Malay P., Torres St. Is, Aust. (C. York), W Indies
Mangifera	40	India, China, Malesia	Litt., ripar., Rf.: lowl. mt., dry f.	Fruit; c, w	Malaysia, Java, Sol.
*Metroxylon	15	Siam to Pac.	Lowl. FW swamps, seas. inundated areas	Pith; cs, ws,	W Malesia, Sol.
*†Musa	40	SE Asia to Pac.	Ripar., Rf.: lowl., lower mt.	Fruit; cs, c, w	Wide
Myristica	100	India to Pac., Aust.	Rf.	Fruit, w	Malay P., Mol.
Nypa	1	Malesia, Sol., Aust.	Estuarine and coastal brackish swamps	Heart "cabbage", seeds; w	Malay P., China, Java
*Pandanus	600	W Africa to Pac.	Litt., swamp f., Rf.: lowl. and mt.	Fruit, nut; c, w	Wide.
Pangium	1	Malesia, Sol., Palau	Litt., Lowl. Rf.	Fruit; w	Malay P., Java
Parartocarpus	2	Malesia, Sol.,	Litt., Rf.	Fruit; c, w	Malay P.
**Pometia	10	Malesia to Fiji	Litt., 1°, 2° lowl. Rf.	Fruit, seeds; c, w	Malay P., Sol., N. Hebr., Fiji
Spondias	9	India, China, Malesia to Pac., trop. Amer.	Litt., ripar., lowl. Rf.	Fruit; w	Wide.
Sterculia	300	tropics	Rf.: lowl. and mt.	Nut; w	Wide.
Syzygium	500	tropics	Litt., Rf.	Fruit; c, w	Wide.
**Terminalia	200	tropics	Litt., swamp f, lowl. Rf.	Nut; c, w	Malaysia

*Important today **Possibly more important formerly †Eumusa section an early introduction from W Malesia

Dry f. = dry forest; FW = freshwater; litt. = littoral; lowl. = lowland; mt. = mountain; Rf. = rainforest; ripar. = riparian; seas. = seasonal; temp. = temperate. c = cultivated as supplementary food; cs = cultivated as staple food; t = transplanted; w = wild; ws = wild form used as staple food.
Amer. = America; Arnhem = Arnhem Land; Aust. = Australia; C. York = Cape York Peninsula; Malay P. = Malay Peninsula; Micr. = Micronesia; Mol. = Moluccas; NG = New Guinea; N. Hebr. = New Hebrides; Pac. = Pacific islands; Phil. = Philippines; Poly. = Polynesia; Sol. = Solomon Is; wide. = widespread.

Table 2. Other indigenous crops

Taxon	No. of species	Area of natural distribution	Vegetation type	Part used in NG and status	Areas of known similar use
Alocasia	70	SE Asia, Malesia, Ceylon	Rf.: lowl. and lower mt.	Tuber, leaves; c, w	India, Malaya, Pac.
Alpinia	250	Asia to Pac.	Rf.: lowl. and mt.	Leaves, rhizome; c, w	Wide.
Athyrium	180	Cosmopolitan	Rf.: wide.	Young fronds; w	
Bambusa	70	Africa, Asia, Amer.	Rf., dry f.: wide.	Shoots; c, w	China, Phil., Java
Commelina	230	Tropics and subtrop.	Grasslands, f. margins, open areas	Leaves, shoots; c, w	India, Java, Malay P
Curcuma	5	Asia, Malesia	Rf.: lowl. and mt.	Rhizome; c, w	Wide.
Cyathea	600	Tropics and subtrop.	Rf.: mt and subalpine grasslands	Young fronds; c, w	Malay P.
Cyclosurus	1	Tropics	Rf.: wide.	Young fronds; w	
Cyrtosperma	18	Indo-Malesia	Rf.: lowl. and lower mt.	Tuber; c, w	Malay Is, Phil., Pac.
Dennstaedtia	70	Tropics, temp. S Amer., Aust., N.Z.	Rf.: wide.	Young fronds; w	
Dryopteris	150	Cosmopolitan	Rf.: wide.	Young fronds; w	
Floscopa	20	Tropics and subtrop.	F. margins, open areas	Young fronds; w	Malay P. (medic.)
Hemigraphis	100	S China, Malesia to Pac., Aust.	F. margins, open areas	Leaves, young shoots; c	Malay P., Java (medic.)
Kaempferia	70	Africa, India to S China, Malesia	Rf.: lowl.	Rhizome, leaves; c, w	Malay P.
Nephrolepis	30	Trop., Japan, N.Z.	Wide.	Roots; w	Java (shoots)
Oenanthe (javanica)	40	Africa, Eurasia, Malesia	Mt. swampland	Leaves, shoots; c, w	Indo-China, Malay P., Sumatra
Polyscias	80	Palaeotropics	Rf.: lowl. and mt.	Young leaves; w	Java, Indo-China
Pteris	250	Cosmopolitan	Wide.	Root, young frond; w	E Indies
**Pueraria	35	Himalaya to Pac.	Mt. grasslands and f. margins	Tuber; c, w	Indo-China
Rorippa	70	Cosmopolitan	Rf.: mt., ripar.	Young plant; c, w	
Rungia (klossii)	50	Africa, India to NG	Rf.: mt.	Leaves, shoots; c, w	

Saccharum (edule)	5	E Asia, Malesia	Open areas, grassl., swamplands	Young inflor.; c	Malay P.
Saccharum (officinale)				Cane sugar; c	Wide.
Setaria (palmifolia)	140	Tropics and warm temp.	Mt. grassl., f. margins	Shoot; c, w	Malay P. (medic.)
Solanum (nigrum)	1700	Tropics and temp.	Mt. grassl., f. margins, open areas	Leaves, shoots; c, w	Wide.
Zingiber	80–90	E Asia, Malesia to Pac., N Aust.	Rf.: lowl. and mt.	Rhizome, leaves; c, w	Wide.

For abbreviations see Table 1.

Table 3. Exotic crops,* areas of origin and time of entry to New Guinea

Taxon	Area(s) of origin	Possible time of entry into New Guinea	Part used and status in NG
Amaranthus spp.	Africa, Asia, Amer.	Early, ? some indigenous	Leaves, young plant; c
Colocasia esculenta	India, Indonesia	pre-4500 BP	Corm, leaves; cs, c
Dioscorea alata	Assam-Burma	? 2000 BP	Tuber; cs, c
D. esculenta	Indo-China	? 1600 BP	Tuber; cs, c
D. bulbifera	Asia, Africa	Early	Tuber, bulbils; c, w
D. nummularia	Asia, Indonesia	Early	Tuber; c
D. pentaphylla	Asia, Indonesia	Early	Tuber; c, w
D. hispida	India, S China, NG	? Indigenous or Early	Leaves, tuber; c, w
D. spp.	SE Asia, NG	Indigenous	Tuber; w
Hibiscus manihot	China	Early	Leaves, shoots; c
Ipomoea batatas	C, S Amer.	Post-15th or 16th Century or ? pre-1200 BP	Tubers, leaves; cs, c
Lablab niger	Asia-India	Post-17th Century	Leaves, seeds; c
Lagenaria siceraria	Africa	Pre-2300 BP (to Mexico 7500–9000 BP) (to Peru 5000–6000 BP)	Leaves, young fruit; c
Phaseolus lunatus	C, S Amer.	Post-15th or 16th Century	Leaves, pods, seeds; c
tetragonolobus Psophocarpus	?Africa, ?Asia	Post-17th Century	Leaves, flowers, pods, seeds, tubers; c

*Some species of Amaranthus and Dioscorea may be indigenous.
For abbreviations see Table 1.

much earlier. The recent claim for pigs being present in highland areas by 10 000 BP (Bulmer 1975, chapter II:3) support the suggestion of an early transfer, since pigs are also apparently exotic to New Guinea. Acceptance of yams and bananas in lowland areas may have been rapid, especially if they were already familiar crops. The establishment of these crops in highland areas would have been somewhat later as the climate of the inter-montane valleys and basins was severe until after c. 10 000 BP.

4. The development of highlands agriculture

Recent archaeological work in the Upper Wahgi Valley of the central highlands of Papua New Guinea suggests that 9000–10 000 BP is, in fact, the time of horticultural experimentation there.

4.1 The archaeological evidence

Several sites have been studied. The earliest investigations, at Manton's site (an open swamp site situated 3 km north of the foothills of the Kubor Ranges, at 1590 m; Golson *et al.* 1967; Lampert 1967; Powell 1970a,b) indicated that stratigraphically there was a zone of disturbed sediments lying between the basal sand or an intermittently developed organic detritus and an undisturbed upper peat. Within this zone a series of ditches, cut from at least 3 different levels were found together with artefacts such as wooden stakes, pointed digging sticks and paddle-shaped spades, and stone axe-blades, sharpening stones, cooking stones and charcoal. The ditches were interpreted as drainage ditches being dug to control water movement during periods of gardening on the swampland itself. Radiocarbon dates indicate the site is of 4800–5000 BP (ANU 252) and the upper undisturbed peat some 400–550 BP (ANU 289). Two of the ditching systems have been dated by associated artefacts to 2300 BP (±120, ANU 43) and to 980 BP (±150, ANU 251).

At the Draepi-Minjigina site (a sinuous swamp at 1890 m within the rolling hilly country at the base of Mt Hagen; Powell 1970a,b, Powell *et al.* 1975) charcoal from a cooking pit embedded in coarse organic detritus dated to 2310 BP (±90, ANU 255) and a more recently cut ditch sealed in by volcanic ash was in use prior to 1190 BP (±50; ANU 1206).

More recent archaeological work has concentrated on the Kuk site. This site lies in a broad, swampy depression on the S side of Ep ridge at 1550 m (Fig. 1). The swamp appears flat but there is a slight rise across the central area, the western end being drained by the Guga river, the eastern end by Tibi creek, a tributary of the Wahgi river. Relatively steep slopes giving way to colluvial fans at the base of Ep ridge border the site to the north while low volcanic-ash-mantled hills are present to the south. Ep ridge rises to 1960 m, the southern hills to 1600 m. Slopes are very gentle, between 1 in 200 and 1 in 400 over most of the site (Powell 1974a). The lowermost organic and inorganic sediments (older than 9000 years) appear to be undisturbed and volcanic ash bands are more or less

Fig. 1. Aerial view looking north across the Kuk archaeological site. Major drainage channel running NW from the S boundary in centre foreground; rectangular-plot patterns obvious; Ep ridge in background.

continuous and horizontally stratified within them; the uppermost, felted peat (formed over about the last 100 years) is undisturbed also. Between lie the sediments disturbed and altered by man's activities: included within them are artifically cut channels and ditches, house platforms and artefacts. Volcanic ashes are horizontally bedded in some cases, while in others they appear as dispersed pellets or fine bands within ditch fills, and in others they seal across the tops of the ditches; they provide an essential marker for correlation of a particular ditching system from one part of the site to another.

Five phases of ditching (dating back to c. 9000 BP) and associated field systems have been investigated and interpreted in terms of agricultural management of the swamp (Golson 1976, 1977). The first two phases (dating to c. 9000 BP and 6000–5500 BP respectively) have field systems characterized by specialized features (including shallow basins, gutters, hollows, pits and stakeholes) which are linked directly to one or more large drainage channels, dug across the swamps. Golson suggests that these systems are associated with the cultivation of a particular plant, such as taro, or possibly some indigenous crops. The phases are separated by a wedge of grey clay (of greatest thickness at the S end of the site) which appears to be "the product of accelerated erosion following the beginning of forest clearance" (Golson and Hughes 1976).

An integrated and increasingly more intensive system of drainage characterizes the more recent phases of swamp use suggesting that the site became wetter and made management more difficult. The field systems also change: the third phase (dating to 4000–2500 BP) lacks the specialized features of the earlier phases and Golson considers a mixed farming regime was the most likely agronomic system being practised. Just before the close of this phase the agronomic practice of tillage seems to have been introduced. The later phases (c. 2000–c. 1200 BP and c. 400–c. 100 BP) reflect a "far more systematic and perhaps specialized use of the swamp" involving cultivation of taro and, in more recent times, sweet potato (Golson 1977).

4.2 The plant evidence

4.2.1 The Wahgi valley sites. The use of the swamplands in the Wahgi valley may have been very localized or more regional; if they were used on a regional scale the pollen diagrams should register any anthropogenic influences on the vegetation there and reflect major changes in land use or agronomic practices.

Altogether 8 pollen diagrams are available for the region: four from the Draepi-Minjigina site, two from the Manton site (Powell 1970a) and one each from Ambra (a lake 110 m long, 70 m wide, 1.5 m deep, fringed with sedge-grass swamp, situated at 1620 m about 2 km west of Kuk; Powell 1974b) and Kindeng (an archaeological site situated 20 km E of Kuk; Powell 1972). All of the diagrams have been drawn up using a pollen sum which includes all forest and woody non-forest pollen taxa. Absolute counts are available only for the Ambra site; these are given in Table 4, expressed as the number of grains/cc $\times 10^3$. The pollen diagrams suggest that:

(1) reduction of forest attributable to human influences had occurred prior to 5000–5300 BP at Draepi-Minjigina (Fig. 2), Manton's and Ambra (Fig. 3). Evidence of earlier vegetation changes is lacking at present due to the poor preservation of pollen in the sediments or a lack of sediments available for the older periods. The stratigraphy of the Ambra site (Fig. 3) suggests that there was a major change in sedimentation rate between 5700–6300 BP* and the site may have been dry or almost dry between 4600 BP and 24 400 BP.

(2) Further clearance of the forest had occurred at all sites by 4000 BP and this was followed by recovery and stabilization of the forests for some 1800–2400 years in most areas. This seems to be correlated with the 4000–2500 BP phase of swamp use for gardening at Kuk. Assuming that a system of shifting cultivation with a long-term fallow was being practised then a general movement to use the swamps of the valley floor may be related to agronomic and/or social factors such as

(a) the introduction of a new crop suited to swamp conditions, for example, taro (if not already present),

(b) the degradation of the soils on the slopes; crop yields would have been lowered unless fallow periods were increased to maintain production,

(c) the altitudinal limit of economic production of the cold-intolerant tropical crops had been reached; if the tree-line had been pushed up to altitudes which

*The radiocarbon date of 4400 ± 130 shown in Fig. 3 is considered to be too young on the basis of sedimentation rates calculated from other dates available.

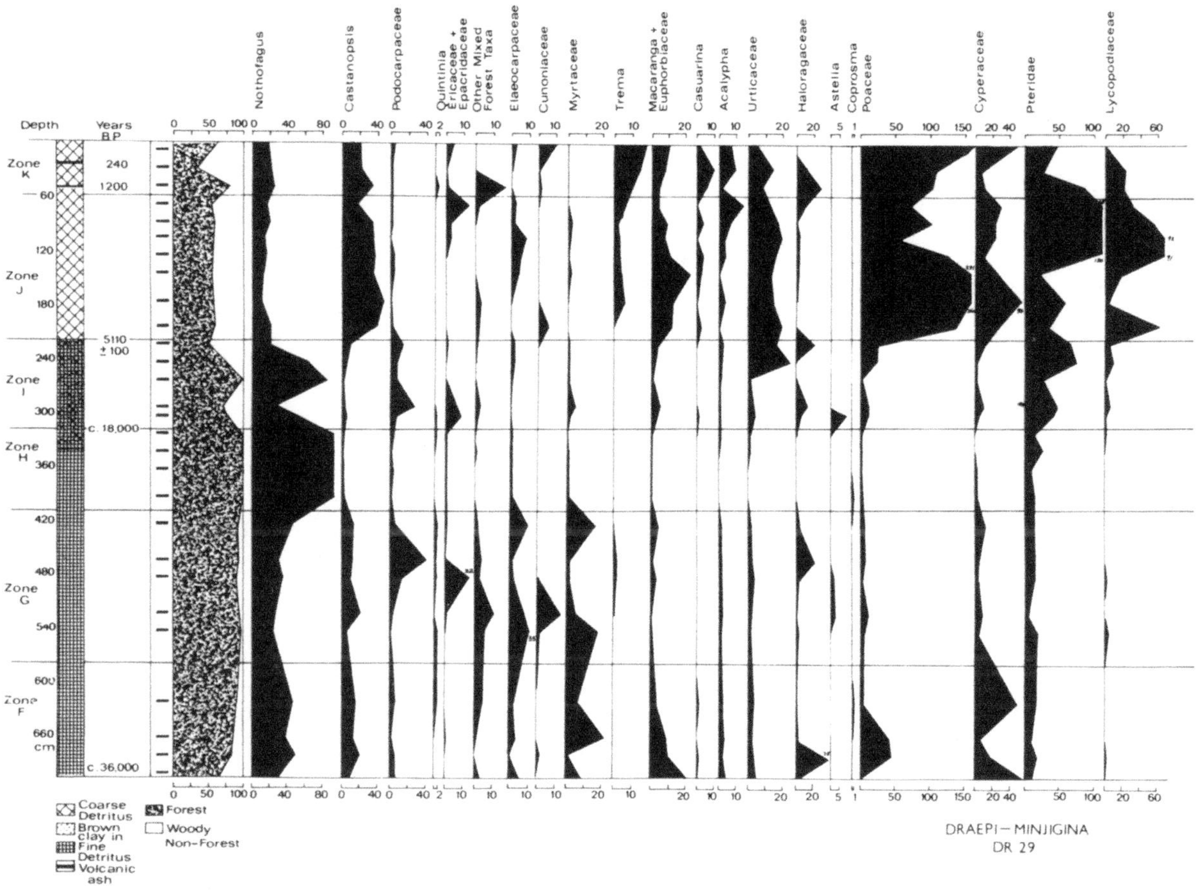

Fig. 2. Pollen diagram for Draepi-Minjigina site, core DR 29. Stratigraphy and radiocarbon dates are shown on the LHS, the summary diagram and selected pollen taxa curves from left to right.

Table 4. Absolute (influx) pollen counts for Lake Ambra core

Pollen	Depth (cm) 141–143	157–158	167–168	175–176	197–198	207–208	217–218	238–239	248–249	258–259	273–274	295–296
Nothofagus	15*	17	51	64	24	30	14	10	9	9	12	3
Castanopsis	16	22	50	73	31	34	17	9	12	17	17	10
Elaeocarpaceae	1	1	3	8	3	1	2	1	+	2	2	1
Podocarpaceae	1	3	6	8	6	3	2	2	1	1	1	0
Casuarina	7	7	3	1	0	+	+	0	+	+	0	+
Trema	7	3	5	4	2	4	3	3	3	3	6	2
Dodonaea	2	1	4	1	0	+	1	1	1	+	1	+
Acalypha	5	3	11	17	6	4	1	2	1	1	2	1
Macaranga	4	5	12	18	13	10	6	2	3	3	4	1
Urticaceae	20	10	25	42	14	16	7	8	5	10	22	8
Charcoal	7300	59400	nc	4540	2750	3200	1170	nc	2440	1060	4430	1160
Age (C-14)		500 BP	1200 BP				2840 BP					c.6500 BP

*Grains per cc × 10^3 nc = no count

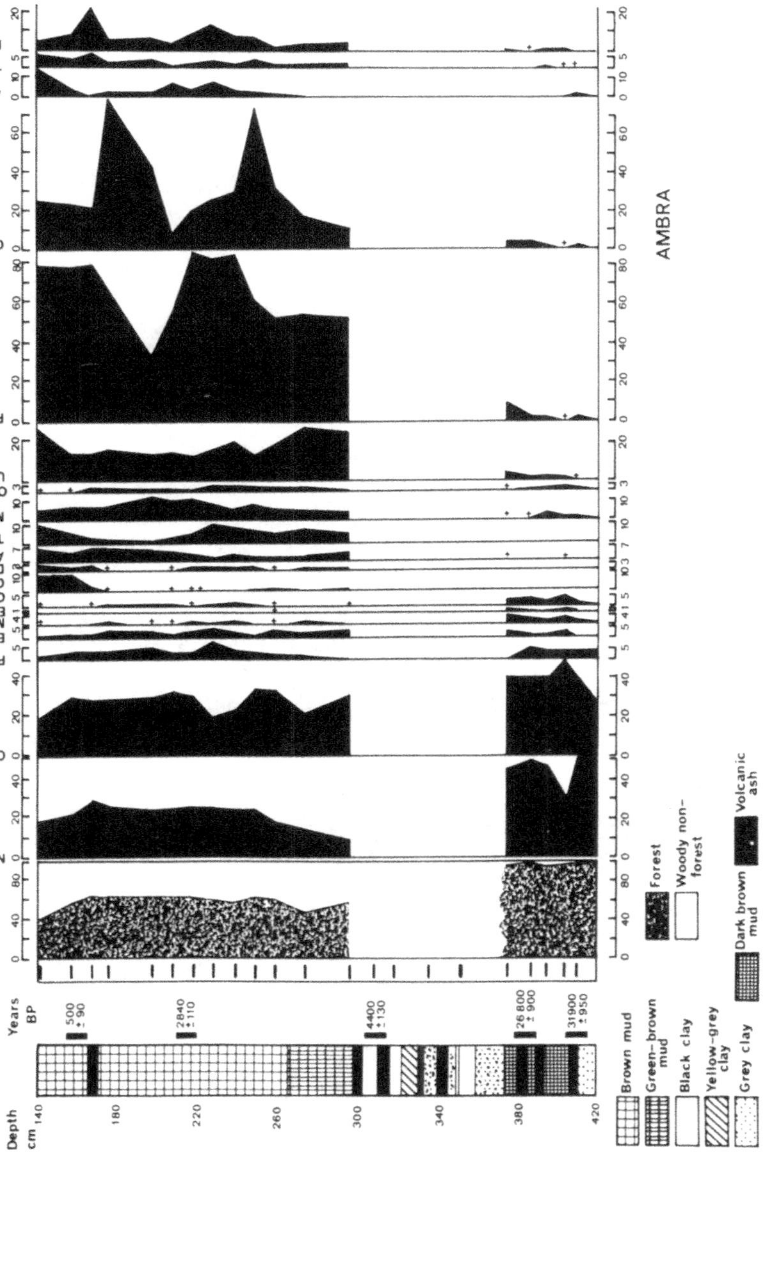

Fig. 3. Pollen diagram for Lake Ambra, core AM 6.

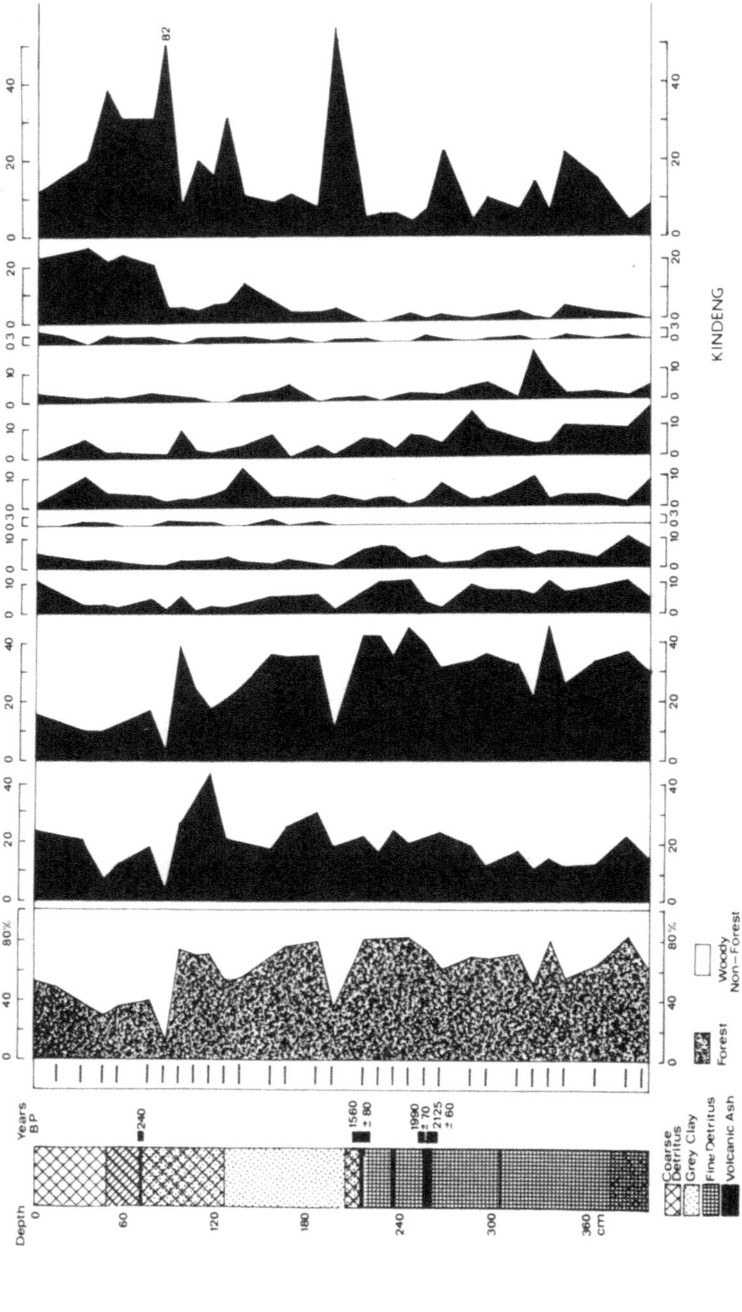

Fig. 4. Pollen diagram for Kindeng site.

were less suitable for the crops being grown then maturity times would have increased until uneconomic.
(d) the greater fertility of the swamp soils had become known through experimentation; crop yields could be maintained with shorter fallow periods there compared with on the slopes,
(e) populations had increased and there was considerable pressure on land.

The use of the swamplands allowed some regeneration of forest to occur on the slopes but these were undoubtedly still used for gardening, and further clearance of forest occurred at least in some parts of the Wahgi as indicated in the Kindeng diagram, and dated to 3000 and 3300 years BP (Fig. 4).

(3) Further disturbance and clearance of forest is registered in the Wahgi valley diagrams at various times after this: between 2000 and 2200 BP at Kindeng and Ambra, between 1500 and 1800 BP at Draepi-Minjigina and again at 1370 BP at Kindeng and between 660 and 750 at both Kindeng and Draepi-Minjigina. Individual pollen curves indicate a reduction in forest ephemerals from about 2250 BP in the Kindeng diagram and in the Ambra diagram the early regrowth shrub, *Acalypha*, assumes importance by 2500 BP.

These changes may reflect the expansion of grassland at the expense of forest regrowth and they could be related to the re-use of the lower slopes for gardening once the technique of tillage has been developed (by about 2500 BP, Golson 1976). Clarke and Street (1967) have shown that economic crop yields can be gained in grassland gardens if complete tillage of the soil is undertaken. The recovery of the forest near Kuk after 2000 BP (as indicated in the Ambra pollen diagram) may be related once again to some of the pressure being taken off the slopes surrounding the site once people started gardening the swampland again, but elsewhere in the Wahgi valley disturbance and clearance of forest seems to have continued.

Changes in individual pollen curves from 1200–1400 BP are important also. *Casuarina* values increase in all diagrams (excluding Manton's where the sediments are mixed), *Trema*, *Acalypha* and *Dodonaea* increase in the Draepi-Minjigina diagram and *Trema* and *Dodonaea* assume importance in the Ambra diagram from about 1200 BP. *Pandanus* appears at about 1270 BP in the Kindeng diagram. It seems likely that these increased values reflect a change in the agricultural system: they may be related to the development of arboriculture and in particular to the adoption of a controlled tree-fallowing system. Whether or not this development was associated with the appearance of a new crop such as the sweet potato, as suggested previously (Powell *et al.* 1975; Golson 1976) is unknown. Golson (1977) has related the abandonment of drainage at Kuk and at Minjigina at about 1200 BP to the adoption of the tree-fallowing practice.

(4) A further reduction in forest values, and particularly in *Nothofagus* and *Castanopsis*, is shown in all the diagrams at 380–400 BP. This is associated with a major increase in *Casuarina*. High values are recorded also for *Acalypha*, *Trema* and *Saurauia* in the Manton diagrams and major increases in *Acalypha* and in *Trema* occur in the Kindeng and Draepi-Minjigina diagrams re-

spectively. These changes suggest there was increased pressure on the forested slopes and an expansion of grassland areas and of areas under controlled tree-fallowing. They may be associated with the introduction or expansion of sweet potato cultivation. The date is closer to that expected for the arrival of sweet potato in New Guinea on the basis of historical and botanical evidence (Yen 1974b) but the possibility of an earlier introduction (at about 1200 BP) cannot be dismissed entirely at present.

4.2.2 Other highland sites. Vegetation changes which may be interpreted in terms of human influences are present in diagram from a number of other highland sites (Walker 1970; Walker and Flenley 1979; Flenley 1967; Hope 1976).

Pollen diagrams from the Sirunki area of the Enga District, some 80 km W of Kuk and at 2500 m, indicate reduction of forest had occurred there by about 4300–4500 BP (inferred age) and further disturbance is registered at about 2000 BP. At Inim (1550 m, near Sirunki) *Trema* assumes importance between 3000 and 2000 BP and clearance of forest occurs between 2000 and 1600 BP. At Birip (1900 m, 50 km W of Kuk) clearance occurred by at least 2000 BP. At both Inim and Birip greater disturbance and further clearance of forests occurred by 450–500 BP. This was followed by a great increase in *Trema*, *Casuarina* and *Dodonaea* at Birip from about 300 BP while at Inim *Casuarina* expands from about 260 BP.

Considering pollen diagrams from sites on Mt Wilhelm, 80 km E of Kuk, Hope (1976) suggests that vegetation changes dating to between 6500 and 5000 BP can be related to climatic and/or anthropogenic influences. After 5000 BP at least some of the changes in the extent of the lower montane forest may be related with more certainty to human impact. The rises in *Casuarina* and *Trema* values after about 1000 BP and the great increase in *Casuarina* from about 300 BP suggest expansion of gardened areas in the Chimbu valley in more recent times.

The pollen diagrams from other areas strongly support some of the Wahgi valley findings and it can be concluded that:

(1) By 4000 BP considerable areas of forest had been cleared throughout the highlands; gardens had been expanded into high altitude areas and swamplands as well as mountain slopes were being used systematically for agricultural purposes. Details of the agricultural system and of the crops being grown remain unknown, but it seems likely that a system of shifting agriculture with long-term fallowing was being practised. Gardens were probably of the mixed-crop type with both indigenous domesticates (bananas, sugarcane, *Setaria*, *Rungia*, *Saccharum edule*) and exotic species (taro, yams, gourds) being grown.

(2) By 2000 BP further pressure was being exerted on the forests throughout the W Highlands and grassland areas were expanding. Gardening of these areas had become possible, however, with the development of techniques of complete tillage. In the Wahgi valley larger areas of swampland were being used at any one time and there may have been a certain amount of crop segregation with taro being grown in the swamp and forest gardens, the other crops in grassland and forest gardens.

(3) By 1200–1000 BP arboriculture and a system of controlled tree-fallowing had developed not only in the Wahgi but also in the Chimbu valley and more intensive use of the dry land sector was then possible. Despite this further forest clearance occurred.

(4) From about 400 BP the extent of forests in all areas of the highlands was reduced even further. At the same time the practice of tree-fallowing had expanded within the Chimbu and Wahgi valleys and been introduced also into the Enga Prov. Just after this the swamplands of the Wahgi valley were drained once again and the archaeological evidence suggests that a system of cultivation using raised plots was being employed. The development of such an agronomic practice would permit frost-intolerant crops to be grown at higher altitudes in grassland or forest gardens as well as allowing the moisture-intolerant crops to be grown on the swampland and it may explain, at least in part, the continued destruction of the forests in more recent times.

Many questions remain to be answered but one in particular overrides others: why is there no information on early forest clearance in the pollen diagrams? The archaeological evidence from Kuk suggests that gardening was being practised in the Wahgi valley some 9000 years ago, while the pollen diagrams indicate only that clearance of forest for agricultural purposes occurred at some time prior to 5300 BP. In the Wahgi valley undisturbed sediments with well-preserved pollen which cover this time period are lacking. Analyses of the ditch infills and of sediments of the Kuk site itself may provide some further information but the evidence gained will have to be interpreted with extreme caution since the deposits are obviously disturbed.

Possibly the earliest indications of gardening would not be shown in the pollen diagrams if the first gardens were very small, short-lived and only very locally distributed. Walker and Flenley (1979) suggest that prior to 4500 BP (inferred age) human influences did not alter the natural regeneration pattern of the forest; later a general degradation of the forests occurred and this is clearly indicated in the diagrams. Certainly the earliest indications would be difficult to separate from the natural successional patterns which occur following climatic change or catastrophic events such as volcanic eruptions or tectonic movements.

Further work (already under way) on a number of other sites in the highlands, together with studies of lower altitude sites may help to resolve some of the problems, and provide us with a clearer picture of the development of agriculture within New Guinea.

References

Allen, F. J. 1977. The hunting neolithic: adaptations to the food quest in prehistoric Papua New Guinea. In: J. V. S. Megaw (ed.) Hunters, gatherers and first farmers beyond Europe. Leicester, Leicester University Press.

Balgooy van, M. M. J. 1976. Phytogeography. In: K. Paijmans (ed.) New Guinea Vegetation. Canberra, Australian National University Press.

Brookfield, H. C. with Hart, D. 1971. Melanesia: a geographical interpretation of an island world. London, Methuen.

Bulmer, R. 1968. The strategies of hunting in New Guinea. Oceania 38 (4): 302–18.
Bulmer, S. 1975. Settlement and economy in prehistoric Papua New Guinea: a review of the archaeological evidence. J. Société des Océanistes 31 (46): 7–75.
Chappel, J. and B. G. Thom. 1977. Sea levels and coasts. In: J. Allen, J. Golson and R. Jones (eds) Sunda and Sahul: Prehistoric studies in southeast Asia, Melanesia and Australia. London, Academic Press.
Clarke, W. C. and J. M. Street. 1967. Soil fertility and cultivation practices in New Guinea. J. Trop. Geog. 24: 7–11.
Flenley, J. R. 1967. The present and former vegetation of the Wabag region of New Guinea. Unpubl. Ph.D. thesis, Australian National University, Canberra.
Golson, J. 1976. The making of the New Guinea highlands. In: J. H. Winslow (ed.) The Melanesian environment: change and development. Canberra, Australian National University Press.
—— 1977. No room at the top: agricultural intensification in the New Guinea highlands. In: J. Allen, J. Golson and R. Jones (eds) Sunda and Sahul: Prehistoric studies in southeast Asia, Melanesia and Australia. London, Academic Press.
Golson, J., R. J. Lampert, J. M. Wheeler and W. R. Ambrose. 1967. A note on carbon dates for horticulture in the New Guinea highlands. J. Polynes. Soc. 76 (3): 369–71.
Golson, J. and P. J. Hughes. 1976. The appearance of plant and animal domestication in New Guinea. In: J. Garanger (ed.) La préhistoire océanienne (IX Congress of Pre- and Protohistoric Sciences, Nice, September 1976): 88–100. Paris: Centre National de la Recherche Scientifique, preprint.
Harris, D. R. 1977. Subsistence strategies across Torres Strait. In: J. Allen, J. Golson and R. Jones (eds) Sunda and Sahul: Prehistoric studies in southeast Asia, Melanesia and Australia. London, Academic Press.
Hope, G. S. 1976. The vegetational history of Mt Wilhelm, Papua New Guinea. J. Ecol. 64: 627–664.
Hope, J. H. and G. S. Hope. 1976. Palaeoenvironments for man in New Guinea. In: R. L. Kirk and A. G. Thorne (eds) The origins of the Australians. Canberra, Australian Institute of Aboriginal Studies.
Kershaw, A. P. 1975. Late Quaternary vegetation and climate in northeastern Australia. In: R. P. Suggate and M. M. Cresswell (eds) Quaternary studies. Wellington, Royal Soc. New Zealand, Bull. 13.
—— 1976. A late pleistocene and holocene pollen diagram from Lynch's Crater, northeastern Queensland, Australia. New Phytol. 77: 469–498.
Lampert, R. J. 1967. Horticulture in the New Guinea Highlands: C-14 dating. Antiquity 41: 307–9.
Löffler, E. 1976. The impact of traditional man on landforms in Papua New Guinea. In: J. H. Winslow (ed.) The Melanesian environment: change and development. Canberra, Australian National University Press.
—— 1977. Geomorphology of Papua New Guinea. Canberra, Australian National University Press.
Nix, H. A. and J. D. Kalma. 1972. Climate as a dominant control in the biogeography of northern Australia and New Guinea. In: D. Walker (ed.) Bridge and Barrier: the Natural and Cultural History of Torres Strait. Research School of Pacific Studies, Department of Biogeography and Geomorphology, BG/3. Canberra, Australian National University.
Paijmans, K. 1976. Vegetation. In: K. Paijmans (ed.) New Guinea vegetation. Canberra, Australian National University Press.
Powell, J. M. 1970a. The impact of man on the vegetation of the Mt Hagen region, New Guinea. Ph.D. thesis, Australian National University, Canberra.
—— 1970b. The history of agriculture in the New Guinea Highlands. Search 1 (5): 199–200.
—— 1972. Report on the Kindeng site, Upper Wahgi valley, New Guinea. Ms. Prehistory Department, Australian National University, Canberra.
—— 1974a. Preliminary report on ecological investigations at Kuk. I. The stratigraphy of the Kuk site. Ms. Prehistory Department, Australian National University, Canberra.
—— 1974b. A preliminary report on the Lake Ambra vegetation history site. Ms. Prehistory Department, Australian National University, Canberra.
—— 1976a. Plants, man and environment in the island of New Guinea. In: J. H. Winslow (ed.) The Melanesian environment: Change and development. Canberra, Australian National University Press.
—— 1976b. Ethnobotany. In: K. Paijmans (ed.) New Guinea vegetation. Canberra, Australian National University Press.
Powell, J. M., A. Kulunga, R. Moge, C. Pono, F. Zimike and J. Golson. 1975. Agricultural

traditions of the Mt Hagen area. Department of Geography, Occas. Paper 12, University of Papua New Guinea, Port Moresby.
Walker, D. 1970. The changing vegetation of the montane tropics. Search 1 (5): 217–221.
Walker, D. and J. R. Flenley. 1979. Late quaternary vegetational history of the Enga Province of upland Papua New Guinea. Phil. Trans. Roy. Soc. Lond. B 286 (1012): 265–344.
Yen, D. E. 1974a. Arboriculture in the subsistence of Santa Cruz, Solomon Islands. Econ. Bot. 28 (3): 247–284.
—— 1974b. The sweet potato and oceania: an essay in ethnobotany. Honolulu, Bishop Museum Bull. 236.
—— 1977. Hoabinhian horticulture: the evidence and the questions from northwest Thailand. In: J. Allen, J. Golson and R. Jones (eds) Sunda and Sahul: Prehistoric studies in southeast Asia, Melanesia and Australia. London, Academic Press.

5. Staple crops in subsistence agriculture

Their major insect pests, with emphasis on biogeographical and ecological aspects

W. C. Gagné

1. Introduction

Research on subsistence agriculture, staple crops and pests of staple crops have generally been neglected by researchers in New Guinea. This could also apply to many other tropical countries. Byrne (1973) ascribes this neglect to: "(a) The attitude that the traditional bush fallow system, which has evolved over many hundreds of years, is difficult to improve upon, and, where it can be practised meets the needs of the country (Papua New Guinea) for subsistence cropping. (b) The development of cash cropping is necessary to meet the increasing demand for money for both the country and the individual. Also it provides a substitute for the traditional male role of protector and fighter which is no longer necessary." Byrne also points out that it is certainly "Possible to improve upon the quantity and quality of crops grown under this [subsistence] system by the application of crop technology. In many cases this improvement is gained with little extra effort."

As a system, shifting agriculture is stressed and eventually falters severely when there is not sufficient land for fallows of necessary length, i.e. at least 15 years in some areas (Scott 1977). In low to medium elevations bush fallows are being shortened because of population pressure with the frequent result that nutrient poor, fire tolerant grasslands are replacing the forest (Gagné 1977, 1979) or yields are dropping critically as on the Gazelle Peninsula on New Britain and parts of the Chimbu and Sepik provinces of Papua New Guinea (PNG) (Byrne 1973). Recent measurements at latitudes and climates similar to New Guinea in South America indicate that a minimum 15-year period is required for fallow aerial biomass to increase nutrient stores sufficiently to maintain a continuous swidden-fallow cycle without causing environmental degradation and savannization (Scott 1977). Scott emphasizes (p. 325) that "Many areas where shifting cultivation is the accepted practice have experienced serious environmental damage because of the shortening of the all-important fallow period in the hope of sustained food production. Savannization and scrublands attest to the unwise, if unknowing, practices used in those very regions where population densities require sustained agricultural yields."

I have chosen to review the major insect pests of staple crops grown in

Monographiae Biologicae, Vol. 42, ed. by J. L. Gressitt.

© 1982, Dr W. Junk Publishers, The Hague.

subsistence agriculture from ecological and biological perspectives as an illustration of applied biogeography.

2. Origin and development of subsistence agriculture

From the excavations at the Kosipe archaeological site, at about 2000 m, we know that people had already gained access to the Papuan Highlands about 25,000 years BP, probably foraging in the area for wild pandanus nuts on a seasonal basis (White *et al.* 1970; and see chapters II:3, II:4). This indicates man's early detailed knowledge of the local environment, for these pandanus are indigenous, and knowledge of the plants needed to be learned after arrival. Coupled with evidence from Australia (Jones 1973), it is now reasonable to assume that humans first reached New Guinea by 40 000 BP. Direct evidence of this extends back 36 000 years (see chapter II:3). We know that intensive gardening was being practised in New Guinea before 9000 BP. Moreover, it is generally accepted that the earliest arrivals into the insular Pacific brought the pig, chicken and dog with them.

Both the locations of archaeological sites and the bone food remains suggest that the subsistence pattern of these early migrants was oriented much towards the sea and the exploitation of sea resources, but that the land hunting also contributed to the diet. The appreciable quantities of pig remains suggest that their animals were probably husbanded rather than hunted in their feral state. If this assumption is correct it is also likely that some gardening was practised, since pigs kept in villages often require feeding. Thus the economy appeared to be a mixed hunting, farming and fishing one (Allen 1977).

If about 1000 years ago coastwards movements of inland groups and coastal migrations by sea took place to the extent suggested by Allen (*loc. cit.*) from archaeological evidence, then we can also suggest with some confidence that population along the coast increased beyond the capacity of the previous localized and general economic systems to maintain them. In such a system of disequilibrium we would expect intensification of the food quest, perhaps by the development of new technology (including not only new tools, but also more intensive gardening techniques, more specialized production of fewer crops, etc.) together with improvements in the systems of distribution and communication, so that any groups' economic catchment would extend well beyond the territory occupied. For example, Motu people in the seasonally dry SE New Guinea savanna could not provide sufficient food for themselves from their own environment, and were dependent upon maintaining an economic catchment that extended inland into more fertile garden areas and along the coast into sago (*Metroxylon* spp.) producing areas (Allen 1977). The sago was exchanged by *Hiri* traders on large sailing canoes called *lakatoi.*

3. Plants, people and environment

New Guinea vegetation is extremely diverse but it can more or less be differentiated into types occupying particular habitats: the coastal and estuarine mangrove and *Nypa* swamps, sago palm fresh water swamp forests, dry eucalypt savanna and grassland, lowland and foothill tropical rainforest, more temperate montane forests, high mountain elfin cloud forest, and alpine grasslands (Powell 1977). All of these are exploited for food, medicines, building materials, etc., but shifting cultivation is carried out in all but the last 2 vegetation types. Sweet potato (*Ipomoea batatas*), the staple grown to the highest elevations, does not appear to be a viable agricultural proposition much above 2700 m.

The vegetation and environment, then, came to profoundly affect the subsistence agriculture mode such that particular staple and supplementary foods were exploited according to the availability of these resources in the various vegetation areas. In a recent ethnobotanical review Powell (1976) lists 650 species (representing 378 genera, 134 families) used for these purposes. Some such as coconuts, bananas, sago, *Pandanus* spp., and bamboos provided basic sustenance and almost all material needs. Powell (1977:13) organized the agriculturalists into 4 basic groups that somewhat mirror the vegetation classification outlined above, and listed the staple and main supplementary foods of each. These (slightly abridged) are:

1. *Coastal and lowland swamp dwellers:* Extensive freshwater swamps are present along both the north and south coasts of New Guinea; mangroves occupy coastal mud flats and river estuaries. Population densities are very low in these areas, 2–4 persons per km^2. Sago is the staple; it is sometimes cultivated but usually harvested from wild stands. Supplementary foods are often traded in from neighboring areas but some may be grown in 'hanging gardens' built on piles among the mangroves or in small gardens situated on higher ground. At least 16 flowering plant species are gathered from the strandline and forests, together with ferns, fungi, and seaweeds.

2. *Savanna and grassland shifting agriculturalists:* The southern plains of Irian Jaya and Papua and some broad valleys of the north coast of New Guinea are covered with open forest, savanna and grassland. Population densities are generally low but may reach 16–30 persons/km^2 in local coastal areas. Yams (*Dioscorea* spp.) are the staple in many of the seasonally dry areas. Both cultivated and wild species are utilized. In coastal areas near Port Moresby and in the Markham Valley bananas (*Musa* spp.) are the staple, while in other coastal areas cassava (*Manihot esculenta*), a recent introduction, is more important; the latter is tolerant of a wide range of environmental conditions and will grow in drier and poorer soils than either yams or bananas. At least 14 other species are cultivated as supplementary crops and a further 9 are available as wild food species.

3. *Tropical rainforest shifting agriculturalists:* Tropical rainforest dominates the coastal plains, foothills, and low mountains where rainfall is higher and more evenly distributed through the year. Population densities vary from 8–16/km^2. Taro (*Colocasia esculenta*) is the staple with bananas, sweet potatoes, and yams being grown as the main supplements. Sweet potato has replaced taro as the staple in some areas (for reasons discussed below). Many other supplementary crops are grown and at least 26 wild species provide edible leaves, fruits and nuts. A number of edible ferns are recorded also.

4. *The highland agriculturalists:* The densest populations, locally up to 200 persons per km^2 and nearly 40% of the total population live at 1400–2700 m in valleys of the central cordillera. Sweet potato is now the staple with taro, yams, bananas, beans, sugar cane, edible pitpits (*Setaria palmifolia* and *Saccharum edule*), and many green vegetables being grown as supplements; 28 spp. are listed. *Pandanus* species are the most important fruit and nut trees planted and gathered. Many other wild foods are also available in the forests, regrowth and grasslands: the 'cabbage' of palms; leaves, fruits and nuts of trees and shrubs; herbs; fern fronds and fungi (58 spp. listed altogether). (After Powell 1977.)

The staples are predominantly starchy: yams, sweet potato, taro (both *C. esculenta* and *Xanthosoma* spp.), sago, cassava and bananas; while the supplementary crops, both cultivated and wild, comprise a wide range of green vegetables, fruits and nuts. Hunting and gathering groups are in the minority. Most are subsistence agriculturalists. The type of gardening practised varies considerably from simple swidden with slash-and-burn clearance, little if any tillage and abandonment after 1–3 years use, to complex systems involving elaborate ground preparation, water and erosion control measures and more or less continuous cropping with controlled short- or long-term fallows.

4. Stress and evolutionary modification in subsistence agriculture

As indicated above, in New Guinea there is a clear relationship between population density, environment and techniques of cultivation. Peoples of the forested and sparsely inhabited fringes practise a long-fallow-slash-and-burn cultivation of a wide variety of crops. Groups of more densely settled peoples living in the anthropogenic grasslands of the Highlands fallow their gardens for shorter periods and substitute various soil-enriching techniques in place of soil regeneration by forest fallowing. There is also the tendency in the Highlands to concentrate on the high-yielding sweet potato (although maintaining many varieties) at the expense of *Colocasia* taro, yams, and other crops that do better in forest-fallowed soil or at lower altitudes. These present-day spatial variations can be arranged to represent a possible historical process wherein there have been continuing adjustments to changes in environment, crops, land availability and pests — changes and adjustments that Golson (1967) and co-workers have now documented archaeologically for part of the New Guinea Highlands for about 9000 years.

This dramatically long period of experimental evolution — out of which have developed the varied techniques for meeting environmental, edaphic and biological constraints — has continued into the historical period. New staple crops of exotic origin such as *Xanthosoma* taro and cassava have spread widely without formal introduction and in some cases have rapidly become important in subsistence economies (Clark 1977).

The tendency towards the polyculture of many species or at least many varieties of single species acts to protect these agro-ecosystems from heavy losses to pests and pathogens. But, like crop plants elsewhere, tropical crops even in a polyculture mix are still subject to attack — perhaps more than crops in non-tropical climates (Janzen 1973:1213), as will be pointed out below.

Other innate biological factors also impact staples and induce the cultivator to alter or abandon a particular crop and/or cropping system. Among those that immediately suggest themselves are: (1) whether the staple is a native species; (2) its length of time in cultivation, either native or introduced; (3) whether the staple has a large or small number of closely related species that do or could serve as alternate hosts of pests and pathogens as well as their geographical proximity to the staple; (4) increases in the staple's area under cultivation exposing it to more potential pests and pathogens (see Southwood 1960); (5) continuities or discontinuities in the area cultivated to a particular crop acting on pests in a manner consistent with the Theory of Island Biogeography as developed by MacArthur and Wilson (1967); and (6) unintentional introduction of pests via staple propagules (tubers, cuttings, seeds) through human commerce.

These same factors impacting a staple crop will also apply to the biological controls (parasitoids, predators, pathogens) of a particular insect pest. The reproductive strategy of a pest, whether r- or K-selected, will also affect the shifting cultivator. But, it is not clear whether he will have greater problems with r-selected species as do "modern" agroecosystems, especially monocultures, than with K-selected species. Pest buildup in root crop staples such as sweet potato, taros and yams appear to be one of the prime reasons for the abandonment of subsistence plots after a few years. These root crop pests, as well as those of bananas and coconuts, have the characteristics of K-selected species: that is they are larger, more likely to be territorial than migratory, and have a long generation time (see Southwood 1977), and these attributes in turn appear due to the durational stability of their habitat (Southwood *et al* 1974; Southwood 1975), or the length of time that the habitat location remains stable for breeding. Species with stable habitats tend towards K-strategies. Coconut groves, bananas and sago habitats can be considered stable, long-term habitats. And indeed, a survey of their primary insect pests shows a high percentage of species that would fit the K-strategist mode. This appears to be the case also with 75% of the major pests of taro and 50% of those of sweet potato. Although these crops are grown in what outwardly appears to be a highly transient, shifting ecosystem ill-suited to K-strategists, invariably subsistence gardens are surrounded by naturalized and/or derelict populations of these crops, often in close proximity to new gardens. So, a continuum of suitable host plants is present for the pest species.

In a wider context, shifting agriculture can be viewed as the evolution of an agricultural system towards the management of crop pests. Swidden plots mimic tropical forest ecosystems in several ways that influence pest problems. For example, the great diversity of crops typically grown, provides a degree of protection because pests seldom are able to build up to destructive proportions on the relatively isolated few crop plants of each species. Also, the partial canopy consisting of some wild trees left standing and tall crops, such as and depending on the elevation: bananas, pandanus, papayas, etc., reduces the severity of pest problems, particularly weeds and plant pathogens. Burning, rotation, multicropping and shading are all practices that, besides their other effects, reduce pest losses. Natural control aspects are maximized because the clearing of small plots permits relatively easy migration of biocontrol agents from the surrounding jungle. Swidden farmers also practice the selection of host resistance simply by using seed and vegetative parts from the most successful crop plants which have survived in this demanding environment (Glass and Thurston 1978).

5. Staples, pest ecology and biocontrol

The reader may have difficulty with the word "pest" as applied to subsistence agriculture. This concept has little to do with economic thresholds since most of these staples do not normally form a part of the cash economy. I (Gagné in press) have used the 90% unuseable damage level (or higher) to denote "major pests" but other authors have not been this precise. If an author cites a species as a "major pest," this was taken into consideration, and if a majority of authors consider an insect species to be a "major pest" of a particular crop, that pest appears in the following discussion.

Lists of insect pests of staple crops (table 1) germane to this region can be found in Hill (1975; in press), Lamb (1974), Lamb and Johnston (1976), and Simon Thomas (1962). Unfortunately the latter does not distinguish between major and minor pests. Biological control of these and other insect pests is most recently reviewed by Wilson (1960), Cochereau (1972) and Young (in press). A bibliography covering these and other New Guinea insects through 1967 is found in Gressitt and Szent-Ivany (1968). Plant pathogens and weeds are dealt with by Firman (1975, 1978) and Henty and Pritchard (1973), respectively.

5.1 Indigenous staples: Sago (*Metroxylon rumphii*, *M. sagu*, *M. oxybracteatum*)

This is the only staple considered to be unquestionably indigenous (Powell 1977). Only in New Guinea and some of the adjacent islands are sago palms used for food purposes, although it is possible their use was discontinued long ago in other regions where these trees grow (Barrau 1958). However, Keleny (1960, 1962) debates the question somewhat. The preponderance of opinion and evidence would indicate an indigenous origin. No primary pests are listed in the literature, although Lamb (1974) specifically mentions the asiatic palm weevil (*Rhynchophorus ferrugineus* [Oliv.]) in conjunction with the most important pests of coconut (*Cocos nucifera*). *R. ferrugineus* and the palm weevil, *R. bilineatus* (Montr.), are stem borers of this and several other palm species (Froggatt 1936). Gressitt (1959a) lists 3 genera with 6 species of hispine Chrysomelidae associated with *Metroxylon* species but these appear to be of minor significance. A candidate pest species of sago is the rhinoceros beetle, *Oryctes centaurus* Sternberg. This indigenous species (known also from New Hebrides: Lever 1969: Endrödi 1973) appears to use *M. rumphii* as its main host plant. It is a minor pest of coconut palms and a few other arecoid palms (Catley 1967; Szent-Ivany in press). Its bionomics on sago remain to be elucidated.

Table 1. Plants used as staple foods in New Guinea (primarily after Powell 1976)

Plant Name	Family	Common name(s)	Part eaten	Area(s) cultivated and other refs.
Alocasia macrorrhiza (cs)	Araceae	Paragum taro	stem (Famine)	Duke of York group (Bourke 1976:83; Pena 1970)
Amorphophallus sp. (cs)	Araceae	—	corm	Louisiade Archipelago: Rossel I. (personal observation)
Cocos nucifera (cs)	Arecaceae	coconut	nut, milk	Widespread lowlands, coastal
Colocasia esculenta (cs)	Araceae	taro true	corm, leaves	Widespread
Cyrtosperma chamissonis (cs)	Araceae	swamp taro	corm	Jimi, New Britain
Dioscorea alata (cs)	Dioscoraceae	greater yam	tuber	Widespread
D. esculenta (cs)	Dioscorceae	lesser yam	tuber	Widespread
Ipomoea batatas (cs)	Convolvulaceae	sweet potato, kau kau	tuber, leaves	Widespread
**Manihot esculenta* (cs)	Euphorbiaceae	cassava, manioc, tapioca	tuber	Widespread
Metroxylon rumphii (cs, cw)	Arecaceae	sago	pith	New Guinea mainland and Manus I., coastal (Barrau 1958)
M. sagu and *M. oxybracteatum* (cs, cw)	Arecaceae	sago	pith	New Guinea mainland, coastal (Barrau 1958)
Musa spp. (cs, cw)	Musaceae	banana	fruit, stalk (famine)	Widespread
**Solanum tuberosum* (cs)	Solanaceae	English potato	tuber	Sparingly in Highlands
**Xanthosoma* prob. *sagittifolium* (cu)	Araceae	kong kong taro	corm	Widespread

cs = cultivated.
cw = wild form used as staple food.
* = introduced in historical times.

5.2 Staples possibly indigenous or introduced by early human colonizers

Here are included crops of undetermined origin but for which there is reason to believe could have originated in New Guinea. In the case of bananas, yams, coconuts and some of the aroids, this will undoubtedly include a mixture of indigenous and introduced species and/or varieties which now pose an almost impossible task for the plant geographer to sort out. Keleny (1962) considers *Colocasia esculenta* and *Musa* spp. to be of Malayan origin, but possibly indigenous. Powell (1977:18) is of the opinion that "by the time the Southeast Asian yams, bananas and taro were transferred to New Guinea some form of subsistence agriculture based on local semidomesticated and domesticated species, was being practised."

5.2.1 Bananas (Musa sapientum *vars.*, Musa *spp.*) The most widespread and longest established banana of Melanesia is *M. sapientum*. There is no particular cultivation technique used for them, though often the fruit are wrapped in thick layers of dry banana leaves to protect them from birds (Barrau 1958) and fruit bats. In the Papuan Highlands the fruit of *M. troglodytarum* are gathered before becoming ripe and buried in mud for ripening. Some species and varieties are eaten raw. New Guinea would appear to have contributed the *fe'i* banana to subsistence agriculture (Simmonds 1962). Most of the bananas cultivated for subsistence purposes in New Guinea are those with high starch content for cooking. They occur wild in the area from India to New Guinea. They are the staple in drier areas of New Guinea such as parts of the Markham Valley, the dry Papuan coast and on Cape Vogel. They are a secondary staple on the Gazelle Peninsula, New Britain, particularly in the drier parts. In villages with acute land shortages they have become the staple crop because of their perennial nature. Triploid bananas can maintain moderate yields in the same ground without any fallow and require quite low labor input. They have become the dominant staple in these and in some grassland regions (Bourke 1976). Bananas with high sugar content (e.g. eaten raw) are mostly recent introductions to cater to expatriate tastes. The major insect pests of New Guinea bananas are listed in table 2, and discussed below: *Cosmopolites sordidus:* The banana weevil also attacks hemp and yams (*Dioscorea* spp.). This is one of the most serious pests of bananas. There are conflicting reports as to whether or not this species occurs in New Guinea (see Hill 1975; Lamb 1974); although Wilson (1960) claims that this species became established in Queensland, possibly from Papua, at the end of the 19th century, and rapidly became the most serious pest of bananas there. It is native to Malaysia whence it has spread by commerce into the Indo-Pacific area (Szent-Ivany in press; Zimmerman 1968b). Attempts at biological control (e.g. with the histerid beetle, *Plaesius javanus* Erichson) have met with little success in New Guinea (Anon. 1977).

Dacus bryoniae and *D. musae*: Local outbreaks of one or the other species of these banana fruit flies occur from time to time. Otherwise, they are not pests of economic importance. This is probably due to the fact that almost all fruit grown is for immediate local consumption and to the fact that a stable ecological balance has developed between the pests, their parasites and

Table 2. Major insect pests of banana (*Musa* spp.) in New Guinea (primarily after Hill 1975; Lamb and Johnston 1976; and Lamb 1974)

Pest species	Order, family and common name(s)	Damage	Economic distribution, biogeographic status in NG and pertinent references
Cosmopolites sordidus (Germ.)	Coleopt.: Curculionidae banana weevil, banana rhizome weevil, banana root borer	Larvae bore into rhizome and pseudostem at ground level allowing pathogen entry	Pantropical including NG according to map in Hill 1975, but not according to Lamb 1974. Exotic (Anon. 1977; Zimmerman 1968b)
Dacus bryoniae (Tryon)	Dipt.: Tephritidae banana fruit fly	Larvae feed in green and ripe fruit	NG, NE Australia. Indigenous (Stibick, 1976; Drew *et al.* 1978; Drew 1975)
D. musae (Tryon)	as above	as above	NG, NE Australia, Solomons Indigenous (see *D. bryoniae* refs.)
Dermolepida nigrum (Nonf.)	Coleopt. Scarabaeidae	Adults chew shoots and defoliate	New Guinea. ?Endemic
Nacoleia octasema (Meyr.)	Lepidopt.: Pyralidae banana scab moth	Larvae feed on fruit	Australia, Indonesia, Bismarcks, Indigenous
Othreis fullonia (Cl.)	Lepidopt.: Noctuidae fruit piercing moth	Adults pierce fruit	Australasia, NG, Pacific Isles (Froggatt 1941; Cochereau 1973). ?Indigenous
Papuana woodlarkiana Montr.	Coleopt.: Scarabaeidae taro beetle	Adults bore into pseudostems	Indonesia, Melanesia, Queensland. Indigenous (Szent-Ivany in press; Endrödi 1971)
Pentalonia nigronervosa Coq.	Homopt.: Aphididae banana aphid	Adults and nymphs suck sap	Pantropical. ?Exotic
Scapanes australis australis (Boisd.)	Coleopt.: Scarabaeidae New Guinea rhinoceros beetle	Adults bore into pseudostems	New Guinea Mainland. Endemic (Endrödi 1957)
S. australis grossepunctatus Sternb.	as above	as above	Bismarcks. Endemic (Endrödi 1957)
Taenaris myops kirschi Stgr	Lepidopt.: Amathusiidae banana butterfly	Larvae defoliate	New Guinea. Endemic (D'Abrera 1971)

predators (Stibick 1976). *D. musae*, however, is a serious pest of triploid bananas throughout the year around Lae and in the Markham Valley. Bunches of green bananas are always bound by villagers with banana leaves to prevent *D. musae* attack (Smith 1977a). Smith (*loc. cit.*) also found varietal differences in susceptibility of attack by this pest, probably due to different hardness of the fruit skins. He also found this species attacking chillies, tomatoes and guavas. An earwig, *Chelisoches morio* F. exerted small predatory influence on larvae and pupae. It is widely distributed up to 1600 m on the New Guinea mainland and on New Britain.

Dermolepida nigrum: This melolonthine scarab beetle has caused regular annual defoliation of bananas in subsistence gardens in the Mekeo area of Papua (Szent-Ivany in press).

Nacoleia octasema: The banana scab moth is considered the worst pest of bananas in portions of PNG (Anon. 1976). Although also present on the New Guinea mainland, *N. octasema* attack there is restricted to species of *Heliconia*, *Pandanus* and *Nypa*. Only in the Bismarcks and N Solomon Prov. does it attack *Musa* spp. and become pestiferous (Byrne 1973). Paine (1961, 1964) in his detailed analyses of distribution, ecology and biological control of this pest, attributes its anomalous pestiferous pattern to the presence of biological races with distinct food preferences. He considers that the banana-feeding habit may have arisen after an initial spread of the species on *Nypa* and/or *Pandanus* in the Melanesian region. *N. octasema* larvae feed on the banana infloresence as it develops and cause a scab on the developing fruit (Hill 1975). The proportion of bunches attacked increases with planting density. Although about 20 species of primary parasitoids were recovered by Paine (1964) over the whole range of *N. octasema*, none of them appeared to play a decisive role in the control of scab moth populations. He considered climate and food supply to play more important roles in its limitation. Attempts to establish an encyrtid (*Litomastix* sp.) which attacks the mainland *Heliconia*- and *Pandanus*-feeding mainland races, on New Britain, failed to establish (Wilson 1960). But Cochereau (1972) still considers *N. octasema* amenable to biological control.

Othreis fullonica: The mouthparts of the adult of this fruit-piercing moth are prolonged into a tube with a sharp point, by means of which they puncture the skin and suck up juice. This allows the entry of such fungi as *Oospora* and *Penicillium* species which then cause rot and fermentation around the puncture. The fruit then falls and becomes inedible (Cochereau 1973). Froggatt (1941) indicates that indigenous species of *Erythrina* are the larval hosts.

Pentalonia nigrinervosa: The banana aphid transmits banana bunch-top virus (Lamb 1974). The direct damage is negligible. The disease results in small and distorted, unsaleable fruit bunches. The aphids are found under the old leaf sheaths at the base of the pseudostem near ground level. Ants always accompany the aphid colonies, and they are responsible for the establisment of new colonies. Winged adults also migrate to new host plants (Hill 1975).

Papuana woodlarkiana: In dry weather adults of this scarab beetle destroy small suckers when freshly planted out by feeding into the growing point. Froggatt (1941) considers it more serious a pest here than *Cosmopolites sordidus*. Other *Papuana* spp. are also implicated here, especially *P. laevipennis*

Arrow (Szent-Ivany and Barrett 1956). Szent-Ivany (in press) considers *P. woodlarkiana* to be an "eurythermal, eurytopous species which must have some indigenous native food plants in the high mountain areas" where introduced staples do not grow.

Taenaris myops kirschi: Large clusters of larvae sometimes defoliate banana plants (Lamb and Johnston 1976) in the eucalypt savanna and monsoon forest areas of Papua (Szent-Ivany and Barrett 1956).

5.2.2 Coconut (Cocos nucifera). The coconut has never been found in the unquestionably wild state. It perpetuates itself and occurs widely in coastal areas and on coral islands, but apparently nowhere that it was not most likely brought by man. Fosberg (1960:74) has reviewed theories of its origin and concludes that most evidence seems to be that it is an Old World plant. He suggests that "the coconut may have been domesticated from a wild species growing somewhere in the present optimum range of the modern coconut ... As the plant became more and more useful, it would have spread, through the agency of man, over a larger and larger area, radiating from the original habitat of the plant. Thus it may well have completely replaced its wild ancestor. This would leave no conceivable way to locate the actual place where the wild plant occurred except by discovery of fossil or archaeological materials." There is no general agreement among authors as to what constitute the major insect pests of the coconut. I selected pest species from the 2 most recent general works from this region (Lamb 1974; Hill 1975), and also made particular reference to Perry (1973). Other publications germane to our area are those of Lever (1934, 1969) and Froggatt and O'Connor (1940, 1941). In 1963, Szent-Ivany stated that there were "no serious regional non-coleopterous insect pests of coconut in the Territory of Papua and New Guinea. Most of our non-coleopterous pests are restricted to relatively small areas ..." (p. 70). The primary insect pests of coconut (see table 3) are discussed below:

Axiagastus cambelli: The coconut spathe bug is most important in parts of the Bismarck Archipelago. High populations cause death of inflorescences Usually, the bug is adequately controlled naturally by the egg parasite *Trissolcus painei* Ferr. (Hymenoptera: Scelionidae), the parasitoid *Pentatomophaga bicincta* de Meijere (Diptera: Tachinidae) and the predaceous ant *Oecophylla smaragdina* F. (Hymenoptera: Formicidae) (Perry 1973) all of which are native to our area. The spathe bug's indigenous hosts are wild betel nut palms (*Areca* spp.) (Lever 1934). There appear to have been no attempts at biological control of this species (Cochereau 1972) with exotic parasitoids and predators, excepting introduction of *P. bicincta* into the Lihir Is from New Britain (O'Sullivan 1973; Young in press).

Brontispa longissima: The coconut leaf hispine is an important pest of coconut nurseries in the Kieta District (Wilson 1960) and is not usually a problem after field planting. Where populations are high the newly expanding leaves may be completely eaten (Perry 1973) by adults and larvae, producing a striped effect on the open leaves (Lamb and Johnston 1976). This species has a number of indigenous parasites, but none are very effective under plantation conditions. Consequently, the larval and pupal parasitoid, *Tetrastichus brontis-*

Table 3. Major insect pests of coconut (*Cocos nucifera*) in New Guinea (considered most serious by Hill, 1975; Lamb, 1974; and Perry, 1973)

Pest species	Order, family and common name(s)	Damage	Economic distribution and biogeographic status in NG
Axiagastus cambelli Dist.	Heteropt.: Pentatomidae coconut spathe bug	Adults and nymphal sapsuckers of spathes and inflorescences	NG, New Hebrides, Solomon Is. Indigenous
Brontispa longissima Sharp	Coleopt.: Chrysomelidae coconut leaf hispa	Adults and larvae feed on inner surfaces of unopened leaves	SE Asia, NG, to New Caledonia. Indigenous
Diocalandra taitense (Guer.)	Coleopt.: Curculionidae Tahitian coconut weevil, lesser coconut borer	Larvae bore into all parts of the plant	NG, Pacific Isles, Madagascar, Hawaii (introduced). ?Indigenous
Oryctes rhinoceros (L.)	Coleopt.: Scarabaeidae Asiatic rhinoceros beetle	Adults are stem borers	India, SE Asia, Pacific Is, Irian Jaya, Manus I., New Ireland, NE New Britain. Exotic
Promecotheca papuana Csiki	Coleopt.: Chrysomelidae coconut leaf miner	Adults are leaf chewers and larvae mine unopened leaves	Manus I., New Britain. Endemic
Rhynchophorus bilineatus (Montr.)	Coleopt.: Curculionidae black palm weevil	Adults and larvae are stem borers	E Indonesia, NE, Solomon Is. Indigenous
Scapanes australis (Boisd.)	Coleopt.: Scarabaeidae New Guinea rhinoceros beetle	Adults are spear and stem borers	NG, Philippines. Endemic.
Segestes and *Segestidea* spp. (5 spp. important, see text)	Orthopt.: Tettigoniidae coconut treehoppers	Adults and nymphs defoliate	Genera occur from Philippines east to NG. Spp. collectively endemic

pae Ferrière, of Javanese origin, has been established in New Britain but is apparently of little importance (Wilson 1960), as compared with the situation in the Celebes (Lever 1969). Eight palm species in addition to coconut are attacked (Szent-Ivany in press). Fifteen of the 20 known *Brontispa* species are associated with Palmae (Gressitt pers. comm.).

Diocalandra taitense: The Tahitian coconut weevil apparently only attacks coconut, but earlier records also implicate nypa palm (*Nypa fruticans*) (Lever 1934). Some controversy exists as to whether this pest does serious primary damage or that its damage is merely secondary (Hill 1975; Lever 1969). It is apparently indigenous to our region (Zimmerman 1968a). No attempts at biological control appear to have been attempted.

Oryctes rhinoceros: Adults of this, the rhinoceros beetle, attack the terminal buds of a wide range of Palmae as well as taro, bananas and others, while the larvae feed in moist, rotting vegetable matter. Palms of all ages are subject to attack, which are sometimes killed directly or by entry of secondary organisms. Much additional damage, and often death of the palms, results from the subsequent attraction of *Rhynchophorus* weevils to the wounds inflicted by *O. rhinoceros*. In the Papuan Subregion this Asiatic species is restricted to Irian Jaya, Manus, New Ireland and the Gazelle Pen., New Britain. It is a serious pest only in the latter area where it appears to have been introduced at the time of the Japanese occupation during World War II (Wilson 1960). The larvae can survive in floating logs on ocean currents. Wartime shipping also aided its spread, and the great number of palms felled or broken during military operations provided it with unprecedented breeding opportunities (Lever 1969). It is native from Pakistan to the Philippines (Garlovsky and Zelazny 1971). A number of biological control agents are active in limiting *Oryctes* larval numbers. The most important is the exotic fungus *Metarrhizium anisopliae* (Perry 1973) which also attacks the pupae and adults. Two species of scoliid parasitoids have been introduced, *Scolia oryctophaga* Coq. from Mauritius and *S. ruficornis* from Zanzibar and attention has been given to the endemic *S. pulchripennis* Cam. (Wilson 1960). Both exotic species were liberated in New Britain where *S. ruficornis* now makes it possible to replant coconuts where previously this was not possible because of *O. rhinoceros* populations. Several species of beetle, bug and mite predators have been introduced, mostly from the Ethiopian region where the genus *Oryctes* predominates (Endrödi 1973). Limited success has been achieved with these. Additional diseases and nematodes (*Rhabditis* sp.) are also implicated in natural controls of *rhinoceros* larvae (Surany 1963). The baculovirus of *O. rhinoceros* has been released on New Ireland, Manus, and New Britain in 1978 and 1979 from Western Samoa. Infected beetles have been recovered from Manus and New Ireland (Young in press).

Promecotheca papuana: The coconut leafminer is an important pest in certain areas of New Britain and Manus I. It also occurs in NE New Guinea, but has not been recorded as a pest there. New Guinea is the center of distribution of the predominantly Palmae feeding (22 of 33 species) genus *Promecotheca* which ranges from Samoa to the Philippines and Indonesia (Gressitt 1959b; pers. comm.). There are several other important coconut pests in this genus outside

New Guinea, as well as a number of other species, like *P. straminipennis* Weise from New Britain which mines the leaves of tree-like *Pandanus* (Gressitt 1958) and *P. pandani* Gressitt from mainland NE New Guinea. At most times all life stages of *P. papuana* occur together; however, at certain times some stages are absent. Then it reaches pest proportions very quickly. The resulting "one-stage" condition has been ascribed to the upsetting effect of the predaceous mite, *Pyemotes ventricosus* (Newp.) on parasite populations (Gressitt 1958, 1959b; Wilson 1960). However, Lever (1969:185) claims that these "one-stage" conditions have arisen in New Guinea "from an incorrect application of insecticides" and that "this development is, therefore, quite apart from the usual external factors such as weather or failure of parasites." Normally, parasites and predators, especially the ant, *Oecophylla smaragdina*, effect control in all of the stages. The preferred foodplant of this beetle is the coconut palm. The original native host, however, may have been the nipa palm (*Nypa fruticans*), betel palm (*Areca catechu*) or the sago palm (*Metroxylon rumphii*). Of these 3, the nipa palm seems to be definitely preferred. It appears unlikely that outbreaks originate in native hosts. Mature coconut palms are preferred (Gressitt 1958; 1959b; Szent-Ivany in press). Natural controls appear to involve a variety of at least 5 native parasitoids and 9 species of ant predators, of which the *kurukum* (*Oecophylla smaragdina*) was considered by Gressitt (1959b) to be the most important controlling factor. There are also a variety of vertebrate predators including lizards and birds. Biological control has been attempted with the internal parasitoid *Pediobius parvulus* (Ferr.) (Hymenoptera: Eulophidae) which attacks late instar *P. papuana* larvae, prepupae and pupae. The parasite is of Javanese origin but New Guinea stocks were introduced from Fiji where this parasite gained considerable success in the biological control of the Fijian coconut leafminer, *P. caeruleipennis* Blanchard, and to some extent *P. opacicollis* Gestro in New Hebrides. *P. parvulus* has been established on New Britain where it is now abundant, but still has failed to prevent *P. papuana* outbreaks (Gressitt 1959b).

Rhynchophorus bilineatus: The black palm weevil is considered one of the most damaging pests of young coconuts. *Rhyncophorus* is a pantropical genus. A complex of closely related species attack coconut and several other palm genera in the SE Asia-New Guinea area (Lever 1969). *R. ferrugineus* (Oliv.) has a similar life history but appears less damaging here. Perry (1973) describes the economic aspect in Papua New Guinea. *R. bilineatus* attack follows dynastid damage or mechanical injury (Szent-Ivany in press). There appear to be no direct attempts at biological control of this pest because more logical control measures are aimed at the coconut rhinoceros beetle and at the New Guinea rhinoceros beetle (see below) which often provide the mode of entry into the palm.

Scapanes australis australis: The New Guinea rhinoceros beetle causes damage similar to that described above for *O. rhinoceros*, although younger palms are preferred (Lamb 1974), especially those in the 2–5 year old range (Lamb and Johnston 1976). Perry (1973) considers it to be the worst pest of coconuts in Papua New Guinea. It also attacks bananas. The worst aspect of *S. a. australis* attack is that entry holes are provided for the black palm weevil which makes the establishment of new plantings difficult, particularly if newly-felled bush country is used. *S. a. australis* occurs throughout most of the New Guinea mainland (Endrödi 1957; Gressitt 1963). Natural enemies do not appear important in its control, although attention has been given to using the native *Scolia pulchripennis* Cam. (Hymenoptera: Scoliidae), a parasite of *Scapanes australis grossepunctantus* Sternb. (Wilson 1960), a lesser pest of coconuts in the Bismarcks and Solomons. Cochereau (1972) considers *Scapanes* spp. amenable to further biological control efforts. No parasites were known of *S. australis* as of 1957 (Gressitt *loc. cit.*).

Segestes and Segestidea spp.: Five species of these large katydids, called coconut treehoppers or Sexavae by economic entomologists, are implicated in damage to coconuts in New Guinea. The species and their distributions are as follows (after Willemse 1977). (See also Perry 1973; O'Connor 1959.)

Segestes decoratus Redt.	Coastal New Guinea; New Britain
Segestidea gracilis (C. Willemse)	New Britain; New Ireland
S. uniformis (C. Willemse)	Admiralty Is.
S. leefmansi (C. Willemse)	New Hanover; New Ireland; Lihir Group
S. novaeguineae (Brancsik)	Coastal New Guinea

All the nymphal stages and the adults are passed in the "foliage." Locally, these species strip the green leaves off coconuts causing a severe setback in growth and loss of production (Szent-Ivany 1963). For example, *S. novaeguineae* periodically causes severe defoliation of coconuts along the Madang coast of PNG (O'Connor 1959). These species also attack bananas and oil palm. There are a number of parasitoids and predators which attack these insects. A tachinid fly helps keep mainland populations under economic thresholds (Wilson 1960; O'Connor 1959). On the other hand, kurukum ants (*Oecophylla smaragdina*) interfere with these natural enemies. An internal parasitoid, *Stichotrema dallatoreanum* Hof. (Strepsiptera: Stichotrematidae) exerts good control of *S. decoratus* and *S. novaeguineae* (up to 60% of the adults of the former species, G. Young in press) on the mainland, but so far it has not been possible to establish this indigenous parasitoid in the Bismarcks (Smee 1965), especially for *S. gracilis* control on oil palm (Young *loc. cit.*) *Leefmansia bicolor* Water. (Hymenoptera: Encyrtidae), bred from *Sexava coriacea* (L.) eggs on Amboina, has been introduced on New Hanover, New Britain and Manus I. (Wilson 1960) for *S. gracilis* control but has achieved little success on the latter island (Smee 1965). On several occasions outbreaks of *Segestidea* spp. have been brought under control by mass liberation of *L. bicolor* in New Britain and New Ireland. Another egg-parasite, *Doirania leefmansi* Water. (Hymenoptera: Trichogrammatidae) controls some *Segestidea* spp. on New Hanover, but, efforts to establish it elsewhere in the Bismarcks have failed.

Smee (1965) dealt with several other species considered important coconut pests within PNG. These are: the rhinoceros beetles (*Oryctes centaurus* Sternb., *Trichogomphus semmelinki* Rits. and *Xylotrupes* spp.), the coconut spathe moth (*Batrachedra arenosella* Walker) and leaf-footed bugs (*Amblypelta* spp.). Sporadically serious nut fall has been ascribed to the feeding damage of

A. gallegonis Lever on Bougainville (Szent-Ivany 1963), where the ant *Iridomyrmex myrmecodiae* Emery interferes with the bugs' natural enemies. Another *Amblypelta* species is also injurious in Milne Bay Prov. (Szent-Ivany 1963). Still other pests of more local or sporadic importance in New Guinea are discussed by Lever (1969). The lesser spathe moth, *Tirathaba rufivena* Walker can cause up to 75% loss in production on coconuts when found in association with high populations of the crazy ant, *Anoplolepis longipes* Jerd. Another localized pest, the toxicogenic lophopid treehopper, *Zophiuma lobulata* Ghauri, causes severe damage to coconuts in the Finschhafen and Popondetta areas of PNG (Young pers. comm.). In some areas *Z. lobulata* eggs are heavily parasitized by *Ooencyrtus malayensis* Ferr. (Hymenoptera: Encyrtidae).

5.2.3 Taro (Colocasia esculenta) *and other aroids* (Alocasia macrorrhiza, Cyrtosperma chamissonis *and* Amorphophallus *spp.*). The familiar taro (*C. esculenta*) has been an important foodstuff in tropical and subtropical regions for more than 2000 years. The plant is highly polymorphic as a result of its extensive cultivation over the centuries (Plowman 1969). A cytological study of the species by Yen and Wheeler (1968) indicated a 2-stage pre-historic introduction from SE Asia. The first introduction may be concurrent with the first introduction of agriculture, while the second introduction may be equally assigned to a pre- or post-European time (Yen 1971). It has been suggested that the Indo-Malaysian taro was probably the main staple (Brookfield and White 1968) before sweet potato introduction (Yen 1963; Conklin 1963; Watson 1965a, b). *C. esculenta* is the most important crop in lowland forest areas where rainfall is well distributed throughout the year, but it is also grown as a minor food in Highlands areas up to 2000 m elevation. The importance of the crop may be seen in the significant place afforded to taro in the legend and ceremony of New Guinea growers. Almost all taro is grown on a subsistence level and is generally cultivated in forest clearings without the use of irrigation, pest control or fertilizers. The number of cultivated varieties is unknown but here probably numbers in the thousands (Rangai 1977). Some authors attribute the decline of *C. esculenta* in the Gazelle Peninsula to sweet potato (e.g. Clarke 1977) and *Xanthosoma* taro (Plucknett 1970) and to the introduction of *Phytophthora colocasiae* leaf blight during World War II. But Bourke (1976) suggests that the main reason for its decline is shortage of land of high fertility because of the shortening forest fallows, coupled with relatively low caloric returns for the labor needed to grow it, the unsuitability of taro for integration with cocoa production, and problems with other pathogens and pests (see below).

Paragum taro (*Alocasia macrorrhiza*) is apparently native to Sri Lanka (Plowman 1969). However, Keleny (1960) considers *Alocasia* (along with *Colocasia*) to have originated in Malaya. Now it is found all over Melanesia, sometimes growing in gardens (Barrau 1958; Powell 1976). It is less important as a staple in New Guinea, except on the Gazelle Peninsula. Another aroid of more sporadic use as a staple here is swamp taro (*Cyrtosperma chamissonis*) which is cultivated in the Jimi Valley and New Britain (Powell 1976). Its nearest wild relatives occur in the swamps of Indo-Malaya as well as in New Guinea. Compared with Micronesia and certain of the Solomon Is, its use here is rather mimimal (Barrau 1958, 1960). This crop is often grown as an emergency food supply in fresh water swamps (Plucknett 1970) when drought affects other dry land staples. It is also salt tolerant (Bourke 1978). Of still lesser importance as a staple are *Amorphophallus* spp. I observed them in cultivation as a staple along with *mami* yam (*Dioscorea esculenta*) on Rossel (Yela) I. in the Louisiade Archipelago in March 1979. Plucknett (1970:131) states that *A. campanulatus* is the main species used for food, especially as an emergency or subsistence crop The major insect pests of *Colocasia* taro, and other staple aroids where noted, are listed in table 4. These are the species that the majority of authors consulted for this region considered to be major pests:

Hippotion celerio: The taro hawkmoth is very common in taro gardens. Severe outbreaks including complete defoliation have been reported at certain localities on New Britain (Perry undated). A parasite and a pentatomid predator, both on larvae, have been reported (Froggatt 1939). Other than these, little appears to be known regarding the natural factors controlling this indigenous species.

Agrius convolvuli (L.), another indigenous pest of taro here, probably has a similar array of natural enemies. Without mentioning either of the above hawkmoth species specifically, Mackey (1977) stated that the sphingid fauna of New Guinea is predominantly of Oriental origin (see also table 5).

Papuana spp.: Eighteen species of taro beetles occur in New Guinea and 11 of these have been recorded as pests. The genus is basically Papuan as only 6 of the 20 species extend beyond our region (Endrodi 1971). The 4 most important species in New Guinea are found in table 4. *P. huebneri* and *P. woodlarkiana* are the most damaging species on New Britain. The adult lays eggs close to the host plant below the soil surface. The larvae feed generally on humus and roots.

Fig. 1. Papuana sp., probably *biroi* Endr., boring into taro, at Wau Ecology Institute (photo B. Gagné).

Table 4. Major insect pests of "true" taro (*Colocasia esculenta*) in New Guinea

Pest species	Order, family and common name(s)	Damange	Economic distribution and biogeographic status in NG; pertinent references
Hippotion celerio (L.)	Lepidopt.: Sphingidae taro hawkmoth	Larvae are defoliators	NG, Pacific Is (except Hawaii); Africa; S. Europe. Indigenous (Riotte pers. comm.)
**Papuana biroi* Endr.	Coleopt.: Scarabaeidae taro beetle	Adults bore into corms	NG, Admiralty Is (?introduced). Endemic (Endrödi 1971)
P. huebneri (Fairm.)	as above	as above	Solomon Is, NG, New Britain. Indigenous
P. tibialis Arrow	as above	as above	NG, Bismarck Archipelago. Endemic.
**P. woodlarkiana* (Montr.)	as above	as above	NG, W to Java, N Queensland, Solomons. Indigenous (Endrödi 1971; Szent-Ivany in press)
Tarophagus proserpina (Kirkaldy)	Homopt.: Delphacidae taro leafhopper	virus vector	Indonesia to Tahiti. Indigenous

*also attacks *Xanthosoma* taro (Gagné, in press).

Adults feed on taro corms and on a number of other staples including *Xanthosoma* taro, *Alocasia* taro, sweet potato, banana, coconut and sago (Szent-Ivany in press). *P. woodlarkiana* and *P. biroi* appear most polyphagous in this respect. Severe damage may kill plants but more commonly losses occur due to declines in quality. Taro variety, trap and barrier crops, garden location and climate appear to influence damage incidence. Also, a number of indigenous control techniques employed by village gardeners have come to notice (Perry undated). Scoliid wasps are larval parasites but do not appear very efficient (Smee 1965). For example, *Scolia punctatissma* parasitizes *P. huebneri* (Endrodi 1971). A tachinid fly, *Formosia* sp., has been reared from *Papuana* larvae in New Britain and Perry (undated) considers that it may help control these pests. Gagné (in press) points to the need to investigate possible biological control of these pests.

Tarophagus proserpina: The entire life history of the taro leafhopper occurs on taro stems. Their direct damage to taro is insignificant during feeding and oviposition. However, they are important as virus vectors of "Bobone" and "Alomae" diseases. The former causes severe leaf stunting and twisting while the latter causes death. Natural controls include mirid egg predators (Perry undated), probably a *Cyrtorhinus* sp. and/or *Tytthus* sp. But this aspect does not yet appear to have been clarified. Sometimes ants appear to interfere with predators. I have also observed parasitized nymphs, probably by dryinid species, at intermediate elevations.

5.2.4 Yams (Dioscorea *spp.*). Of more than 7 species of *Dioscorea* listed by Powell (1976) as being utilized for food in New Guinea, 2 are indicated as cultivated for staple food. These are *D. alata* and *D. esculenta* or "mami." Their origins are obscure but they are probably native to the Old World Tropics (Hill 1975) probably from Malaysia, whence they came via human migrations. Powell (1976:119) indicates that 16 varieties of the former and 6 of the latter species are grown in New Guinea. Other yams locally cultivated as supplementary staples may be indigenous, viz: *D. bulbifera*, *D. nummularia*, *D. pentaphylla* and *D. hispida* (Powell *loc. cit.*). True yam-based agriculture appears to have developed independently in W. Africa and SE Asia around 5000 BP as a result of cultural interaction between preagricultural peoples who exploited wild yams for food, with early Neolithic grain-using agriculturists. Different, although similar, species were brought into cultivation in these 2 areas. The asiatic yams, especially *D. alata*, became disseminated in cultivation throughout SE Asia and around the W. Pacific in fairly early times, probably more than 2000 BP and certainly long before European contact with these areas (Coursey 1970). The primary insect pests of yams are:

Planococcus dioscoreae Williams: This endemic coccid is the only major yam pest in New Guinea. It damages mainly stored yam roots in the Sepik and Milne Bay areas of PNG. Its indigenous hosts are unknown (Szent-Ivany 1960). Likewise, no predators or parasitoids have been recorded. Szent-Ivany (1974) attributes the pest status of *P. dioscoreae* to the collapse of natural enemies and/or unusual climatic extremes (droughts, unusual rainy seasons). He also indicates that the sawfly *Senoclidia purpurata* (Smith) (Hymenoptera:

Fig. 2. Yam leaf damaged by mirid bug: *Harpedona plana* Poppius; nymphs above, adult below, right middle (photo B. Gagné).

Tenthredinidae) and the hawkmoth *Theretra nessus* (Drury) (Lepidoptera: Sphingidae) caused sporadic, heavy defoliation as well as indicating a number of other insect species of minor importance. Gagné (in press) draws attention to the feeding damage of the plant bug, *Harpedona plana* Poppius (Heteroptera: Miridae), a potentially serious pest of *D. alata* (Fig. 2).

5.3 Staple Crops Introduced in Historical times

5.3.1. Sweet Potato (Ipomoea batatas). Yen (1974) has reconstructed the history and prehistory of sweet potato introduction into the Pacific (p. 259 ff). One prehistoric introduction from western S America via the "Kumara Line" through Polynesia to Melanesia is proposed; while 2 historic introductions are postulated: (1) the "Batata Line" from the W Indies to Africa and Europe, thence to Melanesia via Africa and India in the 15–16th centuries, and (2) the "Camote Line" from the Mexican region westwards across the Pacific to regions near Melanesia, especially the Philippines from the 16th century onwards. The complexities of these hypotheses, plus some alternative ones are reviewed. Suffice for our purposes to say that the consensus is that sweet potato is a comparative newcomer to New Guinea, contrasting with the 6 species of staples discussed above. Like Chinese taro, sweet potato has increased at the expense of traditional staples. It is particularly important in grassland areas, but is commonly grown following a forest fallow (Bourke 1976:93). Table 5 lists the pests considered of major importance by the majority of authors consulted:

Agrius convolvuli: The sweet-potato hawkmoth is a major indigenous pest of

Table 5. Major insect pests of sweet potato (*Ipomoea batatas*) in New Guinea

Pest species	Order, family and common name(s)	Damage	Economic distribution and biogeographic status in NG
Agrius convolvuli L.	Lepidopt.: Sphingidae sweet-potato hawkmoth	Larvae are defoliators	Europe, Africa, SE Asia, India, Australia, NG. Indigenous (Riotte pers. comm.)
Bedellia somnulentella (Zell.)	Lepidopt.: Lyonetiidae sweet-potato leaf-miner	Larvae are leaf-miners	Cosmopolitan. ?Exotic
Cylas formicarius L.	Coleopt.: Brenthidae sweet-potato weevil	Adults tunnel in leaves, vines and tubers; larvae tunnel in tubers and stems	Pantropical. Indigenous

Fig. 3. Sweet-potato (kaukau) showing boring by dynastid beetle, *Papuana* sp., probably *biroi* Endr. (photo B. Gagné).

sweet potato in New Guinea. Another hawkmoth, *Hippotion celerio* L. is frequently mentioned as a major pest also. Taro has also been recorded as a food plant of *A. convolvuli* (Smee 1965). No natural controls are recorded among the authors consulted, although I have reared undetermined hymenopterous egg parasites from another sphingid (prob. *Theretra nessus*) on crops growing closely adjacent to sweet potato (e.g. on yams) and I would suspect that these parasites will also be found to attack *A. convolvuli* and *H. celerio*. Larval parasitoids are rarely encountered.

Bedellia somnulentella: The sweet-potato leaf-miner has reached outbreak levels in the New Guinea highlands causing yield reduction. Kimber (1972) states that other insects usually keep it under control, but does not give specific examples. In the Highlands I have reared an undetermined braconid from a sweet potato leaf-miner, presumably another sp., *Acrocercops prosacta* Meyr. (Lepidoptera: Gracillariidae), the predominant leaf-miner there (Gagné unpublished data). This parasite might also attack *B. somnulentella*.

Cylas formicarius: The sweet potato weevil is most serious where there is continuous cropping (Perry 1973), and can be especially severe during droughts (Byrne 1973; Kimber 1972) when soil cracks allow adults to reach tubers more easily (Smee 1965) for egg laying and development. Some authors (Fullaway and Krauss 1945; Essig 1942) believe that *C. formicarius* is of Asiatic origin. Kimber (1972) states that "morning glory" is an alternate host here. This was probably *Ipomoea purpurea* (see Szent-Ivany in press). No predators or parasitoids appear to have been recorded from this pest, nor are there indications of biological control attempts against it. Cultural controls, especially crop rotations, and resistant varieties, are usually recommended as control measures.

5.3.2 Kong kong or Chinese taro (Xanthosoma sagittifolium). No major insect pests have been specifically mentioned associated with this crop, although it probably has the same array of insect associates as has taro (*Colocasia esculenta*), but of lesser severity (Bourke 1976). For example, Szent-Ivany (in press) records *Papuana woodlarkiana* from this host and I have found *P. ?biroi* on it also (Gagné in press). This taro is gaining in importance as a staple in New Guinea because of its higher resistance to plant pathogens which sometimes decimate "true" taro (*C. esculenta*), and especially where a short forest fallow is employed. Its shade tolerance makes it suitable for interculture with banana or coconut (Bourke 1976).

5.3.3 Cassava (Manihot esculenta). This crop seems destined to become an increasingly important staple as land pressures due to population increase and urbanization tendencies increase (Bourke 1976) even though its food value is lower than for other staples. It has the advantage of being comparatively drought tolerant and is capable of withdrawing soil nutients to a greater degree than the others also. The only major pest is the endemic sap-sucking leaf-footed bug, *Amblypelta lutescens papuensis* Brown (Szent-Ivany 1960) and this species appears to be economically restricted to low elevations on the S side of New Guinea where it also causes lesser damage to other plant species, including sweet potato and coconut. It is considered a serious pest of the latter host in the Solomons (Szent-Ivany and Catley 1960). *A. theobromae* Brown is also found on cassava (Lamb and Johnston 1976), but it is of lesser importance. Szent-Ivany and Catley (*loc. cit.*) indicate some natural enemies of *A. lutescens papuensis*. These are all hymenopterous egg parasites: *Hadronotus* sp. (Scelionidae), *Anastatus* sp., (Eupelmidae) and *Ooenocyrtus* sp. (Encyrtidae). None of these parasites are efficient in controlling this pest or in keeping their population below economic threshholds.

5.3.4 English potato (Solanum tuberosum). Although this crop is an increasing component of the diet in the more urbanized centers, it still has not made much inroad in subsistence agriculture. However, there is considerable PNG government promotion of this crop and much work on its entomological pests. The most important pests in PNG (table 6) have been summarized by Smith (1977b) and Thistleton (1977). These consist of 2 native *Papuana* spp. (*P. biroi* and *P. woodlarkiana*, see under *Colocasia esculenta*) as well as 2 aphids (*Aulacorthum solani* and *Myzus persicae*) and the potato moth (*Phthorimaea operculella*) all of which are now cosmopolitan in distribution and are probably exotic to New Guinea.

6. Summary of biogeographic considerations

Most of the staple crops grown in New Guinea are either indigenous or have been grown here for a very long time. Thus, most insect pests are controlled by

Table 6. Major insect pests of English potato (*Solanum tuberosum*) in New Guinea

Pest species	Order, family and common name(s)	Damage	Economic distribution and biogeographic status in NG
Aulacorthrum solani (Klb.)	Homopt.: Aphididae potato aphid	Virus vector	Cosmopolitan. ?Exotic
Myzus persicae (Sulz.)	as above	as above	as above
Papuana biroi Endr.	Coleopt.: Scarabaeidae taro beetle	Adults bore in tubers	NG, Admiralty Is. Endemic
P. woodlarkiana (Montr.)	as above	as above	NG, W to Java, N. Queensland. Indigenous
Phthorimaea operculella (Zell.)	Lepidopt.: Gelechiidae potato moth	Larvae bore into tubers	Cosmopolitan. ?Exotic

natural factors in the environment (predators, parasitoids, pathogens), except for those brief periods when the pests have favorable conditions and they come into prominence (Smee 1965). Were this not the case, some would have been abandoned as a food source by now. The great island of New Guinea, with its adjacent smaller islands (Bismarcks, Bougainville, the Trobriands, D'Entrecasteaux and Louisiades) has a high percentage (36%) of insect pests of staples which are endemic, or at least indigenous (42%) to this region. This also appears true for insect pests of non-staples, cash crops and pasture species (Szent-Ivany 1960).

6.1 *Pest Biogeography*

No major insect pests appear to be indicated in the literature for sago (*Metroxylon* spp.). *Oryctes centaurus* may be a candidate species once its bionomics are more fully known (Szent-Ivany in press). Morphologically, this species is closest to *O. gnu* Mohnike of the Philippines and together with *O. rhinoceros* and 7 other species is grouped in the subgenus *Rykanes* Minck which extends from W Africa eastwards along S Asia to our region. Its larvae are avidly sought for food (Meyer-Rochow 1973).

Of the 11 major pests of banana (*Musa* spp.) listed in Table 2, only 2 species (*Cosmopolites sordidus* and *Pentalonia nigronervosa*) are exotic, and 3 are almost certainly endemic (*Scapanes australis*, *S. grossepunctatus*, *Taenaris myops kirschi*). The remaining species are probably indigenous. *D. musae* appears confined to N. Queensland, PNG, Solomons, and probably extends into Irian Jaya (Smith 1977a; Drew 1975). *D. bryoniae* has a similar distribution, but does not extend into the Solomons. Drew (*loc. cit.*) considers New Guinea to be a center of speciation for the Dacini, especially the subgenus *Bactrocera*, to which both pests belong.

Of the 12 major insect pests of coconut (*Cocos nucifera*) in New Guinea listed in Table 3, only *Oryctes rhinoceros* is unquestionably introduced, 7 are endemic or probably so (*Promecotheca papuana*, *Rhynchophorus bilineatus*, *Scapanes australis* ssp., *Segestes decoratus*, *Segestidea gracilis*, *S. uniformis*, *S. leefmansi* and *S. novaeguineae*), while the remainder are indigenous or probably so (*Axiagastus cambelli*, *Brontispa longissima*, *Diocalandra taitiense*).

Of the 6 major insect pests of "true" taro (*Colocasia escelenta*) 4 are indigenous (Table 4) and 2 are endemic.

Yams (*Dioscorea* spp.) have only one major insect pest, *Planococcus dioscoreae*, and this is endemic.

Sweet potato (*Ipomoea batatas*) has 3 major insect pests (Table 5). One (*Bedellia somnulentella*) is possibly exotic. *Cylas formicarius* and *Agrius convolvuli* are possibly indigenous.

Cassava (*Manihot esculenta*) has only one major insect pest in New Guinea; the endemic coreid *Amblypelta lutescens papuensis*.

Of the 5 major insect pests of English potato, 3 (*Aulacorthum solani*, *Myzus persicae*, *Phthorimaea operculella*) are probably exotic, 1 (*Papuana woodlarkiana*) is indigenous and 1 (*P. biroi*) is endemic.

No major insect pests have been mentioned in the literature specifically for the other staple aroids (*Alocasia macrorrhiza*, *Cyrtosperma chamissonis* and *Xanthosoma* spp.). *C. chamissonis* is probably native to New Guinea, while the 2 other crops are introduced.

Of the 33 major insect pests of staple crops discussed above 12 (36.4%) are endemic, 14 (42.4%) are indigenous and 7 (21.2%) are exotic. The biogeographic affinities of the endemic and indigenous pest species are predominantly with the Oriental region, as demonstrated for the New Guinea insect fauna as a whole (Gressitt 1961, 1967 and chapter IV:13). The geographic origins of the exotic species are now difficult to ascertain; many are now tropicopolitan.

6.2 Biocontrol summary

With respect to the biological control of insect pests on staple crops in New Guinea, Wilson (1960) reported successes on 2 indigenous pests of coconuts. Coconut treehoppers (*Segestes* and *Segestidea* spp.) have been controlled by a parasite of a related host from Ambon, and the coconut leafminer (*Promecotheca papuana*) has been controlled by a Javanese parasite of a related host species. According to Wilson (*loc. cit.*) it appears that, because of the great diversity and limited distribution of the insect fauna in the numerous islands of this tropical region, the possibility exists of improving levels of biological control of some native pests by introducing natural enemies of related species from other islands.

Natural enemies of pests have been introduced into New Guinea from many parts of the world. The area from which biological control material was obtained was often a country where the pest is indigenous, but frequently species have been imported from countries where they have been introduced and established against alien pests. The sources of the established natural enemies are of interest for the bearing that they have on the suitability of particular geographic areas as sources of beneficial insects which have proved of particular value in controlling their hosts, or, at least, have become abundant since introduction. There is an apparent relationship between the source areas and the main area of establishment, and this may well be the reflection, in part at least, of climatic similarity.

Within PNG there has been some movement of natural enemies from province to province and from one island to another. For the control of coconut leafminer, planters have long fostered the green tree ant or kurukum (*Oecophylla smaragdina*) (Wilson 1960), and in New Britain, *Eurytoma* sp. has been bred and distributed when other parasites of the leafminer were slow to appear in the field. There has also been some movement of parasites of coconut treehoppers (*Segestes* and *Segestidea* spp.). An encyrtid and a mymarid were introduced into Manus I. from New Hanover, and the encyrtid was also introduced into New Britain from New Hanover. Unspecified egg parasites of coconut treehoppers were transported from Manus and New Hanover to New Ireland and New Britain (see Wilson 1960 for review of this work).

Nearly all the natural enemies introduced into New Guinea for the control of insect pests have been insects, but, in addition to these, diseases (e.g. of *Oryctes rhinoceros*), several vertebrate and other predators have been introduced. The cane toad (*Bufo marinus*), introduced in 1936–7 from Queensland to New Britain, has been reported to control the sweet potato hawkmoth and other unspecified insect pests (Froggatt 1939; N. G. Dept. Agr. 1939), but modern workers question the effectiveness of this amphibian and point to several detrimental aspects, particularly its impact on native fauna (Schultze-Westrum 1970; Herington 1977; see Bianchi 1977 for background on *Bufo marinus* introduction into the Pacific area).

In New Guinea, the practice of maintaining laboratory stocks of parasites to deal with the sudden outbreaks of pests has sometimes been adopted. *Leefmansia bicolor* has proved of value in controlling coconut grasshoppers in the Bismarck Archipelago, and the laboratory rearing and mass liberation of this species in New Britain and New Ireland has brought outbreaks under control on several occasions. Similarly, *Pleurotropis parvulus*, which is a useful parasite of the coconut leafminer, has been maintained in the laboratory for release during outbreaks of this pest (Wilson 1960).

Table 7. Area and major pest biogeography of some staple crops in Papua New Guinea (after Macewan 1979, in part)

Staple	Area at start of year (ha)	Area planted during year (ha)	No. indigenous/ endemic pest spp.	No. exotic spp.
Ipomoea batatas	71900	42500	2	1
Musa spp.	24400	14500	7	2
Colocasia esculenta	18000	22000	6	0
Dioscorea spp.	7200	7400	1	0
Xanthosoma sp.	4000	2500	(1)	0

6.3 Other biogeographic considerations

There appears to be a relationship between the time that the staple crop has been under cultivation and the percentage of major pests that are indigenous or endemic. Over 90% of the pests of staple crops that are indigenous or prehistoric introductions are themselves endemic or indigenous. This contrasts with 55% of the major pests on the historically introduced staple crops.

The relationship between area under cultivation and numbers of major pests is more difficult to assess. This is because some of the historic and relatively recent introductions have come to supplant the prehistoric introductions. Also, food crops in New Guinea are planted in an intricate admixture which we must convert to "pure" stands for our comparisons. Part of the data presented in Table 7 was derived from Macewan (1979:9). Appropriate data are lacking for the staples not found in Table 7. The low numbers of major pests of sweet potato in spite of its predominance in New Guinea agriculture may indicate recentness of introduction. Other than sweet potato, there is a direct correlation between area and numbers of endemic and indigenous pest species. This correlation would likely be strengthened with the addition of coconut to Table 7 since this crop would go at the head of the table.

Acknowledgements

This contribution has benefited from the critical reviews and comments of G. Young, Entomologist, Agriculture Research Centre, Bubia, PNG; Dr J. L. Gressitt, Director, Wau Ecology Institute, PNG; and Dr F. G. Howarth, Bishop Museum, Honolulu, Hawaii. Also Dr P. Maddison, D.S.I.R., New Zealand.

References

Allen, J. 1977. Management of resources in prehistoric coastal Papua. In The Melanesian Environment. J. W. Winslow (ed.) Australian National University Press, Canberra. xxxiv + 562 p.

Anonymous. 1976. Insects of traditional vegetables. Topic 11. D.P.I. Entomology Conference (Papua New Guinea). 3 p. mimeo.

—— 1977. Pest control in bananas. PANS Manual No. 1. Centre for Overseas Pest Research, London. iii + 126 p.

Barrau, J. 1958. Subsistence agriculture in Melanesia. Bernice P. Bishop Mus. Bull. 219.

—— 1960. The selection, domestication and cultivation of food plants in tropical Oceania in the pre-European era. In: Symposium on the impact of man on humid tropics vegetation. Administration of the Territory of Papua and New Guinea, Goroka. p. 67–72.

Bianchi, F. A. 1977. Cyril Eugene Pemberton, 1886–1975: A biographical sketch. Proc. Haw. Entomol. Soc. 22(3): 417–441.

Bourke, R. M. 1976. Food crop farming systems used on the Gazelle Penninsula of New Britain. In: 1975 Papua New Guinea Food Crops Conf. Proc., Dept. Primary Industry, Port Moresby, K. Wilson and R. M. Bourke (eds.) p. 82–110.

—— 1978. How farmers can reduce the effects of drought on food production in the lowlands. Harvest 4(3): 122–128.

Brookfield, H. C. and J. P. White. 1968. Revolution or evolution in the prehistory of the New Guinea Highlands: a seminar report. Ethology 7(1): 43–52.

Byrne, P. N. B. 1973. Food crops and farming systems. Crop technology course — Lowlands Agricultural Experimental Station, Keravat. 6 p. mimeo.

Catley, A. 1967. Report of the Entomologist (Project Area). In: Semi-annual report of the Project Manager for the period June 1966 — November 1966. UNSF/SPC project for research on the control of the coconut palm rhinoceros beetle. South Pacific Commission, Noumea, New Caledonia. 49 p.

Clarke, W. C. 1977. Progressing with the past: Environmentally sustainable modifications to traditional agricultural systems. Paper presented at the seminar on the Adaptation of Traditional Systems of Agriculture held at Honiara, Solomon Islands, on 3–7 Oct., 1977. 21 p.

Cochereau, P. 1972. La lutte biologique dans le pacifique. Cah. ORSTOM, ser. Biol. no. 16: 89–104.

—— 1973. Les papillons piqueurs de fruits. Ibid. 22: 3–41.

Conklin, H. C. 1963. The Oceanic — African Hypothesis and the sweet potato. In J. Barrau (ed.). Plants and the migration of Pacific peoples. Sym. 10th Pacif. Sci. Congr., Honolulu, 1961. Bernice P. Bishop Mus. Press. Honolulu, p. 129–136.

Coursey, D. G. 1970. The yams as food crops. In D. L. Plucknett (ed.). Tropical root and tuber crops tomorrow. Vol. 1. Proc. 2nd Intl. Symp. Trop. Root Tuber Crops. Univ. Hawaii, College Trop. Agr., Honolulu. p. 87–90.

D'Abrera, A. 1971. Butterflies of the Australian Region. Lansdowne. 415 p.

Drew, R. A. I. 1975. Zoogeography of Dacini (Diptera: Tephritidae) in the South Pacific Area. Pacif. Ins. 16(4): 441–454.

Drew, R. A. I., G. H. S. Hopper and M. A. Bateman. 1978. Economic fruit flies of the South Pacific region. Watson Ferguson and Co., Brisbane.

Endrödi, S. 1957. Rassenkreis von Scapanes australis. Boisd. (Coleopt. Dynastinae). Bonn. Zool. Beitr 8: 71–74.

Endrodi, S. 1971. Monographie der Dynastinae 4. Tribus: Pentodontini (papuanische und pazifische Inselwelt) (Coleoptera: Lamellicornia: Melolonthidae). Pacif. Ins. 13(2): 243–320.
—— 1973. Monographie der Dynastinae 5. Tribus: Oryctini (Coleoptera. Lamellicornia, Melolonthidae). Ent. Arb. Mus. Frey 24: 1–87.
Essig, E. O. 1942. College Entomology. Macmillan, New York, 900 p.
Firman, I. D. 1975. Annotated bibliography of source information on plant disease distribution in the area of the South Pacific Commission. S. Pacif. Comm. Tech. Paper no. 172.
—— 1978. Bibliography of plant pathology and mycology in the area of the South Pacific Commission, 1820–1976. S. Pacif. Comm. Tech. Paper no. 176.
Fosberg, F. R. 1960. A theory on the origin of the coconut. In: Symposium on the impact of man on humid tropics vegetation. Territory of Papua and New Guinea, Goroka, p. 73–75.
Froggatt, J. L. 1936. Coco-nut pests. N. Guin. Agr. Gaz. 2(3): 18–21.
—— 1939. The sweet potato hawkmoth, Hippotion celerio. N. Guin. Agr. Gaz. 5(3): 2–3.
—— 1941. Entomological notes: Fruit-piercing moths N. Guin. Agr. Gaz. 7(4): 298–300.
Froggatt, J. L. and B. A. O'Connor. 1940. Insects associated with the coconut palm. N. Guin. Agr. Gaz. 6(3): 16–32.
—— 1941. Insects associated with the coconut palm. Pt. 2. N. Guin. Agr. Gaz. 7(2): 125–133.
Fullaway, D. T. and N. L. H. Krauss. 1945. Common insects of Hawaii. Illustrated by Yasutaro Oda. Tongg Publ. Co., Honolulu. 228 p.
Gagné, W. C. 1977. Entomological investigations of agro-silviculture using the composted contour mound method in Papua New Guinea. Science in New Guinea. 5(2): 85–101.
—— 1979. Seeking a settlement in shifting agriculture. Dialogue (Winter/Spring 1978/9) 2–12. (with French and Spanish translations).
—— in press. Natural insect control in an intensified Highlands subsistence garden and some biological control needs for subsistence gardeners in Melanesia. Proc. 2nd Papua New Guin. Food Crops Conf.
Garlovsky, D. F. and B. Zelazny. 1971. External morphology of Oryctes rhinoceros. S. Pacif. Comm. Info. Doc. No. 25.
Glass, E. H. and H. D. Thurston. 1978. Traditional and modern crop protection in perspective. Bioscience 28(2): 109–115.
Golson, J., R. J. Lambert, J. M. Wheeler and W. R. Ambrose. 1967. A note on carbon dates for horticulture in the New Guinea Highlands. J. Polynesian Soc. 76: 369–371.
Gressitt, J. L. 1958. Ecology of Promecotheca papuana Csiki, a coconut beetle. Proc. 10th Intl. Congr. Entomol. 2: 747–753.
—— 1959a. Host relations and distribution of New Guinea hispine beetles. Proc. Haw. Entomol. Soc. 17(1): 70–75.
—— 1959b. The coconut leaf-mining beetle Promecotheca papuana. Papua N. Guin. Agr. J. 12(2–3): 119–148.
—— 1961. Problems in the zoogeography of Pacific and Antarctic insects. Pacif. Ins. Monog. 2.
—— 1963. Beetle pests of coconut in the New Guinea area. Proc. IX Pacif. Sci. Congr. 9: 61–62.
—— 1967. The role of the Papuan area in insect evolution and dissemination. Mushi 40(8): 89–95.
Gressitt, J. L. and J. J. H. Szent-Ivany. 1968. Bibliography of New Guinea entomology. Pacif. Ins. Monog. 18.
Henty, E. E. and G. H. Pritchard. 1973. Weeds of New Guinea and their control. Papua New Guinea, Dept. For., Div. Bot., Bot. Bull. No. 7. 2nd ed.
Herington, J. G. 1977. Wildlife introduced and imported into Papua New Guinea. Papua New Guin. Dept. Nat. Resources. Wildlife Branch, Wildlife Publ. 77/2. 9 p.
Hill, D. S. 1975. Agricultural insect pests of the tropics and their control. Cambridge Univ. Press. 516 p.
—— (ed) in press. Handbook of agricultural insect pests of SE Asia. Univ. Hong Kong, Dept. Zool., Occ. Publ., No. 4.
Janzen, D. H. 1973. Tropical agroecosystems. Science 182: 1212–1219.
Jones, R. 1973. The emerging picture of Pleistocene Australians. Nature 246(5431): 278–281.
Keleny, G. P. 1960. Notes on the origin and introduction of the basic food crops of the New Guinea people. In: Symposium on the impact of man on humid tropics vegetation. Administration of the Territory of Papua and New Guinea, Goroka. p. 76–85.
—— 1962. The origin and introduction of the basic food crops of the New Guinea people. Papua N. Guin. Agric. J. 15(1 and 2): 7–13.
Kimber, A. J. 1972. Widespread damage by insect pests in Highlands sweet potato gardens. Harvest 2(4): 117–121.
Lamb, K. P. 1974. Economic Entomology in the Tropics. London, Academic Press. vii + 195 p.
Lamb, K. P. and A. D. Johnston. 1976. Some insects of Papua New Guinea. Univ. Papua New Guin., Dept. Biol. 3rd edit.

Lever, R. J. A. W. 1934. Original food plants of local coconut pests. Brit. Solomon Is Agr. Gaz. 2(3): 8.
—— 1969. Pests of the coconut palm. FAO Agr. Stud. No. 77.
Macarthur, R. H. and E. O. Wilson. 1967. The theory of island biogeography. Princeton Univ. Press, Princeton. xi + 203 p.
Macewan, J. M. 1979? Subsistence agriculture. In: Agriculture in the economy. PNG Dept. Prim. Ind. 56 p.
Mackey, A. P. 1977. The biogeographic relationships of the New Guinea Sphingidae (Lepidoptera). J. Aust. Entomol. Soc. 16: 361–365.
Meyer-Rochow, V. B. 1973. Edible insects in three different ethnic groups of Papua and New Guinea. Amer. J. Clinic. Nutr. 26: 673–677.
New Guinea Dept. Agr. 1939. Annual report for 1938–1939. N. Guin. Agr. Gaz. 6(2): 6–19.
O'Connor, B. A. 1959. The coconut treehopper, Sexava spp., and its parasites in the Madang District. Papua N. Guin. Agr. J. 11(4): 121–125.
O'Sullivan, D. F. 1973. Observations on the coconut spathe bug, Axiagastus cambelli Distant (Hemiptera: Pentatomidae) and its parasites and predators in Papua New Guinea. Papua New Guin. Agr. J. 24(3): 79–86.
Paine, R. W. 1961. Observations on the banana scab moth in the Territory of Papua and New Guinea. Papua New Guin. Agr. J. 14(1): 45–47.
—— 1964. The banana scab moth, Nacoleia octasema (Mey.): its distribution, ecology and control. S. Pacif. Comm. Tech. Pap. No. 145.
Pena, R. S. de la. 1970. The edible aroids in the Asian-Pacific area. In D. L. Plucknett (ed.) Proc. 2nd Intl. Symp. Trop. Root Tuber Crops, Univ. Haw., Coll. Trop. Agr. Vol. 1: 136–140.
Perry, C. H. 1973. The pests and their control: coconut. Crop Tech. Course — Lowlands Agr. Exptl. Stn., Keravat, Papua New Guin. 9 p. mimeo.
—— undated. The ecology and control of some pests in Papua New Guinea. 4 p. mimeo.
Plowman, T. 1969. Folk uses of New World aroids. Econ. Bot. 23(2): 97–122.
Plucknett, D. L. 1970. Colocasia, Xanthosoma, Alocasia, Cyrtosperma and Amorphophallus. In D. L. Plucknett (ed.) Trop. Root Tuber Crops Tomorrow Vol. 1. Proc. 2nd Intl. Symp. Trop. Root Tuber Crops Univ. Haw., College Trop. Agr. Honolulu p. 127–135.
Powell, J. M. 1976. Ethnobotany. In K. Paijmans (ed.). New Guinea Vegetation. Austr. Nat. Univ. Press, Canberra, p. 106–183.
—— 1977. Plants, man and environment in the island of New Guinea In J. W. Winslow (ed.). The Melanesian environment. Australian Nat. Univ. Press, Canberra. xxxiv + 562 p.
Rangai, S. S. 1977. Taro. Papua New Guin. Dept. Prim. Indust. Rural Develop. Ser. Handbook No. 12.
Schultze-Westrum, T. G. 1970. Conservation in Papua and New Guinea: Final report on the 1970 World Wildlife Fund Mission. 46 p.
Scott, G. 1977. The importance of old-field succession. Biomass increments to shifting cultivation. Great Plains — Rocky Mtn, Geogrl. J. 6: 318–327.
Simmonds, N. W. 1962. The evolution of the bananas. Longman, London.
Simon-Thomas, R. T. 1962. De Plagen van enkele Cultuurgewassen in West Nieuw Guinea. Mededlingen van Economische Zaken, Landbouwkundige Serie, No. 1. p. 1–126.
Smee, L. 1965. Insect pests of sweet potato and taro in the Territory of Papua and New Guinea: their habits and control. Papua New Guin. agr. J. 17(3): 99–101.
Smith, E. S. C. 1977a. Studies on the biology and commodity control of the banana fruit fly, Dacus musae (Tryon), in Papua New Guinea. Papua New Guin. Agr. J. 28(2, 3 and 4): 47–56.
——1977b. Potato viruses and potato aphids. In Proc. Potato Seminar H.A.E.S., Aiyura. Printed by Highlands Agric. College. 4 p.
Southwood, T. R. E. 1960. The abundance of the Hawaiian trees and the number of their associated insect species. Proc. Haw. entomol. Soc. 17(2): 299–303.
—— 1975. The dynamics of insect populations. In: D. Pimental (ed.). Insects, Science and Society. Academic Press, p. 151–199.
—— 1977. Entomology and mankind. Amer. Sci. 65(1): 30–39.
Southwood, T. R. E., M. May, M. P. Hassell and G. R. Conway. 1974. Ecological strategies and population parameters. Amer. Nat. 108: 791–804.
Stibick, J. 1976. Fruit flies. Papua New Guin., Dept. Primary Industry, Entomol. Sect. Conf., 1976. 2p. mimeo.
Surany, P. 1963. The nature of diseases attacking Oryctes spp. Proc. 9th Pacif. Sci. Congr. 9: 72–78.
Szent-Ivany, J. J. H. 1960. The zoo-geographical factor in economic entomology on Pacific Islands with special reference to New Guinea. Proc. XI Intl. Congr. Entomol. 1: 534–537.

Szent-Ivany, J. J. H. 1963. Non-coleopterous insects of Cocos nucifera on South Pacific Islands, with special reference to the Territory of Papuan New Guinea. Proc. 9th Pacif. Sci. Congr. 9: 67–71.
—— 1974. Insect pests of yam (Dioscorea spp.) in Papua New Guinea. Folia Ent. Hung. (new series) 26(2): 205–10.
—— in press. Economic insects and mites of Papua New Guinea. Pacif. Ins. Monogr.
Szent-Ivany, J. J. H. and J. H. Barrett. 1956. Some insects of banana in the Territory of Papua and New Guinea. Papua New Guin. agr. J. 11(1): 40–44.
Szent-Ivany, J. J. H. and A. Catley. 1960. Notes on the distribution and economic importance of the Papuan tip-wilt bug, Amblypelta lutescens papuensis Brown (Heteroptera: Coreidae). Papua New Guin. Agr. J. 13(2): 59–65.
Thistleton, B. H. 1977. Pests of potatoes in PNG. In: Anon. (ed.) Potato Seminar. H.A.E.S., Aiyura. Printed by Highlands Agricultural College. 4 p.
Watson, J. B. 1965a. From hunting to horticulture in the New Guinea Highlands. Ethnology 4(3): 295–309.
—— 1965b. The significance of a recent ecological change in the Central Highlands of New Guinea. J. polynesian Soc. 74(4): 438–450.
White, J. P., K. A. W. Crook and B. P. Buxton. 1970. Kosipe: a late Pleistocene site in the Papua Highlands. Proc. Prehist. Soc. 36: 152–170.
Willemse, F. 1977. Classification and distribution of the Sexavae of the Melanesian subregion (Orthoptera, Tettigonioidea, Mecopodinae). Tijdschr. entomol. 120(8): 213–277.
Wilson, F. 1960. A review of the biological control of insects and weeds in Australia and Australian New Guinea. Tech. Comm. Cmwlth. Inst. Biol. Contr., Ottawa 1: 1–91.
Yen, D. 1963. Sweet potato variation and its relation to human migration in the Pacific. In: J. Barrau (ed.). Plants and the migrations of Pacific peoples. Sym. 10th Pacif. Sci. Congr., Honolulu 1961. Bernice P. Bishop Mus. Press, Honolulu, p. 93–118.
—— 1971. The development of agriculture in Oceania. In: R. C. Green and M. Kelly (eds.). Studies in Oceanic Culture History 2. Pac. Anthrop. Rec. No. 12. Bishop Mus. Press, Honolulu, p. 1–12.
—— 1974. The sweet potato and Oceania: an essay in ethnobotany. Bernice P. Bishop Mus. Bull. 236.
Yen, D. and J. M. Wheeler. 1968. The introduction of taro into the Pacific: The indications of chromosome numbers. Ethnology 7(3): 259–267.
Young, G. R. in press. Recent work on biological control in Papua New Guinea and some suggestions for the future. PANS 26: 000–000.
Zimmerman, E. C. 1968a. Rhynchophorinae of southeastern Polynesia (Coleoptera: Curculionidae). Pacif. Ins. 10(1): 47–77.
—— 1968b. The Cosmopolites banana weevils (Coleoptera: Curculionidae: Rhynchophorinae). Ibid. 10(2): 295–299.

PART THREE

Vegetation and flora

1. Late Quaternary vegetation history

D. Walker & G. S. Hope

1. Introduction

At every geographical scale and taxonomic level, the vegetation pattern of New Guinea presents problems, the solutions to which must have substantial historical components. Some of the processes involved doubtless began millions of years ago while some may be only a few centuries, perhaps decades, old. Ideally, explanations of the courses of vegetational change and the forces which have determined them should be based on substantial fossil records and related geological and archaeological phenomena. For the greater part of New Guinea and most of its geological history, however, information of these kinds is entirely lacking. Particularly for explanations of grosser patterns, therefore, recourse must be had to arguments from present day distributions beyond New Guinea, and notions about changing shapes and positions of land masses in the distant past.

For most of the purposes of historical plant geography, pollen analysis provides the best source of positive information. There are now many pollen diagrams and more fragmentary palynological data from the lower Markham valley (Garret-Jones 1979), Owen Stanley Range (Hope 1980a) E Highlands (Powell 1970), Simbu (=Chimbu) (Hope 1976; Corlett 1979), W Highlands (Powell *et al.* 1975), S Highlands (Oldfield *et al.* 1980; Williams *et al.* 1972), Enga (Walker and Flenley 1979; Oldfield *et al.* 1980) and W Sepik (Hope 1980a), Provinces of Papua New Guinea and from the West Snow Mountains (Hope and Peterson 1976) in Irian Jaya. Most of these data are associated with radiocarbon or other independent dating. Little of significance dates before 30 000 years ago (BP).

Particular difficulties attend the use of pollen analysis in the tropics. In the lowlands suitable sites are rare and very few forest trees are sufficiently profligate of pollen as to provide an input to accumulating sediments which accurately reflects their abundances. In the highlands, however, "representative taxa" with adequate pollen production occur in most vegetation types and appropriate basins of accumulation are more common. For these reasons, problems associated with the upland vegetation of the island have been tackled preferentially. Initial field work and the collection of reference pollen began in 1962 but it was not until 1975 that the first significant results reached

© *1982, Dr W. Junk Publishers, The Hague.*

publication stage. The strategy has been first to select accessible sites with deep organic lake deposits from localities which might be expected to have witnessed substantial changes in vegetation of kinds probably clearly reflected in the pollen input to the sites. Such changes are replacement of grasslands by forest, shifts in the balance between major forest pollen producers and the disturbance of forest by human activity. Data collected at this stage of the investigations also expose the full potential of the method for studying finer scale processes of vegetation change in tropical vegetation when new techniques and appropriate sites are found.

What is so far known of the vegetation history of New Guinea is described on three time scales covering the formation of the New Guinea flora, vegetation changes in the late Quaternary, and the development of vegetation communities recognisable in the present.

2. The synthesis of the flora

New Guinea has a complicated geological history which probably accounts for some of the most striking aspects of its flora (Walker 1981). Parts of the island (eg. the Papuan peninsula) were separated from the eastern edge of the Australian crustal plate about 50 million years ago (m.y. BP) (Eocene) at a time when rainforest of Gondwanic origin covered a substantial part of the continent. Drifting eastwards, these fragments came under the influence of the northwestward shear at the Australian-Pacific plates' boundary and moved into their new, and present, positions with respect to Australia about 23 m.y. BP. There, perhaps with other ancient fragments, they formed the leading edge of the Australian plate in its movement towards SE Asia, presumably still bearing Gondwanic plants some genera of which persist in New Guinea to the present day. Fossil evidence shows that some of these plants (eg. *Nothofagus*, *Podocarpus*) have been in New Guinea at least since the mid-Miocene (Khan 1974). When the Australian and SE Asian crustal plates reached roughly their present relative positions about 10 m.y. BP (late Miocene), the former's ancient isolation from invasion was ended and its northern edge, including New Guinea, was exposed to immigration from the enormous Asian floras to its west and north. Throughout the last 5 m.y. the main highlands belt (Fig. 1) and other parts of New Guinea have been rising, probably most quickly at the end of the Pliocene and the beginning of the Quaternary (3–1 m.y. BP), and are still doing so today. Volcanic activity and earth tremors were and are common, and the products of erosion of the uplifting masses accumulate to both north and south. Apart from the Oriomo plateau in the south, which is best regarded as an extension of the stable Australian continent, New Guinea has been unusually unstable geologically for at least 5 m.y., during which time the highest land has become higher thereby entering colder atmospheric layers and modifying the circulation of rain-bearing winds. These are facts fundamental to the understanding of New Guinea vegetation and its constituent flora. The instability has continually created new environments for occupation by oversea immigrants and by the products of their speciation and that of the ancient Gondwanic

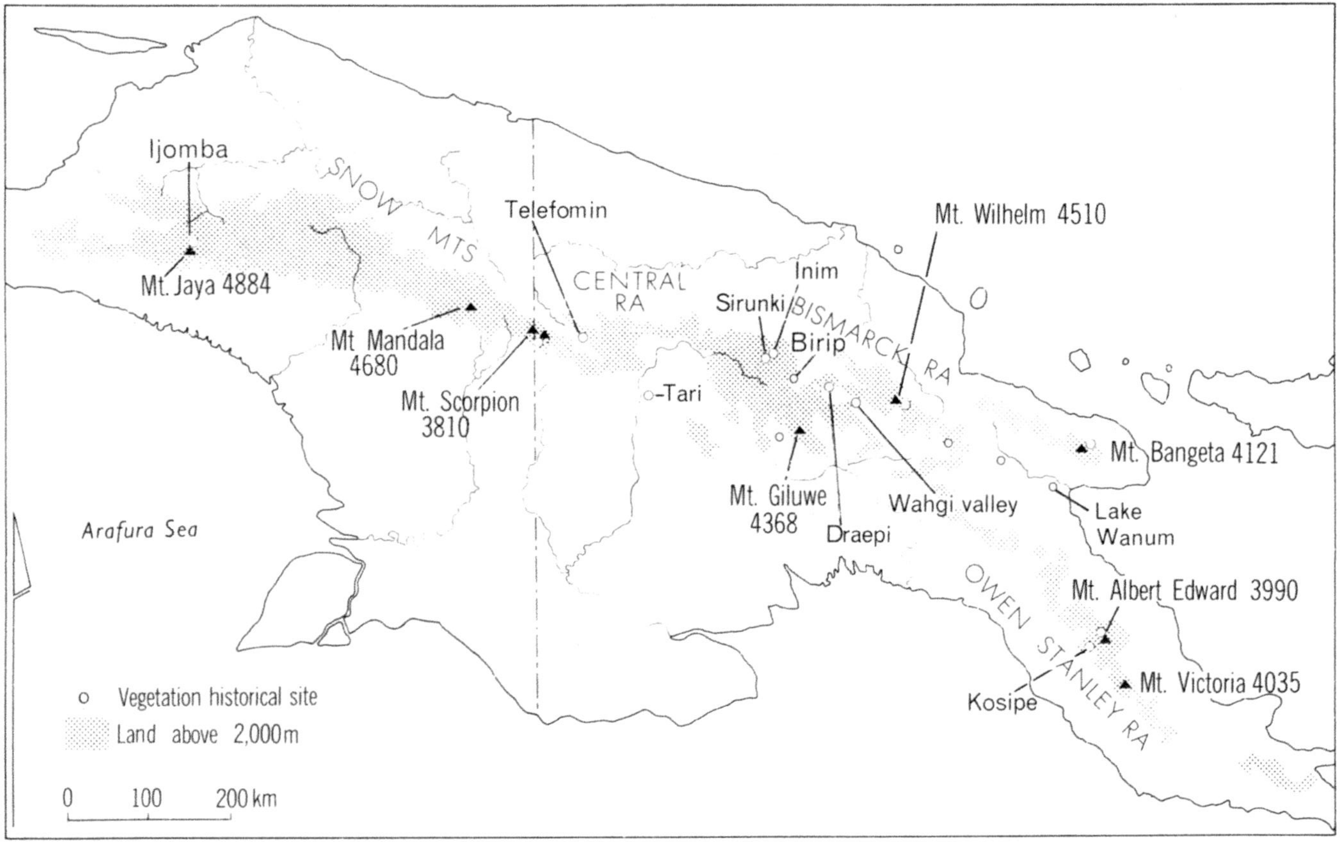

Fig. 1. Location of vegetation historical sites in New Guinea. Some major peaks are given in heights in meters above sea level.

endowment. Indeed, environmental variety seems to exceed the ability of immigrants or new endemics to adapt to it. Compared with other tropical areas of comparable size and usually lesser habitat diversity, New Guinea has a small flora (about 1465 genera and 9000 species), (Balgooy 1976) the evolutionary activity of which is indicated by its 55% species endemicity. On the ecological scale, the very processes which ensure evolutionary opportunity inhibit the development of stability and give rise to pronounced patchiness in the vegetation (White 1975).

During the past few millions of years, therefore, the old Gondwanic stock has been joined by Laurasian immigrants the genera of which now far outnumber those of longer residence. All have speciated actively. On the whole, Gondwanic taxa are best represented in the uplands where some major vicariants live side by side; e.g. in the Fagaceae, the northern *Lithocarpus* and *Castanopsis* and the southern *Nothofagus*; in the Ericales the northern Ericaceae and the southern Epacridaceae. The highest mountains may well have provided the latest major environment to be exposed to immigration, only recently having risen high enough to pass the altitudinal forest limit in even the warmest climatic intervals. Most alpine plants therefore have probably reached New Guinea recently although some have apparently poor migration ability (e.g. *Drapetes ericoides*, *Oreomyrrhis pumila*); some subalpine species (e.g. *Drymis piperita*) may well have evolved in New Guinea from ancestors in the forests (Smith 1977). To the south of the island, Torres Strait and the Arafura Sea, which separate it from Australia, have frequently been dry since glaciation-induced sea level changes began to affect the world. This has permitted the spread of the basically Australian flora into those parts of SE New Guinea (e.g. Port Moresby area, Oriomo plateau) which even today are periodically dry enough to support fire-prone eucalypt-dominated savanna, although immigration across the sea barrier may also be possible (Carr 1972).

An alternative hypothesis about the movement of land masses as a result of the expansion of the earth, wedges parts of New Guinea between SE Asia, Australia, S and C America as long ago as the Early Jurassic (Shields 1979). Thereafter, relatively small movement with respect to Asia and Australia brought New Guinea to its present position. On the face of it, this arrangement might simplify the explanation of the New Guinea flora, including its trans-Pacific affinities. In particular, it would greatly extend the time available for the invasion of New Guinea by SE Asian plants and for their speciation there. The weight of evidence from outside New Guinea, however, still favors the history of crustal plate displacement outlined above.

3. Late Quaternary movements of the main plant formations

The climatic and sea level oscillations which probably began in the Pliocene if not earlier (Shackleton and Kennett 1975), and culminated in the Quaternary, added further complication to an already dynamic system. From the pollen diagrams of New Guinea it is possible to glimpse the vegetation pattern which had developed before the last (Holocene) warming and the changes in the pattern which that precipitated. Comparable vegetation changes probably took place during all the more recent interglacial periods but the continued uplift of the mountains and immigration and diversification of the biota must have imposed long-term trends changing the biological properties of the ecosystems affected by recurrent environmental change. Possibly there was a period during which a dominantly "interglacial" or "Tertiary" climate regime punctuated by cooler phases gave way to one in which the warm times were the less frequent.

There is no doubt that New Guinea was formerly glaciated more widely than it is today. Moraines (Fig. 2) and other glacial features, mapped largely by Löffler (1972) and Hope and Peterson (1976), led Hope and Hope (1976a) to an estimate of 2000 km^2 of ice in the mountains of the island at the maximum of the last glaciation about 18 000–15 000 BP. Today, there are only about 8 km^2, restricted to Mt Jaya (Carstensz), Mt Idenburg and Mt Mandala (Juliana) in Irian Jaya. At glacial maximum, snow lines were highest in the east and west and somewhat lower in the middle (e.g. Mt Scorpion), a feature Hope (1980b) attributes to differential precipitation-ablation regimes rather than the temperature gradients. Throughout the island, however, they lay at least 1000 m and locally more, below present actual or theoretical snowlines implying alpine temperatures 6° to 8°C below present levels. By contrast, sea surface temperature at the time is thought to have been only 2°C below present (CLIMAP 1976). Webster and Streten (1978) conclude that these estimates are incompatible unless cold air incursions from higher latitudes were sufficiently frequent to allow snow lines to be held below the freezing altitude of the ambient tropical air. The course of deglaciation has been dated on Mt Wilhelm where it began about 15 000 and was complete by 8500 BP (Hope and Peterson 1975) and on Mt Jaya, where it probably started slightly earlier (Hope and Peterson 1976); dates for the deglaciation of specific sites on other mountains are not at variance with these two records. There is some evidence from Mt Jaya that overall ice retreat may have been punctuated with temporary advances in its early stages, arguing that substantial and maintained temperature amelioration might have been delayed until about 10 000 BP (Fig. 3).

The main reference pollen diagram of New Guinea comes from Sirunki Swamp, 2500 m and 9 km^2 in the Enga Province, spanning the period from 33 000 to 1500 BP (Walker and Flenley 1979). Much of the land within 5 km of the site now bears grassland and gardens but the surrounding forest is rather undifferentiated lower mountain forest in which *Nothofagus pullei* is strongly represented on the slopes rising northward. A ridge rising directly from the swamp to 3000 m a.s.l. has upper mountain forest at its crest (Walker 1966, 1972; Walker and Guppy 1976; Flenley 1969) (Fig. 4–6). Before about 27 000 BP the pollen analytical data are enigmatic; if they are to be positively interpreted they must suggest hardly vegetated, certainly not forested, conditions yet lacking alpine indicators. These conditions were superseded by the rapid encroachment of trees (e.g. *Rapanea*, *Nothofagus*, *Podocarpus*, *Ascarina*, *Cyathea*) occurring today in both upper and lower mountain forests and the establishment of a subalpine swamp flora. The slopes were not wooded for long and after about 25 000 BP subalpine or alpine grasslands and swamps dominated the region, with perhaps a few enclaves of shrubs. This condition

Fig. 2. Large Pleistocene moraine with subalpine forest, above open grassland with tree ferns, probably cleared by fire. The ice cap on Mt Jaya is visible in the backgroud. Kemabu Plateau 3620 m, Irian Jaya. Photo: G. Hope, February 1972.

Fig. 3. Open glade in *Nothofagus* and conifer forest, probably the result of burning. The area is being invaded by *Vaccinium* and *Papuacedrus*. Near Ilaga, 2850 m, Irian Jaya. Photo: G. Hope, March 1972.

more-or-less persisted in the catchment until about 14 000 BP, although the evidence of far-travelled pollen suggests that the mountain forests had started to expand 2000 years earlier. These data are interpreted as recording a period, certainly extending from 25 000 to 18 000 BP during which the altitudinal forest limit was generally falling. It mostly stood below the altitude of Sirunki but some tree and shrub taxa at least made intermittent forays into the catchment. There is fragmentary evidence from other places in the highlands (Powell 1970, pers. comm; Williams *et al.* 1972) confirming these indications that the forest limit occurred very substantially below its present level for a long time before reaching its greatest depression about 18 000 BP.

The Sirunki data suggest that the forest-limit stood at about 2300 m a.s.l. about 18 000 BP and had begun to move upward between 17 000 and 16 000 BP reaching 2500 m about 14 000 BP. A thousand years later the upper mountain forest was probably no more common than it is today, namely restricted to the 3000 m ridge above the site, the greater part of the catchment being occupied by lower mountain forest rich in *Nothofagus*, *Ascarina*, *Podocarpus* and *Phyllocladus*. Yet this forest dominance was very short-lived for by 12 500 BP the Sirunki catchment was once again occupied by *Cyathea* woodland, sedgey swamps and depauperate grasslands. It was not until about 9000 BP that the catchment became enclosed in a consistently forested belt. Both before and after the greatest depression of the forest, it is evident that 2500 m was an altitude marginal for maintained forest development.

A pollen diagram from Ijomba mire, at 3630 m in Irian Jaya, records at its base alpine vegetation at about 14 000 BP. *Rapanea* and *Coprosma* shrubland invaded the area shortly thereafter and the altitudinal tree line arrived about 11 500 BP. The forest was consolidated by the appearance of *Dacrycarpus* and *Podocarpus* about 1500 years later but there is no indication of any reversion to open communities once the forest had been established (Hope and Peterson 1976).

On Mt Wilhelm, although the ice had begun to retreat by 14 500 and was completely gone by 9000 BP, the pollen diagrams suggest that a consistent rise in forest limit did not begin long before 10 400 BP when it passed 2740 m to reach 3910 m by about 8600 BP. From before 22 000 until 10 500 BP the forest limit was well below the lowest site (2740 m) although its position probably varied (Hope 1976).

The Irian Jaya data suggest a somewhat earlier and less interrupted uphill march of the forest than does the more complete information from Papua New Guinea. In the latter, at least, vegetation response to deglaciation was uncertain for the first 6000 years or so, until, indeed, deglaciation was virtually complete. However, preliminary results (Hope 1980b) agree with the completed pollen data in indicating that forest vegetation extended throughout New Guinea to at least as high as it does today, and probably a little higher by 8500 BP.

During the long periods with altitudinal forest limits close to 2300 m many of the island's great intra-montane valleys lying at or above 1500 m (Fig. 9) must have experienced frequent and severe frosts and cold air drainage effects. They probably harbored extensive grassland extensions from higher altitudes just as many mountain valleys at present carry such vegetation far below the regional

Fig. 4. Kayamanda swamp, 2500 m, near Sirunki, Enga Prov., PNG. *Schefflera* in foreground and swamp sedges behind, with gardens and dissected forest on the slopes. The Sirunki pollen diagram was obtained from this swamp. Photo: G. Hope, November 1973.

Fig. 5. Lake Ipea, near Sirunki, Enga Prov. The slopes are cleared to gardens and grass while mixed mountain forest including *Nothofagus* occurs on the ridges. Photo: G. Hope, November 1973.

Fig. 6. Lake Ipea and Kayamanda swamp, 2500 m, near Sirunki, Enga Prov. Gardens and grassland around the swamp give way to mixed mountain forest on the slopes. Photo: G. Hope, November 1973.

Fig. 7. Dissected Pleistocene lake deposits in the Kaugel Valley, Enga Prov. PNG, at 2000 m. Photo: G. Hope, August 1973.

Fig. 8. Open forest of *Nothofagus* on a ridge crest at 3000 m, northern slopes of Mt Wilhelm, PNG. There are naturally relatively few undercanopy species present. Photo: G. Hope, July 1970.

altitudinal forest limit. Whether such valleys ever became completely forested during the general forest expansion is a moot point.

Holocene pollen diagrams from several highland areas indicate partly open or disturbed forest, from the base of the available record. Certainly the Wahgi valley (Fig. 9) was partially clear of forest 5000 BP (Powell *et al.* 1975) as was the Lai valley near Birip Crater, 2300 BP (Walker and Flenley 1979). The times of initiation of these clearances are not known, but partial clearance took place at Sirunki 4500 BP. However, regeneration after 3800 BP re-established the forest, albeit of different composition, by 3000 BP. Later, about 2000 BP, a more widespread and maintained replacement of forest by open land began in the Sirunki-Inim region (2500 m) which may have been the direct fore-runner of present conditions there (Fig. 4–6). In the absence of evidence of any other force causing the removal of these highlands' forests, human activity must be strongly suspected, particularly since there is some suggestion of horticulture as much as 9000 BP and quite certainly from 6000 BP in the Wahgi Valley (Golson 1977, Powell *et al.* 1975) (Fig. 10).

On the highest mountains it is perhaps less easy definitely to attribute shifts in the position of the altitudinal forest limit to human interference. After all, the tree line is likely to have been as sensitive to climatic shift as to anthropogenic pressures. Thus it is possible that the spread of subalpine grassland at the expense of forests near Ijomba mire during the last 6000 years and the fall of the forest limit on Mt Wilhelm through 200 m to its present level during the last 5000 years have been due to climatic deterioration. On the other hand Hope and Hope (1976a) have clearly demonstrated the presence of hunters at almost 4000 m about 5500 BP and have suggested (1976b) that the beyond-forest grasslands might have been amongst the better mammal hunting grounds since the human settlement of New Guinea.

The use of fire in pursuit of prey might explain the presence of fine charcoal in pollen analysed deposits at times when their sites were close to an accessible forest limit or above it (Hope 1980b; Hope and Hope 1976b). On Mt Wilhelm there is stratigraphic evidence of climatically-induced soil instability at some time within the last 7000 years, which may have correlated with forest changes at lower altitudes. In any event the depression of the forest limit about 5000 BP was followed by burning in the grasslands immediately above it and this has subsequently continued at varying intensities (Corlett 1979). Fire is not necessarily man-made but Corlett has argued that in the generally wet cool mountains, occasional fire-carrying dry spells will only rarely coincide with lightning strikes whereas they would always be exploited for burning by purposive humans (Fig. 11, 12). Corlett (1979) found high frequencies of charcoal particles after about 230 BP at many sites on Mt Wilhelm. He attributes these to human interference with the vegetation while his pollen analyses of the same sites show that, at most, forest was replaced by grassland at the same time that charcoal increased.

While much of the evidence for the destruction of forest by human activity is still circumstantial and likely to remain so, a favorite human instrument, namely fire, has certainly been operating in the highlands for at least 30 000 years and the quantitatively strongest evidence for it is associated with forest-

Fig. 9. Open grassland with very restricted forest, Draepi swamp, 1880 m, Wahgi Valley near Mt Hagen, PNG. Forest had been substantially cleared in this area before 5000 BP. Photo: J.M. Powell, August 1967.

Fig. 10. Kuk tea plantation. The lines of ancient drainage ditches are visible, cut across by modern ditches. The earliest ditches may be 9000 years old. Kuk, near Mt Hagen, W. Highlands, PNG. Photo: J. Golson, 1969.

Fig. 11. Forest of *Nothofagus* spp. killed by fire in 1973, Tel Basin, eastern Star Mts, 2850 m. Photo: G. Hope, June 1975.

Fig. 12. Fires lit in grassland have killed a strip of beech forest surrounding the valley. Tree ferns have resprouted since 1973. Tel Basin, eastern Star Mts, 2840 m. Photo: G. Hope, June 1975.

free, or forest-disturbing, episodes in the pollen analytical record.

The only lowland forest history is from the lower Markham Valley. There, Garrett-Jones (1979) has demonstrated a fluctuating but generally increasing representation of non-forest elements since about 9500 BP. A hypothetical reconstruction of glacial-time vegetation, based on water balance considerations, led Nix and Kalma (1972) to conclude that the lowland closed forest was comparatively restricted in the south of the island, and possibly in the upper Markham and middle Sepik areas. Elsewhere the closed forest would have maintained more or less its present boundaries despite the lower sea levels.

The past 30 000 years therefore have witnessed a single major upward shift of the altitudinal treeline through 1500 m, much of which was accomplished between 14 000 and 8500 BP. Superimposed on the greater stability which preceded and succeeded this change were lesser oscillations of varying magnitude and persistence. Most of these were probably climatically controlled but in some the hand of man was almost certainly involved as he hunted game and forest products and cleared the forests for his gardens.

4. The composition of the vegetation

4.1 Subalpine vegetation

In the present state of knowledge it is very difficult to reconstruct in any detail the vegetation of any part of New Guinea before the beginning of the last major uphill excursion of the forest limit. Nevertheless, it is significant that between 33 000 and 18 000 BP, the appearance of trees in the predominantly herbaceous vegetation around Sirunki did not reflect invasions by integrated forest communities. Instead there seems to have been opportunistic expansions of taxa (e.g. *Rapanea*, *Nothofagus*, *Cyathea*) which today have less obvious individualistic tendencies. This may argue that the highest altitude forests of the times were more varied in composition than those of today, composed of taxa now more differentiated altitudinally.

Although deglaciation, and presumably therefore climatic warming, began about 15 000 BP, it was not accompanied by a comparable movement of the forest limit, at least in the eastern part of the island. Instead, there seem to have been extensive grasslands and herbfields between the retreating ice and snow and the only slightly fluctuating forest edge. The reasons for this hesitant response on the part of the forests are obscure; climatic (Hope 1976; Walker and Flenley 1979) and anthropogenic (Hope and Hope 1976b) reasons have been advanced. However, when the upward spread of forest finally commenced, it proceeded rapidly, suggesting that the climatic limits to tree growth were already at, or somewhat above, the present forest limit. The rather sudden availability of a large area of forestable land seems to have led to much variation between the forests which finally colonised different parts. At Sirunki, which had for some time been only a few hundred metres above the forest limit, a subalpine forest phase is hardly discernible before lower mountain forest taxa (e.g. *Castanopsis*, *Phyllocladus*) as well as those of wider distribution (e.g.

Nothofagus, Podocarpus) became established. For about 1000 years (2 to 4 tree generations) however, the abundances of the contributing taxa fluctuated considerably, suggesting relatively poor overall integration dominated by large more-or-less even-aged patches of gregarious species.

The pollen diagrams from Mt Wilhelm (Hope 1976) relate best to the development of the subalpine forest which, in Wade and McVean's (1969) terms, now extends from 3350 to 3800 m. This ground was occupied by grassland and herbfield until about 10 500 BP when shrubs (e.g. *Styphelia suaveolens, Coprosma* sp., Ericaceae) invaded it together with the tree, or tall shrub, *Rapanea* (probably *R. vaccinioides*). Shortly afterwards *Dacrycarpus* and *Quintinia* arrived and a sort of subalpine forest almost certainly formed. Yet it was substantially different, quantitatively, from that living there today. *Dacrycarpus*, for instance, took about 3000 years to assert the dominance it currently enjoys.

On Mt Scorpion, the process of forest formation seems to have followed a rather different course. At about 3480 m *Rapanea* was already present shortly after deglaciation and was joined by *Papuacedrus* and *Phyllocladus* by 10 500 BP; *Dacrycarpus*, oddly, arrived later (Hope 1980b). On Mt Jaya, by contrast, *Dacrycarpus* was the first forest tree to become established in *Rapanea* shrublands and this occurred about 10 750 BP (Hope 1980b).

Walker and Flenley (1979) have argued that the "full-glacial" highest altitude forests were different from anything known in New Guinea today. The dominant impression from the pollen analytical data is that the course of synthesis of the subalpine forests since this time has varied from place to place and depended very much in its initial stages on the identity of the *opportunistic* forest species locally poised to quickly occupy the newly available land. Indeed, in a flora speciating so impressively, it is likely that ecotypic differences within species differentially affected their success and the associations between them. The subalpine forest as we know it today is diverse in composition for this very reason; it is still being synthesized and does not represent a stable expression of climatic and edaphic conditions.

4.2 Lower mountain forest

Around Sirunki, lower mountain forest taxa, established about a thousand years earlier, achieved a continuing balance about 6500 BP. Even so, this balance was modified on two time scales. Among the canopy trees, although fluctuations with periodicities of less than 2000 years were usually shared by a group of taxa (e.g. *Nothofagus, Ascarina, Castanopsis, Podocarpus, Phyllocladus*), the order of their increases or decreases and their relative rates of change were not the same in all oscillations. Interestingly, those taxa which are most closely associated phytosociologically in living forest (e.g. *Nothofagus* and *Ascarina*) showed the closest and most consistent quantitative associations in the past. Also, amongst the taxa which march together through 2000-odd years of fluctuations are some which frequently fall out of step with one another for shorter periods. Such apparent inconsistencies are readily explained in terms of different responses to temporary environmental variations and differing regeneration times.

Superimposed on this fluctuating, but generally consistent, association between the canopy taxa are their corporate relationships with the forest ephemerals (e.g. *Trema*, *Homalanthus*). Before about 4500 BP at Sirunki, fluctuations of the two groups down to a time-scale of about 500 years were generally parallel, suggesting a constant proportion of ephemerals occupying canopy gaps resulting from the deaths of big trees. Since 4500 BP, however, while peaks of canopy taxa were usually accompanied by peaks of forest ephemerals, the latter were not reduced to the same degree as were the canopy trees between their peaks. This persistence of quantities of forest ephemerals must have been the result of the maintained loss of canopy trees. By comparison with its conditions before 4500 BP the forest may be said to have been "degraded".

Although the data from sites other than Sirunki do not allow this kind of analysis, it is clear that by 5000 BP the forests of the upper Wahgi valley (1500 m) had a high proportion of ephemerals and were discontinuous (Powell *et al.* 1975). Amongst the far-travelled pollen reaching the Mt Wilhelm sites from forested areas below, *Castanopsis* evidently lost some of its earlier importance about 6500 BP (Hope 1976).

There are strong indications, therefore, that, while the qualitative, generic composition of the lower mountain forest (Fig. 13, 14) has been fairly constant for a long time, the relative quantities of its components have continually changed. Depending on altitude, and perhaps other factors, a significant change of mode took place between 6500 and 4500 BP. Although it is certain that a large proportion of today's lower mountain forest is of the "later" degraded kind, there may be areas in which the change did not occur and in which the earlier established forests still persist. If they are found, however, they are unlikely to be uniform in composition. Rather might it be expected that the population fluctuations demonstrated pollen-analytically through time will be expressed by spatial variety in the living forests.

The cause of the major change in this composition of the lower mountain forests is not certain. But at Sirunki it followed the first extensive disturbance of the forest which has been attributed circumstantially to human activity while elsewhere there is positive archaeological evidence for man's presence and horticultural practices at the time of its occurrence. Moreover, its most important features, the thinning of the canopy and the encouragement of the shorter-lived trees, are readily understood as responses to the maintained exploitation of forests for timber, bark, fruits and other useful materials.

5. Conclusions

The history of New Guinea vegetation can be seen as an attempt to fill an increasing diversity of habitats over more than 10 million years. Some of the processes producing habitat diversity have been maintained throughout, for example tectonic change leading to alluviation and landslips, although presumably at rates which have varied from time to time. The kinds of habitats for which these processes are responsible have probably been represented for several million years in New Guinea so that selection of plants to occupy them might be expected to be long past its zenith. Other long-established processes have resulted in the sequential availability of new habitats, as, for instance, uplift has carried land into colder and colder air. The biotic response to these processes must still be very active. Still others, e.g. the general climatic cooling which has affected the earth since the mid-Miocene, though long-established, have been expressed with greater or lesser vehemence from time to time.

The trends associated with these long-present processes have been modified by less ancient changes, some of them of an oscillatory kind. Foremost amongst these have been the periodic climatic fluctuations which have alleviated the overall cooling for thousands of years at a time, particularly during the last 2 million years. These periods, in one of which we now live, have given temporary advantage in some altitudinal zones to plants capable of rapid migration.

Most recent have been man's activities, first as a hunter and gatherer, later as a gardener and pig-keeper and finally as an agent of industrial development. This has very rapidly added further stress to an already dynamic vegetation during the past 30 or more millennia. (See Powell, chapter II:4).

Although there must have been many extinctions resulting from these environmental instabilities, some of them catastrophically sudden, others more attenuated, the overall effect must have been to conserve a large proportion of the variety arising in the flora. This variety was continually produced during the past 10 million years by immigration, particularly from the north and west, and by normal genetic mechanisms in all the components of the flora, old and new. It has been preserved by the very patchiness, spatially and temporally, of the New Guinea environment.

At any particular time in the history of New Guinea's vegetation, the composition and distributions of its parts will be a summation of the effects of all these processes. This level of argument accounts for the relatively small flora and its exceptionally high specific endemicity. But if the vegetation is observed in a place subject to a particular component of the environmental stresses shortly after that component has been intensely active, its composition will probably be dominated by the effects of that stress. So it is that the subalpine forest today is both synthesising as a result of the warming which began 16 000 BP and being destroyed by the hand of man. Similarly, it seems likely that the difficulty in establishing anything but the crudest altitudinal zonation of New Guinea vegetation types results from the localised nature of the responses of a generally rapidly speciating flora to a strong selective agent (climatic warming) recently applied and to the "founder effects" of species (e.g., some *Nothofagus* spp.) which are evidently able to respond quickly and then persist in the areas they occupy.

Figure 15a,b is an attempt to portray the dynamics of New Guinea vegetation by the rather imprecise extension of the notions of Holling (1973) concerning resilience and stability to the sum of the island's vegetation types and to long periods of time. Resilience, in this context, is the capacity of the vegetation to respond to forces of change and still persist. Quiescence, on the

Fig. 13. Mixed mountain forest with *Pandanus*, 2740 m, Mt Wilhelm, PNG. Photo: G. Hope, September 1969.

Fig. 14. Mixed mountain forest with *Pandanus*, 2740 m, Mt Wilhelm, PNG. Photo: G. Hope, September 1969.

other hand, measures the amount of change taking place in the vegetation from time to time. Stability is a measure of the rate at which vegetation returns to its prior state after temporary disturbance. For a given disturbance, the quiescence will be highest in the most stable vegetation. Inquiescence may be the result of inherent instability or repeated or extreme disturbance or all of these. Resilience, which may be attained by stable or unstable systems, is most strongly stressed during periods of inquiescence.

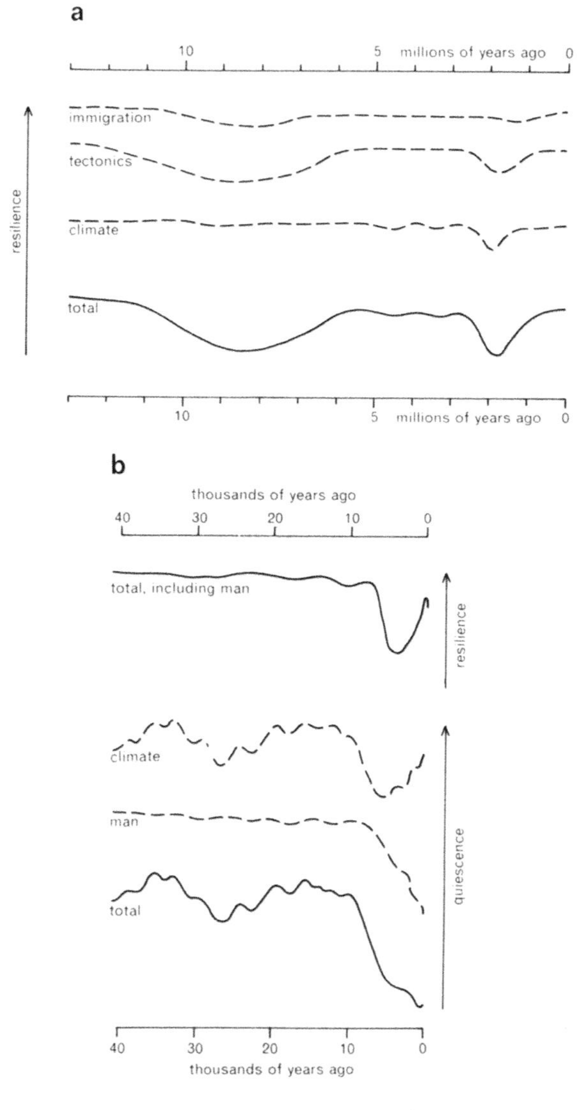

Fig. 15. Supposed trends in resilience and quiesence of New Guinea vegetation, a) during the last 13 million years, b) during the last 40,000 years.

The separate effects of contributory forces are shown by broken lines, their total effects by continuous lines. The same, notional (vertical), scale is used for all resilience plots but there is no relationship between the scales for resilience and quiescence. For further explanation see text.

Immigration of new taxa and environmental changes due to tectonic movements and climate have threatened the resilience of New Guinea vegetation since juxtaposition with Asia. Speciation has probably assisted in the long term maintenance of resilience although sometimes disruptive of it in the short-term. In the early and middle Miocene, tectonic movement probably greatly disrupted a vegetation adapted to only the slowest physical changes. Immigration and the beginnings of climatic change probably contributed to the disintegration of earlier vegetation. The emergence of new vegetation types only slowly established a new resilience capable of undisrupting response to the continuing physical pressures. The increasing climatic uncertainty of the Pliocene and the major tectonic activities and climatic developments at the Plio-Pleistocene transition probably caused very rapid breakdown of the vegetation despite its partial pre-adaptation, although the latter probably assisted it to a more rapid recovery of its resilience. Despite the contribution of the Miocene flora to that of the Late Quaternary, the vegetations of the two periods were ecologically utterly different; that of the later period being capable of withstanding much more environmental change.

Resolving the events of the last 40 000 years more finely requires human activities to be included amongst the factors affecting resilience. The degree of inquiescence can also be read from the pollen diagrams and attributed to climate as well as to human interference. Despite the inquiescence induced by climatic change and continued tectonic movements and immigration during the period, pre-adaptation through the earlier part of the Quaternary or longer, ensured that the resilience of the vegetation was adequate for its recovery. Hunting and gathering probably has but little effect on resilience, species being already available in the vegetation quickly to react to such interference. But the destruction of forests for gardens was a force of a kind hardly felt before, very rapidly applied. Yet, because of the long history of turbulence in the New Guinea environment, vegetation types emerged which adapted to the indigenous horticulture, so that resilience was quickly returned to the vegetation as a whole despite continually falling quiescence. The effects of industrial society on the vegetation, such as the clear felling of large areas of forest over very short periods of time, are sure to impose new stresses on its ability to respond.

New Guinea's vegetation today is evidently adapted to changing physical environments. Such resilience as it has is the result of coming to terms with these forces evolutionarily. Many units of the vegetation as a whole may be stable, in Holling's (1973) sense, but few are quiescent.

Acknowledgements

We are grateful to Dr R. T. Corlett, Dr J. R. Flenley, Dr S. Garrett-Jones and Dr J. M. Powell for their many contributions to the discussion of New Guinea vegetation history and for access to their unpublished data, and to Dr M. M. J. van Balgooy for his revised estimate of species endemicity. Val Lyon prepared the figures for publication.

References

Carr, S. G. M. 1972. Problems of the geography of the tropical eucalypts. In: D. Walker (ed.) Bridge and Barrier: the Natural and Cultural History of Torres Strait. Research School of Pacific Studies, Department of Biogeography and Geomorphology, BG/3. Canberra, Australian National University.

CLIMAP project members. 1976. The surface of Ice Age Earth. Science 191: 1131–1137.

Corlett, R. J. 1979. Human impact on the subalpine vegetation of Mt Wilhelm, Papua New Guinea. Ph.D. thesis, Australian National University, Canberra.

Flenley, J. 1969. The vegetation of the Wabag region, New Guinea highlands: a numerical study. J. Ecol. 57: 465–490.

Garrett-Jones, S. 1979. Holocene vegetation and lake sedimentation in the Markham Valley, PNG. Ph.D. thesis, Australian National University, Canberra.

Golson, J. 1977. No room at the top: agricultural intensification in the New Guinea highlands. In: F. J. Allen, J. Golson and R. Jones (eds) Sunda and Sahul, Prehistoric Studies in Southeast Asia, Melanesia and Australia. London, Academic Press.

Holling, C. S. 1973. Resilience and stability of ecological systems. Ann. Rev. Ecol. and Syst. 4: 1–23.

Hope, G. S. 1976. The vegetational history of Mt Wilhelm, Papua New Guinea. J. Ecol. 64: 627–664.

—— 1980a. Tropical mountain forest in retrospect and prospect. In: J. N. Jennings and G. Linge (eds). Of Time and Place. Essays in Honour of Oskar Spate. Canberra, Australian National University Press.

—— 1980b. Historical influences on the New Guinea flora. In: P. van Royen (ed.) Alpine Flora of New Guinea. Stuttgart, Cramer Verlag.

Hope, G. S. and J. H. Hope. 1976a. Man on Mt Jaya. In: G. S. Hope, J. A. Peterson, U. Radok and I. Allison (eds). The Equatorial Glaciers of New Guinea. Rotterdam, Balkema.

Hope, J. H. and G. S. Hope. 1976b. Palaeoenvironments for man in New Guinea. In: R. L. Kirk and A. G. Thorne (eds) The origins of the Australians. Canberra, Australian Institute of Aboriginal Studies.

Hope, G. S. and J. A. Peterson. 1975. Glaciation and vegetation in the high New Guinea mountains. Bull. Roy. Soc. NZ 13: 155–162.

—— 1976. Palaeoenvironments. In: G. S. Hope, J. A. Peterson, U. Radok and I. Allison (eds). The Equatorial Glaciers of New Guinea. Rotterdam, Balkema.

Khan, A. 1974. Palynology of Neogene sediments from Papua (New Guinea). Stratigraphic boundaries. Pollen et Spores 16: 265–284.

Löffler, E. 1972. Pleistocene glaciation in Papua and New Guinea. Z. Geomorph. Suppl. 13: 32–58.

Nix, H. A. and J. D. Kalma. 1972. Climate as a dominant control in the biogeography of northern Australia and New Guinea. In: D. Walker (ed.) Bridge and Barrier: the Natural and Cultural History of Torres Strait. Research School of Pacific Studies, Department of Biogeography and Geomorphology, BG/3. Canberra, Australian National University.

Oldfield, F., P. G. Appleby and Thompson. 1980. Palaeoecological studies of lakes in the highlands of Papua New Guinea. I. The chronology of sedimentation. J. Ecol. 68: 457–477.

Powell, J. M. 1970. The impact of man on the vegetation of the Mt Hagen region, New Guinea. Ph.D. thesis, Australian National University, Canberra.

Powell, J. M., A. Kulunga, R. Moge, C. Pono, F. Zimike and J. Golson. 1975. Agricultural traditions of the Mt Hagen area. Department of Geography, University of Papua New Guinea, Occasional Paper 12. University of Papua New Guinea.

Shackleton, N. J. and J. P. Kennett. 1975. Palaeotemperature history of the Cenozoic and the initiation of the Antarctic glaciation: oxygen isotope analyses in DSDP sites 277, 279, 281. In: Initial Report of the Deep Sea Drilling Project 29. Washington, U.S. Govt Printing Office.

Shields, O. 1979. Evidence for initial opening of the Pacific in the Jurassic. Palaeogeogr., Palaeoclimatol., Palaeoecol. 26: 181–220.

Smith, J. M. B. 1977. Origins and ecology of the tropicalpine flora of Mt Wilhelm, New Guinea. Biol. J. Linn. Soc. 9: 87–131.

Wade, L. K. and D. N. McVean. 1969. Mt Wilhelm Studies I: the alpine and subalpine vegetation. Research School of Pacific Studies, Department of Biogeography and Geomorphology BG/1. Canberra, Australian National University.

Walker, D. 1966. Vegetation of the Lake Ipea region, New Guinea highlands. I. Forest, grassland and "garden". J. Ecol. 54: 503–533.

—— 1972. Vegetation of the Lake Ipea region, New Guinea highlands. II. Kayamanda swamp. J. Ecol. 60: 297–504.

—— 1981. Speculations on the origins and evolution of Sunda-Sahul rainforests. In: G. Prance (ed.)

Proc. conf. biological model of diversity in tropical rainforests, Caracas, 1979. New York, Columbia University Press.
Walker, D. and J. R. Flenley, 1979. Late Quaternary vegetational history of the Enga Province of upland Papua New Guinea. Phil. Trans. Roy. Soc. Lond. B286: 265–344.
Walker, D. and J. C. Guppy. 1976. Generic plant assemblages in the highland forests of Papua New Guinea. Aust. J. Ecol. 1: 202–212.
Webster, P. J. and N. A. Streten. 1978. Late Quaternary ice age climates of tropical Australasia: interpretations and reconstructions. Quaternary Research 10: 279–309.
White, K. J. 1975. The effect of natural phenomena on the forest environment. Port Moresby, Department of Forests (Papua New Guinea).
Williams, P. W., I. McDougall and J. M. Powell. 1972. Aspects of the Quaternary geology of the Tari-Koroba area, Papua. J. Geol. Soc. Aust. 18: 333–347.

2. Origins of the tropicalpine flora

J. M. B. Smith

1. Introduction

An important focus of interest in New Guinea biogeography continues to be the tropicalpine flora. The closest relatives of members of this flora usually are geographically remote, on other high mountains in the Malesian region or in regions outside the tropics. The visiting biologist from either temperate zone finds plant taxa familiar to him from very distant and different places, inevitably provoking questions concerning the origins of the disjunctions.

2. The tropicalpine environment

In spite of floristic links, there are fundamental differences between mountain environments in New Guinea or elsewhere in the humid tropics, and in the temperate zones. Climatically both are cool, but the diurnal and seasonal marches of temperature are different in the two environments. New Guinea's mountains have a tropical thermal regime, lacking seasonal temperature changes, while experiencing low temperatures because of their altitude. Maximum and minimum temperatures change very little from month to month, except when both are made more equable by cloud cover during a wetter season. On the other hand changes during a single 24 hour period may be substantial, especially in clear weather and at ground level.

Tropicalpine plants are not, therefore, exposed to alternating winter and summer seasons, involving lengthy periods buried beneath snow or exposed to ice-laden winds, alternating with shorter warm periods during which almost all photosynthesis, growth and reproduction take place. Instead plants must cope with a climatic environment in which air temperatures near the ground may both exceed 20°C and fall below 0°C on any day of the year. Their aerial parts may alternately be subjected to intense solar radiation and frozen, while their roots are perpetually at temperatures only a few degrees above freezing.

These and other contrasts have led various authors (e.g. Hedberg 1964; Smith 1975a; Troll 1959) to describe vegetation patterns of tropical mountains in terms unique to these environments, instead of borrowing terms used for mountain vegetation elsewhere. Tropicalpine is used to refer to flora, vegetation

Monographiae Biologicae, Vol. 42, ed. by J. L. Gressitt.

© *1982, Dr W. Junk Publishers, The Hague.*

and environments lying above the forest limit on tropical mountains, or outside the forests at somewhat lower altitudes where these sites are contiguous with or similar to sites above the forest limit. Tropicalpine vegetation can be subdivided into smaller units in any area, but it has been suggested that it is premature to propose such units for more than local use (Smith 1975a).

Many mountains in New Guinea have never been visited by biologists, and very few have received more than cursory ecological study. An exception is Mt Wilhelm, where a research station (Fig. 8) was maintained by the Australian National University from 1965 until its donation to the Papua New Guinea National Parks Board in 1976. A map showing the extent and distribution of high mountains in New Guinea is provided in Fig. 1.

A description of the climate of Mt Wilhelm has been compiled from the discontinuous records of various workers over seven years (Hnatiuk, Smith and McVean 1976). From December to March there is a wetter and mistier season during which light rain falls almost daily. The drier season from May to September still includes rain, but up to 22 consecutive dry days have been recorded. Cloudless days are unusual even at this season. A typical day at any time, but especially during periods transitional between the wetter and drier seasons, has a clear dawn but with cloud resulting from convection up the valleys by about noon; rain develops in the afternoon, the sky clearing again at about dusk.

At the Pindaunde research station (3480 m) annual precipitation, mostly in the form of rain (with a little hail), totals about 3450 mm. At lower altitudes, and perhaps also at higher, precipitation decreases. The Imbukum valley, to the north of Pindaunde, has a wetter climate. Snow falls every few weeks above about 4000 m, but mostly melts within a few hours (Figs. 2, 3, 7, 10).

Maximum and minimum air temperatures at 3480 m average 11.3°C and 3.9°C respectively, with ranges of 9.9°C and 9.5°C. Lapse rate ($\frac{1}{2}$[max and min]) is -0.53°C per 100 m. Temperatures measured at ground level away from sheltering vegetation are more extreme (-9.1 to 29.4°C), but beneath a forest canopy the diurnal temperature range is greatly reduced. There are microclimatic differences between slopes facing east and west. Strong winds are rare.

The climate of Mt Jaya (Carstensz) has been described by Allison and Bennett (1976). It is very similar to that of Mt Wilhelm, though possibly mistier and wetter and with a less pronounced dry season. Data are almost completely lacking from other New Guinea mountains, though there appears to be a general trend for their climates to be less wet and show clearer dry seasons the farther E they are located.

Soils are discussed by Wade and McVean (1969) for Mt Wilhelm, Haantjens (1970) for montane areas generally in the Mt Wilhelm area, and Hope (1976) for Mt Jaya. They are generally acidic and rich in organic matter, and do not clearly exert a controlling influence over vegetation except where they are shallow or waterlogged.

Forests grow up to about 4100 m (with local variation) on Mt Jaya (Hope 1976) and to about 3800 m in the little disturbed Imbukum valley of Mt Wilhelm (Smith 1975a). On smaller mountains, even in the absence of disturbance, forest limits are lower, exemplified by Mt Sigal Mugal in the Kubor

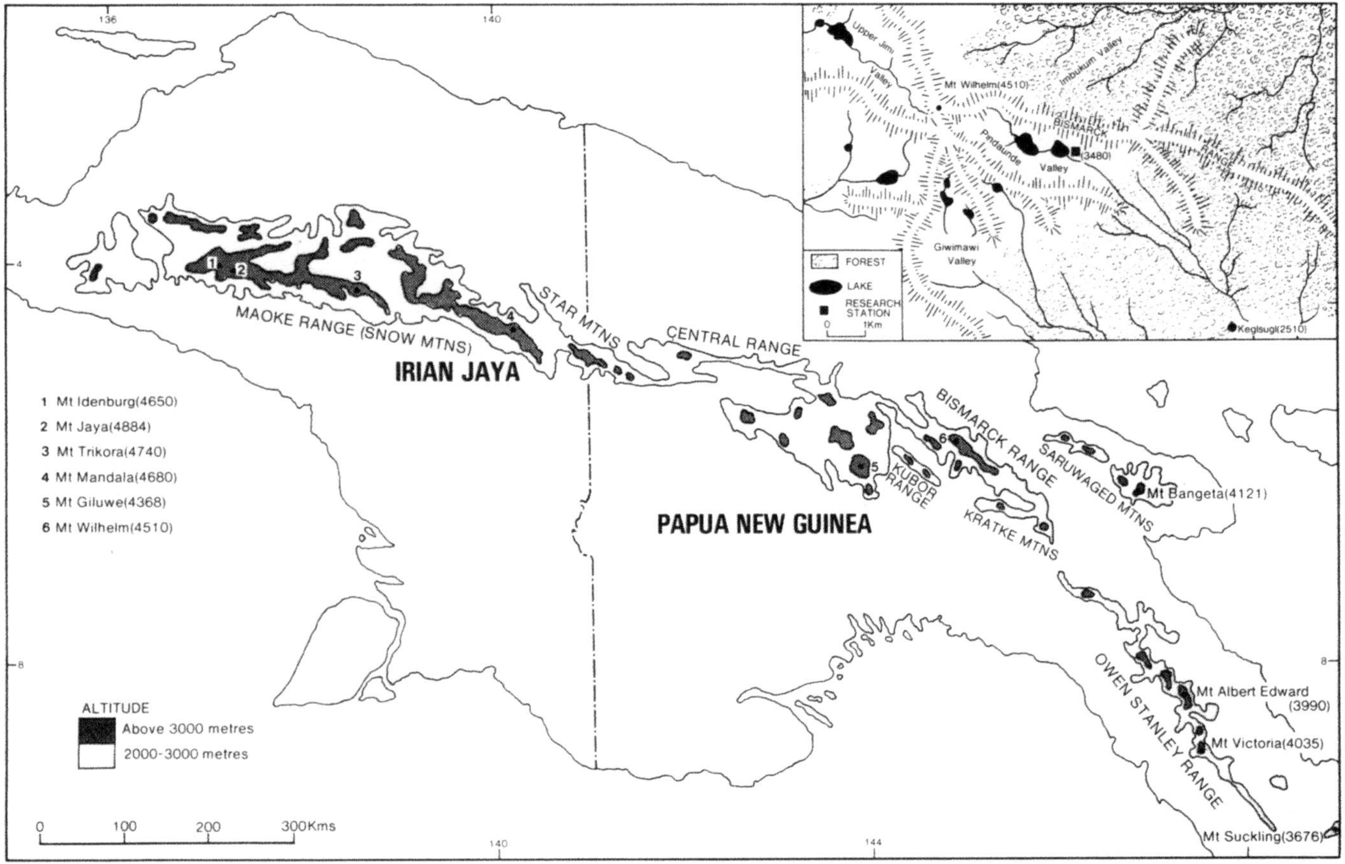

Fig. 1. The high mountains of New Guinea. Tropicalpine environments are now restricted to areas above 3000 m, but extended down to below 2400 m during the lastt glacial period. Not all areas above 2000 m are shown. Inset shows Mt Wilhelm study areas. (Redrawn from Fig. 1 in Hope and Peterson 1975.)

Fig. 2. Snow on the summit ridge of Mt Wilhelm, 4300 m.

Fig. 3. Snow on the summit ridge of Mt Wilhelm, 4400 m.

Fig. 4. Deschampsia klossii Ridl. tussock grassland, with the fern *Papuapteris linearis* C.Chr. and the shrubs *Styphelia suaveolens* (Hook.F.) Warb. and *Coprosma divergens* Oliver, at 4060 m, Pindaunde valley, Mt Wilhelm.

range (Smith 1975a). Man, through burning, has also irregularly lowered the forest limit on most mountains in New Guinea (Paijmans and Löffler 1972; Smith 1975a).

Where forests have been destroyed, tussock grasslands usually take their place, dominated by *Deschampsia klossii, Danthonia archboldii* or species of *Poa*. If burning is not repeated frequently, a variety of shrubs also invades, but especially at lower altitudes the vigor of the tussocks is often such that smaller herbaceous species are not numerous or abundant. In the valley bottoms where these are choked by glacial till and therefore ill-drained, and on very steep, rocky slopes and landslide sites, other non-forest vegetation types are found.

Above forest limit, varied grasslands with some shrubs (Figs. 4, 11, 12) extend upwards to about 4400 m on the highest mountains, with tundra communities above these up to the upper limit of vegetation at 4700 m (on Mt Jaya). There is much local altitudinal and ecological variation in these limits, however, and the distributions of vegetation types on New Guinea mountains are little studied or understood.

A detailed phytosociological study of the vegetation of Mt Wilhelm above 3110 m has been reported by Wade and McVean (1969), extended in less detail to Mt Jaya by Hope (1976). Though associations can be recognised and defined, they frequently cannot be found on other mountains, and include few species of high fidelity. Most tropicalpine plant species occur in several different associations which are often not clearly distinct from each other in the field.

Wallabies are common on a few less accessible mountains (e.g. Mt Suckling) where they exert a noticeable grazing pressure. Generally, large herbivores are not important and insect damage to plants is usually insubstantial. Grazing may have been important before the extinction of some large mammals in the Late Pleistocene, or the reduction of other animal populations through more recent hunting pressures (Hope and Hope 1976).

A fuller review and discussion of New Guinean tropicalpine ecology has recently been undertaken elsewhere (Smith 1980).

3. The history of the mountains

In plate tectonics terms New Guinea is part of Australia. The island is therefore fundamentally southern, part of one of the fragments derived from the disintegration of the early Mesozoic supercontinent Gondwanaland. The other fragments are Africa, Malagasy, India (which all drifted northwards and lost contact with remaining landmasses early), S America, Antarctica and certain W Pacific islands, notably New Caledonia and New Zealand (see Fig. 5).

However, New Guinea has not had its present form, or indeed existed continuously as a land area, since Gondwanaland's fragmentation. Details are not yet clear, and Axelrod and Raven (chapter VI:2) provide a fuller summary of present knowledge. The present land area of New Guinea only emerged from the ocean in late Eocene times, and its high mountains are much younger. Most of the rocks comprising northern New Guinea themselves postdate Gondwanaland's breakup. The continuing northward drift of Australia with

New Guinea led to folding, faulting, uplift, vulcanism and the development of geosynclines along the leading edge of the plate (Davies and Smith 1971). In the Oligocene a large block of arched ocean floor faulted and was uplifted, perhaps to provide New Guinea's first highland area (Thompson 1967). Further folding, vulcanism and uplift continue to the present (Avias 1973; Bain 1973), much having occurred only during the past two million years.

It was during this latter phase that the gradual and uneven cooling of world temperature over the previous 60 million years culminated in a series of glacial periods. Clear evidence has been found in New Guinea only for the last of these, but there is little reason to suppose that previous glacial periods known from elsewhere did not affect the region. Most of the mountains exceeding 3600 m had ice-caps, some with valley glaciers as well (Hope and Peterson 1975; Löffler 1972), and depositional as well as erosional features are easily recognisable (Figs. 6–10) (see Löffler, chapter I:3).

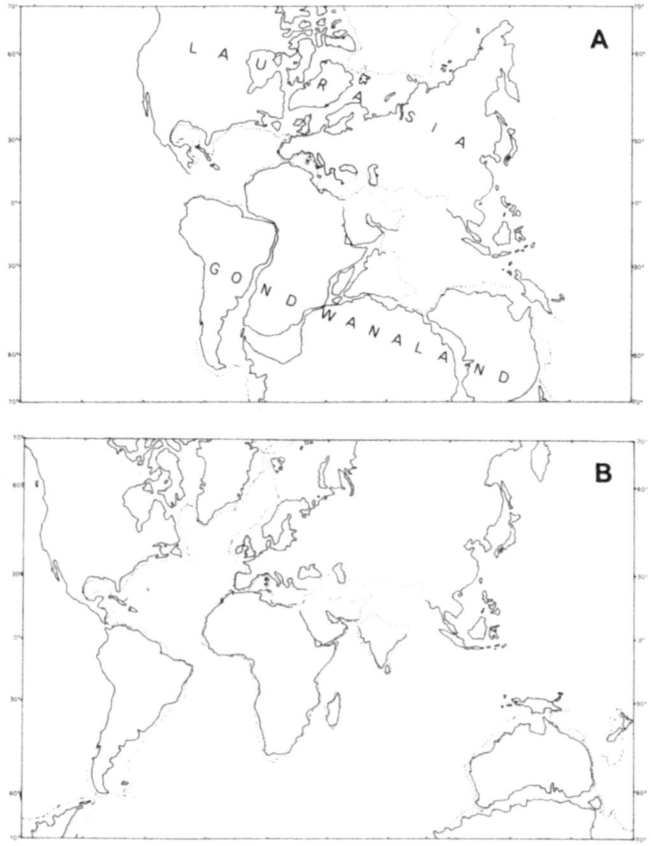

Fig. 5. World paleogeography about 100 million years BP (A) and 50 million years BP (B). Continental margins are shown, but some of the area enclosed by these was probably (as today) shallow sea at any given date. (From Smith 1975, Australian Geographical Studies 13: 7.)

Fig. 6. Terminal moraine in the Imbukum Valley, Mt Wilhelm, 3300 m, incised by stream action.

Fig. 7. Head of the Guraguragugl Valley, Mt Wilhelm, seen from a ridge at 4030 m, showing glacial landforms and a light fall of snow at the higher altitudes.

Fig. 8. The Pindaunde Valley, Mt Wilhelm, seen from 4050 m. Slopes in the foreground are naturally unforested, but near the lakes the grasslands are a fire-disclimax, and the forest shows a relict distribution.

Fig. 9. The Pindaunde Valley, Mt Wilhelm. The lake lies at 3540 m, and the ridgecrest at about 4100 m. The valley is glacially sculpted, and its vegetation modified by man-set fires which have resulted in the replacement of much forest by grassland. The climatic forest limit lies at about 3800 m.

Fig. 10. The Bogunolto ridge and Pindaunde valley of Mt Wilhelm, seen from the research station at 3480 m. Overnight precipitation has settled as wet snow above 4000 m. Grasslands around the lake are a fire-disclimax, occupying formerly forested areas.

4. Origins of the floras of lower altitudes

Taken as a whole, the lowland flora of New Guinea has more in common with areas to the west than to the south, and this led to speculation that the island has only recently come into juxtaposition with N Australia (Good 1957; this and some other early ideas are discussed by Robbins 1971). However it now seems clear that the disruption of plant distributions at or near Torres Strait is restricted largely to taxa whose environmental requirements are only fulfilled on one side of it. The floristic contrast between New Guinea and Australia is therefore an ecological rather than a historical phenomenon (Walker 1972).

The small savanna flora, in areas of monsoonal climate in the Fly delta and Port Moresby areas, has close affinities with northern Australia. These links are explained by the lowering of sealevel by over 100 m during part of the last glacial period, and probably in earlier glacial periods also. The Arafura Sea and Torres Strait were therefore dry land, with similar vegetation extending across the area, until as recently as about 7000 BP (Nix and Kalma 1972).

The lowland rainforest flora shows close affinities, often at species level, with the floras of islands to the west. There has never been a direct land link, but these areas were separated from New Guinea by only relatively narrow straits during glacial periods, and have a similar climate. Before late Cenozoic times, however, New Guinea was far to the south until carried towards its present position by continental drift. These affinities are therefore most easily explained in terms of floristic exchange with the west over moderate sea gaps in late Cenozoic times.

In the higher rainforests, above about 1400 m, a further element becomes important whose affinities lie far to the south, with New Caledonia, New Zealand, SE Australia and temperate S America. These links are not close, being at the generic and family level. Many of the taxa involved (e.g. Cunoniaceae, Monimiaceae, Podocarpaceae, *Drimys*, *Nothofagus*) are known as fossil pollen from late Mesozoic or early Cenozoic sediments in Antarctica as well as other Gondwanaland regions. They are clearly derived from survivors of a late Cretaceous Gondwanaland flora, whose patchy distribution around the S Pacific partly reflects the present distribution of climatically suitable fragments of the supercontinent (Smith 1974).

5. Affinities, characteristics and age of the tropicalpine flora

The tropicalpine flora occupies the insular environments of mountain-tops extending above forest vegetation. Its affinities are various, and its origins cannot be explained in simple paleogeographic terms.

Geographical affinities will be examined by reference to a typical tropicalpine flora which has received detailed study, that of Mt Wilhelm (Smith 1977b). All seedplant species are included which occur commonly or solely in the grasslands of, or contiguous with, the summit area (above 3215 m), or commonly in other local non-forest habitats above 2500 m. One hundred and sixty three native species in 100 genera are involved. Most are shade-intolerant and absent from forests (Figs. 4, 11–15).

Fig. 11. Deschampsia klossi Ridl., the most widespread and important tussock grass species in New Guinea mountains, growing with the low shrub *Gaultheria mundula* J. J. Sm. at 3450 m, Pindaunde valley, Mt Wilhelm.

Fig. 12. Detzneria tubata Diels (Scrophulariaceae), in a monotypic genus close to *Veronica*, *Hebe* and *Parahebe* and endemic to a few high mountains in New Guinea; a shrub about 1.5 m tall, with deep blue cleistogamic flowers; Imbuka ridge, Mt Wilhelm, 4100 m.

Fig. 13. Harmsiopanax ingens Philipson (Araliaceae), a monocarpic soft-wooded species of palm-like habit, growing at 3500 m in the Pindaunde valley, Mt Wilhelm.
a. A group of vegetative specimens; each is about 4 m tall.
b. A flowering specimen; flowers are produced in abundance, and most of the leaves have fallen.
c. Dead individual after monocarpic reproduction.

The global distribution of 42 species extends beyond the Malesian region; 8 to Eurasia, 23 to Australasia, and 11 to both Eurasia and Australasia. 30 of these species are in genera which have bihemispheric or cosmopolitan distributions. Of 121 species endemic to the Malesian region, only 11 are in endemic or nearly endemic genera. 22 are in genera extending to Eurasia; 30 are in genera extending to Australasia, of which 19 also occur in temperate S America; 58 are in bihemispheric or cosmopolitan genera.

Altogether 30% of Mt Wilhelm's tropicalpine seedplant species grow also outside New Guinea, with only 2% endemic to Mt Wilhelm alone. By comparison only 3% of the tropicalpine species of E Africa grow outside that region, and 64% are endemic to single mountains (Hedberg 1969).

The large proportion of widespread species confirms the impression, derived from lack of fidelity of species to vegetation associations, of a low level of specialization and incomplete adaptation to the new tropicalpine environment. Only a few species have narrow ecological requirements as indicated by small altitudinal range (Smith 1975b) or restricted habitats; most are found widely. The total number of species in New Guinea tropicalpine floras is small by comparison with floras of similar environments elsewhere (Wade and McVean 1969), in contrast to the forest flora.

The dearth of local endemism suggests recent attainment of present distribution: had more time elapsed since plant populations were established in their present isolation on New Guinea mountains, more evolution of local taxa would have occurred. This is likely to be especially true in the tropicalpine context of small, fluctuating populations of early-maturing and probably short-lived plants, in which evolution may be unusually rapid as argued for the West African tropicalpine flora by Morton (1972).

A high level of polyploidy in the flora may further suggest recent colonisation. Polyploidy can confer wider ecological tolerance, vegetative reproductive ability and self-compatibility, all characteristics favoring a colonist species in a distant and physically harsh environment. Seventy-five tropicalpine species on Mt Wilhelm were included in a study of ploidy by Borgmann (1962), of which 67% are polyploids. This compares with levels of 49% for the Cameroons Mts and 45% for E Africa Mts (Morton 1966).

However, if immigration of New Guinea's tropicalpine flora is relatively recent, then it is hard to escape the inference that it was from more than one direction and across large tracts of country uninhabitable by the cool-adapted and light-demanding plants concerned. It is not easy to assert positively what dispersal agent(s) may be responsible. Van Steenis (e.g. 1964, 1967) considered plant populations of Malesian tropicalpine environments to be the remnants of a richer flora which occupied a high mountain range extending from Asia to Australia in late Cretaceous times. While removing the necessity of explaining dispersal over long distances, such a hypothesis nevertheless becomes untenable when considered in terms of palaeogeographical reconstructions on the basis of plate tectonics theory. More recently van Steenis (1972) has reiterated his opposition to dispersal of tropicalpine plants across gaps exceeding 500 km. However he also admits that a "peregrine" element may have migrated to Java, from both the NW and SE, in Miocene or Pliocene times.

Holloway (1970) considered the high altitude plants on Mt Kinabalu (Borneo) of Australasian affinity to have an even more recent origin. He pointed out that during Quaternary glacial periods when vegetation belts were lowered, the areas in Malesia occupied by cool-adapted species probably enlarged and became more numerous. Therefore gaps between them become smaller and migration was facilitated. Other recent authors have supported the idea of Quaternary migration of Malesian tropicalpine floras. Raven (1973) stresses the importance of late Cenozoic migrations of plants through Malesian mountain areas, and Gupta (1972) recognises migration between Himalayan and Malesian mountains in the Pleistocene. Schodde (1973) stated that the diverse geographical affinities of New Guinea tropicalpine biotas suggested "itinerant colonization".

Certainly such recent migration agrees better with geological evidence. Australia and New Guinea were separated from SE Asia by a vast marine strait in late Mesozoic and early Cenozoic times, so that dispersal of mountain plants between the two regions must have been very rare until a few million years ago (Smith 1974). Furthermore most Malesian high mountains are of late Cenozoic if not Pleistocene uplift, and it is not clear whether similarly high peaks existed earlier though eroded down today. Even were there earlier high mountains in the region, they are unlikely to have been cold enough to support tropicalpine vegetation until world climate cooled in late Cenozoic times.

Quaternary glacial periods, distributed during the last three million years, provide on the other hand apparently more favorable conditions for migration of tropicalpine floras than those which exist today. Mountain peaks lying above forest limit were more numerous, and could have acted as "stepping-stones" for cool-demanding plant species. In New Guinea especially, geomorphic instability led to the repeated creation of areas of bare ground in the form of moraines, fluvioglacial deposits, volcanic ash and lahars. These could have acted as seedbeds free of plant competition for the establishment of plants from distantly dispersed seeds (Fig. 16). The probable importance of such open sites in the colonisation by plants of the high mountains of East Africa has been stressed by Hedberg (1971) and also by Raven and Raven (1976).

Even if larger areas lying above the forest limit of glacial times are considered, the problem of dispersal is only slightly alleviated. Although many of the Malesian peaks are mutually less isolated, over 1400 km separates the nearest Malesian and mainland Asian sites, and 2250 km lies between the closest Papuan and Australian sites (see Fig. 17). It is important to remember, however, that long time periods are involved. For example, an average of only one successful dispersal and establishment about every 8000 years is quite adequate to account for the immigration of Mt Wilhelm's tropicalpine flora during Pleistocene glacial periods. The dispersal agency need therefore act only very occasionally to be effective. Various agencies can be proposed, among which birds seem the most plausible (Smith 1978).

Fig. 14. Tree-ferns *Cyathea atrox* C. Chr. growing in fire-disclimax grassland on the terminal moraine in the Gwaki valley, Mt Wilhelm, 3200 m.

Fig. 15. A tree-fern *Cyathea gleichenioides* C. Chr. with young fronds, and the tussock grass *Deschampsia klossi* Ridl. at the upper forest limit, 3800 m, Imbukum valley, Mt Wilhelm.

Fig. 16. Pengagl creek, 2950 m, Mt Wilhelm. Streambanks such as this, landslips, and other sites of disturbance are colonized by adventive herbaceous plants. Similar sites may have been the points of initial establishment by plants dispersed over long distances to New Guinea mountains.

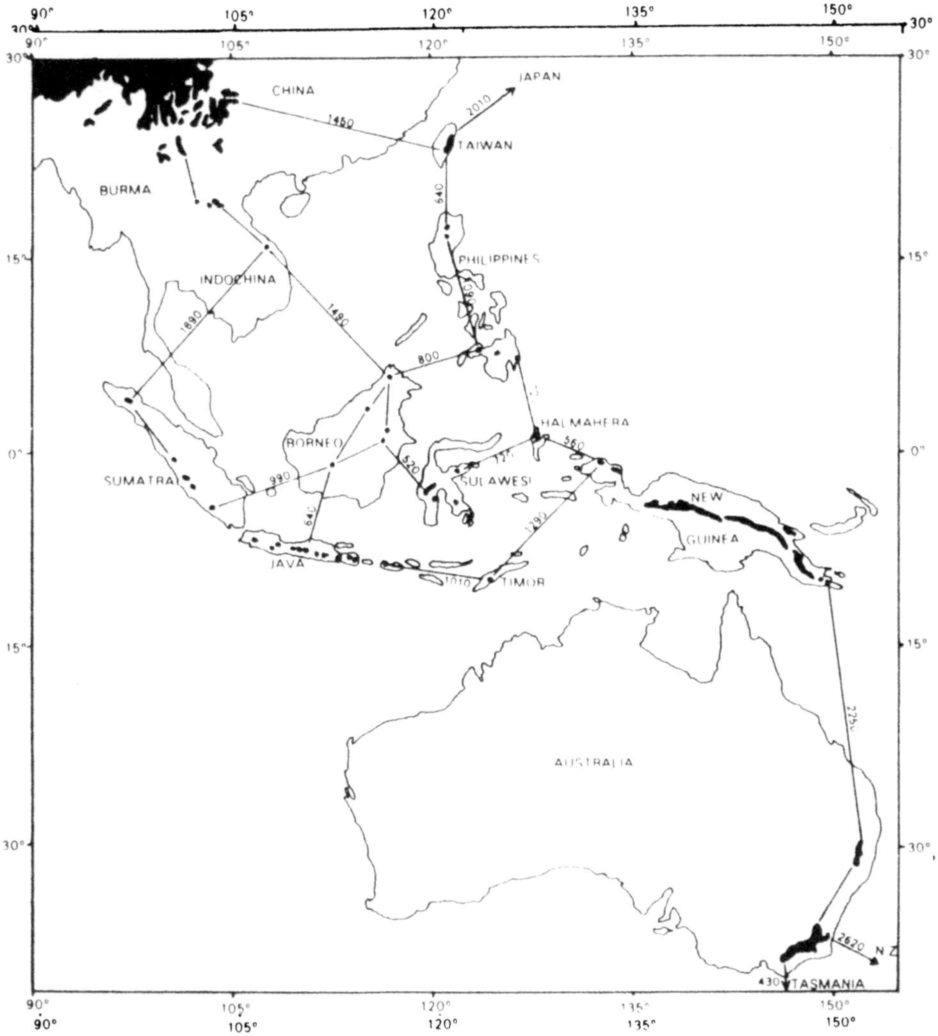

Fig. 17. Land in the Malesian region probably lying above the forest limit at the time of glacial maximum. (From Smith 1977, Biol. J. Linnean Soc. 9: 90.)

6. Ecological indications of migration history

An investigation has been made into some aspects of the ecology of the tropicalpine flora of Mt Wilhelm, revealing differences between floristic elements of apparently differing history (Smith 1977b). The tropicalpine flora was initially divided on objectively applied criteria into 3 main floristic elements. The putatively oldest was named the ancient element, having 35 species. These were species in endemic genera, and also endemic species in genera extending southward to include S America, so that they could arguably be of Gondwanaland origin. The "peregrine" element included all other native species (147), presumed to be of more recent (probably Quaternary) arrival in New Guinea. The third element consisted of 30 species which had been introduced by man to the Mt Wilhelm area, and were therefore of most recent immigration.

The following ecological attributes were assessed for as many species in the flora as was practicable.

1. *Dispersal ability*. This was assessed both by morphological study and weighing of collected disseminules, and by occurrence in discrete, historically created non-forest sites in the Mt Wilhelm area.
2. *Colonist ability* (*adventiveness*). This was assessed by occurrence of species in open, disturbed sites such as landslips, streambanks and eroded pathsides.
3. *Growth rate*. This was measured successfully only for rosette herbs, by periodic description, recording and later weighing of marked individuals.
4. *Phenology*. Species were recorded for flowering behavior in the field over half a year; some were similarly observed over a year in cultivation in Canberra; stems of woody species were examined in section to discover evidence of discontinuous growth revealed by growth rings; species in a small area were examined for damage after heavy frost (Fig. 18).
5. *Response to slope aspect*. East-facing slopes experience higher maximum temperatures than opposite slopes (Smith 1977a), some species growing to higher altitudes on these slopes.
6. *Altitudinal range*.

With regard to dispersal ability, no significant differences could be found between floristic elements. In all other ecological parameters studied, differences were found either between the ancient and peregrine elements, or between native and alien species, or both. In the cases of growth rate, secondary growth of woody stems, and frost vulnerability, the level of statistical significance was only 10%, but this was explicable in each case by rather small numbers of species assessed.

P. F. Stevens (pers. comm. 1978) has pointed out that the proportion of woody species is greatest in the ancient element and least in the alien, so that the observed ecological differences could possibly be explained solely by life-form. However if data for woody and herbaceous species are analysed separately, in spite of smaller numbers significant differences remain in all cases but one, though levels of significance are reduced in some. Significance levels are summarised in Table 1.

The species best able to colonise distant sites are those which are characteristic of transient open or disturbed sites, as produced by rapid geomorphic change or human activity. After migration, evolutionary adaptation to the new environment may entail the transformation of these ruderal species into more competitive but less opportunistic inhabitants of closed vegetation (Baker 1962; Grime 1977; MacArthur and Wilson 1967; Margalef 1968). This may be especially true in the New Guinea tropicalpine situation, where there are no obligately ruderal native species, present natural disturbances probably being too limited or uncommon to permit their long term survival.

Amongst the characteristics typical of ruderals are good dispersal ability, good colonist ability, rapid growth rate, and ability to survive under a wide range of conditions. All these were assessed for elements in the Mt Wilhelm flora, and except for dispersal ability showed differences. The ability to flower in cultivation, and broad and low altitudinal range, indicate ability to survive under a wide range of conditions. Each difference was in the expected direction, indications of migration ability being more apparent in the younger elements.

The other differences revealed also show differences in degree of adaptation to the tropicalpine environment between elements. A lack of synchronous flowering, and continuous growth, may show a failure to have evolved phenological responses to cues of a rather subtle nature in the largely non-seasonal environment. Vulnerability to frost, and failure to grow higher on east-facing slopes, likewise indicate imperfect adaptation to tropicalpine conditions. In these ways too, it is the putatively younger elements which are less well adapted.

In the case of dispersal ability, it is suggested that adaptations favoring effective dispersal over short distances, up to a few km, are irrelevant to occasional dispersal over much longer distances, of several hundred km. In all other cases the bulk of the native flora, and even more the aliens, lend support through their ecological characteristics to the belief that they are of relatively recent immigration.

Combining all the data further confirmed the differences between the floristic elements. For each ecological parameter, every species assessed was given a score of 0, 1 or 2, the higher scores in every case being for characteristics expected in a recent immigrant. When compared, the numbers of each score accumulated by each element are very significantly different (0.1% level). These data expressed as percentages are shown graphically in Fig. 19.

Fig. 18. Needle ice (pipkrake) on damp ground at 3480 m, Pindaunde valley, Mt Wilhelm, after a clear, frosty night.

Table 1.

	All species				Herbaceous species				Woody species		
	No. native spp. assessed	No. alien spp. assessed	Native/alien difference*	Ancient/peregrine difference*	No. native spp. assessed	No. alien spp. assessed	Native/alien difference*	Ancient/peregrine difference*	No. native spp. assessed	No. alien spp. assessed	Ancient/peregrine difference*
Disseminule adaptations	100	13	+	+	79	13	+	+	21	0	+
Local dispersal ability	67	0	—	+	55	0	—	+	12	0	—
Colonist ability	163	30	+	5	131	28	+	5	32	2	+
Growth rate	20	2	—	10	20	2	—	10	0	0	—
Field Phenology	105	7	1	+	77	7	5	+	28	0	+
Flowering in cultivation	42	5	10	2	39	5	+	5	3	0	—
Secondary growth	20	1	—	10	0	0	—	—	20	1	10
Frost vulnerability	35	14	10	+	27	12	+	+	8	2	—
Aspect preference	63	0	—	1	54	0	—	1	9	0	—
Altitudinal range	125	18	5	2	121	18	5	+	4	0	—
Mean altitude	125	18	0.1	+	121	18	1	+	4	0	—

*Numbers refer to per cent significance levels (assessed by χ^2 testing of tabulated data), + indicates no significance (>10%), — indicates insufficient data for testing.

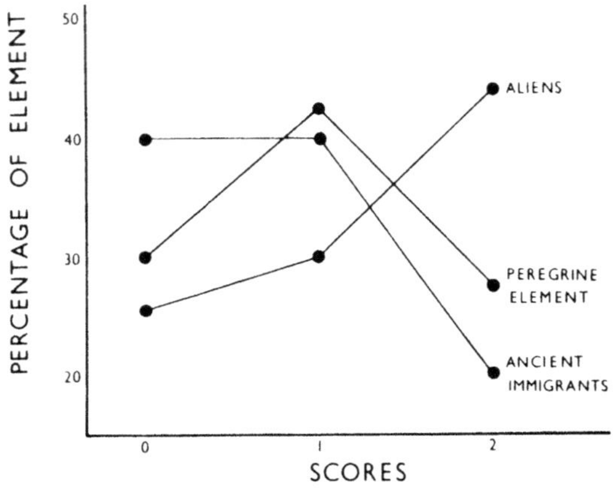

Fig. 19. Numbers of scores 0, 1 and 2 recorded for each floristic element, expressed as percentages. (From Smith 1977, Biol. J. Linnean Soc. 9: 117.)

7. Conclusions: origins of the tropicalpine flora

Geological and botanical evidence both point to the conclusion that the high mountains of New Guinea with their tropicalpine flora are of late Cenozoic age.

A small proportion (less than $\frac{1}{5}$) of the flora may have a longer history in highland New Guinea (or nearby N Australia) extending back to the early Cenozoic when Australia was still attached to Antarctica. Some of these (e.g. *Drimys*, *Pittosporum*, Podocarpaceae) while contributing to the tropicalpine vegetation have their greatest diversity in the forests, and the evolution of species with tropicalpine tolerance is probably fairly recent. Others may have achieved a suggestively gondwanic distribution by late Cenozoic dispersal over long distances. Yet others are in probably young and rapidly evolving families (e.g. *Ischnea* and *Keysseria* in the Compositae) and so are not likely to be as ancient as their generic endemism may otherwise suggest.

The bulk of the native tropicalpine flora probably immigrated by dispersal over long distances in late Cenozoic times, most plausibly during Pleistocene glacial periods. The immigration was from both the S (Australia, New Zealand) and from the N and NW (mainland Asia and Malesian mts). Many of the immigrants have evolved endemic species in New Guinea, and a few may have evolved so far as to be classified as endemic genera (and so categorised as ancient in the study quoted). There has probably been some emigration from New Guinea's tropicalpine regions during the same period; *Parahebe* to Australia and New Zealand and *Tetramolopium* to Hawaii are possible examples. The dispersal agencies involved, which need only have acted very occasionally, remain obscure, though birds appear likely.

A recent, continuing and relatively abrupt wave of immigrant plants is provided by the ever-increasing list of alien species. Nearly all of these are still restricted to disturbed sites, but can be expected to invade closed vegetation as a result of the evolution of local adaptations in the more or less distant future.

Acknowledgements

I am grateful to the Australian National University and the University of New England for financing fieldwork on Mt Wilhelm in 1971–72 and 1975 respectively, and to Prof. R. G. West for providing space in the Cambridge Botany School where this chapter was written during study leave from the University of New England. Many colleagues have influenced my ideas on tropicalpine ecology and phytogeography, amongst whom Prof. D. Walker, Dr N. M. Wace, Dr G. S. Hope, Dr R. J. Hnatiuk and Dr P. F. Stevens deserve special mention. Dr T. P. Bayliss-Smith and Dr E. V. J. Tanner offered helpful criticisms of the manuscript.

References

Allison, I. and J. Bennett. 1976. General Climate. Pp. 61–80 in Hope, G. S., J. A. Peterson, U. Radok and I. Allison (Eds) "The Equatorial Glaciers of New Guinea." A. A. Balkema, Rotterdam. 244 pp.

Avias, J. 1973. Major features of the New Guinea-Louisiade-New Caledonia-Norfolk Arc System. Pp. 113–126 in P. J. Coleman (Ed.) "The Western Pacific. Island Arcs, Marginal Seas, Geochemistry." Univ. Western Australia Press, Nedlands. 675 pp.

Bain, J. H. C. 1973. A Summary of the Main Structural Elements of Papua New Guinea. Pp. 147–161 in P. J. Coleman (Ed.) "The Western Pacific. Island Arcs, Marginal Seas, Geochemistry." Univ. Western Australia Press, Nedlands. 675 pp.

Baker, H. G. 1972. Migration of Weeds. Pp. 327–347 in D. H. Valentine (Ed.) "Taxonomy, Phytogeography and Evolution." Academic Press, London. 431 pp.

Borgmann, E. 1962. Anteil der Polyploiden in der Flora des Bismarckgebirges von Ostneuguinea. Zeitschrift für Botanik 52: 118–172.

Davies, H. L. and I. E. Smith. 1971. Geology of Eastern Papua. Bull. Geol. Soc. Amer. 82: 3299–3312.

Good, R. 1957. Australasian floras. Nature, London 179: 926.

Grime, J. P. 1977. Evidence for the existence of three primary strategies in plants and its relevance to ecological and evolutionary theory. American Naturalist 111: 1169–1194.

Gupta, R. K. 1972. Boreal and arcto-alpine elements in the flora of Western Himalayas. Vegetatio 24: 159–175.

Haantjens, H. A. 1970. Soils of the Goroka-Mt Hagen Area. C.S.I.R.O. (Australia) Land Research Ser. 27: 80–103.

Hedberg, O. 1964. Features of Afroalpine plant ecology. Acta Phytogeographica Suecica 49: 1–139.

—— 1969. Evolution and speciation in a tropical high mountain flora. Biol. J. Linnean Soc. 1: 135–148.

—— 1971. Evolution of the Afroalpine Flora. Pp. 16–23 in W. L. Stern (Ed.) "Adaptive Aspects of Insular Evolution", State Univ. Press, Washington.

Hnatiuk, R. J., J. M. B. Smith and D. N. McVean. 1976. Mt Wilhelm studies. 2. The climate of Mt Wilhelm. Australian National Univ. Press, Canberra. 76 pp.

Holloway, J. D. 1970. The biogeographical analysis of a transect sample of the moth fauna of Mt Kinabalu, using numerical methods. Biol. J. Linnean Soc. 2: 259–286.

Hope, G. S. 1976. Vegetation. Pp. 113–172 in Hope, G. S., J. A. Peterson, U. Radok and I. Allison (Eds) "The equatorial glaciers of New Guinea." A. A. Balkema, Rotterdam. 244 pp.

Hope, G. S. and J. A. Peterson. 1975. Glaciation and vegetation in the high New Guinea mountains. Pp. 155–162 in Suggate, R. P. and M. M. Cresswell (Eds) "Quaternary Studies." Royal Soc. New Zealand, Wellington, 321 pp.

Hope, J. H. and G. S. Hope. 1976. Palaeoenvironments for man in New Guinea. Pp. 29–54, in Kirk, R. L. and A. G. Thorne (Eds) "The origin of the Australians." Australian Institute of Aboriginal Studies, Canberra. 449 pp.

Löffler, E. 1972. Pleistocene glaciation in Papua and New Guinea. Zeitschrift für Geomorphologie, Supplementband 13: 32–58.

MacArthur, R. H. and E. O. Wilson. 1967. "The theory of island biogeography." University Press, Princeton. 203 pp.

Margalef, R. 1968. "Perspectives in ecological theory." University Press, Chicago. 111 pp.
Morton, J. K. 1966. The role of polyploidy in the evolution of a tropical flora. Pp. 73–76 in Darlington, C. D. and K. R. Lewis (Eds) "Chromosomes Today. 1." Oliver and Boyd, Edinburgh, 274 pp.
—— 1972. Phytogeography of the west African mountains. Pp. 221–236 in Valentine, D. H. (Ed.) "Taxonomy, Phytogeography and Evolution." Academic Press, London, 431 pp.
Nix, H. A. and J. D. Kalma. 1972. Climate as a dominant control in the biogeography of northern Australia and New Guinea. Pp. 61–92 in Walker, D. (Ed.), "Bridge and Barrier: the natural and cultural history of Torres Strait." Australian National Univ. Press, Canberra, 437 pp.
Paijmans, K. and E. Löffler. 1972. High altitude forests and grasslands of Mt Albert Edward, New Guinea. J. Trop. Geography 34: 58–64.
Raven, P. H. 1973. Evolution of subalpine and alpine plant groups in New Zealand. New Zealand J. Botany 11: 177–200.
Raven, P. H. and T. E. Raven. 1976. The genus Epilobium (Onagraceae) in Australasia, a systematic and evolutionary study. New Zealand Dept. Sci. Ind. Res. Bull. 216: 1–321.
Robbins, R. 1971. On the biogeography of New Guinea. Australian External Territories 2 (3): 31–37.
Schodde, R. 1973. General problems of fauna conservation in relation to the conservation of vegetation in New Guinea. Pp. 123–144 in Costin, A. B. and R. H. Groves (Eds) "Nature conservation in the Pacific." Australian National Univ. Press, Canberra, 337 pp.
Smith, J. M. B. 1974. Southern biogeography on the basis of continental drift. A review. Australian Mammalogy 1: 213–230.
—— 1975a. Mountain grasslands of New Guinea. J. Biogeography 2: 27–44.
—— 1975b. Notes on the distributions of herbaceous angiosperm species in the mountains of New Guinea. J. Biogeography 2: 87–101.
—— 1977a. Vegetation and microclimate of east- and west-facing slopes in the grasslands of Mt Wilhelm, Papua New Guinea. J. Ecology 65: 39–53.
—— 1977b. Origins and ecology of the tropicalpine flora of Mt Wilhelm, New Guinea. Biol. J. Linnean Soc. 9: 87–131.
—— 1978. Dispersal and establishment of plants. Pp. 207–23 in Walker, D. and J. C. Guppy (Eds) "Biology and Quaternary Environments." Australian Academy of Science, Canberra. 264 pp.
—— 1980. Ecology. In van Royen, P. (Ed.) "The alpine flora of New Guinea, Vol. 1."
Van Steenis, C. G. G. J. 1964. Plant geography of the mountain flora of Mt Kinabalu. Proc. Royal Soc. Lond., B 161: 7–38.
—— 1967. The age of the Kinabalu flora. Malayan Nature Journal 20: 39–43.
—— 1972. "The mountain flora of Java." E. J. Brill, Leiden, 85 pp.
Thompson, J. E. 1967. A geological history of eastern New Guinea. J. Austral. Petrol. Explor. Assoc. 7: 83–93.
Troll, C. 1959. Die tropischen Gebirge. Ihre dreidimensionale klimatische und pflanzengeographische Zonierung. Bonner Geograph. Abhandlungen 25.
Wade, L. K. and D. N. McVean. 1969. "Mt Wilhelm studies. 1. The Alpine and Subalpine Vegetation." Australian National Univ. Press, Canberra. 225 pp.
Walker, D. (Ed.). 1972. "Bridge and Barrier: the Natural and Cultural History of Torres Strait." Australian National Univ. Press, Canberra. 437 pp.

3. Plant zonation

R. J. Johns

1. Introduction

Increased research over the last 20 years in New Guinea has enabled us to gain a much better understanding of the altitudinal zonation of the vegetation. We must accept that a simple classification of vegetation into a series of well defined altitudinal zones is no longer acceptable, although an alternative classification has yet to be proposed. No altitudinal zones can be considered as sharply contrasting with zones either above or below, the transitional forest often occurring over several hundred meters in altitude. The Moss forest considered by Walker (1972) as having the only sharply defined boundary amongst the mountain zones is considered in many cases by Grubb and Stevens (in press) to be a transition forest between montane and subalpine forest.

Because of the impossibility of defining altitudinal zones when one describes a series of "transition" types the approach adopted has been to define a series of nodal forest types, often characterised by physiognomic dominants. The terminology used to define these zones is somewhat modified from that of earlier workers; particularly in reference to the montane zone, where inclusion of both the *Nothofagus* forests and the lower altitude *Castanopsis* and *Lithocarpus* forests in the same zone (Lower Montane sensu Grubb) contributes little of value to our understanding of the vegetation. In consequence the "*Nothofagus*" zone has been defined as Mid Montane and the "*Castanopsis*" zone as Lower Montane. This has additional value because it fits more closely with altitudinal zones shown by the birds of New Guinea as well as enabling the major zone of population pressure to be more easily defined for planning of conservation measures (Fig. 1; Table 1).

2. The lowland zone

The lowlands rainforest of New Guinea remains poorly known both from an ecological and taxonomic standpoint. However considerable advances have been made over the past decade, by repudiation of the accepted concepts of the lowland rainforest as a vast complex and mature (climax) vegetation type. Ecological research has shown that few of the rainforest areas in New Guinea

Monographiae Biologicae, Vol. 42, ed. by J. L. Gressitt.

© *1982, Dr W. Junk Publishers, The Hague.*

have escaped some form of cataclysmic destruction over the past 200 to 300 years (Johns ms.).

The lowland rainforest can be divided into a series of distinctive forest types, based on dominant species (*Pometia pinnata*, *Intsia bijuga*, *Anisoptera* and *Hopea* forest to name but a few. These occur from sea-level to a maximum of 700 m although in many places, especially along the ridges of coastal ranges, the aspect of the lower montane forest, which is dominated by *Castanopsis acuminatissima*, can be found at altitudes below 700 m. Also *Eucalyptopsis papuana* occurs at 200 m as pure ridge top forest on laterite soils to the south of the Huon Gulf. The lowland zone can be divided into a series of distinctive communities. On the coast strand forest occurs. Mangrove swamps occur particularly in the estuaries of the main rivers and extensive swamp forests are found throughout the poorly drained lowlands. Where drainage is adequate and rainfall exceeds 3000 mm per annum, without a marked dry season, the vegetation is mixed lowland tropical rain forest. Under seasonally dry conditions savanna woodlands and monsoon forests dominate. Because of complex historical factors there are great variations in the composition of all these communities within areas with apparently uniform conditions. The lowland zone has been widely influenced by its use for subsistence agriculture. Many forests are seral, or not yet stabilized, and grasslands are of anthropogenic origin. Droughts, fire, volcanic activity, lightning strikes and landslides (particularly during earthquakes) have resulted in extensive destruction of forested areas.

2.1 Coastal vegetation

Below high-water mark, communities of seaweeds and sea grasses occur, Johnstone (1975). In general the tropics do not favor the extensive development of marine plants. Slimes formed by various blue-green algae (Cyanophyceae) occur on accreting mud banks. Coralline algae occur on the coral reefs surrounding the islands. Seagrasses — *Thalassia*, *Halophila*, *Cymodocea* and *Enhalus* occur on mudflats in shallow water. The herbaceous beach vegetation (Fig. 2) growing on dry coral sand and other debris consists of creeping plants such as *Ipomoea pes-caprae* and *Canavalia maritima* with succulents such as *Sesuvium portulacastrum*. There are also grasses, for example *Ischaemum muticum*, and the sedges such as *Fimbristylis* and *Remirea*. Behind this zone is a belt of small shrubs in which are found *Clerodendron inerme*, *Wedelia biflora* and *Morinda citrifolia*. *Pemphis acidula* occurs as a dwarf shrub on coral rock or as a larger shrub on non-coralline rocky shores. *Cassytha filiformis* is a common parasite on many coastal plants (Fig. 2).

Beach forest usually occupies consolidated sand ridges and is characteristically only 5–10 m wide. Common canopy species are *Barringtonia asiatica*, *Terminalia catappa* and *Calophyllum inophyllum*. There is no distinct subcanopy but a number of trees and shrubs including *Vitex trifolia*, *Myristica hollrungii*, *Diospyros sp.*, *Hibiscus tiliaceus*, *Pandanus tectorius* and *P. pedunculatis* are present. *Pandanus dubius* is common on coral. All these are usually

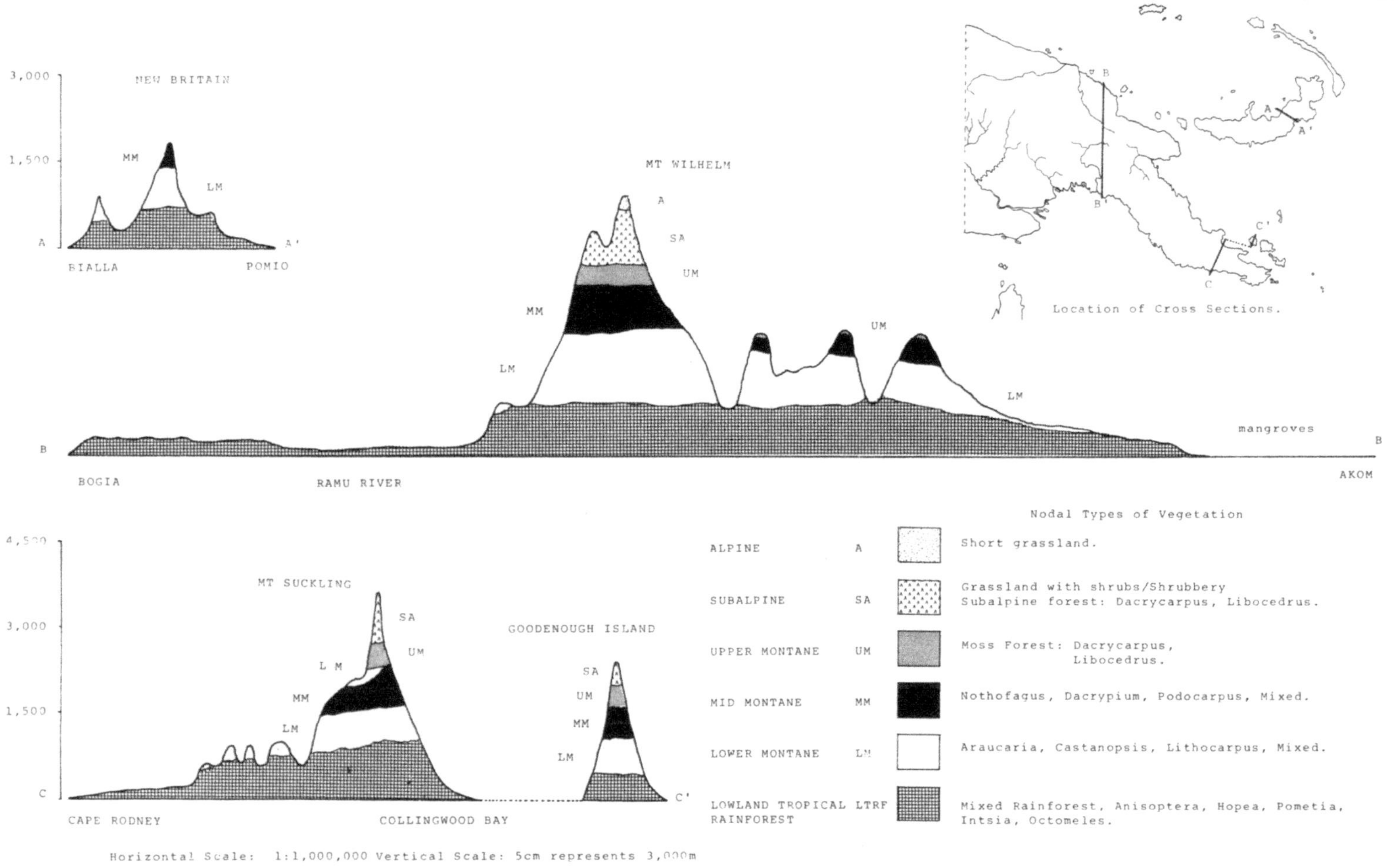

Fig. 1. Idealised cross sections of New Guinea showing the altitudinal zonation of the major vegetation types.

small. Along the south coast of the Huon Gulf, near the mouth of the Mambare River, in the Kokenau area and on Umboi Island the strand forest is dominated by *Casuarina litorea*. In areas around Merauke extensive forests are found dominated by *Cocos nucifera*, the coconut palm: however, it is probable that these were originally planted. *Imperata cylindrica* is often abundant on the beach ridges where they are disturbed. Where tall trees are missing a shrubbery occurs consisting of *Wedelia biflora*, *Crinum asiaticum* and other plants. *Flagellaria indica* is also a common climber in the shrubbery. *Cycas rumphii* is restricted to coral headlands and igneous rocks.

2.2 Mangrove forest

The most extensive mangrove communities occur in the tidal and brackish water of the estuaries of the larger rivers, in particular the delta country of Papua (the Purari and Fly Rivers), and Murik Lakes near the mouth of the Sepik River and areas around the mouth of the Ramu River. Mangrove also occurs as scattered communities along the whole coast, particularly in sheltered bays, inlets and river mouths (Fig. 3) (see chapter III:12).

A marked zonation occurs in the community in the transition from salt to brackish water. The zone inundated by normal and high tides twice daily is dominated by species of *Rhizophora* (*R. mucronata, R. apiculata*, and *R. stylosa*), *Avicennia marina*, *Camptostemon schultzii* (endemic in SW Papua), *Bruguiera parviflora*, *B. sexangula*, *B. gymnorrhiza*, *Xylocarpus granatum* and

Fig. 2. Coastal vegetation dominated by *Pandanus tectorius* with creeping *Ipomoea pes-caprae* (Buso-Morobe Coast).

Sonneratia caseolaris. *Ceriops tagal* occurs in shallow water with a stony bottom. Several species are restricted to the zone which is inundated by spring tides only, for example, *Sonneratia alba*, *Heritiera litoralis*, *Cynometra ramiflora*, *C. inaequifolia*, *Xylocarpus moluccensis* (S coast only), *Lumnitzera littorea*, *L. coccinea* and *Excoecaria agallocha*. On the banks of the tidal streams are found *Bruguiera gymnorrhiza*, *Dolichandrone spathacea*, *Cerbera floribunda*, *C. mangas*, *Aegiceras corniculata* and *Scyphiphora hydrophyllacea*. *Aegialites annulata* and *Osbornia cotondonta* are found on rocky shores in the vicinity of Fairfax Harbor (Port Moresby) as part of the inner edge of the mangrove zone. *Acrostichum* develops as a dense ground cover in older mangrove stands. (See chapter III:12.)

2.3 Swamp forest

These forests include a mosaic of communities varying from tall, well-drained mixed forest on alluvial deposits which are occasionally flooded to extensive areas of seasonally inundated *Metroxylon sagu*, the sago palm, and *Pandanus* swamps. (See Paijmans 1976: Pl. 14, p. 44). Permanently swampy areas are sometimes dominated by *Campnosperma brevipetiolata*.

The lowland swamp has a high canopy at 35 m and the following species are commonly found though usually none dominate: *Albizia moluccana*, *Pometia pinnata*, *Octomeles sumatrana*, *Pterocarpus indicus* and *Intsia bijuga*. Many of the species, for example, *Terminalia sepikana* and *T. complanata*, *Pterocarpus*

Fig. 3. Mangrove forest. A fringe of mangroves occurs along much of the coastline of New Guinea. It is only in the mouths of the major rivers that one finds the extensive development of mangrove forests (Morobe Coast).

indicus and several species of *Ficus*, are deciduous. As the sites become wetter *Metroxylon sagu* can form a distinctive sub-canopy in these forests (Fig. 4).

There are 2 distinctive types of swamp forest in S New Guinea. The genus *Metadina* forms pure open stands up to 15 m tall which are seasonally flooded to depths of up to 1.5 m. Common epiphytes are *Dischida, Hydnophytum, Hoya* and *Cyclophorus*. In the Fly River valley *Barringtonia spicata* and *B. tetraptera* can form extensive pure stands which are flooded seasonally to 3 m deep.

Many additional types of swamp forest occur, some of special interest being the *Terminalia brassii* swamps on New Britain, S New Ireland and Bougainville; the *Sonneratia caseolaris* forest along the banks of freshwater streams, and other scattered forests dominated by *Melaleuca*, *Campnosperma brevipetiolata* or *Pandanus* spp. Another type of swamp forest occurring in the Sepik Valley is well termed "freshwater mangrove." It is composed of species of *Myristica*, *Calophyllum*, *Syzygium*, *Campnosperma*, *Palaquium*, *Diospyros* and *Intsia*.

2.4 Mixed lowland rainforest

The associations of plants making up the mixed lowland rainforest remain the most complex and least understood of the earth's vegetation types. The forests are very diverse in composition and various species can attain local dominance

Fig. 4. A fresh water sago (*Metroxylon sagu*) community in the Gogol River Valley (Photo G. Slater).

for ecological reasons not yet fully understood. Throughout the lowlands the forests have been extensively destroyed or modified by shifting agriculture and many of the present communities represent seral stages developed on old garden sites or on sites where the forest has been destroyed by natural disasters such as wind, landslides or fire. Few detailed studies have been made in these forests, despite their economic importance, because it is difficult to identify species and to analyse the data collected. This broad vegetation type is perhaps best thought of as a mosaic with very varied botanical composition (Fig. 5).

The forests are composed of trees mainly with large buttresses which reach a canopy height of 45 m. In natural forest the canopy is often dense and allows little penetration of light into the lower layers of the forest, resulting in a surprisingly sparse undergrowth despite the over-all luxuriance of the forests. The canopy trees are often covered in a dense mat of epiphytic ferns, mosses and orchids. The most common are *Myrmecodia*, *Lecanopteris*, the staghorn fern *Platyerium wandae* and epiphytic orchids such as *Dendrobium antennatum*, *D. spectabile*, *D. bracteosum* and *Grammatophyllum papuanum*.

The canopy surface is very uneven and species of *Ficus*, which commonly

Fig. 5. Mixed lowland tropical rainforest in the Gogol River Valley (Photo G. Slater).

start as epiphytes, often subsequently form emergent trees. The canopy and sub-canopy of these forests contain more than the 80 genera which include probably more than 1200 tree species.

Along the valley floors (Fig. 5) and the lower valley sides the most frequent genera of trees are *Pometia, Celtis, Pterocarpus, Ficus, Dracontomelum, Vitex, Sloanea, Cerbera, Alstonia, Maniltoa, Cryptocarya, Guioa, Kingiodendron* and *Tristiropsis. Pometia pinnata* is the most common canopy species. *Elaeocarpus sphaericus* is a common canopy tree on disturbed sites and many other species form small recognisable entities within the forest. The sub-canopy of these forests is dominated by the genera *Diospyros, Gnetum, Protium, Pimeliodendron, Myristica, Dendrocnide, Kibara* and *Horsfieldia.* Common species are *Diospyros discolor, Gnetum gnemon, Protium macgregorii* and *Myristica chrysophylla.* Various species of feather palms and the fan palm *Licuala* are common as are several species of *Pandanus* with their striking stilt-roots.

Along the well-drained ridge tops, particularly in the coastal regions of the Morobe Prov., there is a marked change in the composition of the forests. The soils on these ridge tops are either very young and immature acidic brown forest soils or, in more stable areas, red and yellow latosols. The forests of the ridges are often floristically poorer than those developed along the valley floor and are dominated by species of dipterocarps such as *Anisoptera thurifera* and *Hopea iriana* associated with other genera such as *Protium, Intsia, Calophyllum* and *Canarium.* The canopy in these forests tends to be more open than along the valley floor, and the sub-canopy is dominated by the genera *Myristica, Pimeliodendron, Diospyros* and *Gnetum.* Epiphyllous lichens and mosses are less common on the sub-canopy trees of the ridge-top forests.

At higher altitudes the forests show a marked change in physiognomy and floristics with the introduction of an aspect of the lower montane forest. The genus *Lithocarpus* is commonly associated with *Anisoptera* and *Hopea* and a decrease in palms is balanced by the marked increase of tree ferns and the common occurrence of film-ferns, *Hymenophyllum* and *Trichomanes,* on the tree trunks. In the Kui, Paiawa, Maiama areas, *Eucalyptopsis papuana* forms distinctive ridge-top stands, which are in many respects lower montane in aspect.

2.5 Monsoon forest

These forests are found in areas with a markedly seasonal climate and are characterised by the presence of a number of species which remain leafless for prolonged periods. The canopy of these forests is more open than in the rain forest and the lower storeys are not well developed except in areas of high soil moisture where, because of local seepage, dense stands of palms may occur. To the north and north-east of Port Moresby the forest is dominated by *Bombax, Erythrina, Pterygota, Tetrameles, Garuga, Albizia* and *Terminalia.* In SW Papua 2 communities are found, one a forest dominated by *Flindersia, Acacia, Tristania, Mangifera* and *Eugenia,* and the other a scrub community from 1 to 5 m high dominated by *Sinoga lysicephala* with less common species *Melaleuca cajaputi*, *Banksia dentata* and *Grevillea glauca*. These communities show transitions to the savanna woodland.

2.6 Savanna vegetation

With decreasing rainfall (less than 2600 mm/annum and several consecutive months of less than 100 mm) the monsoon forests become susceptible to fire. The extensive areas of savanna woodland around Port Moresby and in W Papua have been created by man-made fires. The dominant trees are *Eucalyptus* species, mainly *E. alba, E. polycarpa, E. papuana, E. confertiflora* and *E. tereticornis* (Fig. 6). These give the forest a distinctly Australian aspect. Savanna vegetation of similar structure but dominated by *Albizia* or *Neonauclea* occurs in the Markham Valley, at Oro Bay and on Umboi Island (Fig. 7). The savanna woodland of Western Papua is richer in species than that in the Port Moresby area and members of the Myrtaceae are particularly abundant, including

Fig. 6. Savanna forest dominated by *Eucalyptus terreticornis* on the edge of gallery rainforest (Photo G. Slater).

Tristania longivalvis, *Xanthostemon crenulatus*, *Melaleuca symphiocarpa*, *M. argentea* and species of *Metrosideros*. Also more and more "Australian" herbs are being discovered. In the Kerema-Vailala area *Melaleuca argentea* forms a very open canopy above the grasses *Imperata*, *Themeda* and *Coelorachis*.

Fig. 7A. Open savanna in the Markham Valley. This is dominated by *Sarcocephalus*.

Fig. 7B. Savanna on Umboi Island dominated by *Albizia*.

2.7 Anthropogenic grasslands

Lane-Poole (1925) recognized that with the exception of the alpine region all grasslands below 3000 m probably originate by a combination of subsistence agriculture and firing. A possible exception are the montane grasslands in the dry region around Henganofi. Regrowth of forest occurs after firing has ceased (Figs. 8 and 9). In the lowlands a tall grassland occurs along the river banks and on the plains up to 300 m. This is composed of *Saccharum spontaneum*, *S. robustum*, *Phragmites karka* (locally dominant) and *Polytoca macrophylla*. A short grassland rich in *Cycas media* occurs on the old gravel fans and screens along the Markham Valley. The principal species are *Themeda australis*, *T. intermedia*, *Ischaemum digitatum* and *Paspalum vaginatum*. Also associated are species of *Arundinella* and *Imperata*.

The montane grasslands are composed of *Imperata*, *Arundinella*, *Sorghum nitidum*, *Ophiuros* and *Capillipedium*. On dry sites *Themeda australis* dominates and *Ischaemum barbatum* covers the moister depressions. In the W Highlands large areas are covered by *Miscanthus floridulus*. (See chapter III:9.)

3. Montane zone

This zone is divided into several distinctive forest types. Mixed broadleaf forest occurs throughout the whole range but varies considerably in composition. Mixed lower montane forest is almost entirely broad-leaved, the mid-montane forest is much reduced in height and characterised by emergent coniferous trees (Table 1). In addition there are forests dominated by *Araucaria*, *Nothofagus* and various oaks such as *Lithocarpus* and *Castanopsis*. The anthropogenous grasslands are dominated by *Miscanthus* and *Imperata*. The montane zone normally occupies an altitudinal range from 700 m to 3000 m.

3.1 Mixed lower-montane forests

This forest differs markedly from the lowland rain forest. The frequency of palms, lianes and trees with buttress roots is much reduced. Tree ferns, filmy ferns and bryophytes (mosses) are uncommon. The canopy of the forest, 20 to 25 m in height, is made up of a number of broadleaf species. There is a wide variation in the composition throughout New Guinea.

Typical canopy genera are *Elaeocarpus*, *Sloanea*, *Elmerrillia*, *Litsea* and *Cryptocarya*. In sub-canopy genera such as *Elaeocarpus*, *Weinmannia*, *Pullea*, *Myristica*, *Daphniphyllum*, *Diospyros* and *Pittosporum* occur. Tree ferns and ground ferns are less common than in the mid and upper montane zones. Several species of climbers and scramblers are present. These include species of the genera *Parsonsia*, *Clematis*, *Smilax* and *Hoya*. Epiphytes include *Schefflera*, orchids and parasitic Loranthaceae.

Fig. 8. Anthropogenic grasslands along the Snake River Valley. These grasslands are regularly burnt (Photo J. E. N. Smith).

Fig. 9. Tall grasslands dominated by *Miscanthus floridulus* and *Saccharum robustum* on disturbed areas along the floor of the Bulolo Valley.

Table 1. Table showing major tree genera of the lower- mid- and upper-montane forests (after Johns 1976)

	Formation type	Major families or genera	Alt. range
Sub alpine zone			
			±3200 m
	Upper montane	*Dacrycarpus* *Papuacedrus* *Schefflera* *Pittosporum* *Elaeocarpus* *Amaracarpus*	
			3000 m 2700 m
Montane zone	Mid montane	*Cunoniaceae* *Nothofagus* *Dryadodaphne* *Podocarpus* *Phyllocladus* *Dacrydium* *Falcatifolium* *Elaeocarpus*	
			2000 m 1500 m
	Lower montane	*Elmerrillia* *Castanopsis* *Lithocarpus* *Araucaria* *Agathis* *Eucalyptopsis* *Podocarpus* *Elaeocarpus* *Syzygium* *Gnetum gnemon*	
			1000 m 300 m
Lowland zone	Lowland tropical rainforest		

3.2 Oak forests

At various altitudes above 610 m ridge-top forests occur which are dominated by *Castanopsis acuminatissima* and species of true oaks for which the generic names *Quercus*, *Pasania* or *Lithocarpus* have been used. *Lithocarpus* is now accepted for the New Guinea species. At the lower levels these species are often mixed with *Anisoptera*, *Hopea*, *Intsia* and *Calophyllum* and a common co-dominant at higher altitudes is *Englehardtia rigida*. *Lithocarpus* spp. occurs through the forests to the upper montane zone and have been found at altitudes of 30 m above sea-level on Rossel I. in the Louisiade Archipelago. At Kini Creek, Mt Michael near Goroka, these forests intergrade with *Nothofagus* forests, or they may be separated by mixed montane forest as on Mt Otto and in the Bundi Gap.

Compound leaves are rare and the average leaf size is much reduced from that in lowland rain forest. Lianes are uncommon and buttressed trees rare. Tree ferns such as *Dicksonia* (rare) and *Cyathea* (many species) are found and the ground cover consists of species of *Hymenophyllum* and mosses such as *Polytrichum*. Brass in 1941 described a typical pure oak forest dominated by *Lithocarpus* and *Castanopsis* as "about 25 m high with slightly spurred greyish trunks sometimes surrounded by coppice shoots the oaks grew well apart and formed a rather thin canopy. In spite of the abundant illumination secondary and sub-stage trees were few." Apart from the oaks the dominant trees are *Pittosporum ramiflorum* and *Calophyllum papuanum*. Typical sub-canopy species are *Phyllocladus hypophyllus* and *Helicia amplifolia*; common shrubs are species of *Rhododendron* and *Vaccinium*.

3.3 Araucaria forest

In the Bulolo-Wau region (Fig. 10), the Hydrographers Range and the Jimi Valley at altitudes from 550 m to 1500 m, *Araucaria cunninghamii* (hoop pine) and/or *A. hunsteinii* (klinki pine) form a pure emergent layer 50 to 70 m in height above a closed canopy of broad-leaf species. The broadleaf canopy is

Fig. 10. Lower montane forest dominated by *Araucaria cunninghamii* in the Bulolo Valley (Photo F. Heath).

Table 2. General characteristics of structure and physiognomy of the montane vegetation: abundance on scale of 0–3 (after Johns 1976)

	Lower montane	Mid montane	Upper montane
Mean height			
canopy	30–40 m	20–30 m	18–20 m
Emergents	70–80 m	to 30 m	to 27 m
(*Araucaria*)			
(*Agathis*)			
Stratification	Diffuse	Diffuse or 2 layers	1 layer
Plank, buttresses	1–2	0–1	0
Surface roots	0–1	1–2	3
Stilt roots	0–1	2, (locally 3)	0–1
Feather palms	2–3	0	0
Fan palms	1–2	0	0
Tree ferns	0 (locally 2)	1–2	3
Pandans	1	2–3	0–1
Pandan lianes	1–2	1	0
Woody lianes	2	1	0
Vascular epiphytes	0–1	3	3
Non-vasc. epiphytes	0–1	3	3
Mistletoes	0–1	2–3	0–1
Epiphyllous lichens and mosses	1	3	3

similar, if not identical, to a lowland forest in composition. Lauraceae and Sapotaceae are abundant. The *Araucaria* component of this forest was considered by Womersley (1958) to be living fossil of a period of more temperate climate. However, it appears that the *Araucaria* forest is seral in nature. *Araucaria* regenerates in forest openings and with increasing age a sub-canopy

Fig. 11. Ridge top stands of *Nothofagus* above the limits of cultivation; Western Highlands.

representing species from the lower montane zone invades the forest. With the death of the emergent *Araucaria*, a slow transition takes place to mixed lower montane forest with hoop and klinki pine increasingly rare. A low altitude stand has been found in Irian Jaya with associated canopy trees of *Phyllocladus* and *Calophyllum congestiflorum*, and other small stands occur at Paiawa (see chapter III:6).

Sub-canopy and shrub layer include varieties of *Drimys piperita* and species of *Sloanea*, *Pullea*, *Schefflera*, *Pittosporum* and *Helicia*. The tree fern *Cyathea perpelvigara* is common and *Cyathea biformis* often occurs as an epiphyte on the lower tree trunks. *Pandanus* spp. and *Perottetia alpestris* are common in the sub-canopy. Terrestrial orchids and epiphytes belonging to the Ericaceae (*Dimorphanthera* and *Rhododendron*) and Araliaceae give the forest a distinctive physiognomy.

3.4 Discussion on the lower montane zone

This zone is quite distinct from the 2 neighboring zones particularly in regard to the physiognomy of the vegetation. While many of the species occur in the mid-montane zone they are rare and seldom dominate this higher zone. In comparison to the mid-montane, Podocarps are less abundant. The common plants of the mid-montane zone, *Nothofagus* and *Cunoniaceae*, are seldom found in this zone. The lower montane is distinctly drier in aspect than the mid-montane, with few species of ferns and mosses. *Pandanus* occurs locally in wet areas, but never forms a regular component of the forest as it often does with *Nothofagus*. Many other forest types are described in detail in Johns (1976).

3.5 Mid-montane forest

These forests are distinct from the lower montane forests both in terms of species composition (Table 1) and in the structure and physiognomy of the vegetation (Table 2). Above 2700 to 3000 m the forest is replaced by one of much lower height, dominated by *Dacrycarpus* (rare at lower altitudes), *Papuacedrus* and various members of the *Araliaceae*. *Harmsiopanax* can form quite a distinctive emergent at higher altitudes.

The main characteristic of this zone is the abundance of terrestrial and epiphytic ferns, mosses and lichens. In this regard the zone is very distinctive from the lower montane zone. Indeed the greatest diversity of most epiphytic and terrestrial genera occurs in these forests. *Nothofagus*, which often occupies secondary sites, occurs throughout the zone.

3.6 Mixed mid-montane forests

This zone is characterized by a higher frequency of conifers than lower montane mixed forest and the high frequency of *Weinmannia*. The main conifer species are *Podocarpus archboldianus, P. schlechteri, Dacrycarpus imbricatus, Falcatifolium falciforme, Papuacedrus papuanus* and *Phyllocladus hypophyllus*. Common understorey species are *Drimys piperita, Pittosporum ramiflorum*, members of the Urticaceae, and the two tree ferns *Cyathea aeneifolia* and *Dicksonia hieronymi*. Where the forest has been disturbed the climbing bamboo *Nastus productus* is abundant. The term broadleaf-gymnosperm forest has been used to describe this type.

3.7 *Nothofagus forest*

Nothofagus dominated forest (Figs. 11 and 12) is common throughout the mid-montane zone. It is dominated either by *N. pullei* at higher altitudes or most commonly at lower altitudes by *N. grandis*. Physiognomically the forests are very similar to the Mixed Mid-montane forests. Epiphytic mosses are abundant. Indeed the greatest diversity of the moss flora comes in this zone, much greater than the upper montane forest which is often commonly called "moss" forest. (See chapter III:5.)

3.8 *Upper montane forest*

This term has been used to include mossy forest, montane cloud forest and cloud forest. (Fig. 13). These forests have been superficially described, but the only detailed ecological work is that of Wade and McVean (1969). The forest is a *Phyllocladus-Xanthomyrtus* association often 20 to 25 m tall; there are 2 recognizable layers. It has been claimed that the upper altitudinal limit of this forest corresponds with the upper limit of prolonged daily cloud but this is open to question.

On Mt Wilhelm an *Amaracarpus brassii-Elaeocarpus azaleifolius-Podocarpus pilgeri* association occurs to 3300 m. This forest is richer in species than the neighboring subalpine forest, with a diversity of life forms and variations in leaf features indicating a forest of some complexity. A similar forest with few species above 10 m high has been found on the Doma Peaks. Most trees are crooked and stunted in growth with the exception of the emergent conifers such as *Papuacedrus* as well as *Elaeocarpus, Schefflera, Saurauia* and *Amaracarpus*. The understorey is dense and composed mainly of tree saplings. Three sprawling shrubs, *Vaccinium keysseri, Coprosma papuensis* and *Dimorphanthera collinsii*, are common. The forest floor is covered with "tangled mass of tree roots, fallen and decaying tree trunks and other forest debris." Although several species of moss are present they are not conspicuous. *Nothofagus* (beech) forests have been found at the same altitude in the Baliem Valley of the Snow Mts in Irian Jaya; no upper montane forest was present. The upper montane forests on Mts Otto, Hagen and Giluwe have a very large number of hepatics covering the tree trunks, true mosses are common, but not abundant in the so-called "moss forest."

Fig. 12. Montane forest dominated by *Nothofagus grandis*; Mt Kaindi.

Fig. 13. Sub-alpine vegetation on Mt Giluwe.

4. Subalpine zone

This zone was first recognized by Brass in 1941 and more recently by Robbins, but both excluded the grassland despite the common occurrence of shrubs in these communities. However this zone has been taken to include not only the floristically poor subalpine forests but also several distinctive grassland communities as well as the bogs and fens which are particularly well developed on Mt Giluwe (Wade and McVean 1969).

4.1 Forests

The subalpine forest, also termed high mountain forest, has a closed canopy at about 10 m with numerous emergents to 15 m in height. Many of the trees branch near the base, and epiphytes and climbing shrubs are less common than in the upper montane forest (Fig. 13). There are two distinct types of forest on Mt Wilhelm. The lower forest with *Amaracarpus caeruleus* and *Pittosporum pullifolium* occurs over an altitudinal zone of some 60 m to 3600 m. Here the tree fern *Cyathea percrassa* is very common in the sub-canopy of the forest and in places, particularly along ridge tops, bryophytes can completely dominate the ground flora. At higher altitudes, up to the forest limits, the forest is dominated by Ericaceae (*Rhododendron gaultheriifolium* and *Dimorphanthera microphylla*) and the Epacridaceae (*Trochocarpa dispersa* and *Rapanea vaccinioides*). The epiphytic fern *Hymenophyllum foersteri* is common, lianes are rare, and dense shrub and fern thickets dominated by *Gleichenia bolanica* are widespread.

Fig. 14. Small tarns are common in the glaciated areas on the summit of the mountains; Mt Giluwe.

On Mt Albert Edward the lowest forest is dominated by *Dacrycarpus compactus* and *Papuacedrus papuanus* with the canopy at 15 m to 20 m. The sub-canopy is dominated by *Vaccinium macbainti.* With increasing altitude the coniferous element decreases and *Dacrycarpus* forms an open emergent layer over a dense canopy dominated by *V. macbainti.* On Mt Wilhelmina *Vaccinium macbainti* is replaced by *V. dominans* and the emergent conifer element in the forest was represented by *Dacrycarpus* and, with less frequency, *Phyllocladus hypophyllus* and *Papuacedrus papuanus.* Two species of *Schefflera, S. megalantha* and *S. falcata* are dominant at above 3800 m.

4.2 Grasslands

The summit grasslands of the mountains have been superficially described by a number of authors: Lane-Poole (1925), Lam (1945), Brass (1941, 1964), Hoogland (1958), Robbins (1960a,b) and Costin (1967). Walker (1968) published the results of a more detailed reconnaissance study of the Pindaunde Valley. With the exception of Lane-Poole all authors refer to the grassland as "alpine." Wade and McVean (1969), after detailed ecological studies on Mt Wilhelm, apply the term subalpine to six distinct grassland associations and to several bog and fen associations (Figs. 13 and 14).

Several shrub species, *Coprosma divergens, C. papuensis, Gaultheria mundula, Haloragis microphylla* and *Leucopogon suaveolens* are common in either small thickets or scattered amongst the dominant grasses *Poa saruwagedica, Deschampsia klossii, Deyeuxia brassii* and *Dichelachne novoguineensis.* Small trees of *Olearia spectabilis* are scattered throughout the grassland.

On the moraines at lower altitudes this association abuts a grassland dominated by *Deschampsia klossii* with *Poa saruwagedica* and *P. languidior* and open groves of *Cyathea atrox.* At higher altitudes, from 3960 m to 4100 m, this association merges gradually into a community characterized by *Danthonia vestita* and *Deyeuxia brassii.*

The most widespread association occurs on the well-drained valley floors and slopes of less than 10 degrees on Mt Wilhelm. Large overlapping tussocks of *Deschampsia klossii* occur occasionally with co-dominant *Hierochloe redolens* and *Poa saruwagedica. Anaphalis mariae* and *Myosotis saruwagedica* equal the tussocks in height, while an abundance of smaller grasses such as *Danthonia penicillata* and *Anthoxanthum angustum* occur. Two other associations of restricted distribution are found. *Danthonia archboldii* and *Gaultheria mundula* are abundant along stream banks and at the base of cliffs with free-flowing water. The other association, restricted to well-drained slopes is dominated by *Poa nivicola* and *Dichelachne novoguineensis.* It is characterized by the absence of *Deschampsia klossii.* A similar association occurs on Mt Giluwe on burnt slopes and there is evidence that this association could be seral to that dominated by *Coprosma divergens* and *Poa saruwagedica.* Several types of grassland were described by van Royen (1967) from the alpine vegetation of Mt Albert Edward, but there seems to be little correlation between the communities he describes and those given by Wade and McVean (1969). It is quite probable that the associations described from Mt Wilhelm are of local distribution and different from those on other mountain peaks.

4.3 Bogs and fens

Only brief mention was made of the subalpine bogs prior to the detailed study by Wade and McVean (1969). They occupy a substantial area on Mt Wilhelm between 3200 m and 4270 m, and nine distinct units have been recognized. Mires are common on Mt Giluwe and several types are found which are poorly represented on Mt Wilhelm. There are considerable differences between the acid bogs developed under conditions of stagnant ground water and the fens which have moving ground water and are less acidic.

The bogs are characterized by hummock plants and shrubs are prominent. Typical examples are the *Agrostis reinwardtii-Gentiana piundenis-Brautelia aristifolia* short grass bog, the *Gleichenia vulcanica-Trochocarpa dekockii* bog (this type occupies large areas on Mt Giluwe and is possibly equivalent to the *Gleichenia vulcanica*-dominated vegetation occurring on Mt Doorman), and the *Astelia papuana* bog which, though limited in extent on Mt Wilhelm, is widespread on Mt Giluwe. Similar formations occur in frost hollows below the alpine zone of Mt Giluwe. Adjacent fens include *Drosera*, *Gentiana* and *Utricularia*.

The fens are characterized by an absence of shrubs and the hummock plants are replaced by grasses and sedges. A typical community is the *Brachyposium sylvaticum-Carex capilliacea* fen, and 3 local types are dominated by *Carex echinata* and *C. gaudichaudiana* and *Scirpus crassiusculus*.

4.4 Alpine grasslands

A short alpine grassland composed of *Festuca papuana*, *Poa callosa* and *Monostachya oreoboloides* associated with dwarf shrubs such as *Leucopogon suaveolens* occurs from 3960 m to 4400 m on Mt Wilhelm. Bog plants such as *Astelia papuana* and *Carpha alpina* are common and emphasize the edaphic nature of this community, particularly at lower altitudes.

Alpine tussock grasslands dominated by *Deschampsia klossii* and *Hierochloe redolens* are confined to deep, moisture-retaining soils between 4190 m and 4420 m. Small shrubs of *Hebe ciliata* and *Leucopogon suaveolens* occur but are not abundant.

Alpine fern meadow is best developed in the alpine zone around 3960 m and is dominated by frequent clumps of the fern *Papuapteris linearis* interspersed with larger tussocks of grass, commonly *Deschampsia klossii*. It is restricted to well-drained sites on steep slopes and ridge crests.

Two distinctive types of heath vegetation occur on Mt Wilhelm. On patches of impervious boulder clays between 4000 m and 4150 m a community dominated by moss *Rhacomitrium lanuginosum* associated with bog species such as *Carpha alpina*, *Oreobolus pumilio* and *Astelia papuana*. A dwarf shrub heath is entirely confined to the alpine zone between 4260 m and 4350 m on the flanks of the rock walls. The soils are shallow and the shrub species *Leucopogon suaveolens*, *Hebe ciliata* and *Drapetes ericoides* form a dense shrub mat 5–10 cm high.

Wet tundra occurs in rock basins above 4420 m near the summit ridge of Mt Wilhelm. *Poa callosa*, *Ranunculus saruwagedicus*, *Potentilla foersteriana* and *Hebe ciliata* are the larger vascular plants. Black mats of the hepatic *Marsupella stoloniformis* are characteristic. At similar elevations on dry sites an alpine tundra dominated by *Poa crassicaulis* and *Psilopilum australe* occurs. Both associations are similar in species composition.

Alpine bog occurs above 4030 m where *Astelia papuana* forms hummocks 10 to 15 cm thick. *Trochocarpa dekockii* is common on these hummocks. Several species, including *Danthonia penicillata*, *Anthoxanthum angustum* and *Keysseria radicans* grow between these hummocks.

References

Brass, L. J. 1941. The 1938–1939 expedition to the snow mountains, Netherlands New Guinea. J. Arnold Arboretum 22: 271–342.

—— 1964. Results of the Archbold expeditions No. 86. Summary of the Sixth Archbold expedition to New Guinea (1959). Bull. Am. Mus. nat. Hist. 127 (Art. 4): 149–215.

Costin, A. 1967. Alpine ecosystems of the Australian Region. In Wright, H. E. and W. H. Osburn, Arctic and alpine environments, 55–62.

Grubb, P. J. and P. F. Stevens. 1979. The forests of the Fatima Basin and Mt Kerigomna and a review of montane and subalpine forests elsewhere in Papua New Guinea. ANU BG/5, Canberra, Aust. Natl Univ.

Hoogland, R. D. 1958. The alpine flora of Mount Wilhelm. Blumea, Supl. 4: 220–238.

Hope, G. S. and J. S. Peterson. 1975. Glaciation and vegetation in the high New Guinea Mountains. Bull. Roy. Soc. N.Z. 13: 155–162.

Johns, R. J. 1976. A provisional classification of montane vegetation of New Guinea. Science in New Guinea 4: 105–127.

—— 1977. The vegetation of Papua New Guinea. Pts. 1 and 2. 64 pp. Office of Forests P.N.G.

—— (m.s.). The instability of the rainforest ecosystem in Papua New Guinea.

Johnstone, I. M. 1975. The sea-grasses of the Port Moresby region. Univ. Papua New Guinea Dept. Biology, Occ. Pap. 7, 38 pp.

Lam, H. J. 1945. Fragmenta Papuana. V. Sargentia. Arnold Arboretum, Harvard University.

Lane-Poole, C. E. 1925. Forest resources of the Territories of Papua and New Guinea. ii+209 pp. Government Printer, Melbourne.

Paijmans, K. editor. 1976. New Guinea vegetation. A.N.U. Press, Canberra.

Robbins, R. G. 1961a. The vegetation of New Guinea. Australian Territories. 1.

—— 1961b. The montane vegetation of New Guinea. Tuatara 8: 121–134.

Royen, P. van. 1967. Some observations on the alpine vegetation of Mount Biota (Papua). Acta Bot. Neerl. 15: 530–534.

Wade, L. K. and D. N. McVean. 1969. Mt Wilhelm studies I. The alpine and subalpine vegetation. Research School of Pacific Studies. Dept Biogeography and Geomorphology. Bio. Geo. series. BG/1. A.N.U. Press, Canberra.

Walker, D. 1968. A reconnaisance of the non-arboreal vegetation of the Pindaunde catchment, Mount Wilhelm, New Guinea. J. Ecol. 56: 503–533.

—— 1972. Bridge and Barrier: The natural and cultural history of the Torres Strait. Bio. Geo. series. BG/3. A.N.U. Press, Canberra.

4. Phytogeography and evolution of the Ericaceae of New Guinea

P. F. Stevens

1. Introduction

The Ericaceae are among the ecologically most important families of the more open woody vegetation from about 1000 to 3750 m especially in the Upper Montane and Subalpine vegetation in New Guinea, but they also occur less frequently in other vegetation types from sea level to over 4000 m. There are over 400 species in six genera in our area alone (Table 1), and the Ericaceae are probably the most diverse woody family in this vegetation: none of the six genera is restricted to New Guinea, but all bar three of the species are. The main purpose of this paper is to examine the morphological and ecological variation of the New Guinea Ericaceae in the context of the variation pattern of the whole family. An attempt is also made to discuss the historical biogeography of the family in New Guinea.

The family is fairly well known taxonomically because of the conspicuous flowers of many of the species and the resultant interest in them in horticultural circles. Sleumer's comprehensive account of the family for the "Flora Malesiana" forms the taxonomic basis of this work (Sleumer 1966, 1967), changes since then being noted below (authorities are not mentioned after species' names except where they are not included, or differ from those given, in Sleumer's accounts). New Guinea as delimited here includes the main part of the East Malesian region (see van Balgooy 1976 for references), but to understand the evolution of the New Guinea Ericaceae one must look at species from the rest of Malesia and SE Asia, as well as tropical America.

The great majority of the family has pollen in tetrads and anthers that invert during the course of development, dehisce by pores, and often lack an endothecium, but, apart from the absence of stipules and the usually axile placentation, there is otherwise great diversity within the family. However, as is quite common (see van Steenis 1978), the coherence of the family is increased because of numerous polyphyletic "trends" or "tendencies" within it, for example, the presence of projections on the anthers, the occurrence of rhizomes and the winging of seeds. The evolution of the New Guinea Ericaceae will be discussed in the light of these and other tendencies, and so placed within the framework of the evolution of the family as a whole. Although the lack of detailed knowledge of cladistic relationships within the family and of the

Monographiae Biologicae, Vol. 42, ed. by J. L. Gressitt.
© *1982, Dr W. Junk Publishers, The Hague.*

Table 1. General distribution of New Guinea Ericaceae and their immediate relatives in Malesia.

		West New Guinea only	East New Guinea only	Widespread New Guinea	Total New Guinea	Rest of Malesia	Pacific and NE Australia
Rhododendron subsection	*Pseudovireya*	8	3	4	15	13	0
	Siphonovireya	4	1	2	7	0	0
	Phaeovireya	19	19	3	41	2	0
	Malayovireya	0	0	0	0	13	0
	Albovireya	3	3	0	6	8	0
	Solenovireya	8	14	3	25	12	1
	Euvireya series						
	Linnaeoidea	7	1	3	11	0	0
	Saxifragoidea	0	0	1	1	0	0
	Taxifolia	0	0	0	0	1	0
	*Stenophylla**	0	0	0	0	1	0
	Citrina	0	0	0	0	1	0
	*Buxifolia/Javanica**	28	19	10	57	73	2
Gaultheria		4	1	1	6	18	0
Diplycosia		13	5	3	21	75	0
Vaccinium section	*Galeopetalum*	0	0	0	0	1	0
	Rigiolepis	0	0	0	0	23	0
	Oarianthe	29	7	5	41	4	3
	Neojunghuhnia	11	2	1	13	0	0
	Bracteata	36	34	4	74	79	1
Dimorphanthera section	*Pteridosiphon*	2	0	0	2	0	0
	Pachyantha	0	6	0	6	0	0
	Dimorphanthera	2	3	5	10	2	0
	Trochilanthe	23	25	5	53	0	0
Agapetes subgenus	*Paphia* section *Paphia*	0	14	0	14	0	4
Total		197	157	49	403	327	11

*For circumscription of these taxa, see text.

polarity of evolution of many of the characters discussed makes this task more difficult, the immediate relationships of most of the genera and supraspecific taxa are becoming clearer, as mentioned below.

It is concluded that New Guinea Ericaceae, and most tropical Ericaceae in general, seem to have evolved very much within the constraints within which the temperate members have evolved. More or less particularly tropical characteristics are the scandent habit, the epiphytic habitat, and some corolla types, especially those probably associated with ornithophily. The specific diversity in New Guinea is in part at least accompanied by the evolution of characteristics similar to those that have developed independently in the family elsewhere, both in the tropics and in the temperate zone.

This chapter is divided into three slightly overlapping parts. The general morphology and taxonomic relationships of the New Guinea Ericaceae and their probable nearest relatives are discussed first, and their general geographical distribution within New Guinea given. In the main section the characters of the Ericaceae in New Guinea are compared in some detail from a basically ecological point of view with those of the family elsewhere. In the third section the evidence presented is discussed in the light of what is known about the geological history of New Guinea.

2. The New Guinea Ericaceae: morphology, relationships, and current distribution

The genera and the main supraspecific taxa of the New Guinea Ericaceae are listed in Table 1. The main species concentrations in the New Guinea taxa are mentioned below, while overall patterns of distribution of the Ericaceae in New Guinea are discussed at the end of this section.

2.1 Rhododendron

All the New Guinea species belong to section *Vireya* which is overwhelmingly Malesian in distribution. A single species of subsection *Euvireya* is known from NE Australia, a few species of subsection *Pseudovireya* grow on the Himalayas and elsewhere in SE Asia, while subsection *Euvireya* is even more sparsely represented on the SE Asian mainland. *Rhododendron* section *Vireya* is currently placed in subgenus *Rhododendron* (Sleumer 1949; Cullen and Chamberlain 1978). It has scales, but no other multicellular hairs. Many members are epiphytes, a habit uncommon in the rest of the genus. The various subsections and series within section *Vireya* are distinguished by differences of indumentum, flower type, habit and leaf size (Sleumer 1966). A reevaluation of the limits of these taxa is necessary, and although this cannot be attempted here, the following points should be noted. Within subsection *Euvireya*, I have combined series *Buxifolia*, *Javanica* and the New Guinea species of series *Stenophylla*. In both flower and leaf type the New Guinea species of series *Stenophylla* differ considerably from *R. stenophyllum*, from Borneo, and are much more similar to species from series *Buxifolia* and *Javanica* (Stevens 1976; see also Woods 1978). These latter differ only in leaf size, and, as Sleumer (1973) noted, closely related species may be placed in different series because of this. The concentration of species of series *Linnaeoidea* and, to a lesser extent, those of subsection *Pseudovireya*, in Irian Jaya is noteworthy; the taxa involved have small leaves. Species limits in subsection *Solenovireya* are exceptionally difficult and the significance of the distributional pattern in this group is unclear.

2.2 *Gaultheria*

There are relatively few New Guinea representatives of this large, circum-Pacific genus, which can be recognized by its fleshy calyx that encloses a capsular fruit and its awned anthers. The New Guinea species seem most related to W Malesian species, with the possible exception of *G. mundula* (Sleumer 1967), which is superficially similar to some species from New Zealand (see also van Steenis 1934). The genus is in need of monographic work to ascertain infrageneric relationships.

2.3 *Diplycosia*

Diplycosia is close to *Gaultheria*, the two having similar fruits, and at least some members of both genera smell of methyl salicylate (wintergreen). *Diplycosia* may be distinguished by its consistently fasciculate inflorescence (very rare in *Gaultheria*), its anthers, which lack spurs but which have tubules, and by the presence of numerous fibers wandering throughout the mesophyll in all species that I have examined (this character is uncommon in *Gaultheria*). *Diplycosia* is also much more commonly epiphytic than *Gaultheria*. All these characteristics are probably derived, although, apart from stamen type, all also occur in *Gaultheria*. The center of diversity in *Diplycosia* is Borneo (Table 1), and the genus has only a foothold on mainland SE Asia; the New Guinea representatives are few and are not distinguished by any characters that set them apart as a group in the genus. The center of diversity of the genus in New Guinea is the Vogelkop Peninsula; there are 7 species there, and 6 are found nowhere else.

2.4 *Vaccinium*

Vaccinium is a large and diverse genus, the New Guinea representatives of which can be characterized by their falsely 10-locular ovaries, small (usually less than 1 cm long) and usually urceolate to globose corollas, and fleshy fruit. (For the circumscription of the genus in Malesia, with section *Pachyantha* being transferred to *Dimorphanthera*, see Stevens 1974). Neither of the New Guinea-centered sections is much different from the polymorphic section *Bracteata*. Section *Neojunghuhnia*, with its very long calyx lobes, is endemic to New Guinea (Table 1); it is poorly known. Section *Oarianthe* is most diverse in New Guinea, only five species not occurring there. It has small leaves, often less than 1 cm long, and basically fasciculate inflorescences; both of these are derived characteristics, although neither is unique in the genus, even within Malesia. Despite the apparent distinctness of section *Oarianthe*, species like *V. summifaucis* and the recently described *V. morobense* Kloet, both placed in section *Bracteata* yet having small leaves and reduced inflorescences, makes one pause for thought.

Species limits in *Vaccinium* will doubtless need emending when more material becomes available, and the local nature of many of the species currently recognized may change. The concentration of species of sections *Oarianthe* and *Neojunghuhnia* in Irian Jaya is perhaps worthy of note, although 11 of the 13 species recognized by Sleumer (1967) in the latter were known to him from only one or two collections, and little more is known about them now.

2.5 Dimorphanthera

Dimorphanthera is characterized by its usually 5-locular ovary; flowers usually over 1 cm long; its relatively short, usually strongly dimorphic, woody stamens, the filaments of which are free (there may appear to be only five stamens, since the larger stamens overarch the smaller); and numerous anatomical features. Available evidence suggests a relationship to *Satyria* and other tropical American genera (Stevens 1972, 1974, ms).

Of the four sections of *Dimorphanthera*, section *Pteridosiphon*, with its conspicuously ribbed calyx, is poorly known. Section *Pachyantha*, with its campanulate, very thick-walled, apparently two-layered corolla less than 1 cm long, is distinct, while section *Dimorphanthera*, with a large, campanulate, usually deeply-lobed corolla, and section *Trochilanthe*, with a basically tubular, usually more shallowly-lobed corolla, are sometimes not easy to distinguish.

Dimorphanthera section *Pachyantha* is restricted to the eastern half of New Guinea. Section *Dimorphanthera* extends to the Philippines and has several wide-ranging species, while section *Trochilanthe* has more numerous, sometimes narrowly distributed species, and further species are still being discovered (Fig. 3).

2.6 Agapetes subgenus Paphia

Agapetes subgenus *Paphia* is anatomically similar to *Dimorphanthera*, but is very different from the other subgenus, *Agapetes*, and may be closer to American taxa (Stevens 1972). It has large, usually rather strictly tubular flowers, and stamens that are almost as long as the corolla, but which are not notably woody or dimorphic. Section *Pseudagapetes* is monotypic, with a single species on the Malay Peninsula, while section *Paphia* is centered in New Guinea, but also occurs in New Caledonia, NE Australia and Fiji. Thirteen of the 14 New Guinea species of subgenus *Paphia* grow on the mountains between Mt Amungwiwa and Mt Victoria in the SE, a distance of only some 250 km, and 11 of them are restricted to this area. Although there is a considerable amount of variation in habit, floral characters, etc., between these species, there are no discontinuities obvious enough on which supraspecific taxa could be based.

2.7 General distribution patterns of New Guinea Ericaceae

About 12% of the species can be called widespread, that is, found along half or more the length of the island (Table 1). Although the figure was arrived at simply by recording as widespread all species which grow in both Irian Jaya and Papua New Guinea, species which just straddle the boundary (e.g., *Vaccinium sanguineum*) are probably balanced by those which have a very wide distribution within one half of New Guinea (e.g., *Dimorphanthera elegantissima*). About 40% of the species are apparently local — restricted to a single mountain or to two adjacent peaks, but this may be an over-estimate (see below). About 26% of the species in the eastern part of the island are local; over a third of these species occur around the Sepik Basin, a very poorly collected area. About 44% of the species from Irian Jaya are local, of which slightly under a quarter are restricted to the Vogelkop.

Ericaceae are poorly represented on the Bismarcks and the Solomons. One species of *Vaccinium* section *Bracteata* occurs on New Britain, one on Rossel Island (and one of section *Oarianthe* on Bougainville) and two species of *Dimorphanthera* on New Britain and one on New Ireland. A few species of *Rhododendron* belonging to several groups also occur on New Britain and New Ireland (Figs. 1–3).

Serious problems arise when discussing details of the distribution of individual species. Although the Ericaceae are among better collected montane families, much more collecting off the beaten track needs to be done, especially in Irian Jaya; novelties are still being collected, and infraspecific variability and species limits and relationships need to be assessed more accurately. Using any species as an example of a particular distribution pattern can be hazardous. Thus *Dimorphanthera calodon* had a restricted geographic range, being known only from the Murray Pass, in the Central Province (Sleumer 1967), until it turned up rather abundantly along the Bulldog Road, Morobe Province (Stevens 1977); *Rhododendron wrightianum* var. *wrightianum* had a disjunct distribution, being known only from Irian Jaya and Milne Bay Province (Sleumer 1966), until it was reported from Western, Morobe and Central Provinces (Sleumer 1973). Hence it is considered premature to analyze species distributions more closely. Kalkman and Vink (1970) discussed the geographical distribution of Ericaceae on the mountains in the well-collected central highlands of Papua New Guinea; they concluded that, although few species there were single mountain endemics, many species occurred on fewer mountains than their overall distribution might have suggested.

3. Characteristics of New Guinea Ericaceae compared with those of the rest of the family

It seems that an explanation of at least some of the variation in New Guinea and other tropical Ericaceae may be sought in terms of adaptive radiation in rather

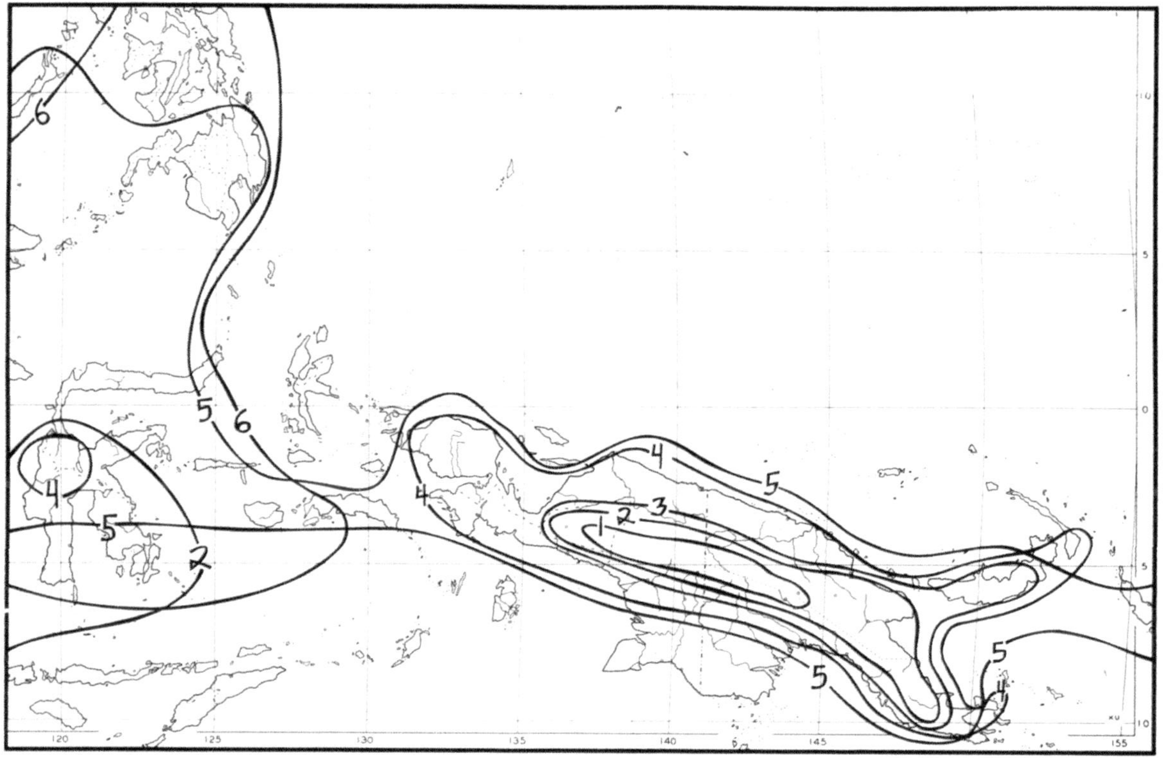

Fig. 1. Distribution of some groups of *Rhododendron* in East Malesia. 1, *Rhododendron* series *Saxifragoidea*; 2, series *Linnaeoidea*; 3, subsection *Albovireya*; 4, subsection *Phaeovireya*; 5, subsection *Solenovireya*; 6, subsection *Malayovireya*.

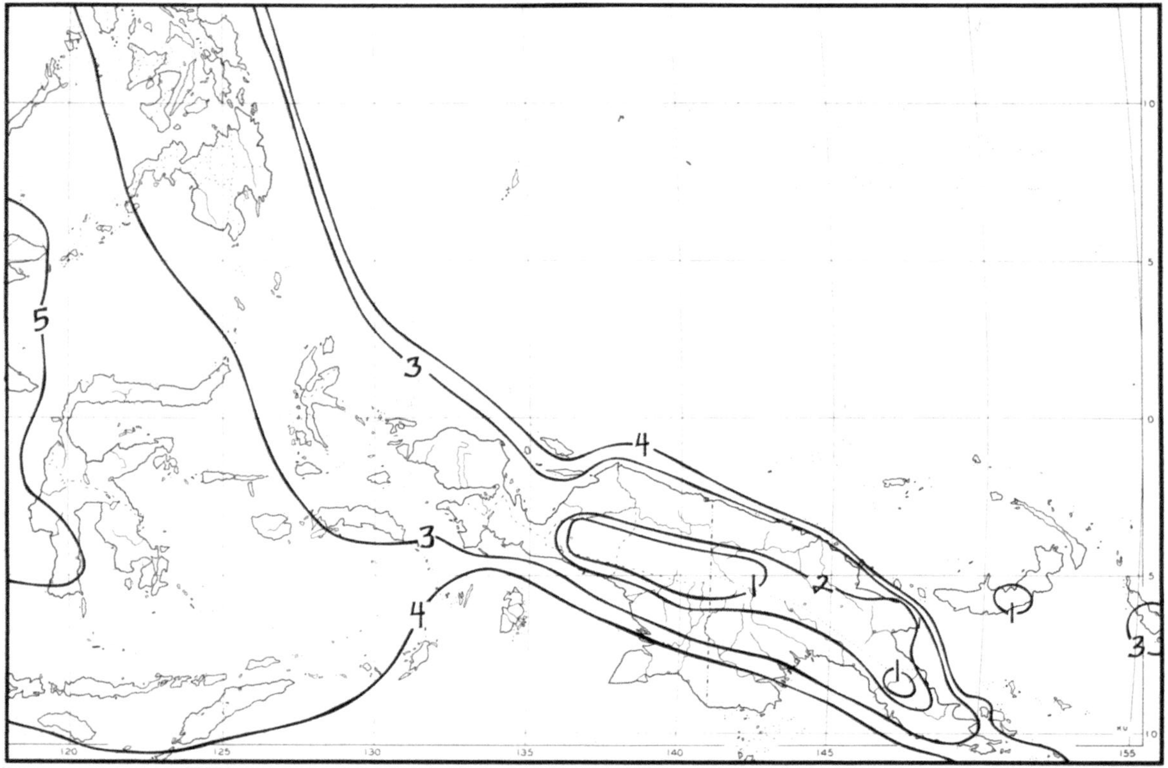

Fig. 2. Distribution of some groups of *Rhododendron* and *Vaccinium in* East Malesia. 1, *Vaccinium* section *Neojunghuhnia*; 2, *Rhododendron* subsection *Siphonovireya*; 3, *Vaccinium* section *Oarianthe*; 4, *Vaccinium* section *Bracteata* (the record from New Britain should be referred to this section); 5, *Vaccinium* section *Rigiolepis*.

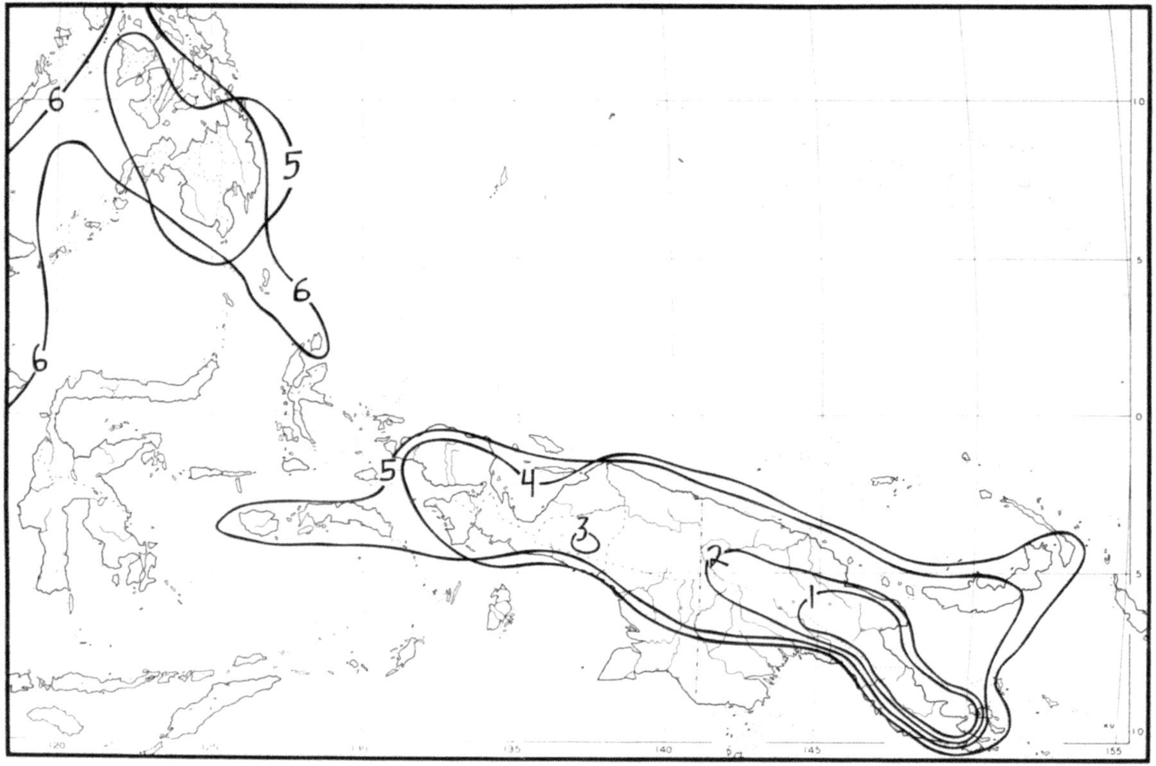

Fig. 3. Distribution of some Vaccinieae in East Malesia. 1, *Agapetes* section *Paphia*; 2, *Dimorphanthera* section *Pachyantha*; 3, *Dimorphanthera* section *Pteridosiphon*; 4, *Dimorphanthera* section *Trochilanthe*; 5, *Dimorphanthera* section *Dimorphanthera*; 6, *Costera*.

particular ecological situations (see also Stevens 1976). In this section, the general ecological constraints within which the New Guinea Ericaceae seem to have evolved will be compared with the ecological characteristics of members of the family. Reference will be made to more specific morphological and anatomical characteristics of the family where necessary. The ecological characteristics are listed below.

1. A preference for cool, rather than hot, climates.
2. A preference for acid habitats.
3. A strong tendency to be heliophilic.
4. A shrubby habit, but with considerable variation in physiognomy.
5. A diversity of leaf size and especially shape.
6. A conspicuous corolla and considerable variation in stamen type.
7. A small fruit with numerous seeds.
8. An accumulation of tannins and sometimes other toxins.

There is a danger of circular argument here, since the New Guinea Ericaceae must have the family characteristics. However, by discussing temperate and other tropical representatives of the family separately, some idea can be given as to how the New Guinea members compare with West Malesian/other tropical/temperate members of the family. Hence a distinction can be drawn between characteristics general in the family, peculiar to the tropical environment, or peculiar to New Guinea.

3.1 Climatic preferences

The Ericaceae are well known for being plants predominantly of arctic to warm temperate conditions (e.g. see Sleumer 1941); in Africa, America and Malesia they occur mostly on mountains as they approach the equator. Throughout Malesia, few species habitually grow below 1000 m (see Stevens 1976, Figs. 2A, 3B–3D for *Rhododendron* in Malesia; van Steenis 1934). In New Guinea, *R. zoelleri* grows from sea level to 2000 m, and a few species, such as *Dimorphanthera tedentii* P. F. Stevens, have not yet been collected above 100 m. The family prefers well-watered conditions, the main exceptions being two warm temperate groups, the Arbuteae in California and the Ericoideae in S Africa.

3.2 Acid habitats

The combination of cool climate and high rainfall often results in the soils becoming acid, even over a base-rich rock (see pp. 118–120 in Whitmore 1975 for a discussion on the montane soils of Malesia), or in the development of peat bogs. The Ericaceae prefer such habitats; as Sleumer (1966, p. 470) notes, it is not clear whether Ericaceae reported as growing on limestone or other base-rich substrates were actually rooted in the rock or growing in a humus pocket. Epiphytic Ericaceae often grow in tree crotches where characteristically often very acid humus accumulates (Grubb and Edwards 1977 recorded a p.H. of

3.9–4.0 for the humus from a tree crotch at Marafunga, Papua New Guinea).

Many temperate Ericaceae are known to have a mycorrhizal association which is believed to help in obtaining nutrients in such base-poor environments (see also Leiser's 1968 report on the mucilaginous root sheath occurring in at least some Ericaceae). However, the study of nutrient cycling in the habitats that Ericaceae favor is only just beginning.

3.3 Heliophilic tendencies

The preferences of most Ericaceae for well-lighted habitats are well known (Camp 1942; Sleumer 1965, 1966). All New Guinea Ericaceae grow in rather open, often stunted forest, clearings, forest margins, or in open habitats such as bogs, grassland, and rock outcrops, or as epiphytes or lianes in the crowns of trees. Sleumer (1965) noted that if they grew in shady habitats they tended to be sterile.

3. 1–3 Combined:
The effects of climate, acidity of the habitat, and insolation are interconnected. Ericaceae are predominantly epiphytic in the crowns of trees at low altitudes in New Guinea, although they sometimes occur on acid sands with open vegetation (more frequently elsewhere in Malesia) or on rock outcrops, i.e., acid, well-insolated, but not notably cool, habitats. Thus in the Kiunga area (Western Province, Papua New Guinea), which is about 300 km from the sea and experiences very heavy rainfall, *Dimorphanthera intermedia, D. tedentii* and *Vaccinium* sp. are not infrequent epiphytes, although they are at less than 100 m elevation. In the same area are species of *Podocarpus*, *Lithocarpus* and *Calophyllum*, all either species of other basically montane genera or species which elsewhere grow at higher altitudes. This suggests that there are factors of the ecology, probably climatic, at Kiunga that favor such apparent altitudinal irregularities. At intermediate elevations the family occurs in similar habitats, and also in the often rather dwarfed and open vegetation on sharp ridges; they also invade road cuttings, grassland, etc. At still higher altitudes, subalpine shrub may be locally dominated by Ericaceae, and the vegetation is low and often partly open.

There is often infraspecific variation in habitat, a species that is epiphytic and/or lianoid (see below) at lower elevations being shrubby and terrestrial at higher elevations. On Mt Strong *Dimorphanthera amplifolia* (F. Mueller) P. F. Stevens was a very common shrub in Subalpine forest growing on peaty soil; at lower elevations it was an epiphyte. On Mt Suckling the variable *Rhododendron truncicola* grew in open vegetation in acid, sandy, riverine flats, on the forest floor in rather open vegetation on sharp ridges, and as an epiphyte in giant *Araucaria cunninghamii* trees which overtopped the closed forest below. Further examples are given by Sleumer (1966, 1967). In the New World tropics, Ericaceae sometimes grow as epiphytes in mangrove forest and also at low altitudes terrestrial on acid, sandy soil. Infraspecific variation in habitat comparable to that described above is reported from New World Vaccinieae (Camp 1945), and in *Rhododendron* section *Maddenia* from the Himalayas (Cant 1937).

Ericaceae throughout the world are generally small to large shrubs, or sometimes smallish trees, as in *Rhododendron*, *Agauria* and *Oxydendrum*. Herbs or near herbs occur in the predominantly temperate Monotropoideae, Pyroloideae and Wittsteinioideae. Growth patterns of New Guinea Ericaceae are dealt with in some detail here; those of the whole family are little known, although the preliminary findings of Temple (1975) indicate considerable diversity (Fig. 4 see Hallé *et al.* 1978 for definitions).

From admittedly casual observations, many, if not all, the species of *Rhododendron*, as well as *Gaultheria pullei* (frequently!), seem to belong to Leeuwenberg's model: there is no trunk, and overall construction is modular, the sympodial axes being terminated by inflorescences. This model has been reported for *R. aurigeranum* by Temple (1975), who notes that it, as well as Scarrone's model, which differs only in having a monopodial main axis, are common in the Ericaceae, intermediates also occurring.

Many other New Guinea Ericaceae are similar, except that the axes are determinate because of the death of the apex, e.g., *Gaultheria mundula*, *Vaccinium amblyandrum* var. *pungens*, *Diplycosia rupicola*, *Dimorphanthera microphylla* and *Agapetes vitis-idaea* (see also below). This growth pattern is probably common in erect Malesian members of these genera.

The lianoid New Guinea Ericaceae also have determinate axes that abort at the apex, but the axes cannot be supported by the lignification that they develop; the sprawling, vine habit is the result. These vines tend to bear their flowers along the leafless, older part of the stem. Plants with this type of growth are probably referrable to the ill-defined Champagnat's model, which has also been reported from a tropical American genus of the Vaccinieae, *Sphyrospermum* (Temple 1975) and probably occurs in *Gaultheria adenothrix* (Miq.) Maxim. (Fig. 4, H).

Mangenot's model, in which the main axes are mixed, the basal part being orthotropic and the apical part plagiotropic, occurs in a temperate shrubby species of *Vaccinium* (Hallé *et al.* 1978), as well as other species of *Vaccinium* and *Leucothoe* (pers. obs.); it should be looked for in lianoid or pendent New Guinea Ericaceae.

There are two complicating factors in the analysis of New Guinea and other Ericaceae in terms of architectural models. The first is that several species have subterranean rhizomes that turn up at the end and produce above ground shoots which, in the cases known, conform to one of the subtypes of Leeuwenberg's model as defined above; however, growth by rhizomes is accommodated neither in currently accepted models nor by reiteration as currently understood. Such rhizomes are known in *Rhododendron anagalliflorum*, *Vaccinium amblyandrum* especially var. *amblyandrum*, *V. crassiflorum*, *V. decumbens*, *V. oranjense*, *Gaultheria mundula* (probable), *Agapetes prostrata* P. F. Stevens, *A. rubrocalyx* and *A. vitis-idaea* from New Guinea, and *G. borneensis*, *G. pernettyoides*, *Diplycosia elliptica* and *D. kinabaluensis*, all from Borneo. They are probably more common, although our field knowledge is poor and collections so often lack the underground parts of the plant that Sleumer (1966, 1967) did not mention rhizomes in his descriptions. Many species layer, especially in moss forest, and ordinary shoots may then become difficult to distinguish from rhizomes, since all Malesian Ericaceae that I know have scales at the bases of the innovations. Short rhizomes occur elsewhere in *Rhododendron* and *Ledum*, and longer rhizomes are common, e.g. *Cassiope*, *Gaultheria*, *Gaylussacia*, *Lyonia*, *Vaccinium* and *Zenobia*.

The second problem is that a number of species may be erect shrubs, trailing shrubs, or vines, depending on the habitat, as anyone who has commonly collected Ericaceae soon realizes. This is also clear from Sleumer's descriptions of the species. Many species of *Agapetes*, *Dimorphanthera*, *Gaultheria* and *Vaccinium* have rather slender, albeit wiry, stems, and the transition to the vine habit given the right genotype or environment seems to be easy. Thus there may be species that are intermediate between 2 models, as well as infraspecific variation in model type. (Hallé *et al.* (1978) note similar examples in their discussions of Champagnat's model; since the final form of the model is not the result of active growth processes, this is to be expected here.)

Thus many Ericaceae, both temperate and tropical, do not lend themselves readily to an attempt at categorization in terms of models; this was also made clear by Temple (1974). It is perhaps more useful to note that New Guinea Ericaceae have determinate axes, and at least the majority have spirally arranged leaves. One can then consider the variation in habit and physiognomy seen and its possible ecological significance, bearing in mind that the particular aspects singled out for attention are part of almost a continuum of variation.

More or less parallel physiognomic series occur in *Rhododendron*, with terminal inflorescences, and the other genera, which, for the purposes of discussion, all have aborting stem apices (Fig. 4). In both there are phanerophytes with more or less steeply ascending branches, either shrubs or small trees. In rather open, Subalpine forest, forms with rather few, steeply ascending branches occur, as well as those with more numerous and widely spreading branches; the branches may be decurved as in *Vaccinium oreites* and *V. kostermansii*.

With increasing plagiotropy of the branches in *Rhododendron*, at least, dwarf, spreading forms may result; these chamaephytes (less than 30 cm tall) may grow as epiphytes, e.g. *R. caespitosum*, or sprawling over rocks, e.g. *R. gracilentum*. The angle at which the branches spread may to a greater or lesser extent be accentuated by environmental conditions. In the chamaephytic species of *Vaccinium* section *Oarianthe* and *Agapetes* subgenus *Paphia* the low stature of the above-ground parts is probably due to environmental pruning and small growth increments; the shoots themselves are usually strictly orthotropic (c.f. the tripartite division of chamaephytes in Raunkiaer 1934). Chamaephytes are not known in *Gaultheria* and *Diplycosia* in New Guinea.

Species of *Vaccinium*, *Gaultheria* and *Agapetes* with rhizomes grow in open, often grassy habitats at high altitudes; the rhizomatous, chamaephytic habit seems suited to such conditions. Thus on Mt Giluwe *V.* c.f. *amblyandrum* grows above the tree limit in tussock bogs; the rhizomes grow in and amongst the dense tussocks, and the exposed, leafy shoot is less than 3 cm long.

Rhododendron anagalliflorum is a plant of lower altitudes. *R. saxifragoides* is the only New Guinea member of the family that forms sometimes quite prominent tussocks in bogs or wet grassland. Growth increments are small, the leaves are steeply ascending, and old foliage persists along the stems. Its architecture, Leeuwenberg's model, is similar to that of other New Guinea species of *Rhododendron*, but its physiognomy and habitat are different. The tendency of the more erect growing species to flop over and layer is best shown in the mossy forest in which so many species grow; branches or fallen trunks may develop roots.

Vaccinium varingaefolium, from W Malesia, persists after a fire because shoots develop from the base (Sleumer 1967). *V. striicaule*, from New Guinea, may also persist after fire (Lane-Poole 1925, p. 179, as *V. blepharocalyx*), and there are *Vaccinium*-dominated shrubberies at high altitudes on the much fired Mt Giluwe. Similar behavior occurs also in the closely related Afro-American

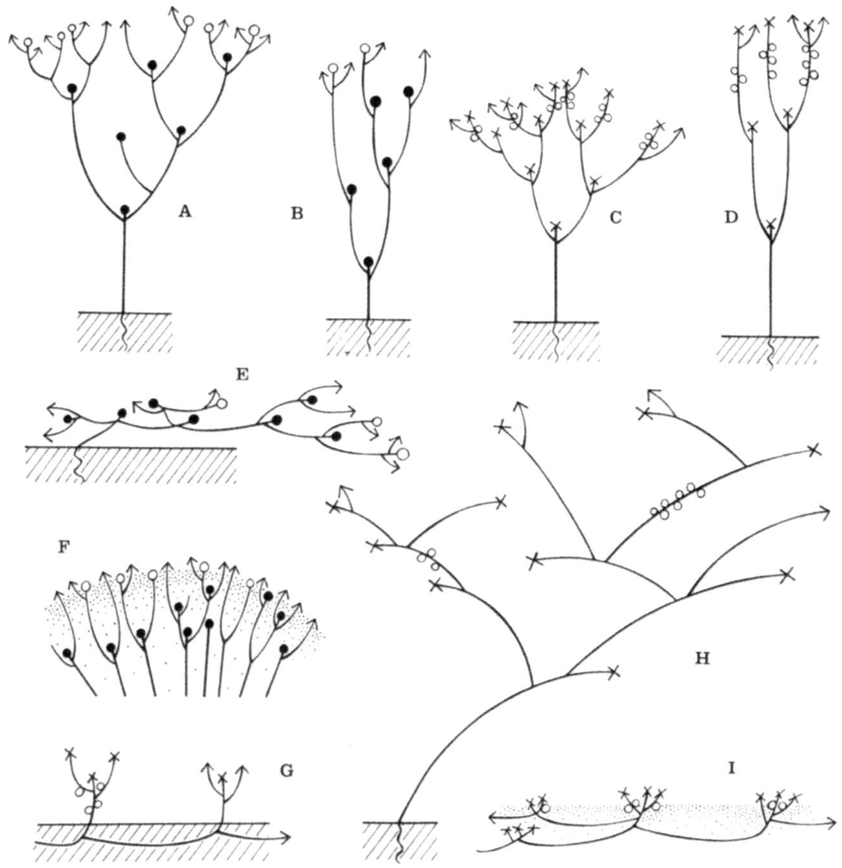

Fig. 4. Growth types in some New Guinea Ericaceae. A, *Rhododendron culminicola*; B, *R. wormersleyi*; C, *Vaccinium striicaule*; D, *Diplycosia rupicola*; *E*, *Rhododendron gracilentum*; F, *R. saxifragoides* — single plant; G, *Agapetes vitis-idaea*; H, *Dimorphanthera anchorifera*; I, *Vaccinium amblyandrum* var. *amblyandrum*. Arrows = growing innovations; x = aborted stem apices; ● = old (terminal) inflorescences; ○ = current inflorescences; diagonal hatching = ground; stippling = organic matter. N.B. Young stages before flowering omitted; those of *Vaccinium*, *Agapetes*, *Dimorphanthera*, and *Diplycosia* probably sympodial, those of *Rhododendron* probably monopodial.

pair of genera *Agauria* and *Agarista*. Roots, burls, swollen stems and associated roots, from which shoots may develop after fire damage to the above ground parts of the plant, occur in genera such as *Arctostaphylos* (Jepson 1916) and *Lyonia* (Judd 1981). The stem bases, and often the adjacent roots, of epiphytic species in all genera are quite frequently more or less swollen. This is common in epiphytic Ericaceae, and is known from tropical American Vaccinieae (Camp 1945), *Rhododendron* section *Maddenia* (Cant 1937), and *Agapetes* subgenus *Agapetes*. Sleumer (1967, p. 476) notes that the roots of *Vaccinium*, *Dimorphanthera*, *Agapetes* (and *Costera*) may be swollen. This swelling is perhaps associated with a swollen stem base.

Scandent forms occur in *Dimorphanthera*, *Agapetes* subgenus *Paphia* (common in both), *Vaccinium* (uncommon); some grow up to 30 m long, e.g. *A. stenantha*, and may be either epiphytic or terrestrial. Some species of *Gaultheria* and *Diplycosia* sprawl, rather than being properly scandent; *Rhododendron* shows the least tendency to be a vine.

It is clear that this whole subject would repay careful study. Differences in Ericaceae growing in cold temperate areas, many similar to those described above, show close correlations between method of growth and habitat (du Rietz 1931).

3.5 Leaf type

Variation in both leaf size and shape within New Guinea Ericaceae is considerable. Generally speaking, the leaves tend to be smaller and thicker, and sometimes with more prominent indumentum, at higher altitudes. Some species of *Rhododendron*, e.g. *R. purpureiflorum*, have narrowly elliptic to sublinear leaf blades, but none has linear leaves like those of some West Malesian species in *Rhododendron* subsection *Pseudovireya* and subsection *Euvireya* series *Stenophylla* and *Taxifolia*. In neither *Gaultheria* nor *Diplycosia* is there great variation in lamina shape; that within *Vaccinium* is moderate, and within *Dimorphanthera* and *Agapetes* subgenus *Paphia* rather little.

The Ericaceae as a whole is noted for diversity in leaf shape, and this sometimes occurs within a genus, as within West Malesian *Diplycosia* and *Rhododendron*, and also the temperate to arctic genus *Cassiope* and the tropical American montane genus *Agarista* (Stevens 1970). Infra- and intergeneric variation in leaf shape in the Vaccinieae is generally less than in other tribes of the Vaccinioideae and Rhododendroideae. New Guinea Vaccineae are as variable as other tropical representatives, but overall, New Guinea Ericaceae are not notably variable in leaf shape. As might be expected, marginal serrations are usually inconspicuous.

The leaves in *Rhododendron* section *Vireya* are scattered or arranged in false verticils. Although prostrate species of *Rhododendron* in New Guinea apparently always have pseudoverticillate leaves, erect species may have scattered to pseudoverticillate leaves. Some species of *Agapetes* subgenus *Agapetes* and the related *Vaccinium* section *Epigynium*, both commonly epiphytic groups, also have pseudoyerticillate leaves, and there is a common tendency in the family for the expanded leaves to be borne mainly in the uppermost part of an innovation.

Another variable of possible ecological significance is the angle at which the leaves are held. Thus, *R. womersleyi* has small, spreading leaves, while the larger leaves of *R. atropurpureum* are often strongly ascending; both species grow in well-lighted conditions. There is comparable variation in the angles at which the leaves are borne in other genera.

3.6 Corolla and stamen

In general stamen type and development, in the occurrence of pollen tetrads in all genera, and in the ubiquitous occurrence of sympetalous corollas, New Guinea Ericaceae agree with the rest of the family, both tropical and temperate. In *Rhododendron*, *Gaultheria*, and *Diplycosia* at least the pedicel is erect in fruit, whatever its position in flower, a change common elsewhere in the family. New Guinea species of *Rhododendron* have viscin threads, like the majority of the Rhododendroideae. However, some details of the variation in corolla and stamen are important in understanding the evolution of the family in New Guinea.

Species of *Gaultheria*, *Diplycosia* and *Vaccinium* have small (usually less than 1 cm long) urceolate to campanulate, rarely tubular, white to red corollas and are more or less similar in this to other members of the genus both elsewhere in Malesia and to many other genera throughout the world. In Java pollination by short-tongued insects or self pollination seems likely for some species of *Gaultheria* and *Vaccinium* (Doctors van Leeuwen 1933).

Most members of *Dimorphanthera*, *Agapetes* subgenus *Paphia* and *Rhododendron* have large flowers more than 1 cm long and tubular to infundibular in shape; corolla colors are red, white or yellow, and there is no asymmetry in the color patterning that occurs in a few of these species. Details of the variation in the Papuasian species of *Rhododendron* has been dealt with in some detail (Stevens 1976). The long, tubular, often bright red, thick-walled corolla so common in *Dimorphanthera* section *Trochilanthe*, *Agapetes* subgenus *Paphia*, and *Rhododendron* (Stevens 1976, Type 1D) seems to be associated with bird pollination (Stevens 1976, 1977), and is known from West Malesia, the Himalayas (*Agapetes* subgenus *Agapetes*), Africa (*Erica*) and the Americas (commonly in the Vaccinieae, but also some members of the polypetalous *Bejaria*, of the Rhododendroideae, where the edges of the broad petals overlap). The diversity in details of shape and coloring of the New Guinea Vaccinieae with this corolla type is less than that of their tropical American relatives, and the variation in mainland Asian representatives of *Agapetes* subgenus *Agapetes* is also considerable. Long, tubular, red corollas occur in some W Malesian species of *Rhododendron*, but they are very uncommon in the Vaccinieae there. Another ornithophilous flower type, with a curved tube and the anthers and stigma positioned on the adaxial side of the tube (Type 1A) is known only in New Guinea, where it occurs in about 38 species of Rhododendron; it seems to be unique in the family.

The large, white, campanulate-infundibular, very sweet-scented flowers found in 10 New Guinea species of *Rhododendron* (Type 2B) seem to be restricted to E Malesia (4 species with a flower type like, or similar to, this type occur in the Celebes-Moluccas region); it is possible that such flowers are pollinated by bats. I do not know of similar flowers elsewhere in the family.

The predominantly white, scented, long-tubular flowers (Type 2A) so common in *Rhododendron* subsections *Solenovireya* and *Siphonovireya* that are probably pollinated by sphingid moths also occur in species of the genus in W Malesia, albeit less frequently. S African species of *Erica* and North American species of *Rhododendron* have a similar floral syndrome, but it is as yet unknown from other tropical areas.

The large, red to orange, campanulate corolla that occurs in *Rhododendron* (Types 1G and 3A) and *Dimorphanthera* section *Dimorphanthera* occur in the former genus throughout Malesia, although the strictly red, scentless flowers occurring in the latter are particularly common in New Guinea. Flowers of this general type are uncommon elsewhere in the family, although they may occur in some species of *Rhododendron* from the hills of SE Asia. The flowers of *Dimorphanthera* section *Pachyantha*, which vary from whitish to red in color and have very thick walls, are also scentless and campanulate, although they are less than 1 cm long. There is circumstantial evidence that both sections of *Dimorphanthera* are visited by birds (Stevens 1976, 1977). Other flower types in both New Guinea and W Malesia are found in only a very few species (Stevens 1976). Nothing is known of their function in pollination; some of these types seem restricted to one of the two areas.

In general terms, much of the variation in corolla characters in New Guinea Ericaceae, especially *Dimorphanthera*, *Agapetes* subgenus *Paphia* and *Rhododendron* seems associated with pollination by large animals. Almost twice as many species with corollas which appear adapted to such pollination occur in New Guinea as in the rest of Malesia.

The stamens of *Agapetes* subgenus *Paphia* and *Dimorphanthera* are rather different to those of Malesian *Vaccinium* and other strictly temperate Vaccinieae, being larger, and in *Dimorphanthera* much more woody and the two whorls more obviously dimorphic. However, the stamens of *Dimorphanthera* are similar to those of the probably related American genus *Satyria*, and the connate filaments that occur in some species of *Agapetes* subgenus *Paphia* are quite common in some American genera and also rarely occur in the not immediately related *Agapetes* subgenus *Agapetes*. A more detailed analysis of the variation of floral (and other) characters in the Vaccinieae confirms this general reticulate pattern of variation (Stevens ms).

3.7 Fruit and seed

Although the general fruit types of the New Guinea Ericaceae are well known, relatively little is known about seed structure, especially in the Vaccinioideae, and practically nothing is known of the details of dispersal.

All genera, except for *Rhododendron*, have colored, fleshy fruits, which are similar to fruits of both tropical and temperate representatives. The seeds, so far as is known, have a single layer of cells of which the outer periclinal walls (and sometimes also the inner) are unthickened, while the rest are thickened, again a common pattern in seeds of fleshy-fruited Ericaceae. Although endozoic transport, probably by birds, of the seeds is presumed, there is little direct evidence that this is so. *Gaultheria* and *Diplycosia* also have capsules, so there may be some local wind dispersal as well. A rather interesting seed type, with testa cells with longitudinally reticulate/scalariform thickening, which become mucilaginous on wetting and an often green embryo, occurs in a number of Vaccinieae from tropical America and mainland SE Asia; it has not been observed in any of the 35 New Guinea species of *Vaccinium*, as well as a few species of *Agapetes* and *Dimorphanthera* examined, although some species have green embryos.

Rhododendron has a capsular fruit and small seeds with long tails. The tails of the seeds are made up of cells similar in shape to those of the rest of the seed. Although such seeds are an extreme type in the genus and family, this is a matter of degree only, and there are other types of winged or tailed seeds. These small, tailed seeds would seem suitable for the epiphytic habitat of many of the species, and also may help dispersal into the open, geologically transient habitats that the genus also favors (see below).

3.8 Chemical constituents

Unfortunately little is known about the chemical constituents of tropical Ericaceae. Methyl salicylate is known from species of both *Gaultheria* and *Diplycosia* in New Guinea (see also Sleumer 1967), but it also occurs in several temperate members of the family (both immediately related, as well as in the Pyroloideae). Casual observation suggests that the Ericaceae are not much eaten, although in the Subalpine region on Mt Strong, Central Division, extensive damage to the leaves of a large population of *Rhododendron culminicola*, probably caused by the weevil *Gymnopholus* (*Symbiopholus*) *symbioticus* Gressitt, was observed (Coode and Stevens 1972). *Rhododendron macgregorii* is toxic to small mammals, and can be used to kill rodents (Henty and Pritchard 1975).

The temperate members of the family are noted for an accumulation of polyphenols and tannins, and sometimes other more specific toxins as well (see Hegenauer 1966 for references). McKey *et al.* (1978) noted that tannins might be implicated in protection against herbivores of vegetation in forest in Africa growing in poor, rather acid conditions, so knowledge of the chemical spectrum of the vegetation types in which the Ericaceae are so prominent would be of more than passing interest (see also Janzen 1974). Tannins may also help to generate the soil conditions that the family likes.

3.9 Other characters of New Guinea Ericaceae

Rhododendron subsection *Phaeovireya* has distinctive scales (Seithe 1960), but the scales throughout section *Vireya* are of the same general type that occurs elsewhere in the genus. The small, glandular hairs of many of the New Guinea Vaccinieae are also similar to those of other Vaccinieae throughout the tropics; the indumentum of New Guinea species of *Diplycosia* and *Gaultheria* is not remarkable.

Anatomical details as far as known also suggest no uniquely New Guinean features. *Vaccinium*, as noted above, is similar to other Malesian representatives of the genus, while *Dimorphanthera* and *Agapetes* subgenus *Paphia* have a number of characteristics in common with American genera. However, in general Ericaceae in tropical (to subtropical) montane areas are more likely to have a single or multiple hypodermis as well as some sort of lignification of the epidermis and/or hypodermis and/or mesophyll. The relatively large vessels and vein endings of *Agapetes* subgenus *Paphia* and *Dimorphanthera* and their American relatives occur in commonly epiphytic groups, however, these characteristics are neither restricted to New Guinea taxa in the Ericaceae, nor, even when considering New Guinea, are they the only anatomical syndrome compatible with the epiphytic habitat. Anatomical characteristics of *Rhododendron* and *Diplycosia* in New Guinea, although distinctive, are found in all Malesian representatives of these taxa as well as in other taxa elsewhere.

4. General discussion

Too little is known about the relationships of New Guinea Ericaceae to make more than suggestions as to their history and origins, and there is little fossil evidence for the whole of Malesia. Also, some of the statements made below are based on questionable assumptions.

Most authors agree over the outline of the plate tectonic history of the New Guinea region (Raven and Axelrod 1972, 1974; Doutch 1972; Nix and Kalma 1972; Walker 1972; and van Balgooy 1976; together with papers cited therein). The Australian plate, New Guinea not yet being recognizable, moved north in the early Tertiary and emergent land on the northern part of the plate entered the tropics somewhat before the Miocene, 25 million years ago. Uplift began as the Australian plate approached SE Asia. Much of New Guinea finally became emergent only in the Pliocene; tectonism probably began in the middle to late Miocene, but mountain uplift may not have been substantial much before the Pliocene, 5 million years ago. By this latter time New Guinea seems to have reached substantially its present position. Mountain uplift and vulcanism are still very active in parts of New Guinea, and Quaternary climatic changes have led to considerable vertical movements in the position of the Upper Montane and Subalpine vegetation which the Ericaceae favor. Despite changes in the sea level, the Moluccas, some islands in Geelvink Bay, and the Bismarck Archipelago remained separate from the New Guinea mainland.

It is unlikely that *Rhododendron*, *Vaccinium*, and at least most of *Gaultheria*, were on both halves of Malesia in the earlier Tertiary. The distinctions between the subsections and series of *Rhododendron* are slight, being based mostly on differences in leaf shape and size, details of scales, and corolla type. It has been known since the days when Malesian rhododendrons were popular as stove plants in the last century that it is easy to cross even the most dissimilar-looking plants and to obtain fertile hybrids (Henslow 1891 — W Malesian species), and this has been abundantly confirmed since then (e.g. Sleumer 1966 (esp. pp. 476, 478), 1973; Stevens 1974 — New Guinea species; E and W Malesian crosses — J. J. Rouse, pers. comm.). Chromosome counts from Malesian species are diploid, 2n=26, except for a single triploid (Jones and Brighton 1972); the species counted came from several subsections and various altitudes. Indumentum and many floral characters in *Rhododendron* section *Vireya* are derived. Although many of the species of *Rhododendron* subgenus *Rhododendron*, to which section *Vireya* belongs, are polyploid (Janaki-Ammal *et al.* 1950), there are also diploids, and *Rhododendron* section *Vireya* is much related to other groups in montane SE Asia.

Rhododendron is clearly very diverse on the SE Asian mainland. The diversity of section *Vireya* in W Malesia is greater than in E Malesia in all characteristics except, perhaps, floral types and habit. Despite the abundance and diversity of *Rhododendron* in New Guinea, only three species occur south and east of the area, although there is no obvious ecogeographic reasons for such an abrupt cut-off (see also van Steenis 1934). Hence it seems reasonable to postulate movement from west to east through Malesia. In Malesian *Vaccinium* as well, the distinctions between the sections are slight, and all sections have substantial similarity in derived floral and anatomical characteristics. Little is known of their cytology and ability to hybridize, although on Mt Wilhelm, Papua New Guinea, *V. amblyandrum* (section *Oarianthe*) and *V. cruentum* (section *Bracteata*) hybridize. Relatives of the Malesian species occur in SE Asia; there is as yet no compelling evidence of a transpacific relationship (c.f. Sleumer 1941).

The diversity of the relatives of the Malesian species of *Vaccinium* is greatest on the SE Asian mainland. Much of the variation there is probably of a derived nature, e.g. inflorescence and flower type, indumentum, leaf arrangement, seed type (Stevens ms., c.f. Sleumer 1941), but there are also less obviously specialized groups. Again, there are few species occurring to the south and east of New Guinea. It is suggested that *Vaccinium* also entered New Guinea from the west.

The history of *Dimorphanthera* and *Agapetes* subgenus *Paphia* in Malesia is unclear. Both have representatives in W Malesia (see above), and they are probably closer to the W Malesian genus *Costera* and to tropical American genera than to Malesian *Vaccinium*. Their history in Malesia may be quite different than that suggested for the above genera, and their immediate ancestors outside Malesia may have been adapted to the tropical montane habitat. (If, as Maguire *et al.* (1978) believe, *Dimorphanthera* and *Satyria* are not close, the problem may be of a different order, but with yet more examples of convergent evolution in characters of anatomy and stamen).

Diplycosia is probably a genus that has evolved wholly in the Old World montane tropics and has also entered New Guinea from the west. As mentioned above, the diversity of the genus is clearly greatest in W Malesia, especially Borneo. In addition, some of the species of *Gaultheria* in West Malesia, notably the curious *G. pernettyoides*, have both the facies and anatomy of *Diplycosia*, although differing in stamen type.

I suggest that diversification of the Ericaceae in New Guinea has occurred since late Miocene at the earliest, some 10 million years ago, and has been dependent on the arrival of probably only a few ancestors, mostly from W Malesia (note that the little evidence we have — Muller 1972 — suggests that in NW Borneo Ericaceae occurred more frequently in Pliocene than in older sediments). The family in New Guinea has always been predominantly montane, since there the conditions that the family favors are most widespread; the occurrence of species in the lowland is a secondary phenomenon. The isolation, youthful topography, and ecological diversity of the abundant montane habitats of New Guinea would seem ideal for speciation. The plasticity shown by many species in their growth form, their frequent ability to grow both as epiphytes and on the ground, and the probable dispersal of their seeds by wind and animals which, on occasion, may result in dispersal over quite considerable distances, might all be expected to give a species rather greater chance of survival than the transient nature of the habitat in which the individual grows might suggest.

Although little is known about the selective value of much of the variation discussed above, it seems clear that much of the variation in form of the plant, coupled with variation in leaf size, has allowed species in *Rhododendron* and *Vaccinium* section *Orianthe* at least to exploit particular micro-habitats on the mountains at the upper limit of woody plant growth; species with these characteristics are concentrated in the Snow Mountains of Irian Jaya, where conditions suitable for their growth are most common. The concentration of species of *Agapetes* subgenus *Paphia* on the mountains of SE New Guinea is accompanied by some similar variation; no other genus in New Guinea has a similar distribution pattern.

Another important factor in the diversification of *Rhododendron* in particular has been variation in flower type, which, coupled with geographical and other aspects of ecological isolation, may have aided speciation. However, there are no obvious internal barriers to hybridization in *Rhododendron*, and the variety of hybrids found in disturbed habitats in the wild suggests that any ethological isolating mechanisms can be readily broken down (Stevens 1974). The adoption of ornithophily seems to have been a factor in the evolution of the Ericaceae in the cool, cloudy conditions of the Upper Montane and Subalpine forest (Stevens 1976), although it should not be forgotten that several species of *Vaccinium*, usually with small, urceolate corollas, also grow there. Clearly other factors must be involved; for example, we know little about how much self pollination and inbreeding takes place or about details of flowering times in different species (for possible differences, see Kores 1978).

It is obvious that the characters discussed above that occur in the New Guinea Ericaceae are paralleled in a general or even quite specific way by Ericaceae elsewhere in the tropics and often in the temperate zone as well. A few characters, like superficial phellogen, may have evolved only once in the family, but this character is found in numerous species through the SE Asian tropics and subtropics. Some of the characters more particularly associated with the tropics, and in part at least involved in the diversification of the family in New Guinea, such as the development of certain flower types, the adoption of the epiphytic habitat, the vine growth form, and possibly some anatomical characteristics, have also clearly evolved repeatedly in the family; it is hard to imagine the monophyletic origin even of the distinctive zygomorphic ornithophilous flower type so common in New Guinea rhododendrons. The possibly bat-pollinated and the zygomorphous bird-pollinated flower types, and the deeply laciniate scales of *Rhododendron* subsection *Phaeovireya*, are the only major characters or character complexes which probably evolved in the New Guinea Ericaceae and nowhere else.

Many of the supraspecific taxa recognized in the Malesian Ericaceae reach their eastern or western limits somewhere between Borneo and New Guinea, although both *Dimorphanthera* section *Dimorphanthera* and *Vaccinium* section *Orianthe* extend to the Philippines (Figs. 2 and 3). That the area between the New Guinea mainland and Borneo and Java is one in which there is considerable quantitative and qualitative change in many elements of the flora and fauna has long been known (e.g. Wallace 1869; Raven and Axelrod 1972). These distribution patterns in the Ericaceae are probably due to factors like the isolation of the two main land masses in Malesia and possible biotic differences on these land masses, which in turn may reflect events associated with plate movements. Biotic differences seem to be an immediate possibility only when considering some pollination syndromes. Even here, although many New Guinea Ericaceae are probably pollinated by the overwhelmingly Australasian group, the Meliphagidae, their floral syndromes do not seem unsuited for pollination by the Nectarindae, which are more common in W Malesia. The two examples of a New Guinea-Philippines distribution (*Dimorphanthera* section *Dimorphanthera* and *Vaccinium* section *Oarianthe*) represent examples of a fairly frequent distribution type which cannot be satisfactorily explained, unless it represents migration along the New Guinea and Philippines montane tracks (van Steenis 1934).

Acknowledgement

I am grateful to M. J. Donoghue and W. S. Judd for their helpful comments on a draft of the manuscript, and to K. Stoutsenberger for drawing the illustrations. The base maps are from Goode's series, copyrighted by the University of Chicago.

References

Balgooy, M. M. J. van. 1976. Phytogeography. pp. 1–22 in Paijmans, K. (ed.) New Guinea vegetation. xvii + 212 pp. Elsevier, New York.

Camp, W. H. 1942. On the structure of populations in the genus Vaccinium. Brittonia 4: 189–204.

—— 1945. The North American blueberries, with notes on other groups of Vacciniaceae. Brittonia 5: 203–275.

Cant, C. M. 1937. Stem structure in the Maddenii series of Rhododendron. Trans. and Proc. Bot. Soc. Edinburgh 32: 287–290.

Coode, M. J. E. and P. F. Stevens. 1972. Notes on the flora of two Papuan mountains. Papua New Guinea Sci. Soc. Proc. 1971. 23: 18–25.

Cullen, J. and D. F. Chamberlain. 1978. A preliminary synopsis of the genus Rhododendron. Notes Roy. Bot. Gard. Edinburgh 36: 105–126.
Docters van Leeuwen, W. M. 1933. Biology of plants and animals occurring in the higher parts of Mt Pangrango-Gedeh in West Java. Verh. Kon. Ned. Akad. Wetensch., Afd. Natuurk., Tweede Sect. 31: 1–278.
Doutch, H. F. 1972. The paleogeography of Northern Australia and New Guinea and its relevance to the Torres Strait area. In Walker, D. (ed.) Bridge and barrier: the natural and cultural history of Torres Strait. xxii + 437 pp. Australian National University, Research School of Pacific Studies, Dept. Biogeography and Geomorphology, Publ. BG/3(1972).
Grubb, P. J. and P. J. Edwards. 1977. Studies of mineral cycling in a montane rain forest in New Guinea. 1. The distribution of organic matter in the soil. J. Ecol. 65: 943–969.
Hallé, F., R. A. A. Oldeman and P. B. Tomlinson. 1978. Tropical trees and forests. xvii + 441 p. Springer Verlag, New York.
Hegenauer, R. 1966. Chemotaxonomie der pflanzen, vol. 4. 551 pp. Birkhaüser, Basel and Stuttgart.
Henslow, G. 1891. Hybrid rhododendrons. J. Roy. Hort. Soc. 13: 240–283.
Henty, E. E. and G. H. Pritchard. 1975. Weeds of New Guinea and their control. Ed. 2. pp. 180. Botany Bulletin 7, Dept. of Forests, Lae, Papua New Guinea.
Janaki-Ammal, E. K., I. C. Enoch and M. Bridgwater, 1950. Chromosome numbers in species of Rhododendron. Rhododendron Year Book, 1950. 78–91.
Janzen, D. 1974. Tropical blackwater rivers, animals, and mast fruiting by the Dipterocarpaceae. Biotropica 6: 69–103.
Jepson, W. L. 1916. Regeneration in manzanita. Madroño 1: 3–11.
Jones, K. and C. Brighton. 1972. Chromosome numbers of tropical rhododendrons. Kew Bull. 26: 559–561.
Judd, W. S. 1981. A monograph of Lyonia. J. Arnold Arb. 62, in press.
Kalkman, C. and W. Vink. 1970. Botanical exploration in the Doma Peaks Peaks region, New Guinea. Blumea 18: 87–135.
Kores, P. 1978. Hunting for New Guinea's elusive rhododendrons. Quart. Bull. Amer. Rhododendron Soc. 32: 12–17.
Lane-Poole, C. E. 1925. Forest resources of the territory of Papua New Guinea. ii + 209 pp. Government Printer, Victoria.
Leiser, T. 1968. A mucilaginous root sheath in Ericaceae. Amer. J. Bot. 55: 391–398.
McKey, D., P. G. Waterman, J. S. Gartlan and T. T. Struhsaker. 1978. Phenolic content of vegetation in two African rain forests: ecological implications. Science (New York) 202: 61–64.
Maguire, B., J. A. Steyermark and J. L. Luteyn. Ericaceae. In: The botany of the Guyana highland. Part 1. Mem. New York Bot. Gard. 29: 139–203.
Muller, J. 1972. Palynological evidence for change in geomorphology, climate and vegetation in the Mio-Pliocene of Malesia. In: Ashton, P. and M. Ashton (eds) The Quarternary era in Malesia. 122 pp. Dept. Geography, University of Hull, and Institute for South-East Asian Biology, University of Aberdeen.
Nix, H. A. and J. D. Kalma. 1972. Climate as a dominant control in the biogeography of Northern Australia and New Guinea. In: Walker, D. (ed.) Bridge and barrier: the natural and cultural history of Torres Strait. xii + 437 pp. Australian National University, Research School of Pacific Studies, Dept. Biogeography and Geomorphology, Publ. BG/3(1972).
Raunkiaer, C. 1934. The life form of plants and statistical plant geography. xvi + 632 pp. Clarendon, Oxford.
Raven, P. H. and D. I. Axelrod. 1972. Plate tectonics and Australasian paleobiogeography. Science (New York) 176: 1379–1386.
—— 1974. Angiosperm biogeography and past continental movements. Ann. Missouri Bot. Gard. 61: 539–673.
Rietz, G. E. du 1931. Life forms of terrestrial flowering plants. Acta Phytogeogr. Suec. 3: 1–95.
Sleumer, H. 1941. Vaccinioideen-Studien. Bot. Jahrb. Syst. 71: 375–510.
—— 1949. Ein System der Gattung Rhododendron L. Bot. Jahrb. Syst. 74: 511–553.
—— 1965. The role of Ericaceae in tropical montane and subalpine forest vegetation of Malaysia. In: Symposium on ecological research in humid tropics. 376 pp. Government of Sarawak and UNESCO.
—— 1966. Ericaceae (part). Fl. Males. 1(6)4: 469–668.
—— 1967. Ericaceae (part). Fl. Males. 1(6)5: 669–914.
—— 1973. New species and noteworthy records of Rhododendron in Malesia (Ericaceae). Blunea 21: 357–376.
Steenis, C. G. G. J. van. 1934. On the origin of the Malaysian mountain flora. Part 1. Bull. Jard. Bot. Buitenzorg Sér. 3. 13: 135–262.

Steenis C. G. G. J. van. 1978. On the doubtful virtue of splitting families. Bothalia 12: 425–427.

Stevens, P. F. 1970. Calluna, Cassiope and Harrimanella: a taxonomic and evolutionary problem. New Phytol. 69: 1133–1150.

—— 1972. Notes on the infrageneric classification of Agapetes, with four new taxa from New Guinea. Notes Roy. Bot. Gard. Edinburgh 32: 13–28.

—— 1974a. The hybridization and geographic variation of Rhododendron atropurpureum and R. womersleyi. Papua New Guinea Sci. Soc. Proc. 1973 25: 73–84.

——1974b. Circumscription and relationships of Dimorphanthera (Ericaceae), with notes on some Papuasian species. Contr. Herb. Austral. 8: 1–34.

—— 1976. The altitudinal and geographical distribution of flower types in Rhododendron section Vireya, especially in the Papuasian species, and their significance. J. Linn. Soc. Bot. 72: 1–33.

—— 1977. Additional notes on Dimorphanthera (Ericaceae). J. Arnold Arb. 58: 437–444.

Temple, A. 1975. Ericaceae: étude architecturale de quelques éspèces. Thesis, Académie de Montpellier, Université des Sciences et Techniques du Languedoc.

Walker, D. 1972. Bridge and barrier. In: Walker, D. (ed.) Bridge and barrier: the natural and cultural history of Torres Strait. xxii + 437 pp. Australian National Univ., Research School of Pacific Studies, Dept. Biogeography and Geomorphology, Publ. BG/3(1972).

Wallace, A. R. 1869. The Malay archipelago. xii + 638 pp. Harper Brothers, New York.

Whitmore, T. C. 1975. Tropical rain forests of the far east. xii + 282 pp. Clarendon, Oxford.

Woods, P. J. B. 1978. Two Bornean rhododendrons. Notes Roy. Bot. Gard. Edinburgh. 37: 157–159.

Note

A specimen of *Dimorphantera, LAE 59446*, was recently collected in the Star Mountains. It apparently represents an undescribed species, and is particularly notable in that the filaments are connate into a tube, i.e., it has the key character of the tropical American genus *Satyria*. This is another example of convergent evolution in this character in the Vaccinieae.

5. The *Nothofagus* Blume (Fagaceae) of New Guinea

J. Ash

1. Introduction

Interest in the *Nothofagus* Blume (Fagaceae) species of New Guinea arises not only from their prominence as canopy trees in the highland forests but also from the distribution of the genus in both tropical and temperate lands around the S Pacific basin. With the accumulation of observations relating to the movements of continents bordering the Pacific Ocean much of the enigma has vanished from the gross biogeography of *Nothofagus* but many details remain uncertain.

In considering the range of ecological conditions occupied by *Nothofagus* in New Guinea it is of interest to place these within the broader context of other tropical and temperate localities. This comparative material indicates both the adaptations to different environments and patterns of specialization and speciation. Comparison of the New Guinea *Nothofagus* with other tropical trees is hindered by their variety and the lack of detailed observations.

Details of the taxonomy, authority, distribution and origins of the New Guinea *Nothofagus* are examined by Steenis (1953, 1971a, 1971b, 1975), Balgooy (1966) and Soepadmo (1972), and their ecology is discussed by Steenis (1971a), Hynes (1973) and Paijmans (1976). This chapter reviews published information and presents some observations from Mt Giluwe and Lake Kutubu in the S Highlands of Papua New Guinea.

2. Biogeography

The genus *Nothofagus* is distributed in southern S America, New Zealand, New Caledonia and the Australian region, including Tasmania, Eastern Australia, New Guinea and New Britain. The dispersal of *Nothofagus* by seed is said to be limited to a few km (Preest 1963) and it may be necessary to postulate adjacent land masses and exposed continental shelves to explain the distribution (but see chapter VI: 2). Reconstructions of continental drift (Griffiths and Varne 1972), Pleistocene sea level fluctuations (Chappell 1974), and records of *Nothofagus* fossils (Steenis 1971a), indicate that *Nothofagus* was probably widespread in Cretaceous Gondwanaland and was subsequently separated on isolated land masses by ocean floor spreading and formation of shallow seas. The gradual

Monographiae Biologicae, Vol. 42, ed. by J. L. Gressitt.

© *1982, Dr W. Junk Publishers, The Hague.*

isolation of New Guinea and the distribution of various sections, subsections and pollen types of *Nothofagus* are indicated in Fig. 1.

The deciduous section is recorded in both S America and Tasmania but all the species of warmer climates are evergreen. The subsection *Bipartitae* occurs in New Guinea and New Caledonia and is the only subsection of *Nothofagus* in these regions.

The pollen of *Nothofagus* is of three types, *brassii, fusca*, and *menziesii* but these types do not coincide with the subsections of the genus. Fossil pollen of all three types has been found in Australia, New Zealand, New Guinea, S America and Antarctica, but there are now no species producing *brassii* pollen in either New Zealand or Australia. It is probable that the abundant *brassii* type pollen from Tertiary Australia was produced by species of the *Bipartitae* which are now represented by the *Nothofagus* species of New Guinea. The identity of the *brassii* pollen from other regions is less certain.

A reconstruction of the history of the *Bipartitae* indicates that this subsection was distributed at least in the Australian and New Caledonian regions, and probably New Zealand, prior to their separation 50–80 million years (m.y.) BP. At this time the northern regions of Australia were about 30° S (McElhinny 1970) with a climate possibly a little warmer than at present. About 45–50 m.y. BP the Australian region separated from Antarctica and began to move towards the Equator. During the early and middle Tertiary the climate of Australia was probably wetter. Fossil pollen indicates widespread forests of *Nothofagus*, including several *brassii* types (Cookson and Pike 1955), and various gymnosperms including *Agathis* Salisb., *Araucaria* Juss., *Phyllocladus* Rich., *Podocarpus* L'Hérit, and *Microcachrydites* cf. Hook f. (Gill and Ingle 1975). The absence of *Nothofagus* macrofossils may indicate that the genus was absent from the swamp forests but abundant on drier ground. During the later Tertiary period Australia became increasingly arid and by the late Pliocene the *Nothofagus*–conifer forest had been replaced by *Eucalyptus* L'Herit and *Acacia* Mill. vegetation over much of the continent. *Brassii* pollen is recorded by Khan (1974) from marine deposits of the upper Miocene in the Papuan Gulf region which was then probably about 15° S. It is likely that *Nothofagus* was established in the New Guinea region, with a lowland tropical climate, prior to the main period of uplift during the Pliocene. Nix and Kalma (1972) use a growth index model derived from present and estimated Pleistocene climates to show that conditions were unsuitable for the migration of rainforest vegetation across the Torres Strait-Arafura Sea during the Pleistocene period with low sea level.

A number of genera share at least part of their biogeographic distribution with *Nothofagus*, e.g., *Araucaria, Agathis, Dacrydium* Soland., *Dacrycarpus, Papuacedrus* Li, *Libocedrus* Endler, *Phyllocladus, Podocarpus, Weinmannia* L. (Balgooy 1966; Couper 1960). In addition to these tree genera there are ferns, mosses, fungi and a diverse fauna (Gill and Ingle 1975; Gressitt 1974). Of particular interest are species and genera of herbivores and parasites specific to *Nothofagus*, e.g., *Sensoriaphis* (Aphididae) (Main 1974; Schlinger 1974).

At some period during the late Tertiary it is postulated that the New Guinea region was in close contact with SE Asia. Many families and genera with Asian

Isolation of New Guinea		Nothofagus taxonomy									
Period of separation	Present Region	No. of species[a]	Section: (Hombr. and Jacq.)	*Calucechinus* (Deciduous)		*Calusparassus* (Evergreen)			Pollen Type[b]		
100 m.y. BP 0			sub-section: (Steen)	*Antarctic*	*Pumil.*	*Quad.*	*Tri.*	*Bi.*	*Brassii*	*Menziesii*	*Fusca*
LAURASIA ———— — — —	SE Asia	0									
GONDWANALAND . .	New Britain	1 } 13						1	1		
. . . .	New Guinea	13 }						13	13		
. . . . .	Australia	2 } 3				2			+	2	+
— — ———	Tasmania	2 }		1		1				2	
— — — ————	Antarctica	3							+	+	+
— — — ————	New Caledonia	5						5	5		
— — ————————	New Zealand	4				3	3		+	3	1
——————————	South America	11		7	1	3			1	2	4
		$\Sigma = 37$		8	1	9	1	19	> 19	> 8	> 5

Key: ———— Oceanic Separation
— — — Oceanic Separation doubtful
. . . . Shallow Sea Separation

[a] Numbers = extant species
[b] +'s = tertiary fossils

Fig. 1. Isolation of New Guinea and Nothofagus Taxonomy.

affinities are now established in New Guinea and form a major part of the forests, especially at lower altitudes. These species may have excluded Gondwanic genera such as *Nothofagus* from the lowlands.

The *Nothofagus* of New Guinea is generally associated with the Lower Montane Forests Zone (Paijmans 1976) but varies in abundance both locally and regionally. In the central ranges the proportion of extant Lower Montane forest comprising *Nothofagus* is 32% in the Wabag-Tararea (Long. 142°–144°) (Saunders 1965), 16% in the Goroka–Mount Hagen area (Long. 144°–146°) (Saunders 1970), and 14% in the E Papua region (S of Lat. 8°) (Saunders 1973). In the northern ranges of the Ramu-Madang area *Nothofagus* comprises only 5% of the Lower Montane Forest (Saunders 1976).

The presence or absence of each *Nothofagus* species in 10 upland regions of New Guinea, together with their altitudinal range, is recorded in Figs. 2 and 3.

A general pattern emerges with most *Nothofagus* species present on the central part of the main range and fewer species present on more isolated ranges and islands. The New Guinea region has been subject to very rapid tectonic movements within the last few million years and ranges such as those of the Huon Peninsula have been isolated by the subsiding Ramu-Markham block-faulted trough while all the northern ranges have been undergoing rapid uplift. The D'Entrecasteaux islands with two species form part of the same continent as New Guinea separated by a shallow sea. New Britain, with a single species of *Nothofagus*, is now isolated on a separate tectonic plate. The adjacent island of New Ireland is situated on a more remote tectonic plate and has no *Nothofagus*. The species occurring in the more isolated ranges have a wide distribution but the poorly distributed species mostly occur within the main range.

If the *Nothofagus* species of New Guinea are derived from species adapted to the lowland tropical environment of the late Tertiary, these Tertiary species may have persisted at low altitudes with little morphological change. The creation of high altitude environments provided the opportunity for the extension of species ranges and the development of high altitude species. Without long distance seed dispersal the high altitude species are restricted to single mountain ranges while low altitude species may be more widespread. In the absence of fossil evidence it is difficult to confirm this hypothesis. Comparison of several morphological characters of the *Bipartitae* in New Caledonia and New Guinea indicates that the triad grouping of flowers and multilamellar cupule are primitive characters (Fig. 4). On the basis of morphological characters and gross aspects of biogeography it is possible to construct a dendrogram of the form shown in Fig. 4 in which close linkage indicates similarity between species. In agreement with the hypothesis we may contrast *N. starkenborghii* which has primitive characters, a low altitude environment and wide distribution, with *N. resinosa* which has derived characters, a high altitude environment and restricted distribution.

Some species share morphological characteristics but they differ in their distribution. *N. rubra* is recorded from the Vogelkop Peninsula to the D'Entrecastaux islands but is rather rare in the eastern ranges. In contrast, the morphologically similar *N. grandis* (Fig. 11) is restricted to the eastern and central ranges, where it is very abundant, but has a more restricted altitude zone

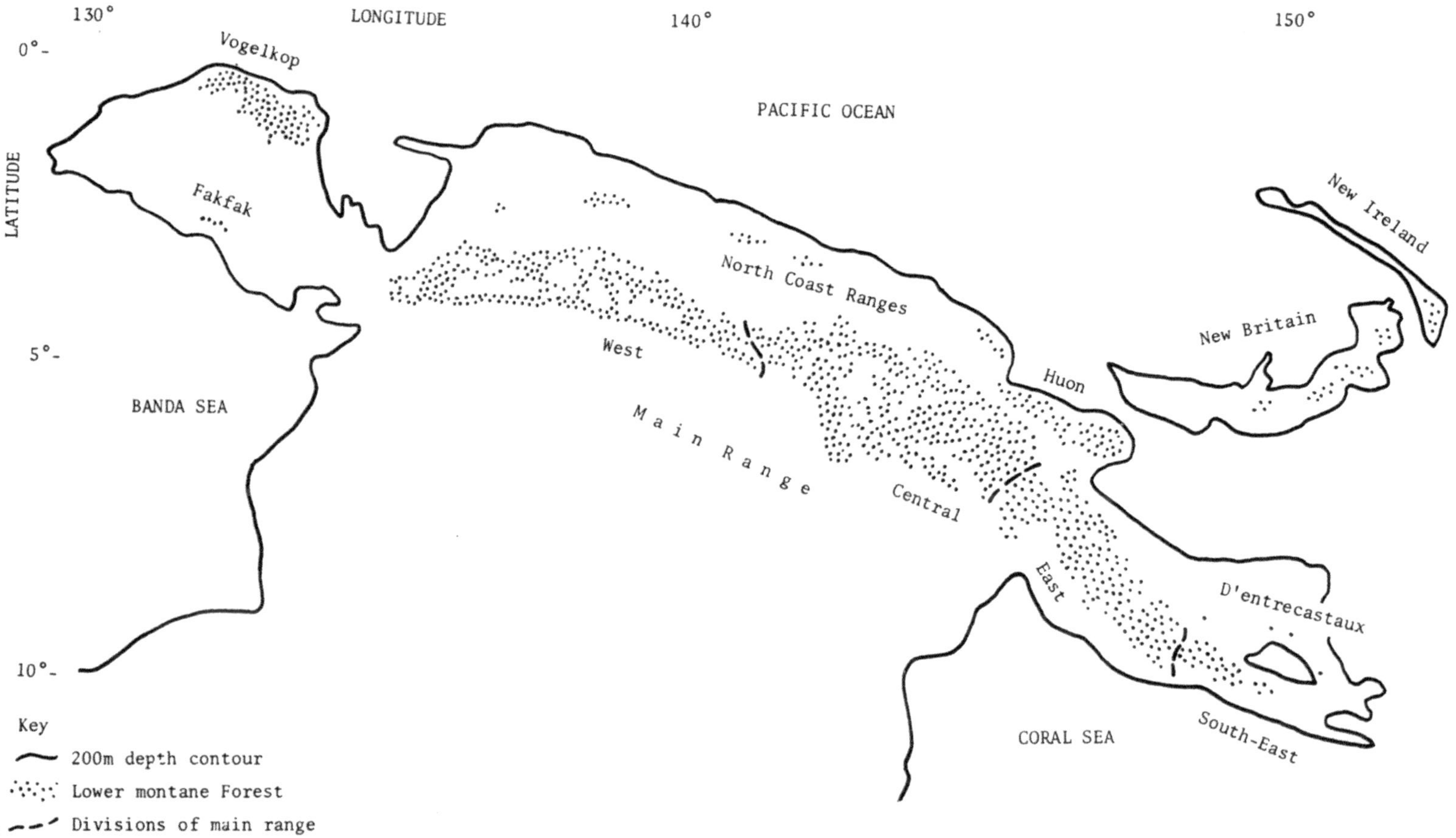

Fig. 2. Map showing lower montane forest in New Guinea (outline is 200 m ocean depth contour).

Species	Altitude (kms) 1 2 3	Fakfak	N. Coast	Vogelkop	West	Main Range Central	Main Range East	S. East	Huon	D'Entrec.	New Britain
starkenborghii	———			+	+	+	+	+			+
brassii	—		+	+	+	+			+		
perryi	—					+	+				
nuda	-						+				
rubra	- ———		+	+	+	+				+	
grandis	——					+	+	+	+		
carrii	- ———			+		+	+		+	+	
flaviramea	————	+	+	+	+	+	+		+		
womersleyi	-			+							
crenata	— -				+	+					
pullei	———			+		+	+		+		
pseudoresinosa	—				+	+			+		
resinosa	—				+	+	+				
Total		1	3	7	7	11	8	2	6	2	1

Fig. 3. Altitude and distribution of *Nothofagus* in New Guinea

Species (after Steenis)	Present in New Guinea	Phyllotaxis Distichous	Flowers in triads ♂	Flowers in triads ♀	Cupule multilamellar	Miscellaneous similarities	Hierarchical classification
aequilateralis			+	+	+	1	
balansae			+	+	+	2	
baumanniae			+	+	+		
codonandra			+	+	+		
discoidea			?		+	3	
starkenborghii	+	+	+	+	+	1	
brassii	+	+	+	+	+		
perryi	+	+	+	+	+	34	
nuda	+	+	+	+		2 4	
rubra	+	+	+		+		
grandis	+	+	+		+		
carrii	+	+	+				
flaviramea	+	+	+			5	
womersleyi	+	+	?			5	
crenata	+	+			+		
pullei	+	+					
pseudoresinosa	+	+				6	
resinosa	+	+				6	

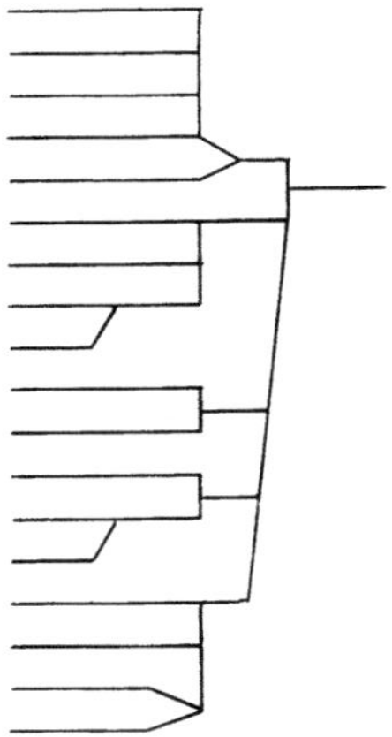

Fig. 4. Morphological variation in the Bipartitae Subsection of *Nothofagus*

than *N. Rubra* (Fig. 8). *N. brassii* is most common in the west, while the morphologically similar *N. perryi* is restricted to the eastern and central ranges. In both cases it is suggested that speciation has occurred through isolation along the main range but the species are ecologically similar and will not cohabit.

Three species are rare or only locally abundant and these each have considerable affinity with a more widespread species, i.e., *N. nuda* with *N. perryi*; *N. womersleyi* with *N. flaviramea*; *N. pseudoresinosa* with *N. resinosa*. It is assumed that these species have evolved recently. In contrast, *N. crenata* (which has been recorded twice); is located in separate areas and is a distinctive species. *N. crenata* was probably once more widespread and abundant at low altitudes. Hybrids have not been recorded in New Guinea but they do occur in S American *Nothofagus* (Donoso Zegers 1976).

2.1. Altitude

The altitude range of the various *Nothofagus* species of New Guinea is recorded in Fig. 3. *N. flaviramea*, *N. starkenborghii*, *N. rubra*, *N. carri* and *N. crenata* are all recorded below 900 m and their upper altitudes are above 2200 m. *N. brassii*, *N. perryi* and *N. grandis* are abundant over the range 1500–2500 m while *N. pullei* extends this range to about 2800 m (Figs. 6, 7). *N. resinosa* and *N. pseudoresinosa* are both abundant above about 2200 m. In any one locality only a few of the species are present and it is unusual to find more than two or three species forming a mixed forest. The distinct altitudinal ranges of *Nothofagus* species are observed in most other regions. In S America Weinberger (1973) characterizes optimum environments of seven species in terms of mean summer temperatures (a correlate of altitude) and mean daily temperature range (an indication of oceanic influence). As secondary factors he discusses the soil, water and shade regimes. In New Zealand there is a reasonably distinct altitude-temperature gradient from *N. truncata* Col. to *N. fusca* Oerst. to *N. menziesii* (Hook f.) Oerst. and *N. solandri* Poole. *N. solandri* occupies rather harsher conditions than *N. menziesii* (Wardle 1970). In Tasmania the deciduous *N. gunnii* Oerst. is locally abundant above 1200 m while *N. cunninghamii* Oerst. is abundant at lower altitudes. In New Caledonia the information describing distributions is incomplete but *N. aequilateralis* Steen, *N. codonanadra* Steen and *N. balansae* Steen are all recorded from 300–1200 m while *N. discoidae* Steen is recorded from lower altitudes and *N. baumanniae* Steen from higher altitudes (Steenis 1971b). The New Caledonian species also differ in their response to soil fertility and water.

Throughout the distribution of *Nothofagus*, the altitudinal gradient provides the primary variable for discriminating the ranges of the local species. Both the altitudinal gradient and some of the factors modifying the local abundance of *Nothofagus* species are a function of the climate. Fluctuations in climate such as those that have occurred during the Pleistocene have shifted the vegetation zones altitudinally and thereby created or removed barriers to horizontal migration.

2.2. Climate

Nothofagus is generally associated with continually high precipitation and cloudiness. The cloudiness reduces light penetration by as much as 70% though average values in highland New Guinea range between 30% in the morning and 50% in the afternoon. Under very cloudy and wet conditions light may be limiting tree growth.

The interception of radiation by clouds reduces both the temperature and the daily range of temperatures. The temperate species of *Nothofagus* will grow competitively provided they experience a few weeks each year with mean daily temperatures in excess of 10°C but they become less successful once the temperature exceeds 16°C for a few weeks (Wardle 1971; Specht and Brouwer 1975). The tropical *Nothofagus* of New Guinea have similar minimum temperature requirements but some will grow competitively in localities with mean daily temperatures as high as 23°C, cf. Fig. 5.

In temperate regions *Nothofagus* species are often the most abundant trees at the timber line but in New Guinea *Nothofagus* only reaches 3100 m while the timber line is usually about 3900 m. Several of the genera occurring at higher altitudes than *Nothofagus* in New Guinea are present at lower altitudes in temperate regions, which probably indicates a different response to the length of the growing season.

The south temperate timber line *Nothofagus* species experience considerable snowfalls and frost during the winter. In New Guinea frost is occasionally recorded on cloudless nights below 1500 m and the frequency increases with altitude reaching about 30–50% of nights at 3500 m (Hnatiuk *et al.* 1976). Brown and Powell (1974) record severe frost damage to *Nothofagus* and other Lower Montane Forest species following prolonged dry and cloudless periods in 1972.

For successful competitive growth, *Nothofagus* requires a fairly continual supply of water derived from reasonably aerated soils. The annual precipitation required to meet this condition rises from about 600–1000 mm in south temperate environments to 1500–1800 mm in New Guinea. The growth of *Nothofagus* in the middle latitudes of both S America and Australia is restricted by the lack of rainfall in summer. Rainfall in New Guinea is generally in excess of the requirements for *Nothofagus* but water deficits may occur during the winter months especially where there is a rain shadow effect. *Nothofagus* is generally absent from localities with regular and sustained water deficits. The coriaceous leaves of *Nothofagus* lose water relatively slowly compared to other Lower Montane species (Hynes 1973) which indicates a resistance to water deficit.

New Guinea is generally very cloudy and certain topographic features such as ridges and passes regularly experience ground level cloud from which the vegetation filters water droplets to provide very high local precipitation combined with low temperatures and low light penetration. In a number of localities, such as the D'Entrecasteaux islands, *Nothofagus* is abundant at relatively low altitudes associated with ground level clouds. The cloud forests, also known as moss forests from the abundant epiphytic bryophytes, often

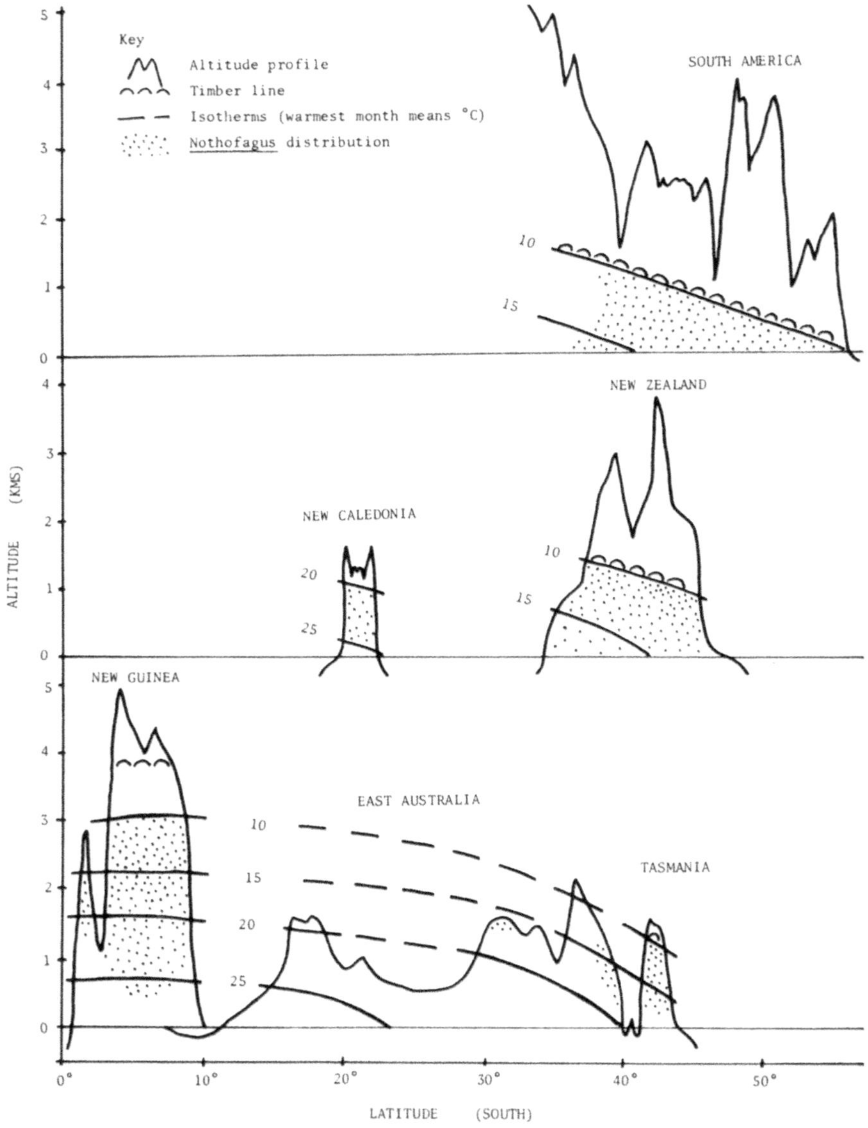

Fig. 5. Altitudinal and Latitudinal distribution of *Nothofagus.*

contain *Nothofagus* but the genus is not restricted to these conditions. In New Zealand Wardle (1970) notes that *Nothofagus* is absent from some very wet and foggy regions of the South Island.

2.3 Wind

Nothofagus pollen and seeds are dispersed by wind. Wind speeds in New Guinea are generally rather low but gusts occasionally cause local damage especially to trees on exposed slopes, ridges and forest edges. Only the SE part of New Guinea is subject to the devastating winds of tropical cyclones but these are relatively frequent in New Caledonia (Lourensz 1977). The regeneration of *Nothofagus* forest after such devastation has not been documented but the response of *Nothofagus* to other catastrophies indicates that it will probably regenerate successfully. The strong west winds in the south temperate regions produce changes in growth form which minimise wind damage.

2.4 Fire

The burning of vegetation depends upon several factors including the morphology and chemistry of the various species, low humidity and a source of ignition. The vegetation of dry regions is generally rather flammable but the species regenerate after fire. In contrast the vegetation of humid regions, such as *Nothofagus* forests, is not very flammable but it will regenerate poorly after fires, especially if the regrowth is also burned. The burning of *Nothofagus* forests is associated with the rare periods of prolonged drought but flammability is increased by excessive tree fall following natural catastrophes or human activity. Many fires are ignited by man and the frequency of forest fires has probably increased during the later Quaternary. In New Guinea disturbed *Nothofagus* forest bordering cultivated areas is frequently subjected to burning but fires rarely penetrate undisturbed forest. Exceptional burning of extensive areas of undisturbed *Nothofagus* forest was recorded in the Star Mountains and Telefomin region (5°S, 14°E) following droughts in 1972. Fires are more frequently recorded in the *Nothofagus* forests of the south temperate regions, especially near their dryer northern limits. *Nothofagus* may regenerate after a single fire from either seedlings or coppicing (Howard 1973; Turner 1976). Repeated fires will eliminate *Nothofagus* and it is replaced by such genera as *Eucalyptus* in Tasmania and eastern Australia (Jackson 1968), *Melaleuca* L. in New Caledonia (Dawson 1966), and *Miscanthus* Anderss. or other grasses in New Guinea.

2.5 Earthquakes, landslides and volcanoes

With the exception of Tasmania and E Australia there is considerable tectonic activity throughout the range of *Nothofagus* especially in N New Guinea, New Britain and parts of the Andes.

Minor earthquakes may be sufficient to damage tree roots and thereby encourage diseases. Rawlings (1964) traces the death of *Nothofagus* trees in a region of New Zealand to the beetle *Nascioides eynsii* Sharp and the fungus *Armillariella mellea* Karsten following earthquake damage.

Major earthquakes will cause trees to fall and they also trigger landslides which both destroy vegetation and provide bare surfaces for recolonisation. Volcanic activity will damage vegetation through burning, toxic gases and burial under ash or lava but it provides a nutrient-rich substrate and light falls of volcanic ash may improve soil fertility over large areas.

Fig. 6, 7. Lower montane forest on SE Mt Giluwe (alt. 2300 m; *Nothofagus* to 2800 m and mountain rising to 4300 m).

Fig. 8. Transition from forest of *Nothofagus pullei* (large pale crowns) and *N. rubra* (large dark crowns) to conifer forest (small conical crowns), SE Mt Giluwe (alt. 2750 m).

Fig. 9. Dieback of *Nothofagus pullei* with *N. rubra* (dark crowns) scarcely affected; SE Mt Giluwe (alt. 2650 m).

Nothofagus forests may be subjected to all these catastrophic events and their response depends to a considerable extent upon whether regeneration from existing individuals is possible or whether the site must be colonised with seeds. In New Zealand Wardle (1970) describes the colonisation of landslide sites by *N. menziesii* and *N. solandri*. In the Valdivian Andes in S America Veblen and Ashton (1978) attribute much of the *Nothofagus* and *Eucryphia cordifolia* Cav. forest to the colonisation of sites following various tectonic catastrophies. In contrast, the *Nothofagus* of New Guinea is rarely recorded from sites of recent landslides or volcanic activity. In New Guinea there are genera with better seed dispersal and faster growth rates than *Nothofagus* and these soon form a dense community on bare ground.

2.6 Substrate

The substrates of *Nothofagus* forests in New Guinea are diverse and range from ultrabasic to acidic igneous rock, calcareous to siliceous sediments and various metamorphics. *Nothofagus* growth is probably most rapid on volcanic substrates but trees may be stunted on shallow soils over limestone. Rock type correlates poorly with *Nothofagus* abundance, which may result partly from the considerable amounts of volcanic ash that have fallen throughout New Guinea. Some variation of *Nothofagus* abundance with substrate has been recorded, such as the local absence from sandstone plateaux near Mt Jaya (4°S, 113°20′E) (Hope 1976b).

Weathering of minerals and decomposition of organic matter is more rapid at high temperatures such that low altitude soils are deep, have a low organic

Fig. 10. Dieback and regeneration of *Nothofagus pullei*, SE Mt Giluwe (alt. 2600 m).

Fig. 11. *Nothofagus grandis* on SE Mt Giluwe (alt. 2300 m); surrounding forest has been logged.

content and contain few nutrients (Rutherford and Haantjens 1965; Haantjens 1970). At extremes *Nothofagus* species are to be found on poor inorganic lowland soils, e.g., *N. starkenborghii*; shallow soils over limestone, e.g., *N. perryi*; and wet peaty soils, e.g., *N. rubra*. In Australia *N. cunninghamii* is associated with a high soil phosphate content but this seems to be a general characteristic of the rain forests (Beadle 1962; Howard and Ashton 1973; Ashton 1976).

2.7 Topography

Much of the local variation in the abundance of *Nothofagus* species is related to topography. At low altitudes in New Guinea *Nothofagus* is usually restricted to ridge crests but at high altitudes various species occur on most types of topography. In New Caledonia Dawson (1966) notes that *Nothofagus* is most common on ridge crests but has a wider distribution at higher altitudes and on ultrabasic rocks. The low altitude species *N. codonandra* is recorded from some riverbank habitats. The distribution in New Zealand resembles that in New Guinea with ridge crest sites in warmer low altitude environments and more varied topography at high altitudes. In Australia *N. cunninghamii* occupies a range of sites but may be restricted to gullies because of their lower fire frequency.

Many factors change along the toposequence catena away from a ridge crest. Microclimatic changes include cloudiness, humidity, solar radiation and wind, while surface processes are dominated by the downslope movement of water, solutes and particles. Slopes are subjected to soil movements which will cause trees to lean and fall while stable ridge crests may support much older trees. Animals (including man) may preferentially travel, feed, disperse seeds and ignite fires along ridges. Certain pathogens, especially soil fungi, may be associated with the wetter conditions on the lower slopes. With so many complex and interacting variables operating upon many tree genera it is not possible to offer a simple explanation of the topographic distribution of *Nothofagus*.

3. Ecology

3.1 Herbivores and Pathogens

There is a diverse fauna and variety of fungi in the Lower Montane forests of New Guinea, but there is little evidence to show their effect upon *Nothofagus*. Several features indicate a resistance to herbivore activity including the hard cupules, coriaceous leaves, hard and durable wood, and the presence of polyphenolic compounds, tannins and triterpenes (Hegnauer 1966; Hillis 1967; Gibbs 1974). Despite these features insect and fungal damage is frequently observed in cupules, seeds, leaves and wood. Leaf litter decays within a few years (Hynes 1973) and even large fallen trunks decay within a few decades. An indication of the variety of organisms associated with *Nothofagus* is apparent from those recorded in New Zealand (Wardle 1967). Many insects feed on New Guinea *Nothofagus* (Gressitt pers. comm.).

Dieback and the deaths of stands of *Nothofagus* is recorded from many localities in New Guinea. Dieback is usually associated with a combination of weather, substrate, pathogens and hosts at a suitable stage of development. Pathogenic soil fungi including *Phytophthora cinnamomi* Rands A.I. and *P. cambivora* (Petri) Buis (Shaw *et al.* 1972; Stamps *et al.* 1972; Arentz pers. comm.; Cartledge *et al.* 1975) and wood boring beetles, e.g., *Xixuthrus*(?) (Cerambycidae), are associated with dieback of *N. pullei* on Mt Giluwe but a causal relationship has not been established. Adjacent stands of *N rubra* are less severely affected and this species occurs on poorly drained sites which generally favor soil pathogens (Figs. 9, 10).

In Australia the pathogenicity of *P. cinnamomi* to seedlings of *N. cunninghamii* has been established (Weste 1975). Dieback in the rain forests of Australia has been attributed to several organisms including the ambrosia beetle

Platypus omnivorus Lea (Platypodinae) and the fungi *Armillariella mellea* and *P. cinnamomi* (Brown 1976).

3.2 Mycorrhizae

Ectotrophic mycorrhizae have been recorded from *Nothofagus* in S America (Singer 1971), New Zealand (Mejstrik 1972) and Australia (Howard 1973). It is likely that they are also associated with *Nothofagus* in New Guinea. There is generally a substantial increase in growth rates associated with mycorrhizal fungi. The successful colonisation of new sites from seed may be limited in the absence of mycorrhizae.

3.3 Epiphytes

The humid environment of *Nothofagus* forests is favorable for epiphytes and many trees support sooty moulds on their leaves, ferns and vascular epiphytes in their branches and bryophytes on the trunks. The detrimental effects of epiphytes may favor the evolution of short leaf longevity and flaking barks. Species with flaking barks, e.g., *N. grandis*, contrast with the moss covered persistent barks of *N. pullei*.

4. Morphology

The various species of *Nothofagus* in New Guinea form canopy trees with maximum heights of 30–40 m and trunk diameters of 80–180 cm. The lower trunk is usually buttressed and the roots extend a few meters from the tree and to a depth of 20–40 cm. The fusion of roots from adjacent trees has not been recorded. The bole is about half the height of the tree and gives rise to several branches which spread at angles of 10–30° from vertical to form a nearly level or domed canopy. The canopy leaves of most species are 4–10 cm in length but *N. pullei*, *N. carrii* and in some localities *N. rubra* and *N. pseudoresinosa* produce smaller leaves. The small leaved species are mostly at higher altitudes but there is not a simple gradient of leaf size with altitude.

Altitude has little effect on the maximum size of *Nothofagus* trees in New Guinea but stunted trees are recorded on shallow soils over limestone, e.g., *N. perryi*, waterlogged soil, e.g., *N. rubra*, and exposed cloudy ridges, e.g., *N. pullei*, *N. brassii*, *N. carrii*, *N. rubra*. In temperate regions, species of *Nothofagus* growing near the timber line may be very stunted (Wardle 1970; Eskuche 1973).

4.1 Wood structure and growth seasonality

Dadswell and Ingle (1954) (cf. Steenis 1971b) have examined the secondary wood from various subsections of *Nothofagus* and they conclude that the genus forms 2 distinct wood types of which one is the *Bipartitae*. The *Bipartitae* is distinguished by the larger and less numerous vessels, bands of parenchyma, and the absence of tracheids. The thick walled fibres and small vessels which characterise the late wood of the cool temperate species are not recorded in the tropical *Bipartitae*. In temperate species parenchyma tissue is most abundant at the end of the late wood growth ring and in *N. moorei* from subtropical eastern Australia (30°S) there is a tendency to form a concentric parenchyma band after the rather indistinct late wood formation. Rings of vessels occur in the early wood of some temperate *Nothofagus* species and these have been observed in samples from New Guinea. Differences in the wood of the temperate and tropical species are not fundamental and may be related to differences in climate (cf. Graaf and Baas 1974).

In temperate species of *Nothofagus* there is usually a flush of leaf and shoot growth in spring and there may be a second flush in summer (Specht and Brouwer 1975; Wardle 1967). The wood anatomy shows corresponding bands of early wood separated by late wood formation (Bussell 1968). In New Guinea the reddish flush of young leaves has been recorded throughout the year and most frequently during June–August and February. A season of parenchyma ring formation and dormancy has not been ascertained.

The growth rings in tropical *Nothofagus* provide information about the growth and age of trees. The parenchyma rings typically have a spacing of 1.5–3.5 mm which is consistent with annual growth rates recorded in N Australian rain forests (Volck 1968). Regularly spaced parenchyma rings are observed in localities continuously favorable for tree growth, e.g., Lake Kutubu (Nix and Kalma 1972), which indicates that the rings are unlikely to be a simple response to dry or cool conditions. In the same environment *Nothofagus* undergoes periodic flushes of leaf growth.

In a more seasonal environment on Mt Kaindi (7°20′S, 146°41′E) *N. grandis* forms annual parenchyma rings (Wardle 1973). A specimen of *N. grandis* from Mt Giluwe collected in 1974 contained 318 parenchyma rings and yielded a cellulose radiocarbon age of 500 ± 50 years B.P. (ANU – 1540). The uncertainty of converting a radiocarbon age to a real age is such that the hypothesis of annual ring formation is not rejected. A very large fallen trunk of *Nothofagus* from near Wabag (5°31′ S, 143°43′ E) yielded a radiocarbon date of 550 ± 85 years B.P. (N.S.W. 4 and 5) (Walker 1966). The information yielded from tree growth rates, radiocarbon dates, and regular parenchyma rings indicate that few trees will exceed 350–550 years. In comparison, the New Zealand species *N. solandri* lives for up to 360 years while *N. menziesii* reaches 600 years (Wardle 1967, 1970).

5. Population dynamics — Life-cycle

Flowers and fruits are recorded from *Nothofagus* trees only a few meters tall growing on open sites but not from shaded subcanopy trees. For each species there is a tendency to flower at a particular season but flowering occurs sporadically throughout the year. Daylength, which is generally an important control of flowering in temperate latitudes, varies by less than one hour and there are few other factors to induce synchronous seasonal development. Seasonality of flowering does not seem to provide a means for genetically isolating the *Nothofagus* species.

Nothofagus produces abundant wind dispersed binucleate pollen which is morphologically variable but the species are not distinctive (Walker 1965). In any locality the greater part of the pollen will be of local origin but occasionally fertilisation will occur with pollen from distant sources.

In 1974 the seed crop in *N. pullei* forest on Mt Giluwe was about 3 million Ha^{-1} while *N. grandis* and *N rubra* each produced about $\frac{1}{10}$ of that number. Annual seed production is said to vary quite markedly and this has also been noted in the tropical *Nothofagus* forests of New Caledonia (Dawson 1966). In New Zealand *N. menziesii* and *N. solandri* are characterised by infrequent seasons with very high seed production while most years have low production (Wardle 1967, 1970). It is thought that unpredictable fluctuations in seed production will allow more seeds and seedlings to escape destruction by herbivores.

On Mt Giluwe in 1974 the caterpillars of a tortricid moth (Tortricidae) were recorded in 74% of the cupules of *N. grandis* and *N. rubra* but the small cupules of *N. pullei* were not parasitised.

Experimental germination of seeds of *N. grandis*, *N. rubra* and *N. pullei*, with moderate shading, yielded less than 1% success. The frequency of seedlings beneath stands of large *Nothofagus* trees is between 50–1000 seedlings Ha'^{1} which indicates that few seeds develop into established seedlings. The majority of seedlings grow upon mossy substrates, especially fallen logs, a characteristic also recorded from New Zealand (Wardle 1970). The advantage of the mossy log substrate probably relates to the reduction of competition with other vascular plants but mycorrhizae, pathogens and herbivores may also be involved. In dense forest seedlings have been recorded with about 2% daylight penetration and the New Guinea species may have a compensation point similar to the 1.15% estimated for *N. cunninghamii* in Australia (Howard 1973). The canopy of *Nothofagus* stands in New Guinea is generally rather open with 5–8% daylight penetration. Under these conditions seedlings will make rather slow growth and few survive to become saplings.

In addition to seedlings, much of the regeneration of *Nothofagus* in New Guinea arises from suckers. The suckers originate from several sources including the undersurface of fallen trunks and damaged or supressed seedlings and saplings. Horizontal suckers may spread along the soil surface for several meters, producing several vertical shoot and root systems. In temperate regions layering is recorded in snow damaged stands of *N. solandri* in New Zealand and probably occurs in the *Nothofagus* Krümholtz of the Andes (Eskuche 1973). *N. moorei* and *N. cunninghamii* may respond to burning by coppicing (Howard 1973; Turner 1976).

The rate of growth of seedlings or suckers is greatly increased by senescence of the tree canopy or the creation of canopy gaps. Compared to other genera, *Nothofagus* grows slowly under shaded subcanopy conditions but rapidly under more open conditions. Similar observations have been made in New Zealand (Wardle 1970) and S America (Veblen and Ashton 1978).

The gregarious nature of many *Nothofagus* populations and the presence of regeneration indicate that allelopathy is not very significant. June (1976) could not detect allelopathy in experiments with *N. fusca* seedlings. New Guinean *Nothofagus* stands normally form a canopy at a height of 30–35 m while other genera generally reach 30–45 m at low altitudes and 20–30 m at high altitudes. The correlation of relative canopy height with relative abundance over the altitude range probably results at least partially from real competitive advantages associated with greater canopy height.

5.1 Population structure of forests

The New Guinea *Nothofagus* forests may be conveniently classified into those which are nearly pure *Nothofagus* stands versus those with a mixture of other genera, and also into those with a restricted range of size classes versus those with a complete range of sizes including regeneration.

Nearly pure *N. pullei* forest with a mosaic of patches each dominated by a single size class is recorded on SE Mt Giluwe. The forest is evidently characterised by a cycle of stand mortality followed by regeneration to form canopy trees which may in turn suffer extensive mortality. Possibly explanations of the mortality have been discussed in relation to various pathogens. There are many genera of competitors in the regrowth but *Nothofagus* apparently grows faster and to a greater height, and thereby dominates the forest. Similar *Nothofagus* dieback is recorded in many regions of New Guinea.

Small populations of *Nothofagus* on ridges may have a single abundant size class of large canopy trees but *Nothofagus* regeneration is often observed where a tree is senescent. The local patches of regeneration are probably sufficient to maintain these populations.

Extensive stands of *Nothofagus* may have all size classes represented but at a small scale the forest resembles that on ridges with local patches of regeneration. The *N. rubra* forest discussed by Kalkman and Vink (1970) is evidently of this type.

Forests in which *Nothofagus* forms only a minor portion of the canopy and is even less abundant in smaller size classes are recorded on Mt Giluwe. The *Nothofagus* trees may be very large and senescent but the regeneration is mostly from other genera. This forest type is apparently successional from *Nothofagus* to a multiple genera forest. In some localities *Nothofagus* regeneration is recorded in multiple genera forest but it is difficult to estimate the stability of the populations from short periods of observation of the individuals.

In New Caledonia Dawson (1966) describes similar ridge top populations amongst *N. balansae* (cf. *Aequilateralis* Steenis 1971b). Hürlimann (1962) also describes poor regeneration in stands of *N. balansae* and *N. codonandra* but in some stands of the latter species a range of size classes were recorded. The *Nothofagus* species in New Zealand also form pure or mixed forest with one or more age classes (Wardle 1964, 1967, 1970). Regeneration is typically within canopy gaps but larger even aged stands may develop after damage to the canopy trees from fire, wind, snow or insect attack. A similar relationship between *Nothofagus* regeneration and local or extensive canopy damage is recorded in S America (Veblen and Ashton 1978). Colonising populations of *Nothofagus* are recorded from bare ground in S America and New Zealand and from grassland in various regions.

Discrete patches of *Nothofagus* forest within a matrix of multi genera forest have been recorded in New Guinea (Robbins and Pullen 1965; Robbins 1970; Kalkman and Vink 1970; Walker 1966), New Caledonia (Dawson 1966), New Zealand (Wardle 1964, 1967) and S America (Mann 1968). The boundary of these patches is usually sharp and in New Guinea *Nothofagus* regeneration scarcely extends beyond the canopy of the *Nothofagus* trees (Walker 1966; Kalkman and Vink 1970). It has been suggested that these patches are relics of more widespread *Nothofagus* forest which has been encroached upon by other genera. The *Nothofagus* stands are therefore in habitats which favor *Nothofagus*, e.g., ridges, or in localities where other genera have not become established for a variety of reasons including their rates of migration.

Palynological records in New Guinea (Walker 1970; Bowler *et al.* 1976; Hope 1976a) indicate that during the Glacial *Nothofagus* was at about two thirds of its present altitude but then migrated rapidly to higher elevations as the climate warmed. These extensive *Nothofagus* forests were then gradually invaded by other genera which together now constitute the Lower Montane Forest. The rates at which *Nothofagus* migrated generally exceed the rates of vegetative spread by suckers and therefore require establishment from seed. *Nothofagus* colonisation of other forest types is rarely recorded in the absence of catastrophies, and grasslands previously associated with cold air drainage may have provided routes for migration. The increase in *Nothofagus* abundance at low altitudes during cooler periods was probably the result of the spread of local populations rather than migration from high altitudes.

5.2 Associated vegetation

The *Nothofagus* species of New Guinea are in competition with hundreds of other genera from many families with diverse life forms. At altitudes above about 2000 m the floristics of forest with *Nothofagus* are generally similar to adjacent forest without *Nothofagus* (cf. Saunders 1965, 1970; Kalkman and Vink 1970). At low altitudes the ridge top forests with *Nothofagus* typically include species and genera at the lower limits of their altitude range while the forest on lower slopes is floristically similar to lowland floras. Forest containing *Nothofagus* changes composition in relation to altitude, topography, location, etc., and there are no species or genera always associated with *Nothofagus*, though some species are usually present in certain enivornments. At altitudes below 2300 m *Nothofagus* is often associated with *Castanopsis acuminitissima* (Bl.) A.Dc. or *Lithocarpus* Blume. These three abundant Fagaceae genera have retained ecological similarities but species development differs, with 13 endemic *Nothofagus*, seven endemic and two non-endemic *Lithocarpus*, and one non-endemic *Castanopsis* species. At high altitudes there are associations with *Phyllocladus hypophyllus*, while *Dacrydium* and *Papuacedrus* occur in some wet and cold habitats. It is notable that these latter associates of *Nothofagus* have similar biogeographic occurrence.

6. Human activity and *Nothofagus*

The *Nothofagus* forests of New Guinea reach an upper altitudinal limit rather higher than the upper limit of cultivation. It has been suggested that Fagaceae forests were preferentially cleared for cultivation (cf. Steenis 1953), but this may be coincidental to other environmental factors. In many localities *Nothofagus* forest is being cleared but higher altitude stands remain which are visited for hunting and gathering timbers, barks and fruits. *Pandanus* may be tended in *Nothofagus* forest and this may involve both clearing undergrowth and ringbarking of canopy trees. Occasionally *Nothofagus* seedlings are planted near villages.

Nothofagus timber is hard and durable with a basic density of 650 kg/m^3 and moderate strength (cf. New Zealand beeches, Anon. 1974). At Lake Kutubu *Nothofagus* is used for house stilts and may be reused in subsequent houses. In the highlands the timber is used in house walls, fences, and for tools such as digging sticks. Prior to the introduction of steel axes the use of *Nothofagus* was probably restricted by its hardness. A large proportion of the timber is burned either in situ or for cooking and warming houses. There are a number of saw mills processing *Nothofagus* for structural purposes and the timber is marketed as New Guinea Beech or under local names such as Gripe. *Nothofagus* stands have a maximum timber yield of about $90m^3/Ha'^1$ from 25–30 trees and the average yields of $45–60m^3/Ha^{-1}$ are comparable with low altitude forests and greater than most high altitude forests.

The effect of logging on *Nothofagus* forest depends upon several factors including the amount and proportions of different species which remain, disturbance to the forest floor, availability of seed and seedlings, and subsequent activity such as cultivation and burning. Secondary forest may contain *Nothofagus* (Brass 1964; Paijmans 1976) and the Kum logging area (Lat. 5°53′, Long. 144°15′) contains regeneration including much *Nothofagus*. In other localities disturbance is too great and grassland or other secondary forest may develop. The present trend is for a reduction in the area of *Nothofagus* forests and their exploitation in accessible localities.

7. New Guinea *Nothofagus* in perspective

On average there are six species in each angiosperm genus in New Guinea but this includes many genera with a single species and a few with many species. The 13 species of *Nothofagus* do not represent exceptional species development though it is greater than in the majority of genera. Since most angiosperm genera are pollinated by animals, the process of speciation may be rather different from that in anemophilous genera such as *Nothofagus*.

The diversity of species and genera in tropical rainforest is greatest on well-drained lowland sites. Dominance by species and genera increases with altitude but the area of habitat declines such that their populations may be similar or smaller. *Nothofagus* comprises 10–20% of the canopy trees of the Lower Montane Forest Zone in New Guinea. Assuming 30–100 canopy trees/Ha^{-1} and a forest area of 50 000 sq.km., the *Nothofagus* population is 15–100 million canopy trees. A few of the species may number several million trees but the rare species may have only thousands of trees.

Nothofagus is numerically one of the commoner trees of New Guinea. In contrast many genera and species associated with *Nothofagus* are scarce. The formation of extensive single species stands is relatively unusual in tropical rainforest but it is characteristic of *Nothofagus*. Pure stands are usually associated with extreme site conditions, especially waterlogging, or regrowth on bare ground, grassland and disturbed forest.

Nothofagus stands are readily identified from aerial photographs and their patchy distribution has been mapped over considerable areas (Saunders 1965, 1970, 1973, 1976). The only other genus to be mapped extensively is *Araucaria* which also has a patchy distribution (Gray 1973; Enright next chapter). Local topographic variation in abundance is recorded from most genera, and several have a characteristic ridge top distribution similar to *Nothofagus*.

The life cycle of *Nothofagus* may be summarised as requiring adjacent trees for effective wind pollination to produce the considerable numbers of poorly dispersed seed. Few seeds become established as seedlings and these require an open tree canopy allowing fast growth to become a tall tree which may then reproduce for many years. Alternatively trees may persist for long periods by suckering. No single aspect of the ecology of *Nothofagus* appears to be exceptional but together they give *Nothofagus* its unique characteristics.

Comparison of the typical *Nothofagus* of New Guinea with the New Caledonian and temperate species indicates both adaptations to particular environments and many similarities. Physiological differences apparently include responses to temperature and daylength. Seasonality is marked in the flowering, leaf production and wood anatomy in temperate regions, but these processes are also periodic in the tropics. Morphological differences result from both phenotypic plasticity in different environments as well as genetic differences. The wood anatomy of tropical and temperate species is distinct but much of the difference is probably related to the environment and not genetic differences. Vegetative regeneration may occur in response to damage caused by heavy snowfalls, fires or tree fall, each of which is characteristic of different regions. The population dynamics of *Nothofagus* are similar everywhere but there are fewer competitors in the temperate regions and *Nothofagus* has a greater ecological range. *Nothofagus* is considered as a colonist of bare ground and a timberline tree in the temperate regions but is excluded from these environments in New Guinea. Overall *Nothofagus* appears to be an ancient genus which has successfully extended its range from humid environments of temperature regions to the cool montane humid regions of the tropics. This transition is not merely climatic but involves adaptation to the prevailing tropical soils and biota.

References

Anon. 1974. Timber properties and uses of the New Zealand Beeches. New Zealand Forest Service, Wellington.

Ashton, D. M. 1976. Phosphorous in forest ecosystems at Beenak, Victoria. J. Ecol. 64, 171–86.

Balgooy, M. M. J. van 1966. Distribution maps of Pacific plants. Blumea Suppl. 5: 53–309.

Beadle, N. C. W. 1962. Soil phosphate and the delimitation of plant communities in East Australia, II. Ecol. 43: 281–88.

Bowler, J. M., G. S. Hope, H. N. Jennings, G. Singh and D. Walker. 1976. Late Quaternary climates of Australia and New Guinea. Quaternary Research 6: 359–94.

Brass, L. J. 1964. The ecology of highland settlement. Watson, J. B. (ed.), New Guinea: The central highlands, American Anthrop. 66: 20–38, Spec. Publ., Part 2.

Brown, B. N. 1976. Phytophthora cinnamomi associated with patch death in tropical rain forests in Queensland. A.P.P.S. Newsletter 5: 1–4, Dept. For. Indooropilly, Qld., Australia.

Brown, M. and J. M. Powell. 1974. Frost and drought in the highlands of Papua New Guinea. J. Tropical Geography 38: 1–6.

Bussell, W. T. 1968. The growth of some New Zealand trees 1–2. New Zealand J. Botany 6: 63–85.

Cartledge, E. G., D. E. Shaw and D. J. Stamps. 1975. Studies in relation to dead patches of *Nothofagus* in Papua New Guinea. Papua New Guinea Dept. Agr. Stock and Fisheries, Res. Bull. 13.

Chappell, J. 1974. Relationships between sea levels, 180 variations and orbital perturbations, during the past 250 000 years. Nature, London 252: 199–202.

Cookson, I.C. and K. M. Pike. 1955. The pollen morphology of Nothofagus Bl., Subsection Bipartitae Steen. Aust. J. Bot. 3: 197–206.

Couper, R.A. 1960. Southern hemisphere Mesozoic and Tertiary Podocarpaceae and Fagaceae and their Palaeogeographic significance. Proc. Roy. Soc. Lond. B 152: 491–500.

Dadswell, H. E. and H. D. Ingle. 1954. The wood anatomy of New Guinea *Nothofagus* Bl. Aust. J. Bot. 2: 141–53.

Dawson, J. W. 1966. Observations on Nothofagus in New Caledonia. Tuatara 14: 1–7.

Donoso Zegers, C. 1976. Nothofagus leoni hybridization and introgression in populations in N. obliqua and N. glauca. Univ. Chile, Fac. Cienc. For Bol. Tec. 36: 1–29.

Eskuche, V. 1973. Estudios fitosociologicos en el norte de Patagonia. Phytoceonologica 1: 64–113.

Gibbs, R. 1974. Chemotaxonomy of flowering plants, 3: 1304–309, Fagaceae. McGill-Queen's Univ. Press. Montreal and London.

Gill, E. D. and H. D. Ingle. 1975. Evolution of Australia's unique flora and fauna in relation to the plate tectonics theory. Roy. Soc. Victoria, Proc. 87: 215–234.

Graff, N. A. Van der and P. Baas, 1974. Wood Anatomical Variation in Relation to Latitude and Altitude. Blumea 22: 101–21.

Gray, B. 1973. Distribution of Araucaria in Papua New Guinea. Res. Bull. 1, Dept. of Forests, Port Moresby.

Gressitt, J. L. 1974. Insect Biogeography. Ann. Rev. Entomol. 19: 293–322.

Griffiths, J. R. and R. Varne. 1972. Evolution of the Tasman Sea, Macquarie Ridge and Alpine Fault. Nature, Phys. Sci., London 235: 83–86.

Haantjens, H. A. 1970. Soils of the Goroka-Mount Hagen area. Lands of the Goroka-Mount Hagen area, Territory of Papua New Guinea, Land Research Ser. 27: 80–103, C.S.I.R.O., Melbourne.

Hegnauer, R. 1966. Chemotaxonomie der Pflanzen. 4: 141–55. Biekhauser Verlag. Basel and Stuttgart.

Hillis, W. E. and T. Inoue. 1967. The Polyphenols of *Nothofagus* species 2, The heartwood of *Nothofagus fusca*. Phytochemistry 6: 59–67.

Hnatiuk, R. J., J. M. B. Smith, and D. N. McVean. 1976. Mt Wilhelm Studies 2, The Climate of Mt Wilhelm. Publication BG/4, Research School of Pacific Studies, A.N.U., Canberra.

Hope, G. S. 1976a. The vegetation history of Mt Wilhelm, Papua New Guinea. J. Ecol. 64: 627–64.

—— 1976b. Vegetation. In Hope, G. S., J. A. Peterson, U. Radok, and I. Allison, (eds), The equatorial glaciers of New Guinea, 113–72. Balkema A.A., Rotterdam.

Howard, T. M. 1973. Studies in the ecology of *Nothofagus cunninghamii* Oerst., 1–3. Australian J. Botany 21: 67–102.

Howard, T. M. and D. H. Ashton. 1973. The distribution of *Nothofagus cunninghamii* rainforest. Roy. Soc. Victoria Proc. 86: 47–75.

Hürlimann, J. H. 1962. The structure of some Biocoenoses of New Caledonia. Proc. 9th Pacific Sci. Congress 4: 89–94.

Hynes, R. A. 1973. Ecology of Nothofagus forest in central New Guinea. M.Sc., thesis submitted to Univ. Papua New Guinea.
Jackson, W. D. 1968. Fire, air, water and earth — an elemental ecology of Tasmania. Proc. Ecol. Soc. Australia 3: 9–16.
June, S. R. 1976. Investigations on allelopathy in red beech forest. Mauri Ora 4: 87–91.
Kalkman, C. and W. Vink. 1970. Botanical exploration in the Doma peaks region, New Guinea, Blumea 18: 88–135.
Khan, A. M. 1974. Palynology of Neogene sediments from Papua (New Guinea), Stratigraphic boundaries. Pollen et Spores 16: 265–84.
Lourensz, R. S. 1977. Tropical Cyclones in the Australian Region, July 1909–June 1975, Dept. of Science, Bureau of Meteorology, Meteorological Summary, A.G.P.S., Canberra.
Main, B. Y. 1974. Occurrence of the lungless spider *Microphokomma* Crosby and Bishop in South-West Australia, (Araneae Symphytognathidae). J. Aust. Entomol. Soc. 13: 79.
Mann, G. 1968. Die Ökosysteme Südamerikas. In Fittkau, E. J. et al. (eds), Biogeography and Ecology in South America, 1: 171–229, Junk, W., The Hague.
McElhinny, M. W. 1970. Formation of the Indian Ocean. Nature, London 228: 977–79.
Mejstrik, V. 1972. The classification and relative frequency of Mycorrhizae in Nothofagus solandri var. cliffortiodes, New Zealand J. Botany 10: 243–53.
Nix, H. A. and J. D. Kalma. 1972. Climate as a dominant control in the biogeography of Northern Australia and New Guinea, in Walker, D. (ed.), Bridge and Barrier: The natural and cultural history of Torres Strait, Publ. BG/3, Research School of Pacific Studies, A.N.U., Canberra.
Paijmans, K. 1976. New Guinea Vegetation. pp. 23–105, C.S.I.R.O.-A.N.U. Press, Canberra.
Preest, D. S. 1963. A note on the dispersal characteristics of the seed of the New Zealand podocarps and beeches and the biogeographic significance. In Gressitt, J. L. (ed.), Pacific Basin Biogeography, 415–24, Bishop Museum Press, Honolulu.
Rawlings, G. B. 1964. A report on the susceptibility of the New Zealand Nothofagus forests to climatic and earthquake damage, New Zealand Forest Service Pathology Branch Report 6, Appendix 2 (unpublished).
Robbins, R. G. 1970. Vegetation of the Goroka-Mt Hagen area, in Lands of the Goroka-Mount Hagen area, Papua New Guinea, Land Research Ser. 27: 104–118, C.S.I.R.O., Melbourne.
Robbins, R. G. and R. Pullen. 1965. Vegetation of the Wabag-Tari area, in Lands of the Wabag-Tari area, Papua New Guinea. Land Research Ser. 15: 100–15, C.S.I.R.O., Melbourne.
Rutherford, G. K. and H. A. Haantjens, 1965. Soils of the Wabag-Tari area, in Lands of the Wabag-Tari area, Papua New Guinea. Land Research Ser. 15, C.S.I.R.O., Melbourne.
Saunders, J. C. 1965. Forest resources of the Wabag-Tari area, in Lands of the Wabag-Tari area, Papua New Guinea. Land Research Ser. 15: 116–124, C.S.I.R.O., Melbourne.
—— 1970. Forest resources of the Goroka-Mount Hagen area, in Lands of the Goroka-Mount Hagen area, Papua New Guinea. Land Research Ser. 27: 119–125, C.S.I.R.O., Melbourne.
—— 1973. Forest resources, in Land-form types and vegetation of Eastern Papua. Land Research Ser. 32: 126–135, C.S.I.R.O., Melbourne.
—— 1976. Forest resources, in Lands of the Ramu-Madang area. Land Research Series 37: 110–125, C.S.I.R.O., Melbourne.
Schlinger, E. I. 1974. Continental drift, Nothofagus and some ecologically associated insects. Ann. Rev. Entomol. 19: 323–43.
Shaw, D. E., E. G. Cartledge and D. J. Stamps. 1972. First records of Phytophthora cinnamomi in Papua New Guinea. Papua New Guinea Agricultural J. 23: 46–48.
Singer, R. 1971. Forest mycology and forest communities in South America, 2, Mycorrhiza sociology and fungus succession in the Nothofagus dombeyi Austrocedus chilensis woods of Patagonia. Hacskaylo, E. (ed.), Mycorrhizae Symposium, 204–15, U.S. Govt. Printing Office, Washington D.C.
Soepadmo, E. 1972. Fagaceae: Nothofagus. Flora Malesiana Ser. 1, Vol 7: 265–94.
Specht, R. L. and M. Brouwer. 1975. Seasonal shoot growth of Eucalyptus species in Brisbane area of Queensland (with notes on shoot growth and litter fall in other areas of Australia). Aust. J. Botany 23: 459–74.
Stamps, D. J., D. E. Shaw and E. G. Cartledge. 1972. Species of Phytophthora and Pythium in Papua New Guinea, P.N.G. Agri. J. 23: 41–45.
Steenis, C. G. G. J. van 1953. Results of the Archbold expeditions Papuan Nothofagus. J. Arnold Arbor. 34: 300–74.
—— 1971a. Nothofagus, key genus of plant geography, in time and space, living and fossil, Ecology and Phylogeny. Blumea 19: 65–98.
—— 1971b. Revision of Nothofagus in New Caledonia. Adansonia 11: 615–24.
—— 1975. Miscellaneous notes on New Guinea plants, 11, (Nothofagus crenata at Lake Kutubu). Blumea 22: 170–72.

Turner, J. C. 1976. An altitudinal transect in rainforest in the Barrington Tops area. Aust. J. Ecol. 1: 155–74.
Veblen, T. and D. Ashton. 1978. Catastrophic influences on the vegetation of the Valdivian Andes, Chile. Vegetatio 36: 149–68.
Volck, H. E. 1968. Silvicultural research and management in North Queensland rainforest, 9th Commonwealth Forestry Conference, New Delhi, 20pp.
Walker, D. 1966. Vegetation of the Lake Ipea region, New Guinea Highlands. J. Ecology 57: 503–33.
—— 1970. The changing vegetation of the Montane tropics. Search 1: 217–21.
Walker, D. and G. Whittman. 1965. Notes on the pollen morphology of Nothofagus Bl. subsection Bipartitae Steen. Pollen et Spores 7: 457–64.
Wardle, J. A. 1970. The ecology of Nothofagus Solandri (1–4). New Zealand J. Botany 8: 494–646.
Wardle, P. 1964. Facets of the distribution of forest vegetation in New Zealand. New Zealand J. Botany 2: 352–66.
—— 1967. Biological flora of New Zealand, 2, Nothofagus menziesii Oerst. (Fagaceae) Silver Beech. New Zealand J. Botany 5: 276–302.
—— 1970. An explanation for alpine timberlines. New Zealand J. Botany 9: 371–402.
—— 1973. New Guinea: Our tropical counterpart. Tuatara 20: 113–24.
Weinberger, P. 1973. The regeneration of the Araucario Patagonic Nothofagus species in relation to microclimatic conditions. Flora (Jena) 162: 157–79.
Weste, G. 1975. Pathogenicity of Phytophthora cinnamomi towards Nothofagus cunninghamii. Aust. J. Botany 23: 277–83.

6. The Araucaria forests of New Guinea

N. J. Enright

1. Introduction

There are two species of *Araucaria* in New Guinea, *A. cunninghamii* D. Don and *A. hunsteinii* K. Schumm. Forests containing these species are of great interest to biologists for several reasons. They are among the most ancient and phylogenetically primitive of the world's surviving conifers, are held in special regard (e.g. superstitious and religious) by the indigenous people in many parts of New Guinea, and are an economically valuable source of timber. Their ecology is of particular interest as most gymnosperms, and most other members of this genus, are absent from wet lowland tropical environments. Competition between species in such areas is generally considered to be intense and one might expect the evolutionary more advanced angiosperms to be advantaged in the struggle for survival.

2. Taxonomy

The araucarias belong to the family Araucariaceae within the order Coniferales. There are two genera in the family; *Araucaria* (19 species) and *Agathis* (about 20 species, Whitmore 1975). *Araucaria* has awl-shaped or lanceolate leaves, and female cones consist of a stout central axis bearing crowded, spirally-disposed cone scales, while *Agathis* has flat, broad leaves and seeds are winged.

The genus *Araucaria* Jussieu has been divided into three sections, although according to Ntima (1968) there is not full agreement on the division among taxonomists. The sections are;

1) *Colymbea*; characterised by flat or broad leaves, large cones, and 2 cotyledons which remain below ground in germination (hypogeal). The section includes the two S American species *Araucaria araucana* and *A. angustifolia* as well as the Australian species *A. bidwillii*. Some authors (e.g. Florin 1963) recognise *A. bidwillii* as belonging to the monotypic section *Bunya*.
2) *Intermedia*; a monotypic section containing only the New Guinea species *A. hunsteinii* K. Schumm ('klinki pine'; Fig. 7). Its characters tend to be intermediate between those of the sections *Colymbea* and *Eutacta*. Juvenile

Monographiae Biologicae, Vol. 42, ed. by J. L. Gressitt.

© 1982, Dr W. Junk Publishers, The Hague.

leaves are awl-shaped, adult leaves flat and broad, cotyledons are normally two and are pushed above the ground in germination (epigeal).

3) *Eutacta*; the largest section, has 15 species including *A. cunninghamii* D. Don in Australia and New Guinea, *A. heterophylla* on Norfolk I., and 13 species in New Caledonia. Leaves are awl-shaped and curved, cones tend to be much smaller than in the other sections, and cotyledons are commonly four and are epigeal.

3. Paleogeography and present distribution

Fossils of *Araucaria* have been reported from as far back as the Triassic, in Arizona (Axelrod, pers. comm.). This is followed by early Jurassic of India (Florin 1963), late Jurassic in Australia and New Zealand, and Mesozoic in S America, all from the section *Eutacta*. This section appears to have died out in India and New Zealand in the Mesozoic, and in S America in the Eocene.

The section *Colymbea* is first known from the middle Jurassic in New Zealand and E Australia, the Mesozoic in S America and Eocene in SW Australia. It still survives only in NE Australia and S America.

No fossil evidence is available for the section *Intermedia* and it is probable that the one species in this section is fairly recently evolved (Ntima 1968). Some Jurassic fossils in N America and Europe appear to be from an *Araucaria*-like group (Seward and Ford 1906; Florin 1963).

At present, the genus is wholly confined to the Southern Hemisphere and, as noted above, species are found in S America, E Australia, New Guinea, New Caledonia and Norfolk I. Similar mechanisms to those suggested for *Nothofagus* by Ash (preceding chapter) must also be invoked to explain the present distribution of the genus *Araucaria*. The most plausible of these suggests a widespread distribution of *Araucaria* in Gondwanaland prior to its break-up about 130 million years BP. The two species of *Araucaria* considered here are both restricted to the Australasian region.

The distribution of *A. cunninghamii* in New Guinea is discontinuous with most stands confined to the central highlands and Owen Stanley Ranges within the altitudinal zone from 500 m to 2450 m (Fig. 1). The areal limits of the species are defined by the Vogelkop, Irian Jaya, in the NW (132° E, 0°30′ S) and Milne Bay, Papua New Guinea, in the SE (151° E, 11° S). Gray (1973) has mapped the distribution of *A. cunninghamii* in Papua New Guinea and notes differences in the appearance of trees presumably related to changes in local environment. Among the sites for which variations are noted are: Bulolo, which has the largest trees; Pai-awa, where trees of variable morphology occur on serpentine at very low altitude; Telefomin, where trees appear stunted; and Erave, where a stand occurs on a swamp. Fergusson I. and Japen I. are the only small islands in the New Guinea region on which *Araucaria* grows. The distribution of this species in Irian Jaya has not been mapped, but Zieck (pers. comm.) notes that sizeable stands occur on the Vogelkop, near Fak Fak, in the central highlands, Cyclops Mts, some of the lower ranges to the north of the central cordillera and on Japen I.

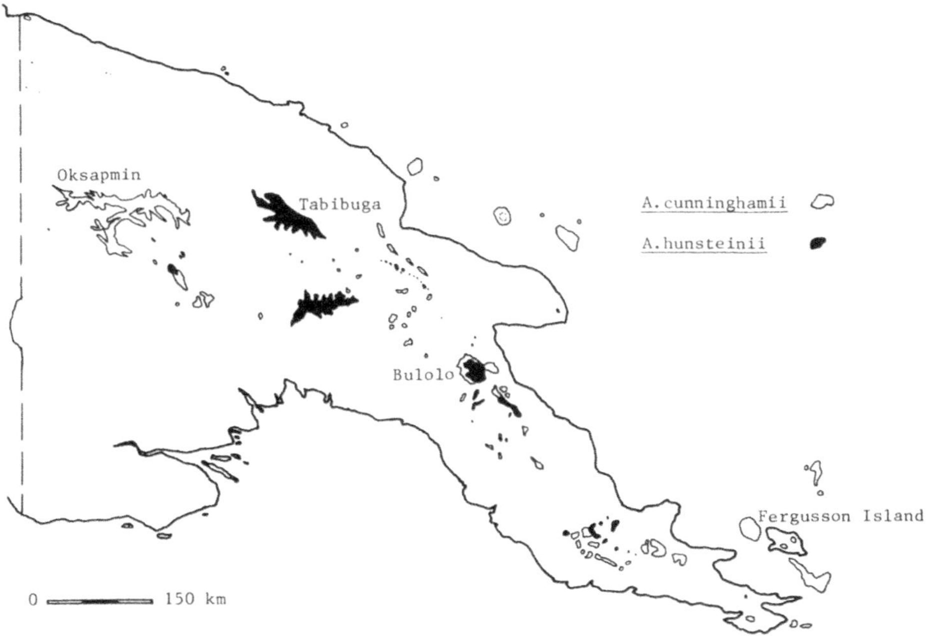

Fig. 1. The distribution of *Araucaria cunninghamii* and *A. hunsteinii* in Papua New Guinea.

The distribution of *A. hunsteinii* is more restricted; the species is endemic to Papua New Guinea. The altitudinal ranges of the two species overlap but *A. hunsteinii* prefers the lower, moister sites of the intermontane valleys, rising to an elevation of about 2100 m (Gray 1973, 1975; Havel 1971). Gray (1975, p. 285) notes that within this zone of overlap "almost invariably one species predominates in a given site."

The range of climatic and soil conditions supporting *A. cunninghamii* and *A. hunsteinii* is large. Rainfall ranges from 800 mm to over 4000 mm per annum (Gray 1973; Webb and Tracey 1967), and mean annual temperature from 11°C with frosts common, to 26°C without frosts. Soils are usually neutral to basic and coarse textured in Australia, but in New Guinea stands on clayey, acidic sites are common.

According to Webb (1959, 1968) *Araucaria* is usually a component of tropical vine forests in Australia, but in New Guinea it occurs also in monsoonal forests, lowland rainforest, and occasionally in cloud forest (rarely above the *Nothofagus* forest). The density of stands also varies greatly, from the very dense stands at Bulolo on both metamorphic and sedimentary parent materials and on limestone sites elsewhere in Papua New Guinea, to the small isolated ('remnant') stands such as those of the central highlands in general.

3. Previous work

Considerable controversy has surrounded the statement by Womersley (1958) that "the *Araucaria* forests of Australia and New Guinea represent a living fossil vegetation which is now slowly being swamped by the tide of broad-leaved forest."

This statement derives partly from the Jimi Valley Patrol (Womersley 1955), a botanical expedition led by Womersley, during which large, non-regenerating stands of *A. hunsteinii* were found. According to the local natives this was due to poor seed production and premature fall of cones. It has since been suggested (Johns pers. comm.) that the area had been cultivated by the natives until the turn of the century, the valley then having been depopulated as a result of the introduction of cerebral malaria. The forest has since passed through a secondary succession leaving *A. hunsteinii* as a remnant of the original forest left for religious or superstitious reasons. A young, closed forest of this kind may well be unsuitable for *Araucaria* regeneration and does not represent a situation typical of *Araucaria* forest. The failure of massive regeneration following logging in the Bulolo area has also been cited as a possible factor influencing the development of Womersley's hypothesis (Havel 1971).

Inherent in the argument of Womersley (1958) is the idea that, in the past, New Guinea and Australia must have harbored large tracts of conifer-dominated forest which have now been all but destroyed. Aubreville (1964, 1965) and Webb and Tracey (1967) also believe that the *Araucaria* forests of E Australia must once have been continuous, but subsequently upset by climatic change and (possibly) human interference during the late Pleistocene and Holocene. This led to the isolation of *A. cunninghamii* dominated forests in fireproof gullies ('remnant' pockets) and 'sub-climax' situations within dry vine forests.

These viewpoints have gained support in recent years through the tropical vegetation history work of Kershaw (1973, 1974, 1976, 1978) who examined pollen sequences one of which dated back to an estimated 100,000 years BP. His sites were a series of volcanic crater lakes on the Atherton Tableland, N Queensland. They covered the altitudinal range from 600 to 670 m and were located in a sensitive ecotonal area between complex mesophyll vine forest and dry sclerophyll woodland dominated by *Eucalyptus* species. Kershaw considers that prior to ca 38,000 years BP the Atherton area was dominated by a conifer forest assemblage which included *Araucaria*, *Podocarpus*, and *Dacrydium*. This assemblage is suggestive of a climate drier (about 900 to 1400 mm rainfall per annum) and possibly cooler than today. After 38,000 years BP these genera disappeared, and were replaced by *Eucalyptus* and *Casuarina* perhaps indicating a further drop in precipitation. By 7000 years BP precipitation is thought to have risen to its present level (2500 mm per annum) and rainforest to have re-invaded the area. However, the new rainforest was dominated by angiosperms rather than conifers. The conifer assemblage present in the lower part of the core is no longer represented in Australia, the closest occurrences of *Dacrydium* being in Tasmania and New Guinea, whilst *A. cunninghamii* is not found in the immediate vicinity of Atherton, but is present in small, disjunct, pockets elsewhere in northern Queensland.

Kershaw (1976) suggests that the reasons for elimination of the conifer dominated forests must relate to climatic change and also, perhaps, to interference by aboriginal man (e.g. fire) who is now known to have been in southern Australia for over 45000 years (Bowler 1976) and New Guinea for at least 26000 years (Powell 1970; White *et al.* 1970). All southern conifers from rainforest environments are extremely fire sensitive, and those being adapted to the drier, marginal types of rainforest would have been seriously affected by any increase in fire frequency.

All the genera of conifers noted in Kershaw's (1973, 1976) conifer dominated assemblage are still present in the montane forests of New Guinea. However, *Araucaria*, *Podocarpus* and *Dacrydium* do not commonly occur together in any abundance. Although both *Araucaria* and *Dacrydium* are often associated with podocarps in different altitudinal zones, no living analogues for Kershaw's pollen assemblages are known. Rymer (1978), Walker (1978), and other palynologists, have acknowledged the problem of attempting to reconstruct the environmental regime of a pollen assemblage no longer represented by a living analogue.

On a timescale of thousands of years, perhaps, Womersley's remarks are justified. While there is no definite indication of the timescale Womersley was envisaging, remarks in the Jimi Valley Patrol (Womersley 1955) and by Havel (1971) suggest that he thought the process to be occurring in the stands he saw in the 1950s. Havel (1971) and Gray (1975) have examined this question and both argue that the two New Guinea species of *Araucaria* are regenerating adequately.

4. Ecology

Havel (1971) presents five altitudinal and topographic transects from the Bulolo-Watut area and, although noting that *A. cunninghamii* tends to occur above *A. hunsteinii* in the altitudinal sequence, summarizes their regeneration ecologies as if only one species is involved:

> "The series presented here illustrate the fact that the optimum development of araucarian regeneration occurs under lower and more open canopy, induced by lower rainfall, steeper topography or disturbance. The increased depth of organic matter on the forest floor and the reduction of light intensity by cloudiness at high altitudes generally does not favour regeneration" (Havel 1971, p. 209).

Gray (1975) examined 8 plots each 50 × 50 m in size; 4 containing *A. hunsteinii* and 4 containing *A. cunninghamii*. Also, 1000 trees >3 cm dbh from each species were enumerated "in the densest natural stands of *Araucaria* known to occur" (Gray 1975, p. 274). From these enumerations, Gray (1975, p. 281) concluded that

> "both species displayed a positively skewed size-class frequency distribution indicative of a self-perpetuating species, not of a declining species."

Such remarks concerning dynamic status of the populations may be true, but seem unjustified in light of the small amount of evidence collected; a positively-skewed distribution need not imply a 'self-perpetuating' species population.

4.1. *Size of individuals*

According to Gray (1975), *A. hunsteinii* is the tallest tree species in the tropics; one individual measured 88.9 m in height and over 2 m in diameter at 30 cm above ground level. Few trees more than 80 m still survive, however, and individuals of this size may always have been rare. Figs. 2 and 3 present data on height and diameter relations of *A. hunsteinii* and *A. cunninghamii* respectively.

Height to the first branch, or first whorl of branches, for mature trees is not related to total tree height, but rather to height of the general angiosperm canopy. Regardless of total height, the first whorl characteristically appears between 20 and 30 m above the ground on emergent trees. This means that most or all available leaf area is above the angiosperm canopy in a high light and relatively low humidity environment. Mature individuals, especially those of *A. hunsteinii*, thus interfere little with the angiosperm forest canopy and represent a spatially separate canopy layer.

4.2. *Density, basal area and biomass*

Density figures for *Araucaria* stands compare well with general figures for tropical rainforest (Paijmans 1970). Mean densities of 510 and 413 trees > 10 cm dbh per ha were found for forests dominated by *A. cunninghammii* and *A. hunsteinii* respectively. Paijmans (1970) obtained densities of from 430 to 691 individuals per ha for the same size range in lowland and hill forest in Papua New Guinea. Values obtained for tropical forests in Borneo (Ashton 1964), Malaya (Wyatt-Smith 1949), Nigeria (Richards 1939, 1952) and British Guiana

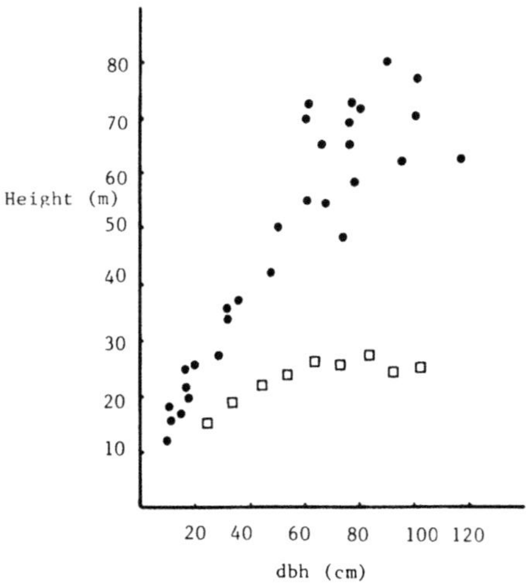

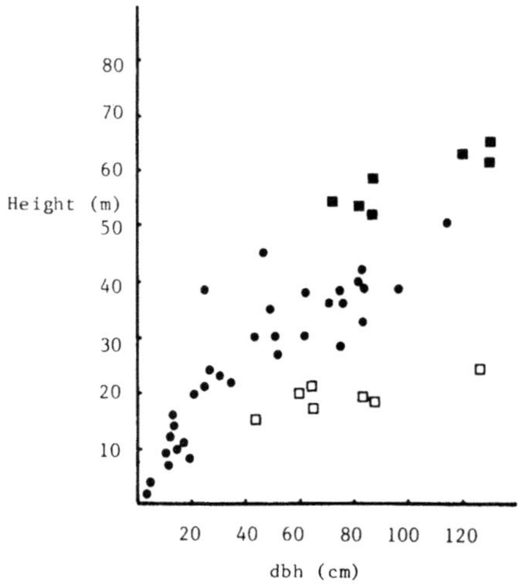

Fig. 2. The relationship between height and diameter in *A. hunsteinii*: total height (●) and mean height to first whorl (□).

Fig. 3. The relationship between height and diameter in *A. cunninghamii*: total height (●), height to first whorl (□) and total height after Gray (1975) (■).

(Davis and Richards 1933) are also mostly within this range (432 to 781).

Basal area and biomass estimates for tropical lowland and lower montane forests are now available from a number of sources (Bray and Gorham 1964; Rodin and Basilevich 1967; Edwards and Grubb 1977). Most basal area values fall in the range from 25 to 45 m^{-2}/ha with the highest recordings being about 60 m^{-2}/ha for cloud forest in S America (Lamprecht 1972). The basal areas of most lower montane forests in New Guinea are also within this range. Plots dominated by *A. hunsteinii* forest approach, and may exceed, 60 m^{-2}/ha in basal area (e.g. one plot examined had a basal area of 53 m^{-2}/ha). As araucarias are the tallest trees in the tropics it is possible that the total biomass of *Araucaria* dominated forests may also be the highest in the tropics.

The regression equation derived by Edwards and Grubb (1977) for biomass estimation of lower montane rainforest in Papua New Guinea has been used to calculate the estimated above-ground biomass of plots which contain *A. hunsteinii* and *A. cunninghamii*. The equation is

$$y = 0.237x - 1.6574$$

where y is the square root of dry weight (kg) and x is the girth of the tree (cm). Edwards and Grubb (1977) derived this equation for all trees >7 cm dbh and multiplied by 1.1 to account for their leaves and twigs and for smaller individuals. Estimates obtained for the *Araucaria* plots were based on individuals > 10 cm dbh and a correction factor of 1.2 has been used. A number of workers have used correction factors ranging from 1.1 to 1.3 (see Edwards and Grubb 1977). Values of 554 t/ha for a sample *A. hunsteinii* plot and 512 t/ha for a sample *A. cunninghamii* plot were obtained. These plots, containing adult *Araucaria* populations, had biomass values above the upper limit recorded for other tropical forests.

Paijmans (1970), in an analysis of four New Guinea rainforest plots, noted that the basal area of a stand dominated by *A. hunsteinii* was considerably higher than that of other plots, and furthermore, that if the *Araucaria* component was removed the basal area was almost identical to that of the plots without *Araucaria*. An examination of 15 plots containing *A. hunsteinii* confirms this observation as statistically significant (Enright 1978) (Fig. 4).

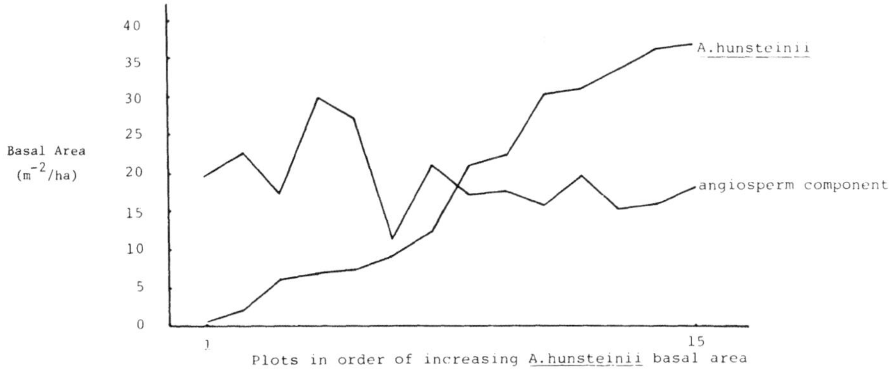

Fig. 4. Basal areas of *A. hunsteinii* and the angiosperm component in 15 × 0.5 ha plots (m^{-2}/ha).

While the variance of *A. hunsteinii* basal area was extremely high (176.8), the variance of the angiosperm component was low (23.6), that is, there was little variation in angiosperm basal area regardless of the amount of *A. hunsteinii* in the plot. Examination of data on *A. cunninghamii* from New Guinea and *Mora gonggrijpii* and *Ocotea rodiaei* from Guyana (Ogden 1966) indicates that this is not a common occurrence in tropical emergent or canopy dominant trees and may be unique. It suggests that *A. hunsteinii* does not compete for resources with the other rainforest tree species. This is probably only true for trees which have reached or penetrated the canopy as the proportion of *A. hunsteinii* basal area made up by subcanopy individuals is usually negligible ($<5\%$). *A. hunsteinii* is a sparsely branched emergent. No branching occurs below, or in, the main 'angiosperm' canopy layer, and the sparse nature of the foliage means that light intensity reaching the broadleaf canopy may be only very slightly reduced. Thus *A. hunsteinii* does not occupy canopy space which would otherwise be occupied by broadleaf species (except for the basal area of the trunk) and may not reduce their photosynthetic efficiency. Moreover it probably causes little disturbance to the canopy layer even as it grows through it, as in the immature form found under relatively low light conditions, its canopy occupies only a small area and is composed of only one whorl of branches. The apparent independence from competition of emergent individuals of *A. hunsteinii* suggests that the population must be checked at an earlier point in its life-history.

In forests containing *A. cunninghamii*, independence between the basal areas of *Araucaria* and angiosperms has not been observed. This reflects the fact that the branching system and foliage of *A. cunninghamii* is denser and often occurs in the angiosperm canopy layer. A reduction in the space available for other species within the canopy may thus occur. In addition, the light intensity reaching the canopy is reduced more markedly than in forests dominated by *A. hunsteinii*. There are also differences in the root systems of the two species which may help to explain how dense stands of *A. hunsteinii* develop with, apparently, almost no effect on the species associated with them. *A. hunsteinii* has a root system comprised almost entirely of 'short-roots' while *A. cunninghamii* has a dense, fibrous root system. Bevege (1972) considers that these short-roots may be indicative of tree species having an efficient, mycorrhiza-mediated mechanism for the uptake of phosphorus and nitrogen.

4.3. The distribution of young and mature Araucaria *individuals*

Regeneration patterns of the two *Araucaria* species differ markedly. *A. hunsteinii* does not regenerate successfully under a closed canopy where light intensity is low and/or the litter layer is very deep. Preferred areas are those of local disturbance (e.g. tree-fall). In the centre of such openings, often approximately 30 m wide, early secondary species such as those of *Macaranga*, *Homalanthus*, *Mallotus* and *Pipturus* grow up rapidly and *A. hunsteinii* seedlings are unable to compete successfully. They do succeed, however, along the margins of the gaps where light intensity has been increased substantially, but not enough to allow invasion of the fast growing species noted above. Here, seedlings survive in large numbers and some are ultimately recruited to the canopy. These groups may not appear even-aged due to the effects of competition and suppression on growth of some individuals. Moreover, if an area larger than 1 hectare were examined one would expect most size-classes to be represented because several such clumps of different age may be included.

In contrast to *A. hunsteinii*, *A. cunninghamii* does appear able to survive under low light conditions for long periods of time. Covariance pattern analyses have shown that *A. hunsteinii* saplings and mature trees are negatively associated (Enright 1978). *A. cunninghamii* populations show no such separation and all size-classes are normally present within small sample plots (0.2–0.5 ha). Segregation analysis has confirmed these differences as significant in a number of plots tested. This method of analysis examines the hypothesis that the size of a tree's nearest-neighbor is independent of its own size. If this were the case it would suggest that saplings of a given species were quite capable of regenerating in close proximity to parent trees.

Sample stands of *A. hunsteinii* were tested for segregation between mature trees (>25 cm dbh), young trees (5–20 cm dbh) and saplings (1–5 cm dbh), and in all cases it was found that there was significant spatial segregation. This implies that regeneration is unlikely to occur in the vicinity of groups of parent trees, or near groups of young developing trees of the same species. It thus seems likely that successful regeneration occurs in patches which grow up *en masse* in areas adjacent to parent stands where light intensity at ground level has been increased through the opening up of the forest by tree-fall.

Examination of sample *A. cunninghamii* stands revealed significant segregation only rarely. It is apparent that, in *A. cunninghamii*, the distribution of all individuals is probably random, and therefore that *A. cunninghamii* is able to regenerate successfully within quite dense parent stands. The patterns expressed may best be seen as 'obligative' in the case of *A. hunsteinii* and 'facultative' in the case of *A. cunninghamii*. Successful regeneration of *A. hunsteinii* is dependent upon relatively high light intensities at ground level, and absence of parent trees in the immediate vicinity (approx. 30 m), while *A. cunninghamii*, although it regenerates under closed canopies and in close proximity to parent trees, may also be found in regeneration clumps in areas of higher light intensity such as tree-fall gaps. This suggests that *A. cunninghamii* is the greater 'generalist' of the two species and this possibility is supported by other known facts about these trees such as altitudinal and latitudinal ranges, soil parent material and rainfall amplitudes.

4.4. Growth rates and age of trees

Average annual increment data and estimated tree age for *A. hunsteinii* based on 18 years of data are presented in table 1. The estimated age of an individual 120 cm dbh and which has grown at the mean rate throughout its life is 465 years. However, when an allowance is made for the time required to reach the height at which diameter measures were made (1.4 m) this estimate is increased to 493 years.

Growth rates tend to be at a maximum during the young-mature and mature stages (i.e. 30–60 cm dbh and 60–90 cm dbh). They are somewhat slower while the tree is beneath the canopy and when individuals become overmature (senescent). Individuals normally attain emergent status (i.e. penetrate the continuous angiosperm canopy) at a size of about 35 cm dbh and it is not until after this that they become reproductively mature. Growth rate increases markedly once the tree is in a full sunlight situation and remains high so long as the individual is healthy. Slowing of the growth rate in large trees is generally credited to senescence, although the causes are not well established. Perhaps, increasing size becomes disadvantageous due to difficulties of translocation within the plant. Long-term (chronic) diseases such as heart-rot may start early in the life of the tree, but only slowly begin to affect its performance. Once the photosynthesis: respiration balance is close to 1, any damage to the cambial layer (through adjacent tree-fall, etc.) which requires the tree to allocate resources for repair is likely to reduce the growth rate.

The confidence limits on mean growth rate per size-class (table 1) are widest in sub-canopy individuals and it is at this stage that major differences in growth response between trees are occurring. While some trees are growing at the mean rate, others may be markedly suppressed, and a few in favorable sites may be 'released.'

Growth data for *A. cunninghamii* are only available for a two-year period and estimates of age are, therefore, initial approximations only. The estimated age of an individual 120 cm dbh is 305 years (including the time taken to reach measurement height). This is considerably less than the estimated 493 years needed for *A. hunsteinii* to reach the same size.

4.5. Insect pests of Araucaria *species in New Guinea*

One of the major causes of death in *Araucaria* populations, besides shade-induced suppression, is attack by insect predators. *A. cunninghamii* is the more

Table 1. Age estimate of Araucaria hunsteinii based on 18 years of annual increment data

Number of Individuals	Size-Class (cm dbh)	Average Annual Increment (cm)	95% Limits	Years in Size-Class	Cumulative Age
2	2–10	.58	±2.00	13.8	13.8
11	10–20	.20	±0.14	50	63.8
11	20–30	.25	±0.15	50	103.8
20	30–40	.26	±0.09	38.5	142.3
18	40–50	.37	±0.11	27	169.3
23	50–60	.24	±0.07	41.7	211
19	60–70	.36	±0.05	27.8	238.8
20	70–80	.24	±0.07	41.7	280.5
19	80–90	.28	±0.05	35.7	316.2
8	90–100	.26	±0.09	38.5	354.7
4	100–110	.16	±0.57	62.5	417.2
2	110–120	.21		47.6	464.8
			Estimated Age at 120 cm dbh = 464.8		

susceptible of the two species to insect attack, suffering especially from infestations by the host-specific scolytid *Hylurdrectonus araucariae* as well as *Vanapa oberthuri*, *Diotimana undata* and *Coptotermes elisae*. Of these, only the last is likely to cause death in trees of *A. hunsteinii* (Gray 1968). Most insect pests are of little significance in natural stands where the low density (compared with plantations) and large disjunctions in population distribution make it difficult for insects to locate suitable host trees. Many, also, for example the Platypodidae, are secondary insects that rely on initial tree injury being caused by other insect species, fire, lightning, etc. before they are able to penetrate the stem. Of the major pests, three are described.

1) *Hylurdrectonus araucariae* is a branchlet-mining scolytid which attacks mostly young trees with juvenile foliage. Attacks on seedlings or mature individuals are rarely successful. Between 1000 and 2000 branchlets on any given tree may be infested, resultant branchlet death leaving some young trees almost completely defoliated.
2) *Vanapa oberthuri* is a weevil which lays its eggs in resin flows on the stem. Upon hatching the larvae tunnel into the cambium and wood allowing secondary invasion by a number of other borer-insects which hastens the death of the tree. Trees <6 cm dbh are generally not attacked.
3) *Coptotermes elisaë* is a termite capable of infesting and killing both species of *Araucaria* in Papua New Guinea. Nests are normally located in the taproot 30–60 cm below ground level with galleries extending as much as 50 m up the trunk of large trees (Gray 1968). These trees are prone to windthrow and to secondary infestation by insect borers. Usually a clump of trees is affected and pockets of dead trees may result.

A number of as yet unidentified members of the Microlepidoptera also cause much wastage in cones of *A. cunninghamii* and *A. hunsteinii*.

4.6. Other predators

Two other groups of organisms may affect survival in natural populations of *Araucaria* in New Guinea; fungi and birds. Woodrot fungi (e.g. *Schizophyllum commune* and *Neurospora* sp.) have been noted in fire-damaged *A. cunninghamii* trees in the Bulolo area (Wylie and Shanahan 1976). Sapstain has been noted in both species and is probably caused by *Botryodiplodia theobromae* (this has been confirmed for *A. cunninghamii* only).

The foraging activity of birds such as the Sulphur-Crested Cockatoo (*Cacatua galerita*) may lead to destruction of half or more of annual cone crops. Department of Forests (PNG) records suggest that over 50% of the *A. hunsteinii* cone crop in the Susu Mountain area, Bulolo was destroyed by cockatoos in successive years (1966 and 1967). *Cacatua galerita* is a notoriously wasteful eater and most seed is lost through the premature fall of green cones which are left to rot on the forest floor.

Wild pigs and cassowaries may be of minor importance in the destruction of seedlings by trampling, but no data are available on this subject.

Ordination of data from a number of 0.5 ha forest plots in New Guinea within the 700 to 1500 m altitudinal zone, using the non-linear ordination technique Continuity Analysis, has differentiated between several types of *Araucaria* forest, each with distinctive floristic and environmental characteristics (Enright 1978). The most marked distinction is between forests containing one of the two species. *A. cunninghamii* is more common above 1000 m in the lower part of the *Castanopsis acuminatissima* and *Lithocarpus* sp. forest zone. Typically associated trees include *Castanopsis acuminatissima*, *Calophyllum* sp., *Schizomeria* sp., *Cinnamomum* sp., *Podocarpus amarus* (= *Prumnopitys amara*) and *Podocarpus neriifolius*. The forest canopy is normally at about 30 m with *A. cunninghamii* emergent to 50–60 m. A thick layer (> 10 cm) of semi-decomposed litter lies on the ground keeping the soil moist almost continuously. Surface feeder roots knit this organic matter together, collecting essential nutrients as they become available from it. The soil below is generally heavily leached and podzolized due to the high rainfall (> 2000 mm/annum) and is consequently low in nutrients.

In the Bulolo valley, *A. hunsteinii* is most common between 700 and 1000 m. Two major forest types containing *A. hunsteinii* may be identified; a dry, semi-deciduous association where rainfall is below 1600 mm/annum, and a moist, tall association with higher rainfall and with many lowland species present. These two types of forest have been noted elsewhere in New Guinea, as for example in the Musa valley (Paijmans pers. comm.).

Forest canopy height in the dry, semi-deciduous forest is only 15–25 m with occasional broadleaved trees reaching 35 m. *A. hunsteinii* rises to over twice the height of the general canopy and is clear boled until well above this zone. Associated species include *Pouteria luzonensis*, *Aleurites moluccana*, *Celtis* sp., *Heritiera* sp. and *Macaranga* sp., all of which may be present in large numbers, and the deciduous species *Garuga floribunda*, *Protium macgregorii*, *Terminalia* sp. and *Sterculia* sp. Less frequent but characteristic species are *Mangifera minor*, *Geijera* sp., *Linociera ramiflora* and *Harpullia* sp.

In higher rainfall areas within the lowland/lower montane forest transition *A. hunsteinii* may be the dominant species of a taller, closed forest type which has many lowland broadleaved species in its canopy. It is in this environment that *A. hunsteinii* attains its maximum height of 70–90 m. In this forest type regeneration is never as evident as in the dry association where the often broken canopy ensures sufficient light at ground level for rapid initial growth (Havel 1971). Commonly associated species of the canopy layer are *Ficus* sp., *Elmerillia papuana*, *Xanthophyllum papuanum*, *Flindersia pimenteliana*, *Acmena* sp., *Pometia pinnata* and *Flindersia amboinensis* all of which may reach heights of 30–40 m. Other species, in the 10–25 m height range, include *Dysoxylum* sp., *Gnetum gnemon*, *Cerbera floribunda*, *Litsea* sp., *Myristica* sp. and *Cryptocarya* sp. (for species lists of sample plots see Johns 1976). Within this association certain areas differ markedly in species densities of the broadleaf components. The two most abundant groupings in the Bulolo area are; (a) *A. hunsteinii* with *Xanthophyllum*, *Gnetum*, *Flindersia* and *Elaeocarpus*, and (b) *A. hunsteinii* with

Pometia, *Myristica*, *Sloanea* and *Dysoxylum*. Whether these groupings represent an environmental difference, or variations in site history is not known. An ordination of species based on quantitative data (basal area) from 26 sites is presented in Fig. 5.

These forest types are not always as clearly defined as may be implied above. *A. hunsteinii* can be found in association with *Castanopsis acuminatissima* above

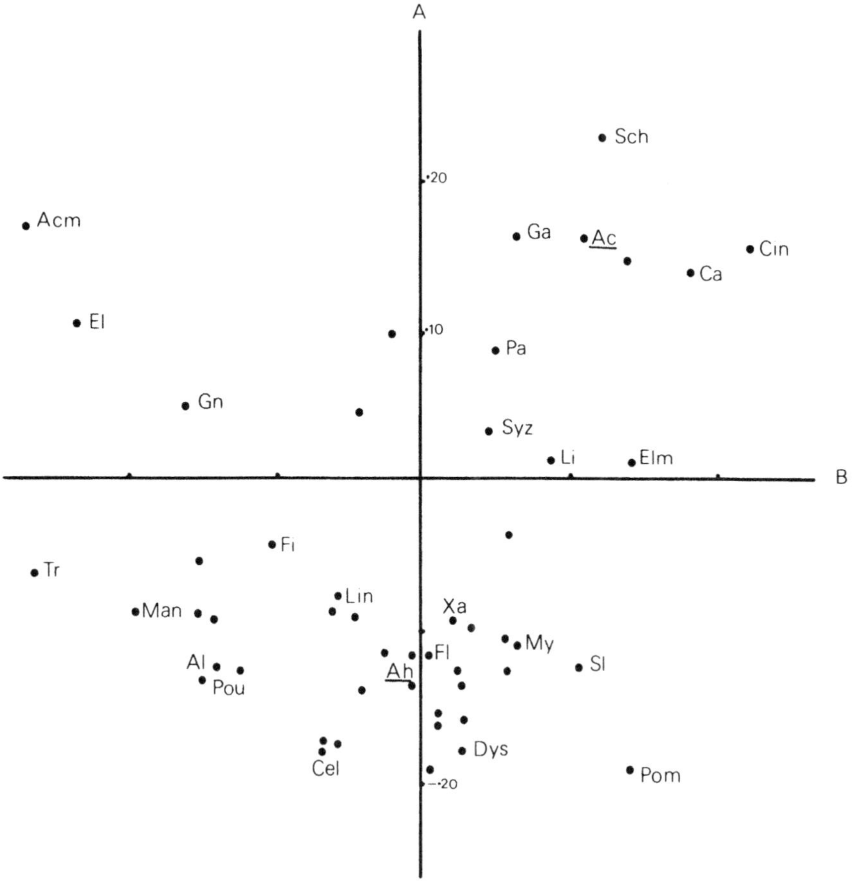

Fig. 5. Non-linear ordination of 50 species in 26 sites using basal area data. Increasing distance between species, or groups of species, indicates increasing dissimilarity between forest types in which they occur. Vertical and horizontal scales are identical.

Key to species:

Ac	Araucaria cunninghamii	Gn	Gnetum gnemon
Ah	Araucaria hunsteinii	Lin	Linociera ramiflora
Acm	Acmena sp.	Li	Lithocarpus sp.
Al	Aleurites moluccana	Man	Mangifera minor
Ca	Castanopsis acuminatissima	My	Myristica sp.
Cel	Celtis latifolia	Pom	Pometia pinnata
Cin	Cinnamomum sp.	Pou	Pouteria luzonensis
Dys	Dysoxylum sp.	Pa	Pandaunus sp.
El	Elmerrillia papuana	Sl	Sloanea sp.
Fi	Ficus sp.	Syz	Syzygium sp.
Fl	Flacourtia sp.	Tr	Trema orientalis
Ga	Gastonia sp.	Xa	Xanthophyllum papuanum

1000 m while *A. cunninghamii* is quite common at altitudes below this in patches thought to represent relic stands, and in obviously secondary forest situations.

There are two major environmental gradients which parallel the distribution of species and assemblages. These are related to rainfall and temperature on the one hand, and soil nutrients on the other. *A. cunninghamii* occurs most commonly above 1000 m in cooler, wetter forests than *A. hunsteinii*. These populations are growing in highly acidic (pH < 5.0) and slightly podzolized soils low in nutrients. *A. hunsteinii* is found in dry *Pouteria luzonensis* forests, where nutrient concentrations, especially of calcium, are high and pH is neutral to basic, and in moist *Flindersia pimenteliana* and *Gnetum gnemom* forests where nutrient concentrations are slightly lower and soils more acidic (pH 5.5 to 6.5). This species does not occur on highly acidic, nutrient-poor, soils and is rarely present in large numbers at altitudes above 1000 m.

Ordination results also provide information on changes in species composition with increasing altitude. Walker (1973) and Walker and Guppy (1976) maintain that there is a convincing floristic break in the generic composition of montane forests at about 3000 m in New Guinea. Other authors (Johns 1976; Robbins' 1961; Paijmans' 1976; Lane-Poole 1925) recognize further structural breaks at lower altitudes and associated, but less clear floristic changes.

The analysis of quantitative data (density, basal area) illustrates that there are major disjunctions between forest types on the altitudinal gradient between 700 m and 1500 m (Enright 1978). The main disjunction divides forests containing many lowland rainforest species from 'Oak' forest which is characterized by the presence of the fagaceous genera *Castanopsis* and *Lithocarpus*. *A. cunninghamii* is more common than *A. hunsteinii* in oak forest and many montane species have their lowest altitudinal occurrence in this zone.

If abundance is ignored and floristic composition only examined, then there is no abrupt change in species composition over this altitudinal range. But, a distinct change is noticeable in the field as the new species which enter the forest at about 1000 m are also those which are the most abundant (i.e. *Castanopsis acuminatissima*, *Lithocarpus* sp., *A. cunninghamii*).

4.8. Succession in Araucaria *Forest*

The two *Araucaria* species have distinctive regeneration strategies and *A. cunninghamii* appears better adapted to stable conditions ('K-selected'). However, there are differences between the species which do not conform to their relative positions on the continuum of regeneration behavior from 'r' to 'K'. *A. hunsteinii* has fewer, larger seeds and a slower growth rate than *A. cunninghamii*, features of a 'K-' rather than an 'r-selected' tree species. The large seeds of *A. hunsteinii* have a low dispersal ability (Havel 1965) but because the frequency of small-scale perturbations (e.g. tree-fall) is high the species is able to take advantage of habitats suitable for regeneration close to parent stands.[1]

[1]The importance of rare climatic events and the role of birds in extending the normal dispersal range of seeds, however, should not be underestimated.

The large food store in the seed allows seedlings to reach a size at which they may survive in shaded habitats for several years while waiting for a gap to be created. Few survive in such habitats for longer than three years. The food store may also have a bearing on dispersal by birds.

A longer reproductive life may provide *A. hunsteinii* with more opportunities to produce successful offspring and guard against the likelihood of local extinction which may occur if the frequency of perturbations falls for several hundreds of years. The likelihood of regeneration success in *A. cunninghamii* on the other hand, does not alter markedly from one year to the next. This fact, combined with its higher seed production means that this species will produce a larger number of potential offspring than *A. hunsteinii* during a shorter reproductive life.

The differences in geographical distribution between the two species may simply reflect differences in their evolutionary histories, but it may be noted that the smaller seeded, more 'ecologically plastic' species (*A. cunninghamii*) is also the more widespread.

The different regeneration strategies of the two species prescribe the types of communities in which they can occur. *A. hunsteinii* is most common in late secondary forests although the successional sequences within those containing it probably differ between climatic zones. Several such sequences are suggested.

In moist areas (rainfall >1800 mm/annum) *A. hunsteinii* may regenerate in small canopy gaps (<0.1 ha) and around the margins of larger gaps. It is excluded from the center of large gaps by fast growing, early secondary species.

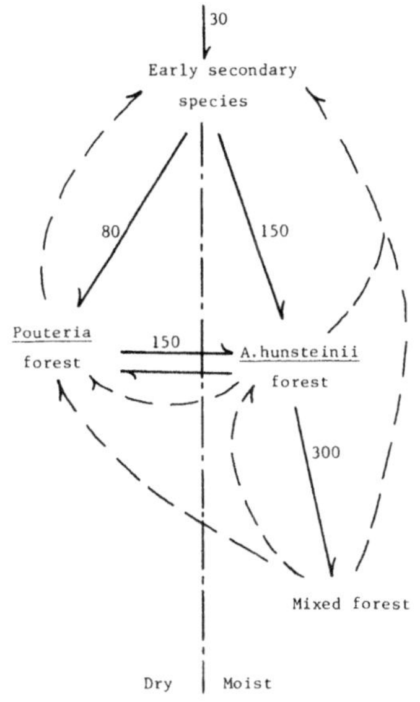

Fig. 6. Possible successional pathways in *A. hunsteinii* forests of the Bulolo area of Papua New Guinea. Approximate timescales in years are included. Three successions are noted:
1) moist forest; early secondary to *A. hunsteinii* to mixed forest (c. 480 years)
2) dry forest; early secondary to *Pouteria* to *A. hunsteinii* forest (c. 260 years)
3) intermediate forest; early secondary to *Pouteria* to *A. hunsteinii* to mixed forest (c. 560 years)
Broken lines indicate perturbation mediated return to an earlier stage.

Fig. 7. Two emergent *Araucaria hunsteinii* ('klinki pine') at Wau. Pale trees in right background are tall emergent araucarias with much *Usnea* lichen, of which some is seen hanging from branches in foreground. (Wau Ecology Institute).

Individuals live for about 500 years so that the composition of the angiospermous component changes beneath a persistent emergent canopy of *A. hunsteinii*. Eventually, as the emergent araucarias die they may be replaced by a mixed species assemblage. However, the creation of canopy gaps seems frequent enough (White 1975) to maintain populations which migrate spatially from one generation to the next over short distances.

In dry areas (rainfall < 1600 mm/annum), and on ridgetops, the successional sequence may end with *A. hunsteinii* dominant or may be cyclical. Forests are dominated by *Pouteria luzonensis* during the early part of secondary succession. *A. hunsteinii* regenerates well within this forest type, probably due to the presence of semi-deciduous species and the more open canopy which allows high light penetration. *Pouteria luzonensis* declines in importance as *A. hunsteinii* becomes the forest dominant, but usually survives at medium densities. Dense populations of *A. hunsteinii* adults cause a marked increase in the depth of litter on the forest floor, increase the water supply reaching the ground, through cloud water interception (Ekern 1964), and slightly reduce light intensity in the sub-canopy. These alterations make conditions less suitable for regeneration of their own offspring. *Pouteria luzonensis* and other species may once again become dominant as the population of *A. hunsteinii* adults declines. Subsequent thinning of the litter layer, reduction in water availability and increase in light intensity will once again favor *A. hunsteinii* regeneration. These two species may thus establish a dynamic equilibrium.

In sites with intermediate moisture conditions, forests may pass through periods of dominance by *Pouteria luzonensis* and *A. hunsteinii*, eventually becoming mixed-forests with no obviously dominant species. These additional species enter while available water supply has been increased by the interception of cloud droplets by *A. hunsteinii* emergents. In the absence of disturbance *A. hunsteinii* declines but the mixed species assemblage may survive because it maintains its own micro-environment. However, after disturbance, such sites will only be suitable for re-invasion by species adapted to drier habitats. These possible successional pathways are illustrated in Fig. 6.

A. cunninghamii can invade sites at an early stage of secondary succession but regenerates well within already established parent populations. There may be a slight change in the floristic composition of the canopy species over time. *A. cunninghamii* appears to be a stable component of the forests above 1000 m, although small shifts in abundances and composition may be expected within any forest purely by chance. Sites on which *A. cunninghamii* is very dense probably reflect its invasion as an early secondary species. Such populations may be expected to decline, but not disappear, as other shade tolerant species enter the area.

References

Ashton, P. S. 1964. Ecological studies in the mixed dipterocarp forests of Brunei State. Oxf. For. Mem. 25.

Aubreville, A. 1964. Les Reliques de la Flore des Coniferes Tropicaux en Australie et en Nouvelle Caledonie. Adansonia 4: 181–192.

Aubreville, A. 1965. Les forets tropicales denses Australiennes et leurs coniferes. Revue Bois et Forets des Tropiques 104: 3–16.
Bevege, D. I. 1972. Vesicular-arbuscular mycorrhizas of Araucaria: aspects of their ecology and physiology and role in nitrogen fixation. Ph.D. Thesis. University of New England. Armidale.
Bowler, J. M. 1976. Recent developments in reconstructing late Quaternary environments in Australia. In: Kirk, R. L. and A. G. Thorne (eds) The Origin of the Australians. A.I.A.S. Canberra. pp. 55–77.
Bray, J. R. and E. Gorham. 1964. Litter production in forests of the world. Adv. Ecol. Res. 2: 101–157.
Davis, T. A. W. and P. W. Richards. 1933. The vegetation of Moraballi Creek, British Guiana; an ecological study of a limited area of tropical rainforest. J. Ecol. 21: 350–384.
Edwards, P. J. and P. J. Grubb. 1977. Studies of mineral cycling in a montane rainforest in New Guinea. I. The distribution of organic matter in the vegetation and soil. J. Ecol. 65: 943–969.
Ekern, P. C. 1964. Direct interception of cloud water on Lanaihale, Hawaii. Soil Sci. Soc. Amer. Proc. 28: 419–421.
Enright, N. J. 1978. The Ecology and Population Dynamics of Araucaria in Papua New Guinea. Ph.D. Thesis. A.N.U. Canberra.
Florin, R. 1963. The Distribution of Conifer and Taxad Genera in Time and Space. Acta Horti Bergiani 20(4): 122–311.
Gray, B. 1968. Forest tree and timber insect pests in the Territory of Papua and New Guinea. Pacific Insects 10: 301–323.
—— 1973. Distribution of Araucaria in Papua New Guinea. Res. Bull. No. 1. PNG Dept. of Forests. Moresby.
—— 1975. Size-Composition and Regeneration of Araucaria stands in New Guinea. J. Ecol. 63: 273–289.
Gray, B. and K. P. Lamb. 1975. Biology of Hylurdrectonus araucariae Schedl., a pest of hoop pine plantations in New Guinea. Bull. Ent. Res. 65: 21–32.
Havel, J. J. 1965. Plantation Establishment of Klinki Pine (A. hunsteinii) in New Guinea. Commonw. For. Rev. 44: 172–187.
—— 1971. The Araucaria Forests of New Guinea and their Regenerative Capacity. J. Ecol. 59: 203–214.
Johns, R. J. 1976. A provisional classification of the montane vegetation of New Guinea. Science in New Guinea 4: 105–117.
Kershaw, A. P. 1973. Late Quaternary vegetation of the Atherton Tableland, northeast Queensland, Australia. Ph.D. Thesis. Australian National University. Canberra.
—— 1974. A long continuous pollen sequence from north-eastern Australia. Nature (Lond.) 251: 222–223.
—— 1976. A late Pleistocene and Holocene pollen diagram from Lynch's Crater, North-eastern Queensland, Australia. New Phytol. 77: 469–498.
—— 1978. Record of last interglacial-glacial cycle from northeastern Queensland. Nature (Lond.) 272: 159–161.
Lamprecht, H. 1972. Einige Strucktur merkmale naturlicher Tropenwaldtypen und ihre waldbauliche Bedeutung. Forstwiss. Cbl. 91: 270–277.
Lane-Poole, C. E. 1925. The forest resources of the Territories of Papua and New Guinea. Gov't. Printer, Melbourne.
Ntima, O. O. 1968. Conifers of the Lowland Tropics No. 3: The Araucarians. Comm. For. Inst. Oxford, and For. Dept. Uni. Qld. Brisbane.
Ogden, J. 1966. Ordination studies on a small area of tropical rainforest. M.Sc. Thesis. University College of North Wales.
Paijmans, K. 1970. An Analysis of Four Tropical Rainforest Sites in New Guinea. J. Ecol. 58: 77–101.
Powell, J. M. 1970. The impact of man on the vegetation of the Mt. Hagen region. Ph.D. Thesis Australian National University.
Richards, P. W. 1939. Ecological studies on the rain forest of Southern Nigeria. I. The structure and floristic composition of the primary forest. J. Ecol. 27: 1–61.
—— 1952. The Tropical Rain Forest, An Ecological Study. Cambridge University Press. Cambridge.
Robbins, R. G. 1961. The montane vegetation of New Guinea. Tuatara 8: 121–134.
Rodin, L. E. and N. I. Bazilevich. 1967. Production and Mineral Cycling in Terrestrial Vegetation. Transl. Scripta Technica. Oliver and Boyd. London.
Rymer, L. 1978. The Use of Uniformitarianism and Analogy in Palaeoecology, Particularly Pollen Analysis. In: (Walker, D. and Guppy, J. C. eds) Biology and Quaternary Environments. Aust. Acad. Sci. Canberra.

Seward, A. C. and S. O. Ford. 1906. The Araucaria: Recent and Extinct. Phil. Trans. Roy. Soc. Lond. Ser. B. 198: 305–411.
Walker, D. 1973. Highlands vegetation. Nat. Hist. 17: 410–414.
—— 1978. Envoi. Chapter 15. pp. 259–264. In: Biology and Quaternary Environments. (ed. by Walker, D. and J. C. Guppy) Aust. Acad. Sci. Canberra.
Walker, D. and J. C. Guppy. 1976. Generic plant assemblages in the highland forests of Papua New Guinea. Aust. J. Ecol. 1: 203–212.
Webb, L. J. 1959. A Physiognomic Classification of Australian Rain Forests. J. Ecol. 47: 551–570.
—— 1968. Environmental Relationships of the Structural Types of Australian Rain Forest Vegetation. Ecol. 49: 296–310.
Webb, L. J. and J. G. Tracey. 1967. An Ecological Guide to New Planting Areas and Site Potential for Hoop Pine. Aust. For. 31: 224–239.
White, J. P., K. A. Crook and B. P. Ruxton. 1970. Kosipe: a late Pleistocene site in the Papuan Highlands. Proc. Prehist. Soc. 36: 152–170.
White, K. J. 1975. The effect of natural phenomena on the forest environment. Dept. Forests. Moresby.
Whitmore, T. C. 1975. Tropical Rainforests of the Far East. Clarendon Press, Oxford.
Womersley, J. S. 1955 unpubl. Jimmi Valley Patrol: report of botanical expedition. In: P.N.G. Dept. Forests. 125/series files.
—— 1958. The Araucaria forests of New Guinea. A Unique Vegetation Type in Malaysia. Proc. Symp. Hum. Trop. Veg., Tjiawa. pp. 252–257.
Wyatt-Smith, J. 1949. A note on tropical lowland evergreen rainforest in Malaya. Malay Forester 12: 58.
Wylie, F. R. and P. J. Shanahan. 1976. Insect attack in fire-damaged plantation trees at Bulolo in Papua New Guinea. J. Aust. Ent. Soc. 14: 371–382.

7. New Guinea Pandanaceae: first approach to ecology and biogeography

Benjamin C. Stone

1. Introduction*

The large island of New Guinea is one of the eight regions in which the family Pandanaceae (Monocot) has become both rich in evolutionary products and dense in individual representation. The other areas are New Caledonia, Australia, Philippines, Borneo, mainland SE Asia, Mascarene Is and Madagascar. It is significant in that all three extant genera of the family occur here; thus New Guinea represents both an area of relictual survival and of phylogenetic diversity and change. In many parts of New Guinea, members of the family play both a significant and conspicuous role in the vegetation, and few of the known vegetation zones or habitats lack at least one species.

The distinctive nature of the genera, and the peculiar and readily recognized plant forms, lend a special character to the family which makes it stand out in environmental classifications, for example interpretations of aerial photography. This is particularly true of *Pandanus* and *Sararanga*, which often occur in exposed aggregations such as beach and savanna vegetation and as a canopy formation in lowland swamp, riverine or gallery forests or in montane forests. Species of *Freycinetia*, less conspicuous from a distance, may nonetheless be obvious and abundant, their climbing stems clothing many forest tree trunks. These salient characteristics, in a family rich in species and represented in so many habitats, make Pandanaceae significant and useful features of New Guinea vegetation for ecologist and biogeographer.

There are however drawbacks to such potential use. Taxonomic understanding of the family is still incomplete, but a good deal has been achieved since the last major contributions by Merrill and Perry (1939, 1940). Still incomplete are species distribution data, many species being known only from one or two collections. Others, however, have now been collected repeatedly in certain localities, yet fail to appear in other areas, and in such cases it is clear what the distributional, and even ecological, limits are. Ecological data are needed for an account of dynamic aspects and roles in successional phases. While it is possible

*I am indebted to J. L. Gressitt (insects), K. A. Pirozynski (fungi) and Peter Raven (biogeography) for help with this chapter.

© *1982, Dr W. Junk Publishers, The Hague.*

to categorize various species according to habitat specialization, it is often still uncertain what roles such species play.

This account is perforce preliminary but sound enough in taxonomic basis to support certain biogeographical interpretations.

2. Taxonomic background

The family Pandanaceae, sole representative of the Pandanales, is a compact, distinctive and disjunct member of the Monocots. For a discussion of the family's relationships, see Stone (1972). Pollen of *Pandanus* is known from Senonian (mid-Cretaceous) sediments in Borneo (Muller 1970), in which pollen also of *Nypa* (Palmae) is found. Much remains to be learned of real relationships, and dated antiquity, of Pandanales and related orders. However Pandanales give overall an impression of primitiveness which is borne out by analysis of a number of characteristics ascertained from investigations of embryology, chromosomes, inflorescence and floral structure, carpel and seed development, vegetative growth and distribution patterns (Stone 1972).

Genera. The three genera are distinct morphologically; no 'intermediate' species exist (Stone 1972). This also indicates antiquity. Significant differences exist also between the generic ranges, which are in a loose sense concentric; this pattern is broken by the extension of *Freycinetia* to New Zealand. New Guinea is part of the central or "core" region where all 3 genera occur (Stone 1976), the rest of this region comprising the Philippines and Melanesia as far east as the Solomon Is.

2.1. Sararanga

This is the smallest genus, with only two species, one endemic in the Philippines (*S. philippinensis* Merr.) and the other (*S. sinuosa* Hemsl.) found in New Guinea (NE Irian Jaya), Admiralty Is (Manus), and the Solomon Is (Santa Ysabel, Malaita, etc.). *Sararanga* is considered the most ancient or at least retains the greatest proportion of archaic features within the family (see Stone 1961, 1974).

2.2 Pandanus

This is the largest, most diverse, and most widespread genus of the family (St. John 1965; Stone 1976). Although a final enumeration of species is not available, and considerable divergence of opinion on species concept exists, all authors agree that there are well over 500 species; a conservative estimate suggests 650–700, another estimate indicates perhaps over 900. New Guinea proper, with about 66 species, constitutes a major element of the genus; and this is true more emphatically when the proportion of endemism, both at species and higher taxonomic levels, is examined.

2.3. *Freycinetia*

This is a genus of about 180 species, from Sri Lanka (Ceylon) and S. Burma, to Taiwan, Ryukyu Is, Central Marianas, Bonin Is, to Hawaii (1 sp.), Society Is (2 spp.), New Zealand (1 sp.) and Queensland (4 spp.). Like *Pandanus* this genus has a major center of diversity in the Philippines, Borneo and New Guinea, and susidiary centers of diversity in New Caledonia, Fiji and Malaya. All of the species are woody root-climbers. There are about 50–56 species in New Guinea. No recent comprehensive treatment of *Freycinetia* is available but my monograph is nearing completion, and the treatment of Pandanaceae for the Flora Malesiana will include all the New Guinea species.

Table 1 lists the currently accepted New Guinea species of the family (excluding synonyms), and shows the distribution outside New Guinea, whether or not the species is endemic, brief mention of habitat when this is a strong limiting factor, and a crude indication of abundance. Cultivated species (whether or not also occurring in the wild) are so marked. A few species limited to the Bismarck Archipelago or smaller islands immediately adjacent to New Guinea are also included.

Table 1. Species of Pandanaceae occurring in New Guinea, with ecological, biogeographic or cultivation status

Species	Distribution	Ecology or forest type)*	Cultivation status
Sararanga sinuosa	NE Irian, Manus, Sol	Lowlands nr coast; tolerant of limestone	Wild
Pandanus adinobotrys	NG/E	Montane; disturbed; 1000 m+	Wild; encouraged in clearings?
aggregatus	NG/E: Papua, Goodenough	Lowland; low montane 100–1600 m.	Wild
antaresensis	NG/E	Montane, mostly 2000 m+	Wild; encouraged in clearings
archboldianus	NG/E: ENG	Montane 1300–2000 m	Wild
atropurpureus	NG/E: Irian	Lower montane forest 1000 m	Wild
amboinensis	NG, Mol	Lowland	C/W
balenii	NG/E	Lowland forest to 850 m	Wild
beccarii	NG, Aru Is	Lowland; lower montane to 1200 m	Wild
boemiensis	NG/E:Irian[1]	Lowland 400 m	Wild
brachyphyllus	NG/E: Irian[1]	Montane; wet gullies, 1800 m	Wild
brassii	NG/E: Papua	Lowland savanna; resistant fire	Wild; occas. planted
brosimos	NG/E	Montane forest above 2000 m, clearings and plantations	C/W, Prob. wild form of P. jiulianettii
castaneus	NG/E: E NG	Lower montane 900–1500 m	Wild
cernuifolius	NG/E: Irian	Lower montane 500–1200 m	Wild

Table 1 (continued)

Species	Distribution	Ecology (or forest type)*	Cultivation status
columbiformis	NG/E: Papua[1]	Lowland under 100 m	Wild
cernuus	NG/E: Irian[1]	1300 m	Wild
concavus	NG/E: Irian	Lower montane 1500 m	Wild
concinnus	NG/E: Irian	Lower montane 1500–1800 m	Wild
conoideus	NG, Bism prob. E Mol	Lowland clearings, cult. ground (edible frt)	CV; Wild and staminate plants unknown
croceus	NG/E: Manus I.	Lowland over limestone	Wild
danckelmannianus	NG, Bism Sol	Lowland 100–1000 m	Wild
daulos	NG/E: Goodenough[1]	Lower montane forest, 1500 m	Wild
daymanensis	NG/E: E NG	Lowland and Hill 700–800 m	Wild
dolichopodus	NG/E: Irian	Submontane 850 m	Wild
dubius	W and E Malesia, NG, Pac	Beaches, coastal rocks	Wild
englerianus	New Ireland/E	Lowland and on limestone	Wild; frt. edible?
eumekes	NG/E: Vogelkop	Lowland 60 m	Wild
exiguus	NG/E[2]	Lowland 100–850 m	Wild
foveolatus	NG/E[3]	Lowland to 1000 m	Wild
galorei	NG/E: Irian[1]	Lower montane 1250 m	Wild
hollrungii	NG/E: E NG	Lowlands, coastal and riverine f.w. swamps	Wild
humicola	NG/E: Irian[1]	Lower montane 1200 m	Wild
hystrix	NG/E	Lowland f.w. swamp to 100 m	Wild
ihuanus	NG/E: Papua	Lowland to 500 m	Wild
jiulianettii	NG/E	Montane above 2000 m also clearings	C/W Wild plants mostly above 3250 m CV. see *P. brosimos*
kaernbachii	NG, Bism	Coastal swamps, estuaries	Wild only
krauelianus	NG; also Bism?	Lowland to 500 m, Common	Wild, but frt occas. eaten
latericius	New Ireland, New Hanover	Lowland below 100 m	Wild
lauterbachii	NG[1], N Qld coast	Lowlands, riverine and estuarine swamps	Wild
leiophyllus	NG/E: Irian, Papua, Rossel I.	Lowland; coastal and riverine swamps	Wild
leptocaulis	NG/E	Low; lower montane; moss for to 1500 m	Wild
leptocarpus	NG/E: Papua	Riverine, estuarine 50 m to tidal	Wild
lictor	New Britain E	Lowland 300 m	Wild
macgregorii	Ferguson I, Louisiades	Prob. lowlands	Wild
meniscostigma	NG/E: Papua: Goodenough	Lowland to 450 m	Wild
microdontus	NG/E: Papua[1]	Low alt	Wild

Table 1 (continued)

Species	Distribution	Ecology (or forest type)*	Cultivation status
minusculus	New Ireland/E	Lowland swamp	Wild
misimaensis	NG/E: Papua, Misima I.	Lowland 300 m Castanopsis forest	Wild
navicularis	New Ireland/E	Lowland on limestone	Wild
Odoardi	NG/E: Irian, Papua	Lowland to 1250 m	Wild
orculaeformis	NG/E: Irian[1]	Lowland	Wild
ornithocephalus	NG/E: Vogelkop	Lowland 100 m on limestone	Wild
papuanus (incl. *pseudopapuanus blakensis*)	NG, Bism N Mol, Aru	Riverbanks, floodplains, secondary forest, to 400 m	Wild
pendulinus	NG/E: Papua[1] ♂only	Riverine under 60 m	Wild
penicillus	NG/E: Irian[2]	Rainforest to 60 m	Wild
permicron	NG/E: Irian	Hill forest, on clay to 1200 m	Wild
plicatus	NG/E: Papua	Lowland at 500 m	CV, prob. of *conoideus*
polycephalus	S and W Malesia, Mol, NG, Aru, Bism Sol	Coastal and estuarine veg.	Wild
pseudosyncarpus	NG/E: Irian[1]	Lowl. to lower montane	Wild
rex	NG/E: Papua[1]	alluvial plain 12 m	Wild
rostellatus	NG/E: throughout	Lower montane 1000–2000 m	Wild
rubrispicatus	NG/E: Morobe	Lower montane 1500 m	Wild? Resembles CV of *conoideus*
scabribracteatus	NG/E	Coastal swamps	Wild
schoddei	Bougainville/E	'cloud' forest 1600 m	Wild
setistylus	NG/E: NE	Lowland	Wild
stenocarpus	NG/E: Arfak Mts[1]	?Rainforest	Wild
subumbellatus	NG, Aru, ?Bism.	Lowland	Wild, ?used, ?planted
tabbersianus (incl. *kivi*)	NG/E: Irian, NE NG, Gulf isls, ?Qld Bism.	Lowland	Wild, occas. used, edible fruit
tectorius	Wide thru Pac isls, E Aust. E. Phil.	Strand; sandy rocky coasts, lowl. 2nd growth savannas	Wild; and planted; CV; ?hybridizes *brassii*
woodlarkensis	NG/E: Woodlark I	Low hill forest	Wild
xanthocarpus	NG/E: Papua	Lowland to 800 m	Wild
Freycinetia			
andajensis	NG/E: Irian[1]	Unknown	Wild
angustissima	NG/E	Montane to 2800 m	Wild
archboldiana	NG/E	700–1500 m	Wild
aruensis	Aru Is	Lowland	Wild
beccarii	NG, Mol	Lowland to 2650 m	Wild
brassii	NG/E: Irian, Papua	Lowland to 640 m	Wild
crucigera	NG/E: Irian	Lowland	Wild
divaricata	NG, Misima, Sol	Lowland to 600 m	Wild
elegantula	NG/E	Montane and submontane 1400 m	Wild
ellipsoidalis	NG/E: Irian, Papua	Lowland to 850 m	Wild
elliptica	NG/E: NE	1500–2500 m	Wild
erythrospatha	NG/E: Irian	Montane 1800–2150 m	Wild
excelsa	NG, Qld	Lowland to 500 m	Wild
fibrosa	NG/E	Lowland to 1000 m	Wild

Table 1 (continued)

Species	Distribution	Ecology (or forest type)*	Cultivation status
flaviceps	NG/E	Montane 1800–2800 m	Wild
forbesii	NG/E but ? = *scandens*	Lowland 800 m	Wild
funicularis	NG, Sudest I, Mol, Sulawesi, Sundas, Sudest I.	Lowland to 750 m	Wild
gibbsiae	NG/E: Vogelkop,	Montane 2000–3000 m	Wild
impavida	NG, Normanby I, Sol, Fiji, Poly	Lowland to 2700 m	Wild
inermis	NG, Solomons	Lowland to 1600 m	Wild
jaheriana	NG/E: Irian	?Lowland	Wild
kanehirae	NG/E	Lowland	Wild
klossii	NG/E?	Lowland	Wild
lacinulata	NG/E	Lowland	Wild
laeta	NG, Sol	Lowland to 750 m	Wild
lagenicarpa	NG, ?Phil	Lowland to 1650 m	Wild
lateriflora	NG/E: Irian	Lowland to 1750 m	Wild
latiauriculata	NG/E	Lowland to 600 m	Wild
latibracteata	NG/E	Lowland to 1500 m	Wild
lauterbachü	NG/E: Prob. = *funicularis*	Lowland	Wild
linearifolia	NG/E	Lowland to 600 m	Wild
linearis	NG/E	Montane at 2500 m	Wild
macrostachya	NG/E	Lowland	Wild
marantifolia	NG, Solomons	Lowland to 1500 m	Wild
marginata	NG/E[1]	Unknown	Wild
multiflora	NG (Vogelkop), Phil	Lowland	Wild
naumannii	Bismarcks	Prob. lowland	Wild
percostata	NG, Qld	Lowland to 1600 m	Wild
perryana	NG/E	Montane at 2000 m	Wild
pleurantha	NG/E: Irian	Submontane 900 m	Wild
polystigma	NG/E	Unknown	Wild
pseudoinsignis	NG/E	500–2400 m	Wild
radicans	NG, ?Mol	Lowland	Wild
rectangularis	NG/E: Irian	Lowland	Wild
rhodospatha	NG/E: Irian	Lowland 450 m	Wild
salamauensis	NG/E	To 1500 m	Wild
scandens	NG, Qld, W to Java	Lowland, rarely to 2500 m	Wild
sogerensis	NG/E	Rainforest 1300 m	Wild
solomonensis	NG, Sol	Lowland	Wild
spinellosa	NG/E	Lowland	Wild
stenodonta	NG/E	Lowland to 500 m	Wild
sterrophylla	NG/E	Montane 2300–2800 m	Wild
tafaensis	NG/E	Submontane 1500–2500 m	Wild
trachypoda	NG/E	Lowland to 600 m	Wild
undulata	NG/E	Lowland	Wild
vulgaris	NG/E	Lowl. and lower mt to 1750 m	Wild

NG/E Endemic to island of New Guinea
CV Represented by more than 1 cultivar
C/W Cultivated but also known as wild plants
* Rainforest unless otherwise indicated
1 Known from a single collection only
2 Known from only 2 collections
3 Known from only 3 collections
Aust = Australia; Bism = Bismarck Archipelago; Micr = Micronesia; Mol = Moluccas; Pac = Pacific islands; Poly = Polynesia; Qld = Queensland; Sol = Solomon Is; wide = widespread.

3. Ecology

3.1. Ecological status of the genera

Although the three genera are rarely found occurring together in the same habitat, this was observed in the Solomon Islands, where on Malaita *Sararanga sinuosa*, *Pandanus danckelmannianus* and *Freycinetia divaricata* were all within a small area near Tantalau village at 300 m. The vegetation was essentially rainforest, locally disturbed. This does not mean that the genera can be equated in ecological terms, however.

3.1.1. Sararanga. The greatest distinction among the genera is undoubtedly the different behavior of those plants which are root-climbers (all *Freycinetia*) and those which are trees and shrubs, i.e. *Pandanus* and *Sararanga* (Went 1895). The two species of *Sararanga* are probably much alike in their ecological needs. They are limited in occurrence, but it seems clear that the limitation is not merely a matter of the physical aspects of the environment. The biological limiting factors are of two kinds, intrinsic and extrinsic; the former includes questions of reproductive capacity; in the latter the chief element is competition from other components of the vegetation. There appears also to exist a biogeographical aspect, partly arising from the particular concomitance of physical and biological factors, and partly distinct from it, arising from gross geological and geomorphological events. These factors have been responsible for the modern distribution of *Sararanga*. Additionally, it is highly susceptible to man-made habitat changes. What little is known of the ecological preferences of *S. philippinensis* correspond well with what is known about *S. sinuosa*.

Sararanga sinuosa (Figs. 1, 2) may form essentially pure stands (Stone 1961). In Tulagi (Solomons), the ground beneath the sararangas is nearly devoid of understorey vegetation. The thick covering of fallen leaves apparently prevents the development of plant cover, the lack of which is also noticeable in *Pandanus* groves. The coriaceous leaves weather slowly, and their continual falling produces a kind of loose pile which may be several feet thick. In this locality, no young plants of *Sararanga* were observed, although not far away several juveniles were seen. It seems entirely possible that this conspicuous and virtually pure stand represents a transitional state between burned-over ground and climax lowland forest.

In 1971, *Sararanga sinuosa* was collected for the first time in the Admiralty Is, on SW Manus Island, where at about 150 m one old tree 15 m high (trunk 22 cm DBH) was found: and in another site, on Los Negros I., on a flat area of limestone, several were found (with ripe fruit in mid-June) in disturbed secondary forest at 3 m altitude. The type collection of *S. sinuosa* came from Fauro I. at 550 m (Guppy 259, in 1884), and this is still the highest elevation known for this species. Most other collections are from altitudes not much above sealevel, such as my collection (*Stone 2579*) from the edge of Tatamba lagoon in Santa Ysabel. In Irian Jaya's Cyclops Mts at about 230 m, van Royen obtained specimens in the foothills north of the road from Hollandia-Binnen (Abepura) to Lake Sentani east of the Koohajoo River (no. *4498*) in 1954, and

Fig. 1. Sararanga sinuosa Hemsl. Habit of mature but as yet unbranched trees. S. Malaita I, Solomons Is. Note pendant infructescence on taller tree. (Photo Stone 1957).

Kalkman obtained specimens (*BW 3666*) along the riverbank at Waisiniwai, Demta, near sealevel, in 1956. Lane-Poole (1925) describes some New Guinea vegetation in the Ramu Valley with pandans lacking proproots; both van Steenis (1954) and I (1961) thought these might be *Sararanga*, but no evidence is at hand to prove this, and it seems probable they were true *Pandanus*.

3.1.2. Pandanus. Following the broad division into major environments and vegetation types in New Guinea as set out by Paijmans (1976) it is possible to allocate many pandan species to these categories.

I. BEACHES including sandy and rocky shores; beach woodlands.
P. dubius, tectorius, polycephalus (Fig. 9).

II. SWAMPS: Saline and brackish swamps near the coast.
P. tectorius.

III. SWAMPS: Freshwater swamps and estuaries near the coast, and swamp woodlands.
P. balenii (once), *hollrungii, hystrix, kaernbachii* (Fig. 8), *lauterbachii,* leiophyllus, *scabribracteatus, tectorius.*

IV. LOWLAND ALLUVIAL PLAINS AND FANS:
Lowland rainforests ('mixed alluvium forest').
P. aggregatus, amboinensis, balenii, beccarii, columbiformis, danckelmannianus, erinaceus, eumekes, exiguus, foveolatus, ihuanus, krauelianus, latericius, leptocaulis, lictor, macgregorii(?), *meniscostigma, microdontus, odoardi, orculaeformis, papuanus* (Fig. 3), *penicillus, pseudosyncarpus, rex, setistylus, stenocarpus*(?), *subumbellatus, tabbersianus, xanthocarpus*(?)
Lowland forest over limestone flats.
P. croceus, englerianus, latericius, navicularis (New Ireland), *ornithocephalus, tectorius* (sometimes).
Riverine forest and gallery forests in lowlands:
P. leptocarpus, pendulinus, papuanus (occasionally)
Savannas and grasslands:
brassii, papuanus (sometimes), *tectorius*

V. HILLS AND LOWER MOUNTAINS; up to about 1000 m.
Mixed evergreen forests:
P. aggregatus, balenii, beccarii, boemiensis, castaneus, cernuifolius, danckelmannianus, daymanensis, exiguus, foveolatus, ihuanus, lictor, meniscostigma, misimaensis, odoardi, papuanus, plicatus, pseudosyncarpus, woodlarkensis, xanthocarpus.

VI. LOWER MONTANE ZONE, approx. 1000–3000 m.; (lower montane zone; montane zone, approx. 1700–3000 m).
Mixed lower and montane forests:
P. adinobotrys, aggregatus, antaresensis (above 2000 m), *archboldianus, atropurpureus, beccarii* (rarely), *brachyphyllus, brosimos* (above 2000 m), *castaneus, cernuifolius, cernuus, concavus, concinnus, conoideus* (not over 2700 m) (Fig. 4), *daulos, foveolatus* (at 1000 m), *galorei, humicola, jiulianettii* (2000–3700 m), *odoardi* (to 1250 m), *permicron* (1200 m), *rostellatus, rubrispicatus, schoddei* (Bougainville only), *xanthocarpus*(?).

VII. UPPER MONTANE ZONE 3000–4000 m.
P. antaresensis, brosimos (Figs. 6, 7), *jiulianettii* (Fig. 5).

Some species appear to have comparatively wide ecological amplitude, while others seem to be very strictly relegated to particular habitats.

Species with wide ecological amplitude. Of these the most obvious at least is *P. tectorius*, which may occur in beach forest on beach sand, on rocky coasts whether limestone or basalt, at the margins of or in higher mounds within some mangrove areas, along small estuaries, or streams into flat hinterlands (as in grasslands of Chipas, Morobe Prov.), in disturbed or secondary lowland forest especially near inhabited sites where they may be remnants of cultivation, and, somewhat more doubtfully, in savannas. The savanna vegetation of Papua in particular is more likely to include *P. brassii*, but the variability of pandans in such sites suggests that perhaps more than one species of Sect. *Austrokeura* may occur in New Guinea; the taxonomy of this group is still tentative. It is possible, however, that hybridization may account for at least part of this variability and that *P. tectorius* may be involved as a contributor to the gene flow. *P. brassii* seems less ecologically variable. Both have a fair resistance to grassland fires and this helps explain their continued presence in both savanna and anthropogenous grasslands.

P. papuanus (fig. 3) is also variable: I include here several synonyms (*P. pistillaris, P. parkinsonii, P. biakensis, P. pseudopanuanus*) indicating this variability, but this mainly refers to fruit dimensions. In New Guinea, *P. papuanus* may occur in primary swamp forest at 3 m (LAE 53432), in disturbed patchy mixed savanna forest at 220 m alt. (LAE 53568), in gallery forest at 180 m (LAE 53550), in freshwater swamp forest at 720 m (LAE 53452), and in regrowth hill forest with *Araucaria* at 900 m (LAE 53033).

It is possible that more species exhibit at rather broad ecological amplitude but the available data do not permit any confident statement, so many species being known from few collections, often with little or no ecological data. In other cases taxonomic problems may conceal ecological amplitude.

Cultivated species. In general there are two major categories of cultivated pandans:

A. Pandans cultivated for their fruits: *P. conoideus*='marita'; *P. brosimos, P. jiulianettii*='karuka'.

B. Pandans cultivated for their leaves: *P. tectorius* var. *laevis* and other varieties or cultivars of P. tectorius; *P. amaryllifolius* (not known in the wild; cult. for its fragrant leaves). Both species may also be utilized for their staminate inflorescences.

A. (1) *Pandanus conoideus* (incl. *P. magnificus, P. ruber*) (Fig. 4).
This is widely known as 'marita' and is chiefly a lowland plant, cultivated in villages as individual trees or in small groves, mostly derived (or entirely derived) from cuttings. The staminate plants are unknown apart from a single very old collection obtained from a plant grown in Buitenzorg (Bogor) in the 19th century and preserved in the Utrecht herbarium (filed as 'P. ceramicus'). It

Fig. 2. Sararanga sinuosa Hemsl. Ripe fruits. From a collection by D. B. Foreman from Manus I, Admiralty Grp (photo Lae Botanic Gard.).

Fig. 3. Pananus papuanus Solms. Lae Botanic Gard.; from beneath showing leaf crowns (photo Stone 1971).

may have come from the Moluccas or W New Guinea. Pistillate plants are widely grown in New Guinea, the fruits ripening but, in those few cases investigated, without functional seeds. The pericarp is a rich dark red and contains a similarly colored oil, which is released upon cooking and mashing to form an oleaginous mush (from which fibers and endocarps are removed) with a mild or insipid flavor. It is used as a sort of buttersauce on starchy foods. The 'marita' (a 'Pidgin' word) may be found up to about 2000 m. Trees or groves probably persist in sites of former habitation and may thus seem to be wild. Other closely related species (in Subg. *Kurzia* Sect. *Kurzia* and Sect. *Microstigma* (formerly Bryantia), the latter section including *P. conoideus*) may be used in the same manner, e.g. *P. hollrungii*, and the question of the origin of *P. conoideus* is made difficult by the existence of several described species which differ only very trivially (e.g. *P. plicatus*). 'Marita' therefore may be a term indicating both *P. conoideus* and these closely related, similarly used species.

A. (2) *Pandanus jiulianettii* (Fig. 5) and *P. brosimos* (Figs. 6, 7).
These species — or as is probable, this species with its cultivars — is known in most of New Guinea as 'karuka' (in 'Pidgin') but individual areas and tribes may have several to many local cultivar names and their own general term as well. The 'karuka' may occur in large groves, as individual cultivated trees, both in primary forest and in anthropogenous grassland or scrub vegetation, always at high altitudes (never below 2000 m and usually about 3000 m or higher). It is one of the very few essentially high-altitude pandans of New Guinea and where abundant gives an unmistakable aspect to the landscape due to the height, the sparsely branched trunks, and in some areas the 'ruffs' of decayed leaves

Fig. 4. Pandanus conoideus Lm. "Marita" fruits on sale at Goroka market (photo Stone 1971).

persistent on the branches (these may be trimmed off in groves of cultivated trees). Below the primary forests (mostly above 3000 m) almost every 'karuka' tree is owned. The fruits are massive subglobose heads composed of numerous large single-seeded drupes. The large and fat-rich endosperm is the edible portion, and the fruits are typically dried over smoldering fires, rendering the drupes storable for some weeks or months. During a period of a month or two each year some at least of the highland tribes depend heavily on 'karuka' seeds, and indeed this plant is one of the 2 or 3 most important food plants in the high montane zone. In the primary forests, 'karuka' trees reach their best development and attain canopy level or nearly do. Both the biology and the agronomics of this significant plant have scarcely been studied but Mr C. Rose is presently carrying out studies which should give us a better understanding of ecology, agricultural status and nutritional value. Staminate trees occur and the floral structure has been described (Stone 1974).

Besides the 'marita' and 'karuka' there are some other species which have edible fruits or seeds, but which are not or hardly ever truly cultivated. These include *P. dubius* (and its var. *compressus*) which have a large drupe or 2-5-seeded polydrupe with comparatively large seeds, of which the tasty white endosperm is sought; *P. kaernbachii* (Fig. 8), which has a thin but flavorsome ('apple-like') pericarp; *P. tectorius*, with a more or less edible pericarp (from which in most cases the calcium ovalate crystals must be removed by boiling before eating); *P. englerianus*, *P. hollrungii*, etc. which are used as substitutes for 'marita,' and possibly others. *Pandanus foveolatus* is mentioned by Powell (in Paijmans 1976) as a 'nut' plant, i.e. providing edible endosperm.

B. Pandans cultivated for their leaves.

(1) *Pandanus tectorius*. In lowland and coastal areas this is the chief species used. The leaves are commonly used in weaving (mats, hats, rainmats, house partitions). Other species may be used. The thick somewhat fleshy broad leaves of *P. dubius* are often preferred for rainmats or roof thatch. In the islands *P. tectorius* var. *laevis*, a cultivar with smooth, spineless leaves, may be preferred to wild forms. Very young leaves may be peeled to free the epidermal/hypodermal tissues, which can be used as cigaret papers.

(2) *Pandanus amaryllifolius*. This unusual species, of which pistillate plants

Fig. 5. Pandanus jiulianettii Mart. Bismarck Mts, W. Clarke 90 (photo Clarke).

Fig. 6. Pandanus brosimos M. and P. Grove of adult trees, Tomba, Mt Hagen (photo Stone 1971).

are unknown, is not known in the wild, but has been collected from villages in Irian Jaya. The fragrant leaves are used in cooking (Stone 1979).

Species with narrow ecological limits. It is probable that many, even the majority, of New Guinea species of *Pandanus* have fairly narrow ecological limits, but detailed ecological studies involving fully identified species are almost nonexistent. Consequently it is appropriate here to discuss only a few selected cases in which there are at least some qualitative data available.

1. A maritime species, *Pandanus dubius*. Although this species may share habitat with *P. tectorius*, it is not coextensive. It is both less common and more precise in its requirements. Its full distribution has recently been discussed (St. John 1975), and if *P. compressus* and *P. bidur* be considered (as I do now) merely subordinate taxa under *P. dubius*, then the full distribution of the species, *sensu lato*, is: from E coast of Malay Peninsula and nearby islands E to New Hebrides and Caroline Is. The habitat is usually a sandy or rocky beach or coastal cliff within direct influence of sea winds. Dispersal is clearly by sea-borne fruits. The species is conspicuously lacking from Polynesia, Marshall Is, Queensland, Lesser Sunda Is, also from the S coast of Borneo and Sulawesi. Such a pattern fits readily in with past and present sea-current and drift patterns

Fig. 7. Pandanus brosimos M. and P. Ripe fruit (LAE 53804), Tomba, Mt Hagen (photo Stone 1971).

in the western and middle region of its distribution but it is not easy to understand its absence in Polynesia. Because of its comparatively large edible seeds the species may have been carried about by pre-European voyagers. It is easily grown in pots and as a cultivated plant (often under the nursery name *P. pacificus*), but it does not seem to be planted or tended by any native peoples, although the fruits and the leaves are often used. *P. dubius* is intolerant of heavy dense clays. *P. polycephalus* (Fig. 9) is also coastal.

2. A riverine species, *Pandanus leptocarpus*. This species deserves, possibly, the appellation of rheophyte. So far it is known only from riverbanks in the lowlands, and though known only from four collections, it appears to be an obligate component of river-bank vegetation and shows, in its fruits, characteristics which seem fairly strongly adaptive to such a habitat. The collections are from the Noord River (Irian), the Vailala River (Papua: 2 collections), and the Purari River (Papua), all near sea level. The cephalium, about 30 × 15 cm, is composed of numerous simple drupes, each of which is a slender and elongated structure (c. 25–30 × 4 mm). Recent evidence from aquatic and riverine species in Malaya indicates that these (i.e. *P. helicopus*), which have similar fruits and drupes, are dispersed upstream by turtles which eat the floating fruits after they begin to break up into the component drupes or small aggregates of drupes. A similar dispersal method is hypothesised for *P. leptocarpus*, which may not be a true rheophyte, but nonetheless must withstand — and probably also requires — proximity to rivers so that when ripe, the fruits fall into the water.

3. A seasonal swamp and savanna species, *Pandanus brassii*. Confined to the Papuan region on both sides of the Irian Jaya-PNG boundary, it belongs with a group of pandans which are prominent in tropical N Australia, and the Papuan habitats are in fact similar to those in Australia where its close relatives occur. It may be found in *Eucalyptus* savanna chiefly along gullies or in depressions, where it may obtain a modicum of water even during the dry season. It is probably rather resistant to fires and is thus often a component of anthropogenous grasslands especially where there is topographical diversity and a mosaic of secondary forest types. As mentioned above, there is a possibility that *P. brassii* may be able to hybridize with *P. tectorius*, or that *P. brassii* is but one of several cryptic species (or subordinate taxa).

4. A species of permanent freshwater swamps, *Pandanus scabribracteatus*. So similar to *P. kaernbachii* that the two may be forms of one species. Habitat preference of both is the same, but *P. scabribracteatus* seems limited to the S part of New Guinea, especially the Papuan coast. It has been observed in shallow swamps with mixed forest not far from the coast and usually close to watercourses, where it is gregarious and may form large tracts (similar to the vegetation shown in Pl. 14 in Paijmans 1976). The trees are fairly tall (to 10–14 m) and in the leaf-axils may be found freshwater crabs. Closely related to *P. pleiocephalus* Martelli ex Fagerlind, of Moluccas (Buru) which may be a synonym, but appears to have a different habitat.

5. A high-altitude species of montane forest, *P. brosimos*. Discussed above: linked with, or merely a form, of the cultivated 'karuka.' It is one of but 2

Fig. 8. Pandanus kaernbachii Warb. Mature Fruit. Stone LAE 52781, from Manus I.

Fig. 9. Pandanus polycephalus Lam. Fruiting tree, cult. in Hortus Bogoriensis, Djawa (photo Stone 1967).

Fig. 10. Freycinetia elegantula Stone. Cult. in Bot. Garten, München, coll. and photo J. Bogner (n. 304). Staminate inflorescence.

species of *Pandanus* in New Guinea, as far as I am aware, which has been found at altitudes approaching 4000 m (other: *P. antaresensis*). On the flanks of Mt Wilhelm it is frequent in forests at 3300–3400 m. Where this species or its close relative or cultivar *P. jiulianettii* occurs in anthropogenous grasslands or scrub forests it is usually shorter than when in undisturbed forest; compare the photograph (fig. 4) in van Steenis 1954. As a cultivar, this may be found as low as about 2000 m, apparently lower than wild plants. At the upper extreme of its altitudinal range it drops out before the level of the elfin forest or foggy, windswept vegetation dominated by *Cyathea* is reached.

3.1.3. Freycinetia. The precise ecological characteristics of most of the many species of *Freycinetia* are poorly known, often virtually or actually nil. As can be seen from Table 1, there are in general only two categories which can be recognized when all the New Guinea species are taken into consideration: lowland and montane; with a division of the latter category into two;

Lowland species: *F. aruensis, brassii, crucigera, divaricata, ellipsoidalis, excelsa, fibrosa, forbesii, impavida, funicularis, inermis, kanehirae, klossii, lacinulata, laeta, lagenicarpa, lateriflora, latiauriculata, latibracteata, lauterbachii, linearifolia, macrostachya, marantifolia, multiflora, naumannii, percostata, pseudoinsignis, radicans, rectangularis, rhodospatha, salamauensis, scandens, solomonensis, spinellosa, stenodonta, trachypoda, undulate, vulgaris.*

Montane species: (A) Mostly or entirely found below 2000 m alt.: *F. archboldiana, beccarii, elegantula, elliptica, erythrospatha, pleurantha, pseudoinsignis, scandens, sogerensis, tafaensis, vulgaris.* (B) Mostly or entirely found above 2000 m alt. to about 3000 m alt.: *F. angustissima, beccarii, elliptica, erythrospatha, flaviceps, gibbsiae, linearis, perryana, pseudoinsignis, scandens* (rarely), *sterrophylla, tafaensis.*

As may be noted a few species have to be included under both (A) and (B), and some also under 'Lowland' and (Montane-A) groups, e.g. *F. pseudoinsignis, F. vulgaris*; of these the former also is in (Montane-B) and thus seems able to occur at virtually any altitude up to the maximum which among specimens I have examined appears to be just over 3300 m.

Besides making this crude division into altitudinal categories, one may record certain known habitat preferences for a very few of the better known species: *F. funicularis* and *F. lauterbachii* (which are possibly synonyms anyway) are found in lowland strand forest; *F. angustissima* is a characteristic species of the high montane moss forests: as appears to be the case also with *F. gibbsiae* and *F. sterrophylla. F. macrostachya* may occur in lowland swamp forest.

Because freycinetias are dependent on forest environment they are usually lacking in open environments, certainly so in grasslands and usually in savannas (but they may occur in remnant gallery forests). Their ecology is conditioned importantly by the forest type and secondarily by the generalized climatic and other physical factors. Because their tiny seeds are distributed by birds and small mammals with arboreal habits their mode of germinating includes both terrestrial and arboreal types, and in the latter case root contact with soil may result from elongation downward of roots from well-established plants. All the freycinetias form clasping roots. Those which germinate on the ground usually must climb up at least a few meters before their vegetative parts reach adult size and form (seedlings are as yet virtually unidentifiable to species), and in many cases flowering may not occur until the climber reaches canopy level or at least subcanopy level with a considerably higher light intensity than that near the forest floor. As a result, flowering is not nearly so well documented as for pandans, and a number of species are known as yet only from one or at most 2 or 3 fertile specimens. Although there are some useful vegetative characters and also some anatomical (leaf) characters it is as yet impossible to identify all from sterile materials.

Clearly the large number of species of *Freycinetia* in New Guinea reflects in some measure the environmental diversity of the island but it is also clear that there are a good many species which have largely overlapping ecological capacities, and the distribution factors must include geographic isolation and various sympatric assortments. Freycinetias are basically plants of primary forest or of the later seral phases of mixed natural regenerating secondary forests. They probably do not play a major role in succession, unlike some pandans. On the other hand they may possibly in some areas constitute an important food resource for some of the bird and small mammal fauna, not to mention invertebrates.

Useful species. Unlike *Pandanus* there are no cultivated species of *Freycinetia* (outside of botanical gardens). The stems or roots of some species may be used for cordage, immediately if slender, or after retting; various species may be used (Powell gives no identifications; in Paijmans 1976). There are vague reports of a medicinal use of the roots, and of the fruits for abortion, and some substantiated records of use for bark cloth and as ceremonial or ornamental leaves and flowers. There is no evidence, however, that freycinetias are ever actually planted.

3.2. Diseases and animals

Pandans and disease. Fungal parasites and saprophytes associated with pandans. In the following list (Table 2) ascomycetes and their conidial forms (Fungi imperfecti: Coelomycetes and Hyphomycetes) make up the bulk of such records, and represent both the unitunicate and the bitunicate groups. Basidiomycetes are relatively few, chiefly saprophytic agarics and polypores, and there is one slime-mold of the Physaraceae (Myxomycotina). I am indebted to Dr Pirozynski of Canada National Museums for correcting and augmenting the list I extracted from literature available to me.

Viral or bacterial diseases of pandans. A disease called aurigo (or yellowspot) is reported by Sorauer, Lindau and Reh (1922) to occur in *Pandanus*; it may be caused by a virus. Smith (1915) listed no bacterial diseases of Pandanaceae.

Galls in pandans. Docters van Leeuwen (1926) describes root-galls which form on aerial roots of *Freycinetia*, also on those of *Pandanus nitidus* in Java and the

Table 2. Alphabetical list of fungi reported from pandans.

On *Pandanus*	Origin	Tissue
Boninogaster phalloides Kobayasi	Chichijima	stump
Coniothyrium pandanicola Petr.	Hawaii	leaf
Coniothyriopsiella perottianum Verona	Hawaii	leaf (dead)
Ceuthospora pandani Verona	Samoa	fruit (dead)
Didymella pandani Höhn.	Java	leaf (dead)
Dimerosporium pandani Verona	Samoa	leaf (dead)
Dothidasteromella pandani Höhn.	Java	leaf (living)
Echidnodes sandwicensis Petr.	Hawaii	leaf (dead)
Filosporella aquatica Nawawi	Malaysia	submerged leaf
Gloeosporium pandani Verona	Samoa	fruit (dead)
Glomerella cingulata (Stonem.) Spauld. and Schenk.	Philippines	leaf (dead)
Haplaria pandani Gutner	greenhouse, USSR	leaf
Haplosporella pandanicola Roy	India	leaf
Hardnessia pandani Sawada	Taiwan	leaf
Heraldoa pandani Bat.	Malaysia	leaf
Hypocrea hemisphaerica Doi	New Guinea	log
Johansonia pandani Müller	Java	leaf
Lactarius pandani Heim	Madagascar	wood
var. amantiacus Heim	Zaire	wood
var. intermedius Heim	Zaire	wood
Leptosphaeria pandanicola Sawada	Taiwan	leaf
Macrophoma pandani Verona	Samoa	fruit (dead)
Melanconium pandanicola Sawada	Taiwan	wood
Meliola pandani Syd.	Borneo	leaf (living)
Meliola pandani Sawada (non Syd.)	Taiwan	leaf (living)
Meliola pandanicola Hansf. and Deighton	Sierra Leone	leaf (living)
Meliola juttingii Hansf.	Java	leaf (living)
Mycospharerella pandani Tilak	India	leaf
Oxydothis pandanicola (Syd.) Petr.	Philippines	leaf
Parastigmatea pandani Hugeunin	Loyalty Is	leaf (living)
Pestalotia pandani Verona	Rarotonga	fruit (dead)
Pestalotia pandani (non Verona)	Taiwan	leaf
Pestalotia peyronelii Verona	Australia	fruit (dead)
Paecilomyces iriomoteanus Mats.	Japan	leaf (dead)
Phoma fructus-pandani Verona	Samoa	fruit (dead)
Phoma martellii Verona	Samoa	fruit (dead)
Phoma pandani Verona	Marquesas	leaf
Phomatospora cylindrotheca Petr.	Hawaii	wood
Rhabdospora pandani (Höhn.) Petr.	Philippines	
Schizochora pandani Stev.	Hawaii	wood
Solheimia kamatii Mani Varghese and Rao	India	wood (dead)
Thielavia neocaledoniensis Moreau	New Caledonia	wood (dead)
Volutellaria fuliginea Hino and Katumoto	Japan	

On *Freycinetia*	Origin	Tissue
Chaetosphaeria aoteazoae Hughes	New Zealand	wood
Crepidotus parietalis Horak	New Zealand	leaf (dead)
Clypeosphaeria stevensii Syd.	Hawaii	stem
Hugueninia freycinetiae Bez. and Barros	New Caledonia	leaf (living)
Metasphaeria christophersoni Verona	Sava Is.	leaf (dead)
Mycosphaerella freycinetiae Stev.	Hawaii	leaf
Nectria freycinetiae Samuels	New Zealand	leaf (dead)
Odontia flexilis Cunningh.	New Zealand	stem (dead)
Peltosoma freycinetiae Syd.	Philippines	leaf (living)
Pyrenocyclus ambiguus Petr.	Hawaii	stem (dead)

lesser Sunda Islands. These are caused by gall-midges such as "*Trishormomyia*" *pandani* Felt, which produces leaf-galls in *P. furcatus*.

Leaf feeders. Pandans in New Guinea and surrounding regions often show traces of leaf predation or mining by larval stages of Hispine beetles (Gressitt 1963), which are found also on other Monocots (see Coleoptera chapter). Rarely occurrences are prominent enough to severely harm the plant hosts. On the other hand the abundance of pandans in New Guinea is important for the hispine beetles. Tettigoniid and acridid grasshoppers and phasmids are important leaf predators, and certain lepidopterous larvae are important predators of new leaves (Gressitt pers. comm.). One such moth is *Aeolarchis spheotoma* Meyr. Another making holes ("keyholes") in older leaves is *Trissodoris honorariella* (Walsingham) (see Zimmerman, 1978, p. 1035).

Banana Scab Moth disease in pandans. The Banana Scab Moth (*Nacoleia octasema* (Meyrick) has been intensively studied by Paine (1964) who found that *Pandanus* was an important food plant for the species, which also feeds on banana, *Heliconia*, and *Nypa*. It feeds mainly on inflorescences, especially floral bracts. While the moth is a sufficient pest of bananas to be called a disease, it may be more accurate to refer to the relationship as predation. Adults oviposit on the lower quarter of the leaf or leaves immediately below a very young inflorescence, though they may also be on older leaves. The majority of oviposition sites was found to be on leaf undersurfaces. In only one instance was a larva found actually feeding on leaf-tissue; the usual food is bracts. The larvae are xerophobic and photophobic. Larvae preferred *Pandanus* bracts to banana when experimentally offered the choice, suggesting that pandans are an older host plant and that the banana-feeding habit may have arisen in certain situations after an initial spread of the species on *Pandanus* or *Nypa*. *N. octasema* is not known to feed on banana in New Guinea proper, and here its pandan host usually is *P. tectorius* or similar members of Sect. *Pandanus*, and also some other species.

Wood-boring beetles may be found in pandan trunks and fruits, where they may damage tissues to the extent that branch-loss or fruit-fall may occur. *Anomocaulus fulvovestitus* Fairm. (dynastine) is an example known from Fiji (Lever, 1945). *Oryctes rhinoceros* L. (dynastine) can kill stands of pandans by adult boring, and larvae breed in dead trunks (Gressitt 1953). Weevils are common associates of pandans, e.g. *Diathetes pandanae* Zimm. (Zimmerman 1939).

Pandan leaf-axils as microhabitats, and other insect-pandan associations. The pandan leaf axil is of considerable interest. It is usually stocked by the infall of detritus from above, usually retains water to the extent of a cupful or more, or a thick solution of decaying debris, and may incorporate dying fragments of the endo-axillary rootlet system. Its chemical parameters remain to be documented, but it is known to be a breeding location for mosquitoes such as *Aedes pandani* A. Stone (which may be a carrier of the microfilarial disease *Filaria immitis*), and thus may have a role in human disease ecology. Other mosquitos also breed in pandan leaf axils, though not necessarily limited to them. The most ubiquitous insects found in pandans are predacious earwigs. They run freely over the plant in search of prey.

Other invertebrates found in pandans, especially leaf-axils, are: Acari (mites); Coleoptera (particularly Nitidulids, but also Curculionids and Staphylinids); Cyclorrhaphan fly (larvae): Psocopterans (which are usually fungus-feeders); and Drosophilids (two identified species found in Malaya are *Drosophila funebris* Fallen and *Drosophilella seminigra* Duda). Also amphipods, isopods, centipedes, scorpions, land-crabs, and earthworms. Earthworms can be found in leaf-axils of very tall pandans (e.g. in the Western Highlands); it seems likely they enter the leaf-axil system when the pandan is very short as a juvenile, and are "carried" upward by normal growth of the pandan.

The developing inflorescences almost always contain Nitudulid larvae (chiefly *Amystrops* spp.) and thrips (esp. *Docidothrips* spp.). These are extremely abundant on staminate inflorescences, less so on pistillate. A wasp of particular interest is *Bootania* (Hymenoptera Torymidae), which develops in *Pandanus* seeds (Boucek ex Parker, *in litt.*)

Larger animals associated with pandans outside the pollination syndrome are likely to be concerned chiefly with fruits. Flying-foxes are important in this connection, and so are mammals such as cuscus, rats, and possibly some birds. The overseas distribution of *Freycinetia* seeds endozoically and epizooically, and the same for at least some *Pandanus* drupes, is extremely probable. Hornbills eat certain types of pandan fruit. Most pandan seeds are well protected from this kind of predation and only the fleshy pericarp is digested.

Those pandans close to rivers and streams may provide food, in the form of fruits, to turtles, as is known to be true in Malaya (*P. helicopus*) and is probably also the case in Queensland for *P. aquaticus*. The latter species also attracts, when with ripe fruit, white cockatoos.

3.3 Pollinators and dispersal

3.3.1. Pollinators. There is no verified case of pollen transfer in Pandanaceae. The possible or plausible vectors include a number of visually assessed "candidates" and it is possible that pandans, or various species of them, are pollinated by a number of vectors simultaneously. Bees (melipones) have been seen visiting male inflorescences in Micronesia and elsewhere, but have not been seen on pistillate inflorescences. Also beetles, chiefly Nitidulidae (*Amystrops*) but perhaps also Cetoninae (*Lomaptera* etc) visit pandans (Willemstein, *in litt.*). *In Pandanus* Sect. *Maysops* the pollen grains of several species are when freshly released lightly coated with oil and thus adhere in small clusters of a greatly variable number of grains each: the oil is apparently volatile, and of a very faint orange tinge; this suggests high edibility of the pollen (personal obs. in Bot. Gard. Lae, 1971).

Wind-pollination is believed to act in those *Pandanus* species which are components of coastal, riverine, or low vegetation where they are at the edge of or form the top of the canopy, as in *P. tectorius* and *P. brassii*, but is unlikely or quite impossible in species restricted to the understorey or interior of dense forests. It may be relevant that those species capable of wind-pollination are often widespread and may possess oceanic current fruit dispersal syndrome as well. In *Sararanga*, there is no data whatsoever, bearing on pollinators.

In *Freycinetia*, bats, birds, and rats are all implicated fairly strongly as suitable pollinators. We will learn something more about this from the work of Paul Cox, presently in Samoa. *Freycinetia* floral bracts are particularly soft, fleshy, slightly sweet, often showing signs of having been partly eaten. In *Pandanus* it is not likely that mammals feed on the bracts, but they may be palatable to various insects. Degener also reported that bracts of *F. banksii* in New Zealand are eaten by the Maoris; he further states that rats (presumably *R. exulans*) pollinate *F. arborea* in Hawaii. By implication this may be true also for other Polynesian areas in which both this plant and the same rat occurs, e.g. Marquesas and Tahiti. Birds, however, are also pollinators.

3.3.2. Fruit dispersal in Pandanaceae. In both *Sararanga* and *Freycinetia*, the fruit is a berry; the seeds are very small (under 3 mm long) and are embedded in a soft, pulpy or fleshy fruit. In *Sararanga* these fruits are globular-reniform, not or rarely over 2 cm broad and long, and ripen to a vivid candy red or orange. Undoubtedly they are distributed by birds and fruit-bats. In *Freycinetia*, although the berries are aggregated into small round heads or cylindric spikes, they are soft, easily torn apart when ripe, and thus most probably sought by the same groups of animals mentioned above. In *Pandanus* this type of fruit presentation is found in a few species also, such as *P. polycephalus*; in such cases the individual drupe (carpel) is usually simple, 1-seeded, and fairly small. In *P. danckelmannianus* the long apical corpus and style of the drupe abscisses when ripe just above the basal seed, leaving the latter, which is bright yellow, attached to the receptacle, where it is highly visible. In these species, the endocarp is firm and if the vector animal swallows the seed it is probable that it is excreted whole and viable. The same may be true for *Freycinetia*, but their seeds are not protected by an endocarp; instead there is a slightly indurated seed-coat. The effectiveness of this structure is unknown. Other species of *Pandanus* have very different modes of dispersal. The *aquatic* modes may be divided into *fresh-water* and *marine*; but the former is difficult to specify when alternatives are possible, as is the case with e.g. *P. leptocarpus* (endozoic, turtles) or *P. leiophyllus* (possibly bird or bat?). The latter is demonstrable by the finding of pandan "keys" or phalanges in beach-drift, and seedlings established in coastal and strand areas from some drift fruit. Yet even in *P. tectorius* the possibility of occasional alternative dispersal by fruit-bat is to be considered, as observations in Micronesia have shown. Dispersal by humans is of course also possible even when the species of pandan is not a particulary useful one; ornamental pandans and useful cultivars are however more commonly spread by cuttings.

Circumstantial evidence based on species distribution suggests that throughout much of the tropical E. Pacific, dispersal of pandan-fruits is by ocean-current; but in high islands, especially larger ones such as Fiji and New Guinea, non-coastal species will evince other mechanisms.

4. Biogeographic status of the genera

4.1. Sararanga

The foregoing discussion of *Sararanga* included details of the distribution, and the "central" position the genus has in terms of the distribution of the family (Fig. 11). We will revert to this question below.

4.2. Pandanus

A review of the distribution of the genus has been provided recently by St. John (1965) (Fig. 12). More recently a longer essay on the biogeography of the genus was published (Stone 1976) based on the infrageneric taxonomy earlier proposed (Stone 1974). The eight subgenera and 61 sections (which have subsequently been added to by Huynh and Stone) are mainly natural taxa and have been mapped. (Stone 1976). In this review only those infrageneric taxa which include New Guinea are listed. Table 3 shows the New Guinea species attributed to each subgenus, section, and subsection. Four of the eight subgenera of *Pandanus* occur in New Guinea.

4.2.1. Distribution of subgenera and sections of Pandanus in New Guinea (see Table 3). In subg. *Lophostigma* (Fig. 13, 14), five sections are endemic to New Guinea; the other two sections which occur in New Guinea are found only in the immediately adjacent regions, east of Wallace's Line, and not extending east of the Solomons nor south of N Queensland. The only curious occurrence is the local endemic species *P. patina* of Ponape, allocated to Sect. *Megastigma*; but this section (consisting of but three species) is perhaps unnatural. Unfortunately very few collections of these three species are known, and no staminate specimens. However the occurrence of *P. cominsii* var. *micronesicus* in Ponape and Truk (Sect. *Kurzia* of subg. *Kurzia*) suggests that extension of a group to the Caroline Islands is not so unlikely.

Of the five New Guinea sections of subg. *Kurzia*, three are endemic; two others occur only E of Wallace's line and extend to larger islands of E Melanesia; none extend to Australia; the last extends northward to the Philippines where it is extensively developed, and has species in both Borneo and Sulawesi.

Of the five sections of subg. *Pandanus* only one, sect *Excavata*, is endemic, Sect. *Austrokeura* extends the genus into the harsher, drier frontier of N Australia, to its maximal tolerances. *Excavata* (which is monotypic) is limited to New Guinea mountains above 2000 m. The New Guinea species of Sect. *Pandanus* are like the majority of others in this section primarily coastal and rely mainly on ocean fruit dispersal, supplemented by estuarine and zoochoric modes (flying foxes).

In subgenus *Acrostigma* (Fig. 15) only one subsection is endemic to New Guinea. Interestingly 2 subsections appear to show disjunctions separated approximately by Wallace's line, but perhaps they indicate imperfect taxonomic

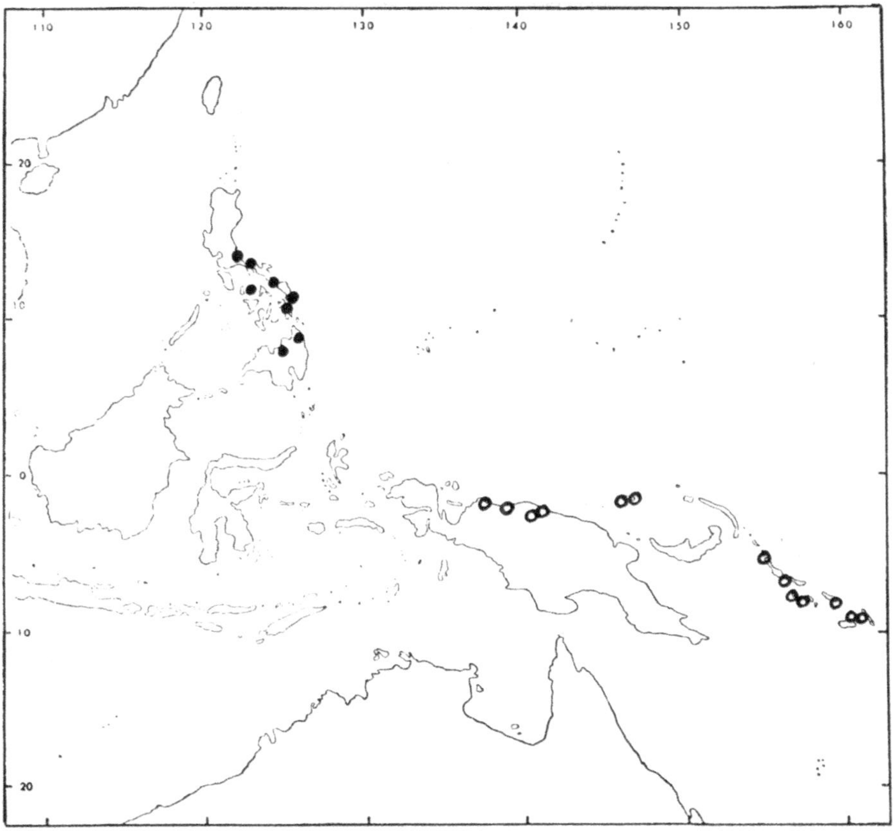

○ Sararanga sinuosa ● S. philippinensis

Fig. 11. Distribution of genus *Saranga* Hemsl. Both species are shown: all herbarium collections known are mapped.

conclusions. The sole representative of subgenus *Rykia* in New Guinea is *P. dubius*, of Section *Hombronia*, and it is a coastal species.

4.2.2. Species endemism in New Guinea Pandanus. To determine how many species are endemic in New Guinea involves sifting through many binomials, not all of which can now be properly evaluated. Considering New Guinea proper — excluding the surrounding islands — there are probably 50 endemic species of *Pandanus*, not allowing for future discoveries. For many species staminate plants and inflorescences are unknown, collections so far made being of pistillate plants with fruits. An exception is *P. pendulinus*, which is known only from the staminate plant. Lack of associated material representing both sexes is an impediment to classification as well as to knowledge of field biology and ecology. However, our greatly improved knowledge of vegetative characters (including anatomical data) will permit association of pistillate and staminate specimens even when they are not obtained from the same locality. There is hardly any study indicating real diversity within a species except for the commonest, such as *P. tectorius* and *P. dubius*. Thus our understanding of the New Guinea pandans is quite incomplete. Despite the imperfect data, it seems

Table 3. Subgenera, sections, and subsections of *Pandanus* occurring in New Guinea (with included species).

Subg. RYKIA: Sect. *Hombronia*; P. dubius.

Subg. LOPHOSTIGMA
- Sect. *Stonedendron* Huynh. (Endemic)
  - *P. atropurpureus, concinnus, columbiformis* (Fig. 14)
- Sect. *Megastigma* Stone
  - *P. schoddei* (Bougainville); *rex*. Also in Ponape (*patina*).
- Sect. *Karuka* Stone (Endemic)
  - *P. brosimos, jiulianettii*
- Sect. *Maysops* St. John
  - *P. cernuifolius, floribundus, krauelianus, microdontus, xanthocarpus, beccarii, amboinensis, castaneus, croceus*, etc. Also Solomons, Moluccas, N Queensland (Fig. 13).
- Sect. *Metamaysops* Stone (Endemic)
  - *P. brachyphyllus, galorei.*
- Sect. *Perrya* Stone (Endemic)
  - *P. aggregatus, archboldianus*
- Sect. *Liniobtutus* St. John (Endemic)
  - *P. hystrix, foveolatus, boemiensis, orculiformis, penicillus, hooglandii.*

Sub. KURZIA
- Sect. *Kurzia* Stone
  - *P. englerianus* (New Ireland), *exiguus, hollrungii, latericius* (New Ireland), *macgregorii, minusculus* (New Ireland), *plicatus, subumbellatus*. Also in Solomons, New Hebrides, and Carolines.
- Sect. *Microstigma* Kurz
  - *P. conoideus, ruber, erythros, magnificus*. Also in Moluccas.
- Sect. *Jeanneretia* (Gaudich.) Stone
  - *P. polycephalus*. Also Indonesia, Philippines, Solomons
- Sect. *Involuta* St. John (Endemic)
  - *P. leiophyllus.*
- Sect. *Marginata* St. John (Endemic)
  - *P. meniscostigma.*
- Sect. *Paralophostigma* Stone (Endemic).
  - *P. balenii, limbatus*. (note: this Section was formerly placed in Subg. *Lophostigma*).

Subg. PANDANUS
- Sect. *Pandanus* (from Polynesia to E. Africa).
  - *tectorius* (including many binomials/'microspecies' such as *novo-hibernicus, pedunculatus, polyacris*, etc.)
- Sect. *Austrokeura* (Stone) Stone
  - *P. brassii*. Also in N Australia. Some additional Papuan species may exist.
- Sect. *Fagerlindia* Stone
  - *P. kaernbachii, scabribracteatus, dolichopodus*(?). Also in Philippines, Moluccas.
- Sect. *Intraobtutus* St. John (almost endemic).
  - *P. papuanus* (and synonms): Also in Biak, Bismarcks.
- Sect. *Excavata* Stone (Endemic)
  - *P. antaresensis* (incl. *bowersae*).

Subg. ACROSTIGMA
- Sect. *Acrostigma* subsect. *Acrostigma*:
  - *P. adinobotrys, lauterbachii, odoardi, setistylus, pseudosyncarpus* (Fig. 15)
- Sect. *Acrostigma* subsect. *Rostellati* Stone
  - *P. eumekes, permicron, rostellatus, ihuanus, odoardi* (Fig. 15)
- Sect. *Acrostigma* subsect. *Dimissistyli* Stone
  - *P. danckelmannianus, erinaceus* (Buka), *lictor* (New Britain), *misimaensis, stenocarpus, humicola*. Also in Halmaheira.
- Sect. *Pseudacrostigma* Stone
  - *P. ornithocephalus*. Also in Borneo.

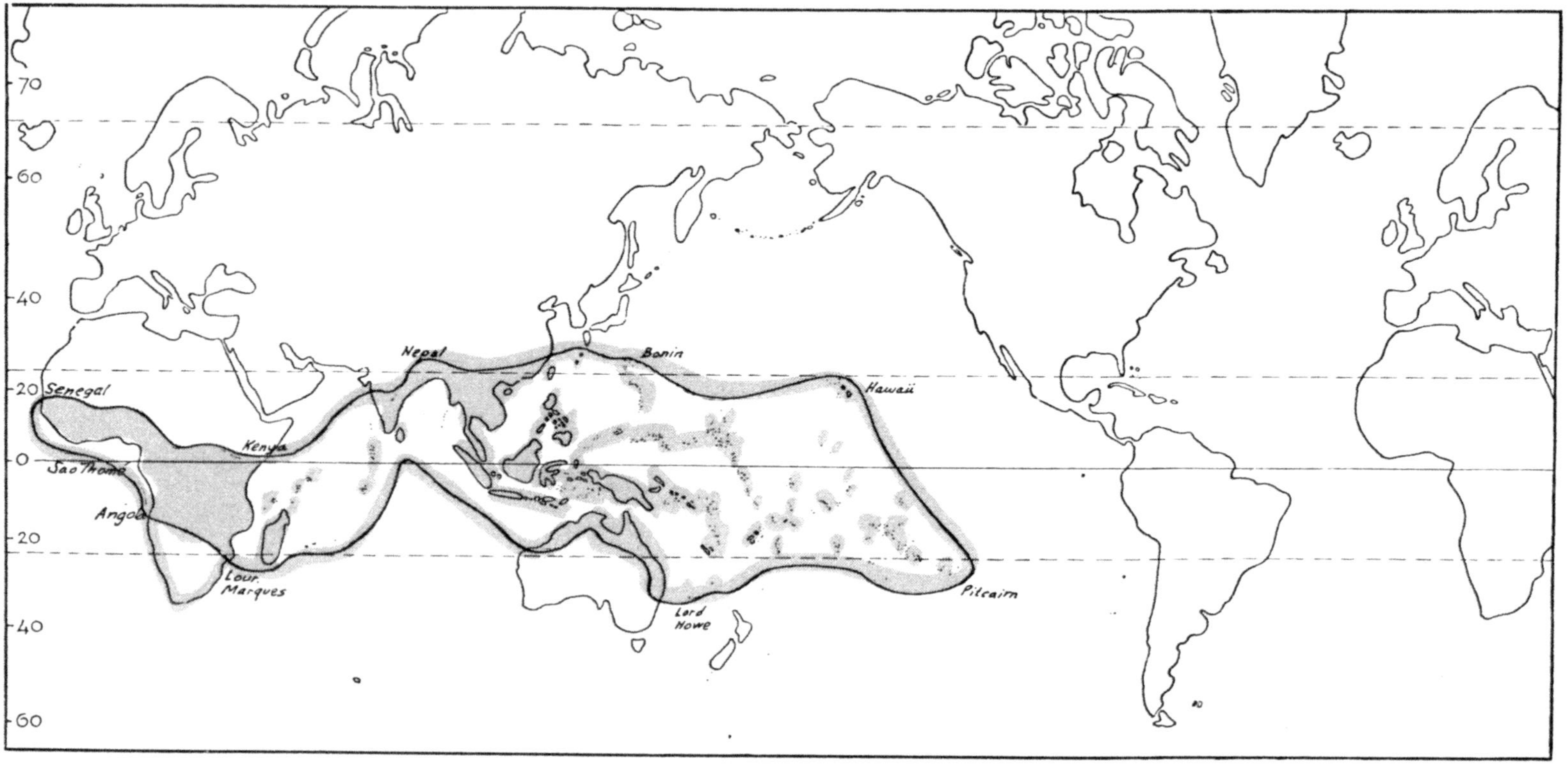

Fig. 12 Distribution of the genus *Pandanus* Park. (after St. John 1965).

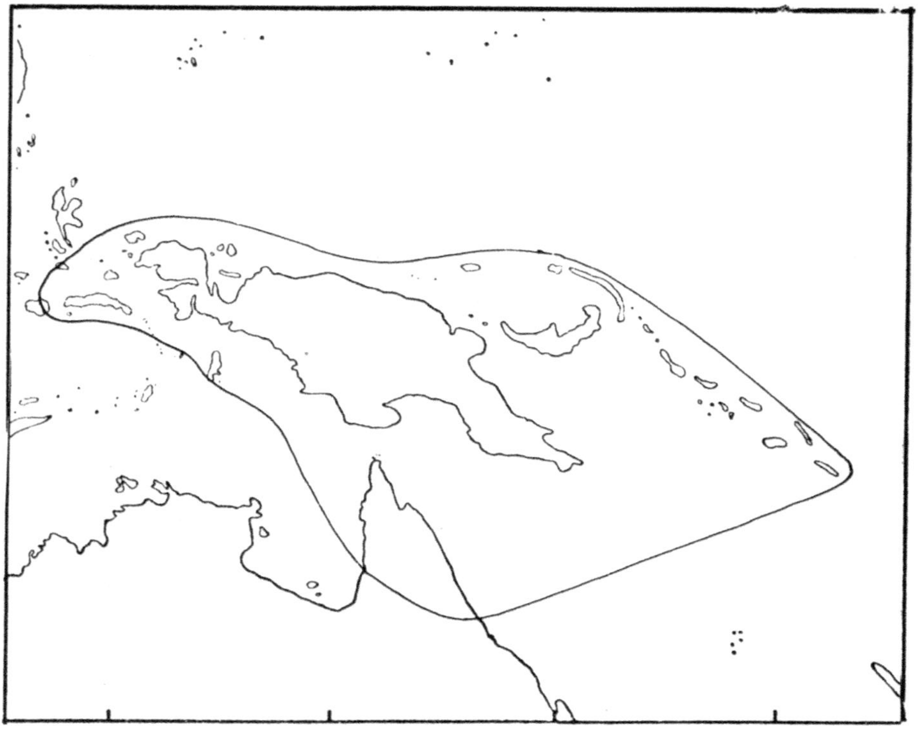

Fig. 13. Distribution of *Pandanus* subg. *Lophostigma* sect. *Maysops* (after Stone 1976).

clear that a strong and distinctive endemic element of *Pandanus* occurs in New Guinea. In terms of species, something like 80% endemism occurs — a figure which may perhaps rise as studies continue. New Guinea has no endemic subgenera, but has nine endemic sections (of 20). In Sect. Acrostigma, one of the three subsections present is endemic.

4.3. *Freycinetia*

Sections have been established but, as yet, no subgenera. There are also still some species *incertae sedis*. However, a resumé of the infrageneric classification can be presented for its interim value. A sectional classification was published (Stone 1968) but has been added to subsequently (Stone 1969, 1970, and *in prep.*) I now recognize 21 sections of which 12 occur in New Guinea. (Fig. 16).

4.3.1. Species endemism in New Guinea Freycinetia. As in *Pandanus* there are still several binomials which have not yet been evaluated taxonomically; however we may tentatively accept 56 species for New Guinea. Of these 40 have been recorded only from New Guinea; apparent endemism is thus 71%. The majority of the remaining species have been found outside New Guinea in Melanesia, especially Solomon Is, which have seven of their total of 22 in common with New Guinea (Stone 1970). One species is found outside New Guinea only in

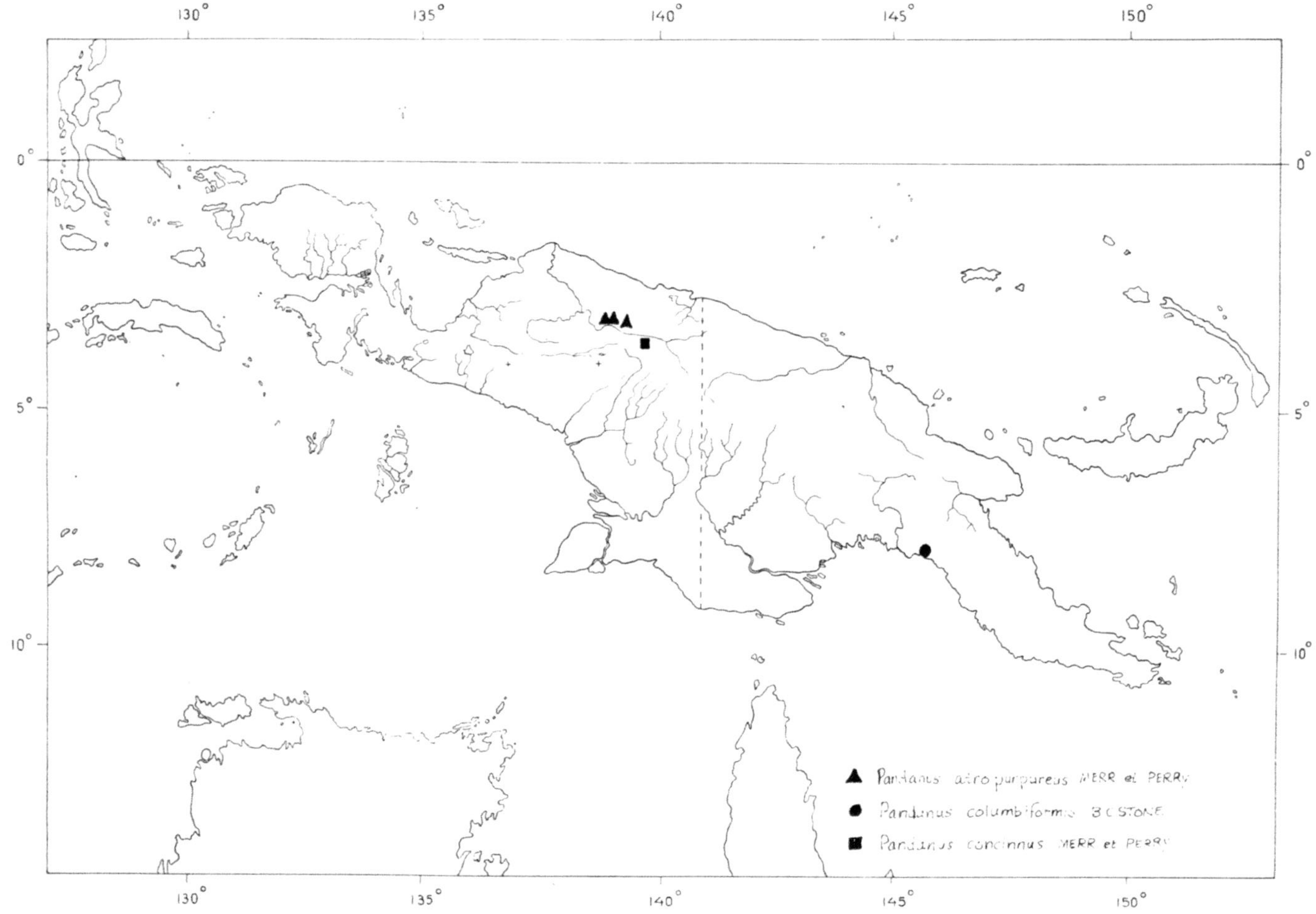

Fig. 14. Distribution of *Pandanus* subg. *Lophostigma* sect. *Stonedendron*. Note sparsity of collections. Each symbol represents a single collection

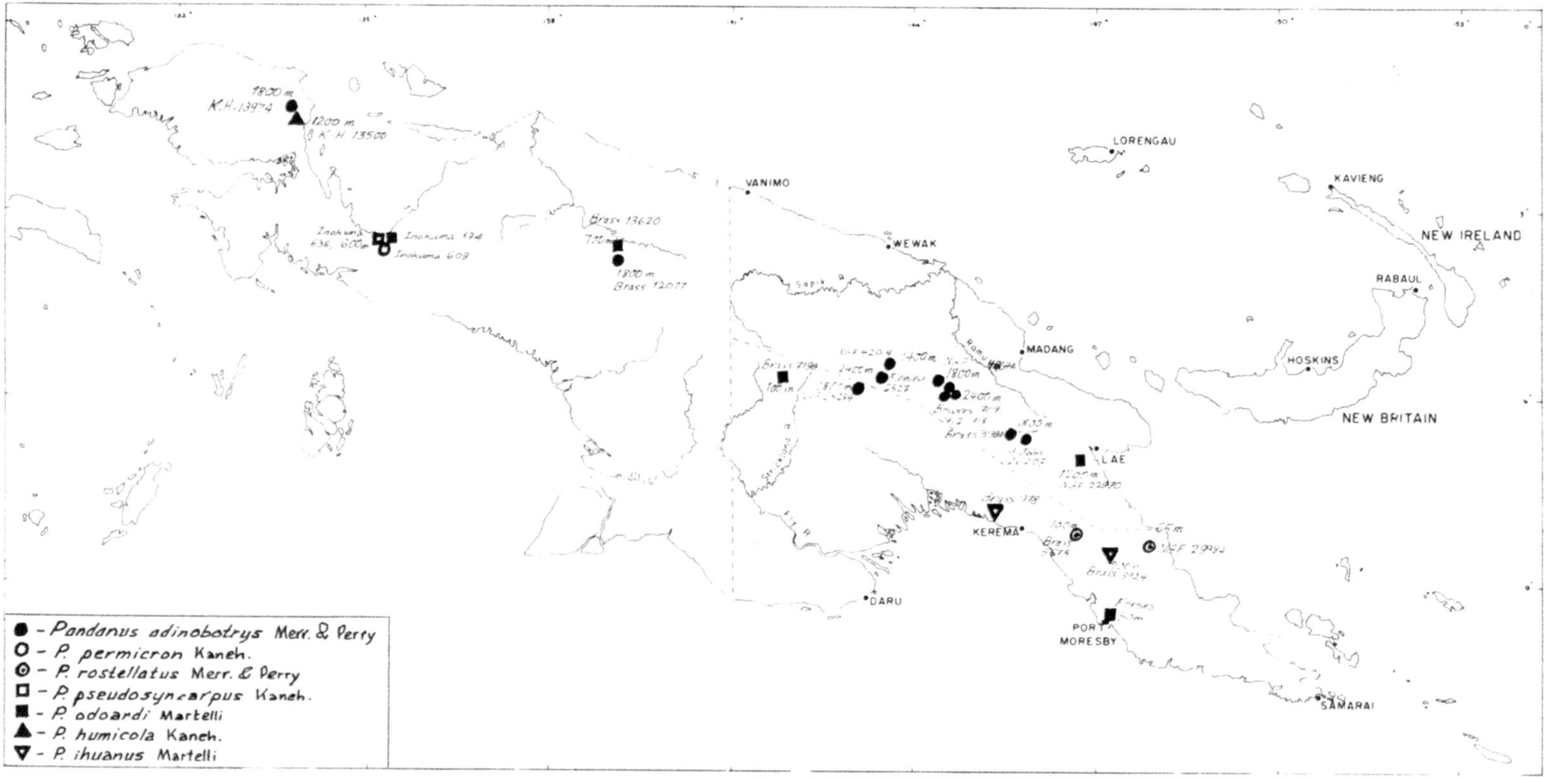

Fig. 15. Distribution of *Pandanus* subg. *Acrostigma* sect. *Rostellati* (part) and *Acrostigma* (part) to show insufficient sampling. Herbarium and collector's numbers are indicated, and montane collections show altitude in meters. The poverty of the material data-base is evident.

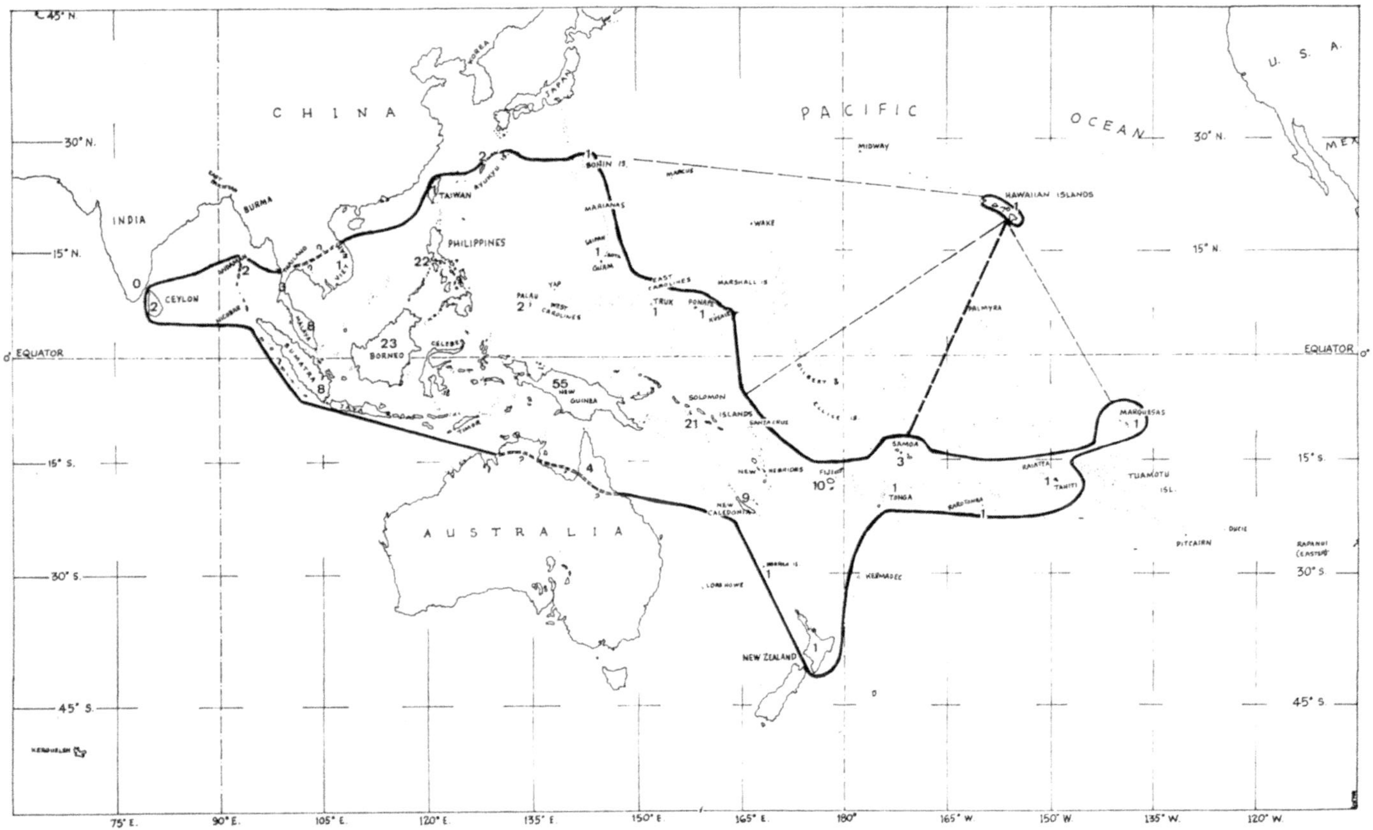

Fig. 16. Distribution of the genus *Freycinetia* Gaudich. (after Stone 1968). A few localities on New Zealand South Island are now known (*banksii*) and a slightly more nort locality in S. Burma (*sumatrana*). The doubtful localities in N. Terr., Australia, have not been substantiated, and all records are from N. Queensland.

Australia: *F. excelsa*; three (*multiflora*, *percostata*, and *lagenicarpa*) display either strong affinity to the Philippine-Moluccan flora or actually occur there; two (*radicans*, *marginata*) are linked closely with the Moluccan-Sunda Is flora pertaining to Sect. *Blumeella*, while another (*scandens*) occurs in these areas and also in Queensland (where it has been called *propinqua*). The Australian *australiensis* (Sect. *Filiformicarpae*) is very similar to, and perhaps identical with New Guinea *F. marginata*. *F. funicularis* (Sect. *Lateriflorae*) also occurs in E Moluccas, and continues eastward beyond New Guinea to the Solomons. A few species in the above list (Table 1) have not so far been found in New Guinea proper but are for convenience included here as "New Guinea" species, e.g. *naumannii* (Bismarck Archip.) and *aruensis* (Aru). It is quite possible that these will turn up in New Guinea. Some other species such as *keyensis*, so far known only from the Key (or Kei) Is, fit the same category but have not been included because their taxonomic status is doubtful and they are tentatively (unpubl.) considered synonyms.

The New Guinea *Freycinetia*-flora is well marked and includes several unusual and distinctive species, such as *angustissima* and *elegantula*, which appear as some of the smallest plants in the genus; and others such as *kanehirae*, which are among the largest. Several species have a very memorable ap-

Table 4. Sections of *Freycinetia* in New Guinea with their distribution and included species*

Section	Additional distribution	Species
Lateriflorae Stone	E. Mol., Sal.	*funicularis, lauterbachii, pleurantha, rhodospatha, sterrophylla, gibbsiae, erythrospatha*
Solmsiella Stone	N Caledonia	*elegantula, lateriflora*
Polystachyae Stone	Phil., Mol.	*multiflora*
Filiformicarpae Stone	Queensland, Micr (except atolls), Mol., Phil.	*macrostachya, kanehirae, undulata*
Oligostigma Warb.	Sol., Queensland, some Sunda Is	*scandens, beccarii, crucigera, elliptica, ellipsoidalis, nervosa, forbesii, divericata, rectangularis, vulgaris* (and *oblanceolata, globiceps,* if not synonymys)
Warburgiella Stone	Sol., N Cal., N Hebr.	*marantifolia, decipiens*
Ridleyella Stone	Queensland	*angustissima, flaviceps, linearis, linearifolia, linearis, fibrosa, brassii, tafaensis, excelsa, klossii, spinellosa,? lacinulata*
Polygynae Stone	Sol.	*laeta*
Blumeella Stone	Phil., Sunda, Sulawesi, Ceram, Ambon, Timor	*radicans, marginata, archboldiana, pseudoinsignis, latiauriculata, latibracteata*
Pristophyllae Stone	Phil., Mol., Queensland	*percostata*
Devrieseella Stone	Phil., Mol.	*lagenicarpa*
Gaudichaudiella Stone	SE Polyn., Fiji, Samoa, Sol., Bismarcks	*impavida, solomonensis*

*Species *incertae sedis*: jaheriana, andajensis, naumannii, salamauensis, sogerensis pearance, e.g. *salamauensis* and *marantifolia*. Others are very obscure and impossible to interpret until further collections are made, e.g. *jaheriana*.

4.3.2. Section endemism in New Guinea Freycinetia. As shown in Table 4, 12 sections occur in New Guinea, of which none is endemic. The nearest approach to an endemic section is Sect. *Ridleyella* (New Guinea and Queensland) but biogeographically this is entirely satisfactory. The extension to the Solomons through the Bismarcks of Sect. *Polygynae* (monotypic: *F. laeta*) and Sect. *Lateriflorae* is also normal. The links between New Guinea, the Moluccas (sometimes including Sulawesi) and the Philippines are also evident in the Sections *Polystachyae*, *Pristophyllae*, *Devrieseella*, and *Filiformicarpae*. A small but evident tie to the New Caledonian flora is evinced in Sections *Solmsiella* and *Warburgiella*. The only extension of *Freycinetia* into Micronesia from the New Guinea center is shown by Sect. *Filiformicarpae* (2 spp., *F. ponapensis* and *F. reineckei*).

5. Biogeographical conclusions

Pandanaceae make up a significant element in the New Guinea flora: approximately 123 species in three genera. As individuals certain species may be very heavily represented, locally dominating large tracts of terrain within their habitat. They are often conspicuous and characteristic. They must play, in some situations, important ecological roles. They interconnect in myriad ways with the flora, fauna, and local aspects of their environment. They provide several species which, at least in primitive societies, play an important or even vital role.

Biogeographically the Pandanaceae are entirely paleotropical; yet there are persistent, if dubious, reports of fossil palynomorphs said to pertain to the family from the New World (Elsik 1968; Leffingwell 1970; Jarzen 1978). Until these records (e.g. *Pandaniidites*) can be satisfactorily demonstrated to pertain to the family, they cannot be a controlling factor in hypotheses attempting to locate the origins of the family.

Except for its multiovulate characteristics *Freycinetia* is structurally more advanced in most respects than the other two genera. *Sararanga*, with its complex inflorescence and trichomes of stellate hairs (in *S. philippinensis*). quadrifarious phyllotaxy, and lack of large proproots, seems the most ancient, yet has perhaps some advanced features (berries, small seeds) as well. *Pandanus*, obviously an extremely successful group, may still show cases of current evolutionary change, but seems to have become an essentially fixed genus of very numerous, often relictual species. There are situations in Australia and some Pacific islands, including New Guinea, where putative hybridization indicates a genetic elasticity in certain species-groups that imply a current adaptiveness and plasticity. Extinction in such cases applies only to genetic entities of less than species rank. But in other parts of the world species are on the threshold of extinction (e.g. *P. vandermeerschii* in Mauritius).

Clearly much evolution in all three genera has taken place in what we now call New Guinea. Possibly, a good deal of extinction may also have occurred.

The endemics in the New Guinea flora stand out as evidence, but at present there seems no easy way to date them either absolutely or on a comparative basis. Stratigraphic exploration in New Guinea as yet throws no light on this problem. The pandan flora of New Guinea apart from its local endemics is clearly regional; lines of affinity extend E, W, N and S. The southern affinity — i.e. with Queensland and N Australia in general, brings us to the climatic margin of the family. Extension to the N — into Micronesia — is in general explainable either by oceanic drift (*Pandanus* Sect. *Pandanus*) or bird distribution (*Freycinetia*); but the occurrence of the strange *Pandanus patina* limited to Ponape and related apparently to *P. schoddei* of Bougainville and *P. rex* of New Guinea demands a different interpretation. It is not a unique pattern, however, as it is shown also by the apocynaceous genus *Lepinia* (van Steenis 1962). The affinities to species occurring E of New Guinea and the Bismarcks are for the most part relationships to species of the Solomons, Samoa and Polynesia, or (to a smaller degree) to species of New Caledonia, where there is strong endemism in *Pandanus* (esp. in subg. *Lophostigma*) and in *Freycinetia*. The affinities linking New Guinea with areas to the W are primarily exemplified by Moluccan and/or Philippine linkages, very few relationships extending W of Wallace's line. In short, the New Guinea pandan flora is related outward in a manner which corresponds with our general understanding of the relationships of other components of the New Guinea flora.

Since the Pandanaceae clearly originated in upper Cretaceous time, and may well have originated earlier, direct migration between Asia and Australasia via Africa would still have been possible (Raven and Axelrod 1974). The relatively poor representation of the family in W Africa probably is related to the sorts of massive extinction in that continent discussed by the same authors. Certainly some of the complexes of species in New Guinea, New Caledonia and other parts of Australasia may have evolved relatively recently in relation to changes in topography and climate, but other groups, e.g. *Sararanga* appear to be much older.

References

Balgooy, M. J. van. 1976. See Paijmans, 1976.
Brass, L. J. 1934. Bull. Amer. Mus. Nat. Hist. 68: 534.
Degener, O. 1930. Plants of Hawaii National Park. Publ. by the Author. (Reprint 1945).
Docters v. Leeuwen, J. and W. 1926. Zoocecidia of the Netherlands East Indies. 'S Lands Plantentuin, Buitenzorg, Drukk. de Vrie.
Elsik, W. 1968. Palynology of a paleocene Rockdale lignite, etc. Pollen et Spores 10: 314.
Froggatt, J. L. 1936. Entomological Notes. New Guinea Agric. Gaz. 2: 10–14.
Gressitt, J. L. 1953. The coconut rhinoceros beetle (*Oryctes rhinoceros*) with particular reference to the Palau Islands. Bishop Mus. Bull. 212: viii + 157 pp.
—— 1963. Hispine beetles (Chrysomelidae) from New Guinea. Pacif. Insects 5: 591–714.
Jarzen, D. M. 1978. Some Maestrichtian palynomorphs and their phytogeographic and paleoecological implications. Palynology 2: 29–38.
Laing, F. 1927. Insects of Samoa. Brit. Mus, Nat. Hist. 2(1): 34–45.
Lane-Poole, C. E. 1925. Forest Resources of the Territory of Papua and New Guinea. Parliament of the Commonwealth of Australia: Government Printer, Victoria.
Leffingwell, H., D. A. Larson and M. L. Valencia. 1970. Study of fossil pollen Wodehousia spinata I, II. Bull. Canad. Petrol. Geol. 18: 238–262.
Leefmans, S. 1927. Korte Med. Inst. Plantenziekten nos. 72, 73, 74. Buitenzorg, Java.

Kanehira, R. 1940. A summary of our knowledge of Papuan Pandanus. Bot. Mag. (Tokyo) 54: 249–261.
Lever, R. W. 1945. Agric. J. Fiji 16(4): 98–103.
—— 1947. Bull Entom. Res. 38: 140.
Martelli, U. 1929. Pandanaceae collected for Arnold Arboretum by L. J. Brass in New Guinea. J. Arnold Arb. 10: 137–142.
Merrill, E. D. and L. M. Perry. 1939. On the Brass collections of Pandanaceae from New Guinea. J. Arnold Arb. 20: 139–186.
—— Plantae Papuanae Archboldianae II. J. Arnold Arb. 21: 163–200.
Muller, J. 1970. Palynological evidence on early differentiation of Angiosperms. Biol. Rev. 45: 417–450.
Paijmans, K. (Ed). 1976. New Guinea Vegetation. Elsevier Scientific Publishing Co., Amsterdam. 1–208.
Paine, R. W. 1964. The Banana Scab Moth Nacoleia octasema (Meyrick): its distribution, ecology, and control. South Pacific Commission, Techn. Pap. n. 145: 1–70.
Powell, J. M. See: Paijmans, 1976.
Raven, P. H. and D. Axelrod. 1974. Angiosperm biogeography and past continental movements. Ann. Mo. Bot. Gard. 61: 539–673.
St. John, H. 1960. Revision of the genus Pandanus Stickman, Part I. Pacif. Sci. 14: 224–241.
—— 1961. Ibid. Part 6. New Pandanus species from Queensland, Australia. Pacif. Sci. 15: 563–575.
—— 1962. Ibid. Part. 12. Queensland Pandanus. Pacif. Sci. 16: 291–346.
—— 1965. Le distribution mondiale du genre Pandanus. Adansonia ser. 2, 5: 307–308.
—— 1967. Revision of the genus Pandanus, Stickman, Part 21. The Pandanus monticola group in Queensland, Australia. Pacif. Sci. 21: 272–281.
—— 1968. Ibid. Part 29. New Papuan species in the section Microstigma collected by C. E. Carr. Pacif. Sci. 22: 514–519.
—— 1968. Ibid. Part 30. The New Section Marginata from Papua. Pacif. Sci. 22: 520–522.
—— 1968. Ibid. Part 32. The New Section Involuta from Papua. Pacif. Sci. 22: 532–535.
—— 1973. Ibid. Part 35. Additional Pandanus species from New Guinea. Pacif. Sci. 27: 44–01.
—— 1975. Ibid. Part 39. Pandanus of Rotuma Island, Pacific Ocean, Pacif. Sci. 29: 371–406.
Smith, E. F. 1915. Ann. Mo. Bot. Gard. 2: 377.
Sorauer, P., K. Lindau and F. Reh. 1922. Manual of Plant Diseases, ed. 3. 1: 434. Berlin.
Solms, H. Grafen zu. 1883. Über die von Beccari auf seiner Reise nach Celebes und Neu-Guinea gesammelten Pandanaceae. Ann. Jard. Bot. Buitenz. 3: 89–104.
Stone, B. C. 1961. The genus Sararanga (Pandanaceae). Brittonia 13: 212–224.
—— 1968. Materials for a monograph of Freycinetia (Pandanaceae) IV. Subdivision of the genus, with fifteen new sections. Blumea 16: 261–372.
—— 1969. Materials for a monograph of Freycinetia (Pandanaceae) VIII. A revised list of Philippine species, with critical notes and some new taxa. Webbia 23: 597–607.
—— 1970. Materials for a monograph of Freycinetia (Pandanaceae), IX. Species of the Solomon Islands. Kew Bull. 24: 355–375.
—— 1971. Materials for a monograph of Freycinetia (Pandanaceae) XII. Fedn. Mus. J. (Kuala Lumpur) p.s. for 1968, 13: 155–165.
—— 1972. A reconsideration of the evolutionary status of the family Pandanaceae and its significance in monocotyledon phylogeny. Quart. Rev. Biol. 47: 34–45.
—— 1972. The genus Pandanus in the Solomon Islands with notes from adjacent regions. Part I. Malays. J. Sci. 1A: 93–132.
—— 1973. The genus Pandanus in the Solomon Islands with notes on adjacent regions. Part II. Malays. J. Sci. 2A: 59–80.
—— 1973. Studies in Malesian Pandanaceae, XIII. New and Noteworthy Pandanaceae from Papuasia. Contrib. Herb. Austral. No. 4: 7–40.
—— 1974. Towards an improved infrageneric classification in Pandanus (Pandanaceae). Bot. Jahrb.Syst. 94: 459–540.
—— 1976. The Pandanaceae of the New Hebrides, with an essay on Intraspecific variation in Pandanus tectorius. Kew Bull. 31: 47–70.
—— 1976. On the biogeography of Pandanus (Pandanaceae). C. R. Soc. Biogeogr. (Paris) Seance 458: 69–90.
—— 1979. Studies in Malesian Pandanaceae XVII. On the taxonomy of 'pandan wangi,' a Pandanus cultivar with scented leaves. Econ. Bot. 32: 285–293.
Steenis, C. G. G. J. van. 1954. Pandanus in Malaysian vegetation types. Advance reprint from Flora Malesiana ser. I, 4(1): 1–12.
Steenis, C. G. G. J. van and M. J. van Balgooy. 1966. Pacific Plant Areas, vol. 2. Blumea, supplement volume V. Cfr. p. 56, and Maps 28–29.

Warburg, O. 1900. Pandanaceae, in Engler, A., and K. Prantl, eds., Das Pflanzenreich 3, IV. 9: 1–99.
Went, F. A. F. C. 1895. Über Häft und Nahrwurzeln bei Kletterpflanzen und Epiphyten. Ann. Jard. Bot. Buitenz. 12: 1–72.
Zimmermann, E. C. 1939. Proc. Haw. Entomol Soc. 10: 335–338.
—— 1978. Insects of Hawaii. vol. 8 (I, II), 1903 pp.

8. Semi-deciduous scrub and forest and Eucalypt woodland in the Port Moresby area

Petrus C. Heyligers

1. Introduction

Woodlands dominated by *Eucalyptus* species are widespread in the seasonally dry lowlands along the NE coast of the Gulf of Papua. Denser woody vegetation is often juxtaposed with sharp boundaries without an obvious difference noticeable in land form or soil.

During the Torres Strait Symposium (Walker 1972) the special position of these lowlands was discussed with respect to the Quaternary climatic circulation and its consequences for plant and animal life. However, lack of distributional data on biota was evident then as it is today. This chapter aims to fill part of this gap by giving a general description of some of the vegetation types near Port Moresby.

Following a general survey of most of the lowlands (Mabbutt *et al.* 1965) and some plant-soil relationship studies in the eucalypt woodlands (Heyligers 1966) the woodlands and adjacent forests were investigated in more detail to determine floristic relationships and to find clues about the causes of the sharp boundaries. Three areas showing a diversity of vegetation types on similar topography were selected from air photos. Fieldwork was carried out in the mid and late 1960s. Additional information has been gathered from literature and herbarium collections. The basic information is available as Technical Memoranda from the Division of Land Use Research, CSIRO (Heyligers 1972b,c; 1979).

This study was a reconnaissance, and hence replicated observations were restricted. It should also be noted that the expression "Port Moresby area" pertains only to the combination of the three areas investigated. There are undoubtedly more species and other dry-land vegetation types in the Port Moresby region, as well as in the rest of the seasonally dry lowlands (Heyligers 1965, 1972a).

The terminology for vegetation structure follows Specht (1970). It describes growth-form, height and projected foliage cover of the tallest stratum. The term "savanna" is therefore not used and has been replaced by "open-woodland" with the addition of "shrubby" or "grassy" to indicate the understorey structure.

The selected areas (Fig. 1) were spread over various land systems, all

© *1982, Dr W. Junk Publishers, The Hague.*

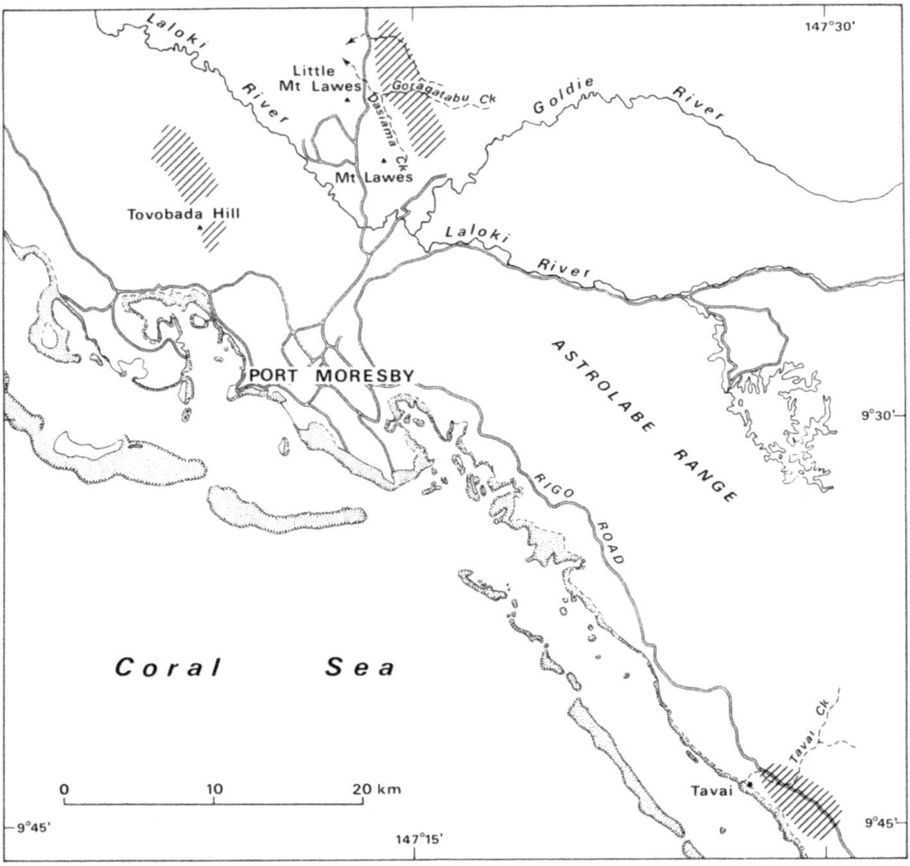

Fig. 1. Locality map of the investigated areas (hatched).

belonging to the coastal hill zone (Mabbutt *et al.* 1965). The rainfall is in the order of 1000–1200 mm per annum with a strong seasonal contrast. During the driest part of the year, usually between May and August, the passing of whole months without rainfall or with amounts less than 12 mm is not uncommon. September generally is a month of higher rainfall, followed by a shorter dry period before the arrival of the wet season in November.

2. Study areas

2.1 The Goragatabu area

This area situated about 25 km NE of Port Moresby has a basin-shaped topography (Figs. 1 and 2). Relatively flat interfluves between incised mostly ephemeral streams are backed by ridges of various heights.

The area is underlain by gabbro and minor sedimentary rocks of Paleogene age. On the interfluves texture-contrast soils are dominant: a sandy loam surface horizon, often with quartz and chert gravel, overlies at a depth of 20 to 50 cm a sandy to heavy clay. Slight depressions have a coarse sandy layer over the heavy clay subsoil and become waterlogged during the wet season. Soil reaction varies from neutral to slightly acid in the surface horizon and is medium acid in the subsoil. Brown clay soils and neutral brown lithosols, both derived from gabbro and tuff, are found on the low ridges. They are sandy clays to clay loams, usually with some gravel, overlying weathered rock at varying depth.

Grassy open-woodlands dominated by *Eucalyptus confertiflora* and *E. alba* (Fig. 3) cover most of the broader interfluves and extend over some of the ridges. There, the large tussock grass *Themeda novoguineensis* dominates the understorey, but on the interfluves it is sparser and associated with a floristically richer herb layer. The depressions are characterized by a tussock grassland dominated by the annual grasses *Eriachne squarrosa* and *Eragrostis cumingii*, while in the adjacent woodland there usually is a transitional zone where *Imperata cylindrica* is dominant.

Fig. 2. An aerial view over the S section of the Goragatabu area with Little Mt Lawes and teak plantations in the middle distance and the N part of the Tovobada area in the background. Eucalypt open-woodland and open-forest can be seen on the interfluves.

Fig. 3. Eucalypt open-woodland with a *Themeda novoguineensis* ground storey before and after a mild fire. The charred tussocks demonstrate the protection afforded to the growing tips by the tightly bunched leaves.

The woodlands and open-forests are found on narrow interfluves, footslopes and low ridges. They have irregular, open canopies up to 18 m tall with emergents reaching 25 m. Canopy and undergrowth are always of mixed composition, but lianes are infrequent. Some common species are *Protium macgregorii*, *Colona discolor*, *Acacia auriculiformis*, *Terminalia archboldiana*, *Psychotria lolokiensis* and *Guioa* sp. *Eucalyptus confertiflora* and *E. alba* also occur, but young eucalypt trees are only found in the most open stands. The forest floor is usually covered with leaf litter and fallen, decaying trunks. *Oplismenus hirtellus* and/or *Scleria lithosperma* form an open ground cover. They are joined by *Eriachne squarrosa* and other moisture-loving herbs on sandier soils. If these are waterlogged for longer periods, they form open spots in the forest comparable to those in the open-woodlands.

Taller open-forests with a more conspicuous storey of emergent deciduous trees occur on the slopes of ridges to the north and east of Mt Lawes on brown clay soils. *Albizia carrii*, *Garuga floribunda*, *Bombax ceiba* and *Gyrocarpus americanus* are the more common species of emergents. The canopies are of mixed composition and comprise many species not or only sporadically found in the woodlands and other open-forests, but eucalypts are not present. The understorey is usually open, but rich in species, with *Syzygium* sp. or *Jossinia desmantha* common.

Transition zones between open-woodlands and open-forests, some tens of meters wide, are not uncommon, especially in the eastern and southern sections of the area where low ridges are more frequent. Eucalypts decrease in numbers and species such as *Terminalia archboldiana*, *Protium macgregorii*, *Grevillea pinnatifida*, *Acacia auriculiformis*, *Guoia* sp. and *Rhus taitensis* become interspersed. The cover of *Themeda novoguineensis* thins out and *Scleria lithosperma* and other forest floor species appear.

Some ridges are covered with a shrubby open-woodland which resembles this transitional vegetation in floristic composition, but which has a denser shrub layer dominated by *Guioa*. After an extremely dry season the undergrowth in the transition zones as well as in this shrubby open-woodland was found to have drastically changed. *Trema cannabina* and *Urena lobata* were now predominant, while a large number of the original undergrowth shrubs had died. There were no traces of fire. It would be interesting to know whether or not eucalypts can establish from seed after such a season.

Larger forest remnants on the interfluves retain their open-forest structure, but smaller ones become shrubby open-woodlands. Boundaries with the grassy open-woodlands are usually sharp. As a rule, the canopy has the more common species from the woodlands and open-forests, but with *Litsea glutinosa* more prominent. In the understorey *Psychotria lolokiensis*, *Diospyros "littorea,"* *Glochidion* sp. and *Antidesma ghaesembilla* are usually found amongst the common species, but there is considerable variation between remnants. The ground cover is characterized by moisture-loving herbs. The sites of these remnants have sandy surface horizons with ground water lying at the surface during the wet season.

2.2 The Tavai area

This area is centered on the Rigo Road bridge over the Tavai Creek, about 45 km SE of Port Moresby (Fig. 1; Fig. 4). Short, branching ridges with crests to about 200 m above sea level and rather steep slopes dominate the landscape. Broad, concave footslopes are extensive and form a trough-shaped depression between the hills along the coast (just visible at the left on Fig. 5) and those further inland.

The underlying rocks are of Eocene age and comprise thinbedded mudstone, tuff and limestone, locally intruded by gabbro. The soils on the ridge crests are usually strongly alkaline, shallow, dark lithosols; those on the spurs and higher parts of the footslopes neutral brown lithosols or red gravelly clay soils with sandy loam topsoils. Their reaction varies from moderately alkaline to slightly acid. Texture-contrast soils with an alkaline topsoil over a mildly acid heavier subsoil are found on the lower footslopes.

Forest and scrub are dominant in the area, with open-woodland common on the coastal hills. Many of the woody species are different from those found in the Goragatabu area, but most of the herbaceous species are common to both areas.

Eucalyptus papuana, up to 16 m tall, is the only important species in the canopy of the open-woodlands. *Desmodium umbellatum* and *Cycas circinalis* are common in the shrub layer. The ground cover is dominated by *Heteropogon contortus* or by *Themeda novoguineensis* with *Panicum macranthum* common between the tussocks. The eucalypt storey is well-developed on neutral and mildly acid soils, but is sparser on alkaline soils. Where these occur on narrow ridges and steep upper slopes *Eucalyptus papuana* is absent, but shrubs spread into the grassland from small patches of open-scrub. Here, an occasional *E. confertiflora* can be seen.

The scrub has a very irregular structure, mainly due to a certain degree of clustering of various growth-forms: patches of small trees and tall shrubs, locally with a low shrub understorey, alternate with concentrations of "shrub-lianes" (i.e. shrubs of which the branches ramble through and over other shrubs and trees) which usually are of lesser height than the tree-and-shrub patches. Other lianes are common throughout the vegetation and are often spiny or thorny. Emergents, up to 18 m tall, are common: *Albizia carrii* and *Gyrocarpus americanus* are dominant (Fig. 5). Besides *Jossinia desmantha* which in patches reaches almost single species dominance, the most numerous species in the canopy are *Drypetes australasica*, *Bridelia monoica*, *Ehretia dichotoma*, *Securinega melanthesoides*, *Zizyphus oenoplia* and *Alectryon repando-dentatus* (Fig. 6). A large group of other species is found in smaller numbers. The more common shrub-lianes besides *Securinega* and *Zizyphus* are *Harrisonia brownii*, *Capparis quiniflora* and *Carmona retusa*. *Glycosmis pentaphylla* and *Murraya paniculata* are common shrubs in the understorey. The ground cover usually is fairly dense with *Oplismenus hirtellus* and *Cymaria acuminata* as common species.

Open-forest is commonly found in patches surrounded by scrub or bordering open-woodland. However, it also occurs on low ridges and spurs of higher hills,

Fig. 4. An aerial view over the area S of the Tavai Creek, which flows at the foot of the ridge in the background. Bare *Gyrocarpus americanus* stands out in the vegetation of the plain and low ridges.

Fig. 5. This picture, taken 7 May 1967, features *Gyrocarpus americanus*, which has already shed its leaves and bears clusters of winged fruits, and *Albizia carrii*, which is still in leaf at this time.

both with alkaline sandy loam soils. It has a canopy with irregular gaps, about 12 m tall with scattered emergents up to 20 m. The understorey is rather open, but denser in gaps; lianes are common, but shrub-lianes are considerably less numerous. Many species are shared with the scrub, but *Celtis philippensis* is dominant and all other species are subordinate. *Atalaya papuana* takes a prominent place amongst the emergents, with *Gyrocarpus* and *Albizia* subdominant. The ground cover is sparse with *Oplismenus hirtellus* common.

Open-forest or woodland types with a strong eucalypt element are conspicuously lacking in the Tavai area. *Eucalyptus confertiflora*, *E. alba* and *E. papuana* occur in the open-woodlands especially of the coastal hills and valleys to the north of the area, but except for the occasional occurrence of *E. confertiflora* in depauperate open-scrub, none of the investigated scrub and forest types contained any eucalypt species.

There has been some localized indigeneous cultivation in the area and some

Fig. 6. Some of the growth-forms in the scrub of the Tavai area. The specimen with two grooved trunks is *Jossinia desmantha* of which the growth-form is a small tree or tall shrub. *Parsonsia* sp., at the left, is a single-stemmed woody liane with a thick corky bark. *Celtis philippensis*, at the right, is one of the few tree species with low buttresses. In the background, visible between the *Jossinia* trunks, is *Drypetes australasica*, a small tree. All four species are evergreen; *Celtis* is notophyll, the other three are microphyll.

attention was paid to the regrowth of what were apparently old gardens. *Securinega melanthesoides* was one of the most important colonizers, accompanied by *Bridelia monoica*, *Ehretia dichotoma* and many other species found in the scrub, i.e. the abandoned gardens reverted back to the original vegetation without an intermediate stage of light-demanding secondary growth species as is the case in wetter environments. If small areas of grassland and open-woodland are old gardens then their succession to scrub must have been deflected by frequent burning soon after abandoning (*vide* Eden 1974), because at least the species just mentioned occur in *Eucalyptus* open-woodland in the other areas and are to a certain extent fire-resistant. If the grasslands are precursors of grassy open-woodlands, the question is at what stage do eucalypts move in? It was noticed that usually there is a grassland zone between scrub and open-woodland; also that narrow tongues of grassland are often devoid of eucalypts. It could be that an area has to exceed a certain threshold size before *E. papuana* can establish.

2.3 The Tovobada area

This is an area of high limestone hills and ridges; Tovobada Hill, just 11 km north of Port Moresby, reaches 380 m alt. (Fig. 1). Vegetation types range from grassland to semi-deciduous forest with frequently sharp boundaries between types (Fig. 7). Along the NE margin a suite of lower, more rounded ridges is found on mudstone, limestone and tuff, intersected by incised valleys with ephemeral streams which drain the higher country. Smaller valleys arise from the rounded ridges in a strongly dendritic pattern; hence, interfluves in this section are quite variable in size and slope. Ridges and the interfluves generally carry eucalypt open-woodland, but areas with a scrubby vegetation become more common towards the N.

All rocks in the area are of Eocene age, comparable to those of the Tavai area. Alkaline dark lithosols are found on the higher limestone ridges and hills, neutral red lithosols on the lower ridges. The interfluves have texture-contrast soils: neutral to slightly acid sandy loams overlie a medium acid sandy to heavy clay at a depth of 20 to 40 cm.

Floristically, the Tovobada area takes a position intermediate between the other two areas: open-woodland, woodland and open-forest are relatively strongly related to the Goragatabu area, scrub and closed-forest to the Tavai area.

Eucalyptus confertiflora and *E. alba* dominate the tree storey of the grassy open-woodlands in the NE section. They are usually not more than 10 m tall and often have crooked stems. Lower trees are *Acacia auriculiformis*, *Timonius timon*, *Glochidion* sp., *Antidesma ghaesembilla*, and *Exocarpos latifolius*. Common shrubs are *Canthium suborbiculare*, *Securinega melanthesoides*, *Acronychia rhytidocarpa* and *Bridelia monoica*. The very gravelly, loamy, slightly acid, red lithosols are well-drained and very drought-sensitive: therophytes ("annual" grasses and herbs) dominate during the wetter period. *Sorghum laxiflorum*, for instance, can grow to 2 m tall and obscure many of the other plants in the undergrowth. Weeds such as *Eleutheranthera ruderalis* or *Hyptis suaveolens* may locally cover large tracts. When these plants die off early in the dry season, perennial grasses such as *Panicum viale* show up. There is a rich assortment of associated species in this ground cover community.

In the southern part tussock grasslands cover large areas. *Themeda australis*, *Sehima nervosum* and *Heteropogon contortus* are usually more or less co-dominant, but any one of these can locally form almost pure stands. Other species are infrequent. Scattered *Cycas circinalis* and an occasional shrub or small tree may be present (Figs. 7, 8). This grassland may abut forest or continue as the ground storey of open-woodland in which *E. confertiflora* is the dominant species, with occasional *E. papuana*. Tree height is 10–20 m. There do not seem to be significant differences in the soils under grassland and open-woodland, and soil reaction varies from alkaline to acid without noticeable change in the vegetation. Near Tovobada Hill, however, the boundary between grassland and open-woodland on the one hand and closed-forest on the other did coincide with a soil boundary. A mildly acid redbrown sandy clay loam merging into weathered rock at about 50 cm was found under grassland and open-woodland, but a shallower mildly alkaline soil of rather similar color and texture occurred under the forest.

Low open-forest and shrubby low woodland (Fig. 9) occur on ridges, spurs and interfluves as bush remnants in open-woodlands as well as in more extensive tracts. The irregular canopy is 6–8 m high with emergents to 15 m tall. Some of the common species are *Canthium suborbiculare*, *Eucalyptus confertiflora*, *E. alba*, *Glochidion* sp., *Flacourtia zippellii*, *Exocarpos latifolius*, *Psychotria lolokiensis*, *Allophylus cobbe*, *Desmodium umbellatum*, *Bombax ceiba* and *Elaeocarpus arnhemicus*. The species dominance varies between stands. *Guioa* sp. and *Turraea pubescens* are common in the floristically rich understorey. Usually there is a variety of lianes present, e.g. *Dalbergia densa*, *Harrisonia brownii* and *Abrus precatorius*. Some of these, e.g. *Capparis quiniflora* and *Zizyphus oenoplia*, have thorny branches. The ground storey is often sparse, but weeds such as *Bidens* sp. and *Achyranthes aspera* are common in patches where the overstorey is lacking or very open, e.g. due to treefall.

The shrubby woodlands with *Eucalyptus confertiflora* and *E. alba* dominant in the canopy are transitional to eucalypt open-woodland. These types, also in their ground layers, are floristically closely related. It seems that drought and fire are the agents which change the shrubby woodland into a grassy open-woodland due to progressive killing of woody undergrowth plants and promoting the growth of therophytes. In 1967, after the severe dry seasons of the preceding years, it was observed that drought alone can be responsible for the killing of above-ground parts of a variety of shrubs.

The closed-forests vary in canopy height, 8–16 m, with mostly deciduous emergents occasionally reaching 28 m. They are found on higher ridges with red or dark lithosols. *Bombax ceiba*, *Garuga floribunda*, *Albizia carrii* and *Gyrocarpus americanus* are the most frequent emergents; *Drypetes australasica* is always predominant in the canopy with a variety of other species, some of which are codominant in a particular stand, e.g. *Celtis philippensis*, *Harpullia* sp., or *Bridelia monoica*. The understorey is also of mixed composition;

Fig. 7. A view in the Tovobada Hills. *Themeda australis* dominates the grassland, with eucalypts coming in on some of the lower slopes. Forest with emergent, bare *Bombax ceiba* is visible in the valley reentrants. Note the sharp boundaries between intact and burnt grassland and between grassland and forest.

Fig. 8. Cycas circinalis developing a new crown of leaves after a fire, which consumed the old leaves and the pinnae of the living ones. Near the scrub border the fire was milder and only scorched the leaves of the taller plant. *Themeda australis* is sprouting from its charred tussock bases.

Glycosmis pentaphylla, *Murraya paniculata* and *Canthium* sp. are amongst the predominant species. There is a great variety in the lianes, which are especially common in the canopy where they twine crowns together and cover branches of dead trees. Openings in the canopy occur due to the blowing over of such trees or the breaking of their branches.

In 1967, on the southern ridges, several examples were seen of woodlands and forests which had degenerated into open-woodlands. The likely cause for this is a combination of fire and drought. Fire-resistant trees, e.g. *Eucalyptus confertiflora*, *E. papuana*, *Bombax ceiba* and *Albizia carrii*, were still standing, often with some lower trees, such as *Desmodium umbellatum* and *Bridelia monoica*, and shrubs surviving from the original vegetation. The ground was completely covered by weeds, erect ones as well as creepers, with *Passiflora foetida* and *P. aurantia* also covering shrubs and climbing trees. *Panicum macrocladum* and *Heteropogon contortus* may be common in the fringes where this vegetation grades into eucalypt open-woodland with a *Themeda australis* ground cover.

Fig. 9. Eucalyptus confertiflora in shrubby woodland. The diameter of the trunk is about 30 cm. The shrub in the foreground is *Turraea pubescens*, the grass *Oplismenus hirtellus*.

3. A geographical analysis of the vegetation types

During the Torres Strait Symposium it was demonstrated by several contributors that a comparison of similar vegetation communities or habitat types can yield more useful pointers to the history of an area than a general analysis of the flora (e.g. Hoogland 1972, Webb and Tracey 1972; *vide* Heyligers 1972a). More than a decade earlier, such an analysis had been made of the vegetation types of Arnhemland (Specht 1958), in which the differential geographical distribution of the species of the various vegetation types was clearly shown.

Table 1 summarizes the results of a geographical analysis of the vegetation types of the Port Moresby area. The ground flora was treated separately from the over-storeys.* For sake of brevity, vegetation types with similar spectra have been grouped and only average spectra are given. Finally, the table shows some high frequencies of unidentified species. Some of those may be undescribed species, e.g. *Guioa* sp., others — often lianes — were only found in the vegetative state and could not be satisfactorily matched with herbarium collections.

The forest, scrub and woodland spectra show that the total frequency of New Guinea endemics in the Goragatabu area (32%) is twice that of the other areas; that species, in New Guinea restricted to SE Papua, are relatively frequent in the Tavai and Tovobada areas, but that apart from the local endemics, these are virtually absent from the Goragatabu area; that species distributed over Australia and monsoonal Asia or with a wider distribution are more prominent in the Tavai and Tovobada areas than in the Goragatabu area (39%, 35% and 14%, respectively); that species occurring outside New Guinea only in Australia are relatively frequent (13%) in the Goragatabu area, and that in the Tavai area the frequency of species restricted to SE Papua and Australia is relatively high (11%). In the eucalypt open-woodlands Australian and widespread species are strongly represented, the Australian element (29%) being more than twice as frequent as in the denser woody communities, while the endemic element (14%) is only slightly less frequent than in the scrub and forest types of the Tavai and Tovobada areas.

The differences can at least partly be explained by the geological histories of the areas. The Tavai and Tovobada areas are underlain by sedimentary rocks of Eocene age, but the Goragatabu area is developed over an early Oligocene intrusion of gabbro which lies in a broad band along the NE margin of the Eocene deposits. It was towards the end of the Oligocene that the orogen which eventually would form the central mountain ranges of New Guinea began to emerge (Pieters 1978). It is probable that the area intruded by gabbro emerged before the adjacent sedimentary regions. Associated deep-sea sedimentation in

*It is a commonly observed fact that at least in eucalypt woodlands there often is little correlation between the floristic composition of the canopy storey and that of the ground layer (Heyligers 1966; Specht 1958, p. 422). Also, the modes of dispersal differ significantly in these storeys: of the 268 species in the tree and shrub strata 65% has fleshy fruits or arillate seeds, 12% propagules adapted to wind-dispersal, 9% large and 4% small or very small seeds without obvious structural adaptation for dispersal, and only 3% propagules with hooks or sticky glandular hairs. However, for the 163 ground storey species, the figures for these classes are 7%, 3%, 4%, 50% and 31%.

Table 1. Geographical distribution spectra. The spectra are based on the frequencies of occurrence of the species to reflect to some degree the commonness of species; frequency being the number of times a species was represented in the total of observations for a particular vegetation type. The figures for the classes have been converted to percentages to facilitate comparison of spectra.

	Species distribution area*											
	e	e′	E	a	A	m	M	b	B	w	W	U
Canopy and shrub storeys												
Forest, scrub and woodland												
Goragatabu area	11	10	11	0	13	1	7	0	5	0	9	32
Tavai area	10	2	4	11	1	1	5	6	9	9	15	28
Tovobada area	7	3	6	4	7	2	7	7	8	6	21	21
Eucalyptus open-woodland	7	3	4	3	26	0	5	3	8	2	27	11
Ground storeys												
Oplismenus	0	2	0	0	6	6	4	0	4	3	49	26
Themeda australis — Heteropogon	0	1	0	1	6	1	9	2	20	7	46	6
Eriachne — Eragrostis	1	5	0	3	6	5	4	3	15	4	43	11
Themeda novoguineenis — Sorghum	3	7	1	7	9	4	6	0	27	4	29	3

*Species distribution area

e restricted to SE Papua
e′ restricted to seasonally dry areas in New Guinea
E restricted to New Guinea
a SE Papua and Australia.
A New Guinea and Australia.
m SE Papua and monsoonal Asia.
M New Guinea and monsoonal Asia.
b SE Papua, Australia and monsoonal Asia.
B New Guinea, Australia and monsoonal Asia.
w as b, but also occurring outside monsoonal regions.
W as B, but also occurring outside monsoonal regions.
U not known due to unidentified species.

the Port Moresby trough has continued unbroken up to the present day with the late Tertiary and Quaternary sediments likely to have been provided by the Fly River system (Ewing *et al.* 1970). It is therefore possible that the SE Papuan lowlands became isolated from ecologically similar environments some time during the Miocene.

The differences between the spectra of the Tavai and Tovobada areas and the spectrum of the Goragatabu area could be due to a longer history of exchange with the rest of New Guinea and Australia by the predecessor of the Goragatabu area. The high frequency of species in New Guinea restricted to SE Papua could mean that the vegetation of the Tavai and Tovobada areas developed when there was a considerable barrier to movement of plants across New Guinea except for species with a wide tolerance to climate. It is likely that most of the eucalypt open-woodland elements came in during the earlier phases of colonization as very few open-woodland species are restricted to the Tovobada or Tavai area, and that the Australian and widespread species became more prominent due to selective fire pressure.

The *Oplismenus* ground cover, which comprises the groundstoreys of scrub and forest, is dominated by widespread species indifferent to seasonally dry conditions. The ground covers of *Themeda australis* and/or *Heteropogon contortus*, either occurring under eucalypt open-woodland or as grassland, also have a large widespread element, but widespread species restricted to seasonally dry areas form an important subsidiary group. In the *Eriachne — Eragrostis* tussock grasslands there is a higher frequency of species which in New Guinea are limited to SE Papua. Lastly, groundcovers either dominated by *Themeda novoguineensis*, by annual grasses such as *Sorghum laxiflorum* and *Pseudopogonatherum irritans*, or by mixtures, found under woodlands and open-woodlands, show a further increase in New Guinea endemics and in widespread species restricted to seasonally dry areas, with a concurrent decrease in widespread indifferent species.

The interpretation of these spectra with regard to their historical significance seems less possible than for those of the overstoreys. The preponderance of widespread species seems to be correlated with the large number of species which have very light seeds or spiny, hooked or bristled propagules, which are easily dispersed by animals. The *Themeda novoguineensis — Sorghum* groundcover has the highest frequency of endemic species among the ground storey communities as well as an important group of widespread species restricted to seasonally dry regions. This could indicate that this groundcover type contains the oldest elements, possibly contemporary with the woody species of the Goragatabu area. It is in this area that this groundcover type is most widespread, although it occurs also in the other areas.

Vegetation types physiognomically similar to the eucalypt open-woodlands of the Port Moresby area are widespread over N Australia. However, these woodlands are floristically richer and show a large range of communities (Webb and Tracey 1972). The flora of these woodlands contains a considerable widespread element. Several species belong to genera which are best represented in rainforest (Hoogland 1972), Specht (1958) demonstrated that 17% of the species found in the "*Eucalyptus* tall open forests" of Arnhemland can be found as far west as the Indian peninsula. He also showed that this element has a wide distribution across the N part of the Australian continent.

Small areas of vegetation characterized by non-eucalypt species occur throughout the seasonally dry regions of tropical Australia. Their structure ranges from closed-forest to open-scrub and there is a considerable variation in floristic composition. This "monsoon forest formation," analysed by Specht (1958) for Arnhemland, shares 58% of its 81 species with New Guinea, 49% with the Malayan Archipelago and 35% with the Indian peninsula. Its floristic list has 10 species and 24 genera in common with the forest and scrub types of the Port Moresby area. *Vice versa*, various seasonal vegetation types in India (Legris 1963) show close connections at generic level with N Australia and SE Papua. However, more information on floristic composition and species dominance in the various types would be required before a detailed comparison could profitably be made.

4. Deciduousness

The shedding of leaves at the onset of or during the dry season, sometimes accompanied by profuse flowering or seed dispersal, is a conspicuous characteristic of the woody vegetation. Sixty of the 275 species (22%) found in the tree and shrub layers are deciduous. Deciduousness amongst the lianes is similar to the general average. Shrubs are usually evergreen, but five of the nine species which are "flexible" in their shrub-habit, i.e. they may also develop as trees, are deciduous including *Guioa* sp., *Acronychia rhytidocarpa* and *Turraea pubescens*. Among the tree species 43% of those growing more than 8 m tall is deciduous but only 17% of the smaller ones.

Deciduous species more often have compound leaves than evergreen species: 37% as against 17%. The leaf-size distribution of deciduous and evergreen species with simple leaves is not significantly different, with notophyll and microphyll species about 30% each and small mesophylls 25%.* However, 41% of the leaflets of deciduous species with compound leaves is microphyll and only 9% small mesophyll, while these figures for evergreen species are 22% and 33%. An example of each of these species is: *Desmodium umbellatum*, *Erythrina* sp., *Jagera serrata*, and *Polyscias elegans*.

Deciduousness is either obligatory or facultative, and the adaptive significance is different for various species. For example, in a species like *Bombax ceiba*, which sheds its leaves even in ever-wet climates, it is probably correlated with the exposure of flowers and fruit, while in *Eucalyptus alba* and *E. confertiflora* the severity of the dry season determines how much of the foliage will be shed. The proportion of deciduous species varies between types as well as between the storeys of each type. The situation, based on counts of individuals, can be summarized as follows: open-forest in the Goragatabu area has about equal numbers of obligate deciduous, facultative deciduous and evergreen emergents

* Leaf-size classes are Raunkiaer's as amended by Webb (1959), with a further half-way division of the mesophyll class in small and large (Heyligers 1972c).

and proportionally fewer obligate and more facultative deciduous individuals in the canopy. In forest and scrub in the Tavai and Tovobada areas up to 80% of the emergents are obligatory deciduous, with a further 10% or so facultative deciduous, while in the canopy 64% or more of the individuals are evergreen and 20% or less obligatory deciduous. Facultative deciduous species, e.g. *Guioa* sp., *Harrisonia brownii* and *Acronychia rhytidocarpa*, become more prominent in the woodlands of the Goragatabu and Tovabada areas, a trend culminating in very high proportions of this class, up to 85%, in the open-woodlands due to the dominance of *Eucalyptus alba* and/or *E. confertiflora*.

Longer-term observations than were possible during the present study would be necessary to establish the phenological patterns for the various vegetation types in the region, which could yield not only information on adaptation strategies (cf. Frankie *et al.* 1974), but also on the evolutionary history of SE Papua (cf. Westman 1978).

5. On the origin of the open-woodlands and the forest/open-woodland boundaries

Eden (1974) has investigated the impact of shifting cultivation in the vicinity of the Tovobada and Tavai areas. He compared air photos taken at least 10 years apart and found that after being abandoned, 78% of the gardens reverted back to forest, while the remainder were converted to grassland or savanna (=open-woodland). From inspection in the field and interrogation of the local population he concluded that only if the initial regeneration cover of grasses is burnt is the succession deflected with the result that a grassland dominated by *Imperata* becomes established. If not burnt within this period the combustibility markedly declines as the invading forest shrubs take over and the grasses are shaded out. His conclusion is that at the present time the maintenance of grassland and savanna and their further spread is attributable to man's activities, but that it is not clear whether they originated wholly in this way, as there is some archaeological evidence that the effective start of deforestation by man in S Papua dates back only to about 2000 years ago. Also, their development may have been influenced by past climatic fluctuations, more specifically during drier phases in the Quaternary.

Indigenous cultivation practice could have been responsible for sharp boundaries between forest and grassland in several localities in the areas investigated in the present study. Gardens are established in the forest or forest edge rather than in the grass areas because of the difficulty in preparing the root-ridden topsoil (Eden 1974, Ward (*vide* Golson 1972)). Eden also states: "in general, villagers seek to prevent such forest replacement [by grasslands] in order to safe-guard their future cultivation land, but ... their success is only partial". However, our observations on this subject concentrated on situations where influence of past cultivation was not evident and often unlikely to have played a role, e.g. on steep slopes or on very stony soils.

It seems plausible that the eucalypts and a variety of other Australian and Malesian species found only in S Papua, have become a component of the vegetation after the upthrusting of the main ranges in New Guinea in the later part of the Tertiary. As eucalypt species do not regenerate under forest conditions drought and fire must have created niches for their entrance into and continued existence in the area. At present fires due to lightning are of limited significance, because lightning is commonly associated with thunderstorms (Eden 1974), and most if not all of them are lit by man. However, climatic variations have undoubtedly played a role and natural fires will have been at one stage or another just as much a part of the scene as they have been and still are in Australia (Mount 1979; Webb 1968).

On the low ridges and interfluves in the Goragatabu and Tovobada areas open-forests and woodlands occur which have eucalypts as a conspicuous component amongst a rich variety of other species. Transitions between these vegetation types and eucalypt open-woodlands occur, as well as remnants of these types surrounded by open-woodland. With minor exceptions, the woody species of the open-woodlands are also found in the woodlands and open-forests. They survive in the open-woodlands owing to their ability to regenerate after fire. The evidence from the shrubby woodlands and the transition zones suggests that occasional fires are not sufficient for the development of grassy open-woodlands, but that a regular fire regime is necessary to develop and maintain these open-woodlands.

The survival of bush remnants in the grassy open-woodlands could be a result of fire-behavior. Originally, these remnants were probably denser shrubby undergrowth patches in the open-forest or woodland possibly related to slightly more favorable soil conditions. They formed a barrier to mild fires as there was not sufficient fuel on the floor, while the green leaves would not burn. In course of time positive feedback would reinforce this barrier as the surrounding vegetation became more open and the shrubs in the patch developed more branches and leaves around its perimeter. In this way a sharp boundary would result between bush remnant and open-woodland. Grass patches on ridge crests and in depressions of interfluves could have formed the initial foci for the fires.

Not only in the southern part of the Tovobada area (Fig. 7), but also on the high ridges nearer to Port Moresby sharp grassland/forest boundaries are common and appear to be quite stable. A picture of Port Moresby Harbor, taken in 1910 (Fig. 10), shows at least eight forest remnants sharply delimited against the grassland on Burns Peak and the ridge to the north. These remnants were still there in 1967, and their boundaries did not appear to have changed at all. The coincidence of such boundary with a change in soil conditions was found near Tovobada Hill, where the soil under forest was shallower than under grassland and mildly alkaline instead of slightly acid. It is possible that at some stage in the past the hills were all covered with woody vegetation, open on shallow droughty soils, denser where growing under somewhat less xeric conditions. Fire could have led to the development of open-woodland and grassland as postulated above, and positive feedback effects would have promoted the sharp boundaries with the surrounding denser woody vegetation. However, the forest/open-woodland boundaries are not always correlated with soil differences, as was also demonstrated by Eden (1974), and it is likely that topography is a more important factor. A clue is provided by the fact that many of these boundaries run across the contour rather than along. Due to strong

Fig. 10. Two views of the high ridge backing Port Moresby Harbor. The top picture is a detail from a postcard of c. 1910, the bottom picture was taken in 1963. The persistence of the forest remnants and their sharp boundaries against the grassland contrasts sharply with the changed townscape.

relief and uniform wind regime fires burn most strongly in uphill direction, whilst the sides burn only relatively mildly. Hence, the least fire pressure would generally be in a lateral direction, enhancing through positive feedback effects the interface of forest and grassland across the direction of the contours.

In conclusion, then, it appears likely that fire has been part of the local scene long before the arrival of man and that the interplay of environmental factors such as slope, soil condition and wind direction, with the various ecological characteristics of the individual species as well as those of the vegetation communities, has given fire a chance to establish sharp boundaries without the hand of man.

6. Summary

Vegetation types on well-drained soils ranging from grasslands and open-woodlands to forest and scrub are described for three areas, all of which are subject to a dry season lasting from June till October. Deciduousness is most strongly developed amongst the emergents and varies considerably in the canopies and understoreys. Endemism, either local or island-wide, is relatively high in the woody layers, but is generally less pronounced in the ground storeys. In the eucalypt open-woodlands, species restricted to New Guinea and Australia are more than twice as frequent as in the other vegetation types.

Forest and woodland types with a significant eucalypt component occur in two of the areas. They form the transition between forest without eucalypts and eucalypt-dominated open-woodlands. However, there are also instances of very sharp boundaries between open-woodland or grassland and forest or scrub without eucalypts. These boundaries are attributed to slight differences in slope and soil factors, of which it is surmised that they may lead to positive feedback effects in the fire history. In the examples studied, man's activities do not appear to be the direct cause of these boundaries.

Acknowledgments and Dedication

I am indebted to many persons and various organizations which in one way or another have been involved with this study. I like to mention especially the help of my wife Yantine during the first field trip and that of Mr R. Pullen, then botanical collector with the Herbarium Australiense, during the second one; also the unstinting work of technical assistants and typists during the desk-stage of the project.

Finally, I wish to dedicate this paper to the memory of Frodo, our first-born and only son (1964–1978).

References

Eden, M. J. 1974. The origin and status of savanna and grassland in southern Papua. Inst. Br. Geogr. Trans. 63: 97–110.

Ewing, J. I., R. E. Houtz and W. J. Ludwig. 1970. Sediment distribution in the Coral Sea. J. Geophys. Res. 75: 1963–1972.
Frankie, G. W., H. G. Baker and P. A. Opler. 1974. Comparative phenological studies of trees in tropical wet and dry forests in the lowlands of Costa Rica. J. Ecol. 62: 881–919.
Golson, J. 1972. Land connections, sea barriers and the relationship of Australian and New Guinea prehistory. In "Bridge and Barrier" (D. Walker ed.), p. 375–397.
Heyligers, P. C. 1965. Vegetation and ecology of the Port Moresby-Kairuku area. CSIRO Aust. Land Res. Ser. No. 14: 146–173.
—— 1966. Observations on Themeda australis — Eucalyptus savannah in Papua. Pacif. Sci. 20: 477–489.
—— 1972a. Analysis of the plant geography of the semi-deciduous scrub and forest and the eucalypt savannah near Port Moresby. Pacif. Sci. 26: 229–241.
—— 1972b. Reference lists for distribution data of taxa occurring in the Port Moresby area. CSIRO, Division of Land Research, Technical Memorandum 72/1, 17 pp.
—— 1972c. Reference list for life from data of species occurring in the Port Moresby area. CSIRO, Division of Land Research, Technical Memorandum 72/2, 43 pp.
—— 1979. Reference material on some vegetation types in the Port Moresby area. CSIRO, Div. of Land Use Research, Technical Memorandum 79/16.
Hoogland, R. D. 1972. Plant distribution patterns across the Torres Strait. In "Bridge and Barrier" (D. Walker ed.), pp. 131–152.
Legris, P. 1963. La vegetation de l'Inde; ecologie et flore. Inst. Fr. Pondichery, Trav. Sec. Sci. et Tech., VI, 389 pp.
Mabbutt, J. A., P. C. Heyligers, R. M. Scott, J. G. Speight, E. A. Fitzpatrick, J. R. McAlpine and R. Pullen. 1965. Lands of the Port Moresby-Kairuku area, Territory of Papua and New Guinea. CSIRO Aust. Land Res. Ser. No. 14, 182 pp.
Mount, A. B. 1979. Natural regeneration processes in Tasmanian forests. Search 10: 180–186.
Pieters, P. E. 1978. Port Moresby-Kalo-Aroa; Papua New Guinea. 1:250 000 geological series — explanatory notes. (AGPS: Canberra.) 55 pp.
Specht, R. L. 1958. The geographical relationships of the flora of Arnhemland. In "Records of the American-Australian Scientific Expedition to Arnhemland". 3: 415–478. (Melbourne Univ. Press.)
—— 1970. Vegetation. In "The Australian Environment", 4th Ed., ed. G. W. Leeper, pp. 44–67. (CSIRO Aust. and Melbourne Univ. Press: Melbourne.)
Walker, D. (ed.) 1972. Bridge and Barrier: the natural and cultural history of Torres Strait. Aust. National Univ., Research School of Pacific Studies, Publication BG/3, 437 pp.
Webb, L. J. 1959. A physiognomic classification of Australian rainforests. J. Ecol. 47: 551–70.
—— 1968. Environmental relationships of the structural types of Australian rainforest vegetation. Ecol. 49: 296–311.
Webb, L. J. and J. G. Tracey. 1972. An ecological comparison of vegetation communities on each side of Torres Strait. In "Bridge and Barrier" (D. Walker ed.), 109–129.
Westman, W. E. 1978. Evidence for the distinct evolutionary histories of canopy and understorey in the Eucalyptus forest-heath alliance in Australia. J. Biogeogr. 5: 365–376.

9. Grasslands and grassland succession in New Guinea

E. E. Henty

1. Introduction

The vegetation of New Guinea is mostly forest, which, from its situation within the humid tropics, may be expected. However, an appreciable area is under grassland. The occurrence of some of this grassland may be attributed to natural environmental conditions, for example the subalpine grassland of the higher mountains, and perhaps in "frost-pockets" at slightly lower altitudes. There is also tall grassland on the flood-plains of rivers, and swamp grassland, some of it floating, at both low and median altitudes. Part of the grassland/savanna country along the S coast of the island may also be accepted as "natural" (see preceding chapter). The balance, and by far the larger area, occurs in regions where neither soil nor climate would preclude the growth of forest (see Fig. 9). This grassland is believed to have developed as a result of the destruction of forest by man, mainly in the course of food-gardening; it is anthropogenic grassland (Robbins 1960).

Food gardening in New Guinea has, to a large extent, traditionally followed a system of shifting cultivation or bush fallow; forest is felled and burned, the resulting clearing planted with food crops, cultivated for a few years, and then abandoned, for trees to re-establish themselves. Provided that the rotation period is long enough for trees to become re-established between cultivations (probably a minimum period of the order of 20 years) the system can be maintained. If, however, due to population pressure or other factors, the period of rotation is made shorter, a stage is reached where woody plants no longer regenerate to a complete canopy and grass becomes established. The grass will, after a time, become inflammable, and fire will further inhibit the growth of woody plants. The nucleus of an area of grassland is formed.

Once formed, grassland is maintained by regular burning; the fires are mostly of human origin, deliberately lighted. The area will increase, over the years, by fires burning into the grassland/forest margins, particularly uphill, or in the direction of the prevailing wind. In general, in New Guinea, undisturbed forest or regrowth will not burn, although there is evidence that forest has burned, presumably in very dry seasons; in the Madang Province, near Usino, fire-scars have been found deep inside large trees in numbers which suggest extensive fires. Such fires were not, however, sufficient to destroy the forest or to allow grassland to develop.

Monographiae Biologicae, Vol. 42, ed. by J. L. Gressitt.

© 1982, Dr W. Junk Publishers, The Hague.

Fig. 1. Tall lowland grassland; *Imperata* with some *Saccharum spontaneum*.

Fig. 2. Imperata/Saccharum grassland, similar to that shown in Fig. 1, after grazing; *Saccharum* has increased and *Imperata* decreased.

Fig. 3. Short lowland grassland, almost pure *Themeda australis.*

Fig. 4. Themeda australis; recovery after burning.

Fig. 5. Grassland on hillside; fires have burnt uphill, denuding the ridges, without much lateral effect.

Fig. 6. Tall highland grassland, *Miscanthus floridulus* dominant.

Fig. 7. Phragmites karka in a swampy highland valley.

Fig. 8. Short highland grassland; *Themeda australis* and *Arundinella setosa* dominant.

In the change from forest to grassland, the dominant factor is the destruction of forest by human agency, and soil type does not appear to be important. The change has occurred on all major soil types, and Robbins (1960) found that forest-grassland boundaries were commonly independent of soil boundaries. Gardens are usually made on the better soils in any case, and it is evident that forest on such soils is quite susceptible to degradation. The effect of soil may, however, soon appear by its influence on the type of grassland which develops.

The effect of climate is more marked than that of soil, but mainly in its extremes. In very wet localities, gardening may be conducted with minimal burning. In the NE corner of the Western Province (Ningerum to Ingembit) the rainfall approaches 6000 mm. Clearings are rarely burned and regrowth is already appearing among the bananas and sugar cane which are the second year food crops. The forest soon re-establishes itself, but near villages it is seen to be modified, with coppicing species (notably Myrtaceae) numerous (Henty 1967). However, near Lae, Morobe Province, with a rainfall a little below 5000 mm, intensive gardening has produced grassland, on land previously occupied by forest, within 15 years. Certainly forest cannot withstand disturbance so successfully, under a drier or sunnier climate. This is demonstrated in parts of the Huon Peninsula, where mornings are usually clear, but afternoons cloudy. Opposite slopes may be observed, where that facing east bears grassland, while that facing west retains more or less degraded forest.

Since the floristic make-up of grassland, like that of other vegetation, varies according to altitude, grasslands are here considered as (a) lowland grasslands, up to 1500 m, and (b) highland grasslands, from 1500 m upwards. The division is somewhat arbitrary and overlapped by a number of species, but it is convenient for the purpose of discussion.

2. Lowland grassland

The main pioneer grasses, colonising abandoned garden land, are the two *Imperata* species, *I. conferta* and *I. cylindrica*; usually there are other, more robust species: *Coelorhachis rottboellioides*, *Ophiuros tongcalingii*, *Pennisetum macrostachyum* and *Polytoca macrophylla*. The sprawling species, *Microstegium spectabile* and *Eulalia leptostachys*, may also be present. In wetter areas, the introduced, creeping, *Paspalum conjugatum* may cover the ground at first, but it is finally overgrown by taller species. In some localities, *Saccharum spontaneum* may be prominent.

In the early stages of grassland establishment, a variety of woody plants survive for a while, but these are reduced as the years pass, with regular burning. A few woody species are sufficiently fire-resistant to persist. The most common of these, widespread throughout the island at low altitudes, are *Nauclea orientalis*, *Alphitonia incana*, *Antidesma ghaesembilla*, *Cordia dichotoma* and *Albizia procera*. They appear as scattered, smallish trees, often rather misshapen. The large grasses, *Polytoca* and *Pennisetum*, do not usually persist in extensive grassland, but are confined to forest margins, ravines and stream banks.

2.1. Tall lowland grassland

This, the earliest stage of grassland succession, is usually dominated, in the lowlands, by *Imperata*. The two species present in New Guinea are similar in habit; rhizomatous grasses, with the stems close together, producing dense stands 1–2 m high. *I. conferta* may be as frequent as *I. cylindrica* on recently-abandoned garden land, but it seems to decrease as the grassland ages; examination of *Imperata* grassland about Lae has shown a predominance of *I. cylindrica* on the older stands. Both species form strong rhizomes, deep enough to be protected from fire, and recovery after burning is rapid. The grass flowers at this stage, and strong vegetative growth follows until the next burn.

On well-drained soil, without serious moisture stress, *Imperata* grassland contains few other species. *Coelorhachis rottboellioides* and *Ophiuros tongcalingii* are usually present; these both form dense erect tussocks, somewhat taller than the *Imperata*. *Sorghum nitidum* is occasional, often on patches with somewhat marginal soil conditions, with poorer internal drainage. Legumes and herbs are scarce; scattered plants of *Crotalaria montana* may occur, and a few species of *Desmodium*.

Where drainage is somewhat impeded, and water-logging occurs at some season of the year, *Saccharum spontaneum* may become prominent, even dominant. This may be seen in parts of the lower Markham valley, and in Papua on the east coast, particularly near Popondetta, inland from Collingwood Bay, and on the S coast of the Central Province. The *Saccharum* grows to 3 m high, and may be mixed with *Imperata*, or in pure stands forming a mosaic with *Imperata* which is somewhat shorter, usually 1–2 m (Heyligers 1965). The stands are usually dense, with few other species, but on somewhat less fertile soils, as in southern Papua, more open communities are found, with *Cymbopogon globosus* and a number of minor species of grasses, herbs and sub-shrubs (Taylor 1964).

2.2. Transition of grassland types

Once grassland is established, soil deterioration follows; loss of organic matter is the most noticeable change in the early stages. The organic content of the soil has already been depleted by gardening activities. Under grass, less organic matter returns to the soil, than under the original tree cover. The ground cover is removed, at least annually, by fire, leading to higher soil temperatures and consequently faster decomposition of organic matter; at the same time surface litter is destroyed, reducing the sources of organic replenishment. With reduced organic content, soil structure begins to fail, which, combined with loss of litter, leads to decreased absorption of water and depleted water-holding capacity. Runoff increases, and erosion is accelerated on sloping land. The ultimate result is thin, infertile soil, subject to severe moisture stress during dry periods. Under those conditions, *Imperata* does not persist, but is supplanted by *Themeda australis* and other species, forming short grassland.

On land with minimal slope, soil structure is likely to deteriorate in the same

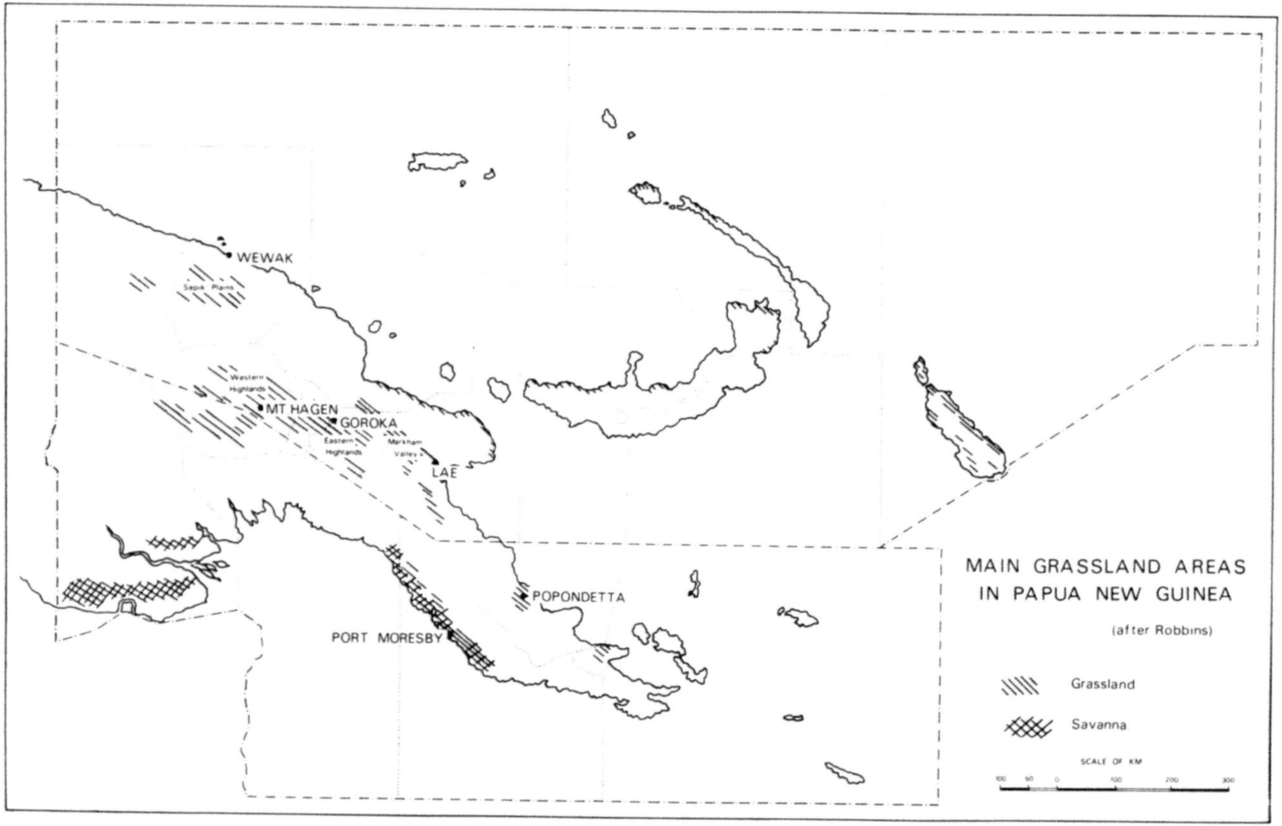

Fig. 9. Map of main grassland areas in Papua New Guinea (after Robbins).

way, and the grassland which ultimately develops will be influenced by soil texture and other features, and by climate. Observations indicate that *Imperata* may be long persistent on such land, if the soil is of sandy texture, with good, but not excessive, internal drainage. Organic content is not crucial here; the soil remains permeable, even though its structure may deteriorate. Good examples may be found in the Markham-Ramu valley (see below). On soils of finer texture, deterioration may lead to pan-formations, with the soil above the impermeable layer alternately sodden and parched (see Sepik Plains, below).

2.3. Short lowland grassland

Themeda australis is usually dominant. At low altitudes, on well drained soil with a rather dry seasonal climate, it may form over 90% of the vegetation. It is a tussocky grass, forming clumps to 20 cm across at the base, sometimes larger, irregularly spaced but mostly 10–20 cm apart, so that only about 20% of the soil surface is occupied, although the fully-grown foliage forms a closed canopy about 1 m above the ground (Fig. 10). The soil between the clumps is mostly bare. Associated grasses (which may be very infrequent) are *Capillipedium parviflorum*, *Sorghum nitidum*, *Heteropogon contortus* and small plants of *Eragrostis* spp. and *Digitaria* spp. A number of legumes may be present, mostly sub-shrubs with a woody rootstock, including *Christia subcordata*, *Crotalaria alata*, *C. ferruginea*, *C. montana* and other small species, *Desmodium gangeticum*, *Indigofera linnaei*, *I. linifolia* and *I. trifoliata*, *Pycnospora lutescens*, *Uraria lagopodoides* and *Zornia* spp. Trailers and twiners are *Atylosia scarabaeoides*, *Glycine tomentella* and *Vigna* spp. Plants of other families include the trailing *Boerhaavia diffusa* (Nyctaginaceae), *Euphorbia serrulata* and the introduced *E. hirta*; another introduction, *Tridax procumbens* (Compositae) has become common in places.

Short grassland is much richer in species than tall grassland, but although

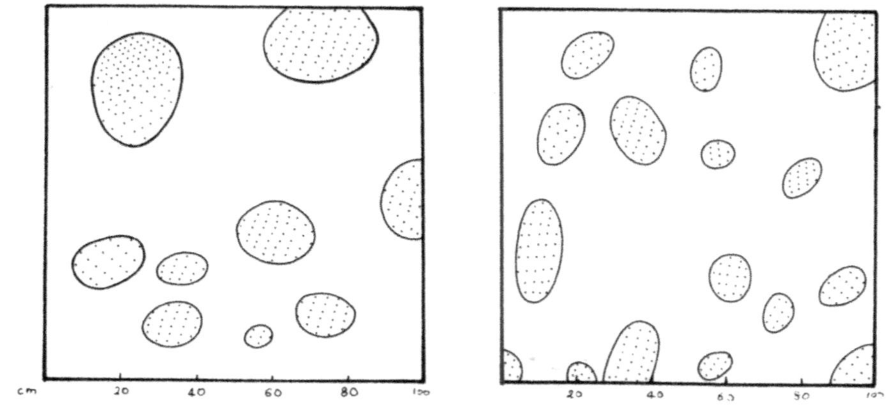

Fig. 10. Quadrats 1 m square, showing basal area of tussocks of Themeda australis (dotted). Unshaded area is bare ground. Nadzab, lower Markham valley, on level ground, 1 month after burning.

such a variety of plants occur they are not prominent among the grass. After burning, the legumes make rapid growth, and produce flowers and fruit.

At somewhat higher altitudes (500–1000 m) more grasses appear, including *Arundinella setosa*, *Alloteropsis semialata*, and *Hyparrhenia* spp.; *Capillipedium* may become more frequent. The legumes, *Cassia mimosoides*, *Desmodium heterocarpon* and *D. gyroides* may be common, and *Melastoma malabathricum* may be found as a shrub and *Osbeckia chinensis* as a sub-shrub.

Grassland dominated by *Themeda australis* is characteristic of steep hillsides, where the soil has presumably followed a course of degradation as outlined above, and of land of easy slope with soil of coarse texture and good internal drainage, containing gravel or stone to the extent that severe moisture stress occurs during dry periods. Such conditions may be seen in the Markham-Ramu valley (Morobe and Madang provinces). The soil is made up of alluvial deposits, both fine and coarse-textured, so that the surface is a mosaic of stone and gravel, sandy or silty soil, and areas of heavier, poorly-drained soil. *Themeda* flourishes on the gravelly soil, and in places where a thin layer of finer soil overlies the gravel; where a deeper layer of fine-textured soil occurs, *Imperata* dominates, and in poorly-drained areas, *Saccharum spontaneum* or, in a few patches, *Phragmites* or *Saccharum robustum*. On some of the land occupied by *Themeda*, the soil is deep enough to cultivate. After cultivation, *Imperata* colonises the land, but after a few years *Themeda* is re-established.

On the Sepik plains, south of Wewak, are large areas of short grassland of an unusual type. The soils are infertile, with negligible internal drainage, and large tracts have very low relief; the soil is either saturated with water or extremely dry, according to the season. Strips of relict forest, along stream-cut ravines, separate the grassy areas. The grassland is a complex of *Themeda australis* with associated *Arundinella setosa* on the one hand, and *Ischaemum barbatum* and sedges, on the other (Robbins 1961). Where the relief is sufficient to give some surface drainage, *Themeda/Arundinella* occupies the drier crowns and *Ischaemum* the shallow depressions. The vegetation is rich in species of grasses and other herbs. The grasses include *Alloteropsis semialata*, *Capillipedium*, *Centosteca latifolia*, *Dimeria* spp., *Eragrostis* spp., *Eremochloa bimaculata*, *Eriachne* spp., *Paspalum* spp., *Sacciolepis* spp., *Sclerandrium truncatiglume*, and *Sorghum nitidum*. In the wet patches with *Ischaemum barbatum* are *Apluda mutica*, *Isachne* spp. and *Leersia hexandra*. Legumes include *Crotalaria calycina*, *Desmodium* spp. and *Indigofera linnaei*. Among the sedges are *Cyperus* spp., *Eleocharis* spp., *Fimbristylis dichotoma* and *F. littoralis*, *Fuirena ciliaris* and *F. umbellata*, *Rhynchospora rubra* and *Scleria* spp. Other herbs include *Eriocaulon* spp. (*E. australis* was formerly burned and used as salt), *Osbeckia chinensis*, species of *Euphorbia* and *Phyllanthus*, Polygalaceae and Scrophulariaceae, with species of *Buchnera*, *Lindernia* and *Striga*.

In the SE part of the island (Northern, Milne Bay and Central provinces), particularly where the climate is markedly seasonal, a mixture of species commonly occurs in short grassland and the sharp division between *Imperata/Saccharum*-dominated tall grassland and *Themeda australis*-dominated short grassland becomes less distinct. Taller species, such as *Coelorhachis rottboellioides*, *Cymbopogon globosus*, *Ophiuros tongcalingii*, *Panicum* spp., *Sehima nervosum* and *Themeda novoguineensis*, occur both in tall grassland, and intermittently among short grassland. *Heteropogon contortus*, which is sporadic throughout the lowlands, becomes prominent in some of the drier areas; this grass has been characterised as a drought indicator (van Steenis 1965). It is sufficiently common in parts of the Western Province to be of some importance as grazing for deer (Gillison in annot.).

Along the southern coast, grassland grades into savanna (see preceding chapter) or forms a mosaic. Part of this is anthropogenic grassland, but part is certainly natural, the result of edaphic and climatic conditions. On undulating land in the Central Province, tall grassland with *Imperata* and species such as *Cymbopogon globosus* and *Themeda novoguineensis*, occurs on the deeper soils, with short grassland, dominated by *Themeda australis*, on lighter soils and steeper slopes. Further west, particularly in the Trans-Fly area, flat terrain with an impermeable subsoil leads to extensive waterlogging during the wet season. Grasses form a dense ground cover among the trees; *Imperata* occurs in better-drained areas, other species range from tall *Vetiveria* to prostrate forms such as *Thaumastochloa rariflora*. On badly-drained soil, *Ectrosia leporina* colonises disturbed situations and land left bare after burning, normal behavior for *Imperata* elsewhere.

3. Highland grassland

As the term is used here, highland grassland lies between altitudes of 1500 m and 2500 m; the latter is about the upper limit of food gardening.

3.1. Tall highland grassland

The main pioneer grass, after gardening, is *Miscanthus floridulus*, which may develop into dense stands 2–3 m high, often mixed with trees of both fire-resistant and forest species. Walker (1966) found that *Imperata* may develop on abandoned gardens before *Miscanthus*, and may become established if burning is practiced and pigs are excluded. The absence of burning, and the activities of pigs, seem to be the most important factors favoring *Miscanthus* in competition with *Imperata*.

Less important tall grasses are *Phragmites karka* (swampland), *Saccharum robustum* (stream banks) and *Pennisetum macrostachyum* (hillsides and forest margins). *Coix lachryma-jobi* and *C. gigantea* are sporadic, sometimes locally common, probably as escapes or relicts of cultivation; the fruits of *C. lachryma-jobi* are used as beads, and *C. gigantea* has been used to make salt (Freund *et al.* 1965).

Themeda intermedia, a tall grass forming tussocks to 3 m high, is not usually common or dominant, but has been, in parts of the Goroka valley, co-dominant with *Imperata* in tall grassland (and in places in almost pure association). This type of grassland is becoming scarcer, since it usually occupies well-drained, agriculturally useful soil.

Several species are usually prominent, including *Themeda australis*, *Allopteropsis semialata, Arundinella setosa, Capillipedium parviflorum, Ischaemum barbatum*, *Ophiuros tongcalingii* and *Sorghum nitidum*.

On dry ridges, *Themeda* is usually dominant, with associated *Alloteropsis*, and often *Hyparrhenia filipendula*. On wetter sites, *Ischaemum polystachyum* occurs, often with *Sacciolepis indica* and in depressions, *Leersia hexandra*. A very robust species of *Ischaemum*, taxonomically similar to *I. polystachyum*, occurs in gregarious patches, apparently always on hillside soaks, in various localities in the Eastern Highlands. The leaves are a glaucous blue-green color, and patches of the grass are conspicuous from the air.

The weak-stemmed grass, *Apluda mutica*, occurs in the wetter grassland. Several decumbent or creeping species, of which the most common is *Arthraxon ciliaris*, are found in all but the driest situations, with several species of *Isachne*. At the higher altitudes are such temperate species as *Microlaena stipoides*, *Echinopogon ovatus*, *Polypogon monspeliensis*, *Dichelachne novoguineensis* and *D. crinita*.

A variety of other herbs and shrubs is scattered through the grassland. Legumes include several *Crotalaria* spp., *Desmodium heterocarpon*, *D. microphyllum*, *D. velutina*, *Indigofera trifoliata*, *Pycnospora lutescens*, *Uraria lagopoides* and the tallish shrub *Tephrosia vestita*. Another widespread shrub is *Melastoma malabathricum*; in the same family, the sub-shrub *Osbeckia chinensis* is sometimes common. There are various sedges, with *Fimbristylis dichotoma* usually present. Several ferns occur, and *Pteridium acquilinum*, bracken, may be common.

A few trees persist in the open grassland; *Alphitonia incana* is present in the lower parts of the Eastern Highlands and *Dodonaea viscosa* may be common, particularly in the Western Highlands. This small tree is found in both tall and short grassland, and is to some extent an indicator of recent disturbance. The introduced shrub, *Tephrosia vogelii*, is spreading, particularly in the Eastern Highlands.

Short grassland is more widespread in the Eastern Highlands than in the more western provinces, Chimbu, Western and Southern Highlands and Enga. This has been interpreted as indicating that the Eastern Highlands Province was settled earlier than the others, and that the grasslands are more mature (Robbins 1960). This is a reasonable hypothesis, but grassland type is not necessarily a reliable indicator of age. The short grasslands of the eastern Highlands are often in areas of lower rainfall. As proposed above, progression from *Miscanthus* grassland to short grassland may not, in wetter areas, be inevitable; even reversion to forest is possible. In drier areas, with more frequent burning, degradation to short grassland could be expected to proceed at a faster rate. It is even possible that, in the lower and drier parts of the Eastern Highlands, the pioneer grass was *Imperata*, not *Miscanthus*.

4. Mountain grassland

Areas of grassland exist above 2500 m, and so above the range of gardening activity, but still below sub-alpine grasslands. The origins of this grassland are obscure; some no doubt occupies frost-pockets where woody vegetation has been killed. Human activity was probably a factor, but nothing comparable to the disturbance of gardening could have occurred. Whatever their origin, these grasslands are maintained by fires, mostly deliberately lighted. Their area may be increased when fire destroys surrounding shrubbery; on the other hand, shrubbery, mainly by vegetative propagation, is constantly invading grassy areas, so that there is a dynamic transition zone between the forest and the grassland (Gillison 1970).

The composition of these mountain grasslands, with microtherm, festucoid species, contrasts sharply with the panicoid bias of grasslands at lower altitudes. Dominant species may be *Anthoxanthum angustum*, *Chionachne* (*Danthonia*) *archboldii*, *Deschampsia klossii*, *Hierochloe redolens* or *Poa saruwagedica*. These are tussocky grasses, and usually occupy the better-drained peaty soils. Many smaller species, of *Brachypodium*, *Danthonia*, *Festuca* and *Poa*, occur among the tussocks or in heath vegetation or in bogs, among Cyperaceae and mosses. A number of cushion-forming plants occur, among them the grasses *Danthonia* (*Monostachya*) *oreoboloides* and *Poa papuana*.

5. Grazing

In recent years (mainly during this century) a new factor, grazing by ungulates, has begun to affect the grasslands of New Guinea. The native grasses are not well adapted to grazing and the initial effect is usually destructive.

In *Imperata* grassland, the immediate effect of grazing by cattle is to reduce the dominance of *Imperata*. After burning, the young grass may be kept low for long enough (at least in patches) to allow the entry of low-growing grasses and other herbs. Burning is reduced, by the reduction of inflammable material, also to the disadvantage of *Imperata*. Under moderate to high rainfall, several regrowth tree species, unpalatable to cattle, may establish themselves, and it seems that grazing could initiate the re-conversion of grassland to forest.

In stands of mixed *Imperata* and *Saccharum spontaneum*, the unpalatable *Saccharum* is less affected by grazing; a patchwork of pure *Saccharum* communities develops, interspersed with patches of degraded *Imperata* with weeds and low grasses.

Under drier conditions, natural grassland seems to be more stable; this applies both to long and short grassland. *Imperata* grassland, however, often contains isolated tall tussocks of *Coelorhachis tongcalingii*. This grass is apparently quite unpalatable; it is also vigorous and resistant to trampling, and under grazing it may increase to an extent which greatly impairs the value of the grassland as pasture.

Themeda australis-dominated grassland has, in various localities, survived moderate grazing under dry conditions, but it is no better able to withstand heavy sustained grazing than *Imperata*.

Under particular conditions (moderate, rather seasonal rainfall; somewhat immature soil with fair internal drainage; easy topography) grassland of medium height, with *Imperata* and *Themeda australis* co-dominant, has been established in a few places (Gusap-Suliman Rivers, middle Ramu valley; Baiyer River valley). This may represent a transitional stage, between *Imperata*-dominated tall grassland and *Themeda*-dominated short grassland. As pasture, this grassland sustained grazing well, and judging by appearances, both cattle and horses were well nourished by it. However, even this has been degraded by grazing within 25 years.

Highland short grassland has generally been more stable under grazing than lowland grassland. *Arundinella setosa* is persistent, partly, no doubt, because it is not very palatable. *Ischaemum barbatum, Capillipedium parviflorum, Sorghum nitidum* and *Ophiuros tongcalingii* all persist fairly well, not necessarily in the same localities. *Ischaemum polystachyum* is less persistent, and *Alloteropsis semialata*, which seems very palatable and is considered nutritious in Australia, has almost disappeared from parts of the Eastern Highlands.

Modification by grazing has resulted in the emergence of various species, both native and introduced, which may partly or totally replace the dominant species of the original grassland. The increase of the undesirable *Coelorachis rottboellioides* in grazed *Imperata* grassland, has been cited above; a few other instances are given below.

In the SW part of the Western Province, between the Bensbach and Morehead Rivers, is an appreciable area of short swamp grassland, alternately flooded and dry. The short dense turf (in the dry season) is almost exclusively composed of *Pseudoraphis spinescens*, which elsewhere is known as a coloniser of bare ground, left by the retreating water of drying swamps. Here the grass is heavily grazed by deer, which appeared in the area after 1920, and are now present in great numbers. Probably the area was originally a reed-bed of *Phragmites karka*, which has been eliminated by grazing. A few patches of *Phragmites* remain, and regeneration by rhizomes on fenced quadrats has been demonstrated (J. Waithman pers. comm.).

In the same area, on slightly higher ground, *Imperata* grassland has been degraded by grazing and the creeping grass *Chrysopogon aciculatus* has gained entry. This species spreads by a prostrate creeping stem, and produces erect flowering stems, to 30 cm high. The spikelets have a sharp callus by which they penetrate and cling to animals' coats and the clothing of humans; grazing animals avoid patches of the grass, with their sharp-pointed fruits, while suppressing other herbage. Under uncontrolled grazing complete replacement of *Imperata* by *Chrysopogon* seems probable (Henty 1973).

In the lower Markham valley, another grass, which threatens even moderately grazed pasture, has appeared. *Digitaria insularis* forms tussocks 1–1.5 m tall, increasing by short stolons. It spreads by the light seeds, which are windblown or cling by the soft hairs to mobile objects. Native to tropical South America, it was first noticed in Morobe Province in 1952. Since then it has become firmly established in various localities, or land formerly occupied by *Imperata*. It is quite unpalatable to cattle and so has an advantage over useful grasses. It has not made headway in degraded *Themeda* grassland.

Several species of *Dichanthium* are established in the Central Province, where a dry seasonal climate prevails. Probably none of them is native, or at any rate normal in natural grassland, but they follow disturbance by grazing or other means; for instance, they move along roads. *D. annulatum* is a prominant species, which has also spread to other parts of Papua New Guinea; it was noticed near Lae in 1952. Since then it has colonised degraded *Themeda* grassland in the lower Markham valley, particularly those parts where degradation was far advanced, where the *Themeda* had almost disappeared, giving way to low grasses and taller weeds (such as *Sida cordifolia*), of negligible grazing value. The invasion of *Dichanthium annulatum*, as a short dense turf, has already rehabilitated appreciable areas, and its further spread may be expected.

The above examples indicate the extent to which modification of grassland by grazing is influenced by the species available. These may be present in the original grassland, or may arrive as immigrants. The history of extensive grazing in Papua New Guinea is short, and it is probable that many more changes will be observed as the years pass.

References

Freund, A. P. H., E. E. Henty and M. A. Lynch. 1965. Salt-making in inland New Guinea. Trans. P. and N. G. Sci. Soc. 6: 16–19.

Gillison, A. M. 1970. Structure and floristics of a montane grassland/forest transition, Doma Peaks region, Papua. Blumea 18: 71–86.

Henty, E. E. 1967. Border Survey collections, Western District (internal report).

—— 1973. Vegetation of the Bensbach Wildlife Reserve (internal report).

Heyligers, P. C. 1965. Vegetation and Ecology, in C.S.I.R.O. Land Research Series No. 14, Lands of the Port Moresby-Kairuku area, Papua New Guinea: 146–172.

Robbins, R. G. 1960. The anthropogenic grasslands of Papua and New Guinea, in Symposium on the impact of man on humid tropics vegetation, Goroka, UNESCO: 313–329.

—— 1961. Vegetation, in C.S.I.R.O. Division of Land Research and Regional Survey, Divisional Report No. 61/2, Lands of the Wewak-Lower Sepik area, New Guinea: 121–153.

Steenis, C. G. G. J. 1965. Concise Plant-geography, in Backer and Bakhuizen van den Brink, Flora of Java, vol. 2: (61).

Taylor, B. W. 1964. Vegetation, in C.S.I.R.O. Land Research Series No. 12, Lands of the Wanigela-Cape Vogel area, Papua New Guinea: 69–83.

Walker, D. 1966. Forest, Grassland and Garden in New Guinea. J. Ecol. 54: 503–527.

10. Ecology of fungi in New Guinea

Dorothy E. Shaw

1. Introduction

The fungi whose ecology will be considered in this paper occur mainly on the island of New Guinea itself, but also on adjacent islands off W New Guinea, and in the Bismarck, Louisiade and d'Entrecasteaux Is off E New Guinea. This area corresponds to the phytogeographical region of Balgooy (1976) except that the Solomons are not included here, unless passing reference is made to the North Solomons. Reference to the political divisions of "Papua New Guinea" and "Irian Jaya" is occasionally inevitable, as collection and study zones have usually occurred within these boundaries. Also, in recent decades, quarantine barriers have conformed to the political divisions. Breaches of such barriers, of course, may affect the biota of various ecological niches to a greater or lesser extent.

The land mass under discussion extends roughly from the equator to about 12°S lat., and from sea level to about 5000 m at Carstensz in the central cordillera. Paijmans (1976) gave the main environments of Papua New Guinea as: coastal beach ridges and flats, $\frac{1}{2}\%$; coastal saline and brackish swamps, $1\frac{1}{2}\%$; lowland fresh water swamps, mainly 0–50 m, 11%; lowland alluvial plains and fans, mainly 0–500 m, 15%; foothills and mountains below 1000 m, 43%; lower montane zone 1000–3000 m, 25%; and upper montane zone 3000–4000+ m, 4%. He considered that the vegetation of Irian Jaya appeared to be similar to that of Papua New Guinea, but with a somewhat higher proportion of climax forest, and lacking the eucalypt savanna of the central SW coast region of Papua New Guinea; also, the climate is progressively wetter in the mountain ranges at least, from east to west. New Guinea lies within the heavy precipitation belt of the humid tropics, with well over half of E New Guinea receiving 2450 mm rainfall or more per annum (Hart 1970) with recordings up to 9604 mm mean annual rainfall (McAlpine, Keig and Short 1975).

With its latitudinal position and high rainfall, and such a range of land forms, together with associated biota including man, as well as man's artifacts, New Guinea provides a variety of macro- and micro-habitats for fungi. Park (1968) pointed out that environmentally fungi are notorious and versatile opportunists. The large surface:volume ratio inherent in hyphal growth gives a large contact area with the environment and greatly affects fungal biology. Not only

© *1982, Dr W. Junk Publishers, The Hague.*

is the area of contact large, but there is no great distance between any point in the protoplasm itself and the environment. This is different from higher multicellular organisms, much of whose protoplasm is protected from the environment by physical and physiological barriers. The environment, therefore, plays a much more immediate and direct part in fungal behavior than in the behavior of many other organisms. Consequently, a fungus may in turn influence an environment rapidly, by depletion of foodstuffs, or by the accumulation of waste products (Park 1968).

Fungi are also notorious for their extremely prolific reproduction. Large numbers of dispersal spores are often produced very rapidly, giving dispersal *in space*, and equally important, large numbers of residual propagules are left behind on substrata where the mycelial growth took place, giving dispersal *in time* to withstand unfavorable environmental conditions or large discontinuities such as drought, heat, cold etc.

There are many reviews and papers on the ecology of fungi generally, and of specific groups in particular, but in this chapter only brief mention of some of the ecological niches occupied by fungi in New Guinea will be possible, with most emphasis on fungi associated with living plants. The examples given, with a few selected references only, may serve to illustrate the contention that fungi are versatile opportunists of the environment with an ability to exploit varied ecological niches.

In this chapter, names of fungi have been brought up to date where possible. In other cases, as where names are under review, the name used by the original author is given without comment.

2. Fungi on man and other animals

Man himself is an ecological niche, and many species of fungi are now known which are parasites or pathogens (parasites which induce disease in the host) of *Homo sapiens*. Some of these cause pulmonary disabilities, others are allergens, some are dermatophytes, others cause deep-seated malformations to organs and some are toxin producers. Those species recorded in New Guinea include *Cryptococcus neoformans* which causes cryptococcosis (Champness and Clezy 1962; Frey and Durie 1964) with worldwide distribution; *Sporothrix schenckii* which causes sporotrichosis (Campbell 1965) with distribution in overseas countries, and various unidentified fungi causing mycetomas as well as well known dermatophytes. Some of the human-infecting fungi, however, are not confined to man, but are also recorded on other animals including birds. Others have an even wider range. For example, dermatophytes were until fairly recently isolated only from man and other animals and their taxonomic affinities were obscure. Since the development of the "hair-baiting" technique in soil, and also since the discovery that dermatophytes are heterothallic (that is, that the perfect states can be obtained by mating appropriate isolates) they are now known to be widely distributed in soil, and they are securely classified as gymnoascaceous fungi (Ainsworth 1968). As is usual with pathogenic fungi on animals, the perfect states are not associated with the pathogenic phase. In New Guinea,

Microsporum gypseum (with the perfect state in *Nannizzia*), one of the ringworm group which has been recorded on every continent, has been isolated from soil (Morahan 1970 and others) as have other fungi such as *Trichophyton terrestre*, *Ctenomyces serratus* and *Chrysosporium keratinophilum* capable of attacking keratin of feathers, hair, hoof and horn (Frey 1959, 1968; Frey and Griffin 1961; Morahan 1970). Some of the dermatophytes in New Guinea have affected a considerable number of people. *Trichophyton concentricus*, causing tinea imbricata, for example, was estimated to affect about 150,000 people in the lowlands of E New Guinea (Schofield, Parkinson and Jeffrey 1963), while Peters (1960) reported that in one area 20.8% of those examined suffered from tinea imbricata and of these over half were almost uniformly covered with lesions. McLennan and O'Keefe (1975) found that various species of *Trichophyton* were affecting 48.3% of the people in one lowland region, but only 14.6% of the nearest high valley population in an area of lower temperatures.

Gregory (1961) was of the opinion that outdoor airborne allergens such as *Cladosporium* and *Alternaria* appear to originate from vegetation above the ground, but not significantly from the soil. Species of *Cladosporium* he stated, are ubiquitous, and dominate the day-time air spora in temperate regions and in most of the tropics. Species of both genera are recorded for New Guinea (Johnston 1961; Shaw 1963a). Allergic reactions in man may also be induced by other fungi, for example, *Aspergillus fumigatus*. This species may also cause acute and chronic infections of the respiratory and other systems of man and other animals. The fungus has an optimum temperature of 45°C or higher, and is common in compost and other organic materials undergoing decomposition. Some species of *Aspergillus*, such as *A. fumigatus* and *A. flavus*, have strains which may produce mycotoxins, as do strains of *Penicillium* such as *P. islandicum, P. patulum* and *P. citrinum*. All of these (and others) have been recorded in New Guinea (Shaw, Brown and Cunningham 1972) while mycotoxins have been recorded by Wookey (1972) and Collins, Beard and Still (1975) on peanuts (*Arachis hypogaea*) and by Still and Munro (1978) on winged bean (*Psophocarpus tetragonolobus*).

Relatively few fungi have been recorded causing disease of the larger animals in New Guinea as yet. However, species reported do include *Candida albicans* and *Aspergillus fumigatus* on fowl and *Microsporum canis* on cat and dog (Egerton and Rothwell 1964), while the fungus causing swamp cancer of horses in Papua New Guinea has been discussed by Austwick and Copland (1974).

Other fungi are well known as insect parasites. Some of these are mainly superficial on the host's body, with penetration apparently haustorial. The great majority of these fungi belong to the Laboulbeniales, and many are highly host specific (Madelin 1968). Other fungi are endoparasites, with the fungal mycelium living within the host's body after, in many cases, penetration of the host's integument, and eventually causing death. In some cases diseased insects commonly climb to high places, for example, on vegetation, or if subterranean, rise to the soil surface; death in such situations facilitates dispersal of the parasites' spores (Madelin 1968). Some of the early records of entomogenous fungi in New Guinea were made by Hennings, Sydow and Petch, and more recently by Kobayasi and Shimizu (1976) and Prior (unpublished), including such genera as *Cordyceps*, *Hymenostilbe* and *Isaria*, and involving some new species. *Beauveria bassiana* and *Metarrhizium anisopliae* are frequently collected, and there is one record of *Entomophthora* sp. on eri silkworm (Shaw unpublished). Other fungi enter through integumental wounds and others live on decomposing hosts, but these records are not listed here.

Fungi parasitic on insects which are closely associated with plants in New Guinea were recorded by Shaw (1963a) and include species in the genera *Aschersonia*, *Hypocrella*, *Myriangium*, *Pseudomicrocera* and *Septobasidium*. In many cases the parasitism is so severe that very little of the insect host remains, and the fungal colonies may be mistaken by collectors for parasites on the plant itself. In these cases the remains of the insect host are usually unidentifiable.

An interesting example of an unusual ecological niche was reported by Gressitt, Sedlacek and Szent-Ivany (1965), who found that fungi were present on the backs of more than 250 live specimens of 11 species of *Gymnopholus*, large flightless weevils feeding on leaves of woody plants in high moss forest on various New Guinea mountain ranges. Fungi were also found on weevils of the subfamily Cryptorrhynchinae on the same mountains. The fungi were said to include Ascomycetes (Ascomycotina) and Fungi Imperfecti (Deuteromycotina), although the specific identifications have not been published as yet. The authors stated that mutalistic symbiotic relationships seemed to be clearly indicated.

Another ecologically natural yet taxonomically mixed group recorded overseas are the fungi which are predaceous on protozoans, soil amoebae and on nematodes. These latter fungi especially have evolved many unusual morphological adaptations such as sticky networks, knobs or hyphal branches, or short hairs ending in rings, which by the sudden swelling of the cells, capture nematodes by constriction (Duddington 1968). Besides the truly predacious fungi, there are others which are internally parasitic. Very little detailed work has been done on these fungi in New Guinea as yet, although species in the genera of *Arthrobotrys*, *Dactylella* and *Dactylaria*, which include some nematode-attacking fungi, have been recorded (Matsushima 1971) from soil.

3. Fungi on plants

Many fungi have close associations with living plants. Those associations may range from true epiphytes to parasites and to pathogens, where disease is induced in the host. A few New Guinea examples, with selected references only, in some of these ecological niches, are given below.

Many of the yeasts, yeast-like fungi and some other species are known as epiphytes on leaves and other surfaces of plants in many parts of the world. They live on substances deposited on the plant surfaces, although they are also able to use exudates from the leaves and other plant parts, especially during the senescent stage of the "host" part. In New Guinea, yeasts from slime fluxes on plant parts (and other substrata) have been recorded by Soneda and Uchida (1971) and false yeasts of the family Sporobolomycetaceae isolated from various living plant surfaces by Shaw and Cartledge (unpublished).

The surfaces of leaves, and to a much less extent petioles and twigs but only rarely fruit, form an important and extensive habitat for many fungi with brown or black pigmentation (Fig. 1, A). The colonies may range in size from 1 to 20 mm in diameter, or be confluent and cover a much larger area, and even completely festoon trees. The fungi belong to a diverse range of families in the Ascomycetes and Fungi Imperfecti, and include the sooty moulds, sooty mould-like fungi, the "dark mildews" (Meliolales) and other species of various fungal groups. They occur as epiphytic saprobes, hyperparasites on other fungi or as superficial leaf parasites. Some like the Meliolales, are obligate parasites, giving no growth on artificial media. This latter group particularly are considered to be common in "shrub and open parklands" (Hansford 1961) being comparatively rare on undergrowth in the rainforest (though perhaps quite common on the inaccessible top canopy) unless breaks in the canopy allow more light to reach the forest floor. The superficial leaf parasites penetrate the host by means of small haustoria which develop in the epidermal cells of the host, but no disease is produced. Some, especially the Meliolales, are frequently so parasitized by other fungi that the production of the meliolaceous fruiting bodies may be prevented altogether. The fungi in all the above groups are especially common

Fig. 1. A. Black colonies of a superficial fungus on leaf of *Syzygium* sp. B. Black tumorous outgrowths of maize cob (*Zea mays*) caused by the smut fungus *Ustilago maydis*.

in the tropical and sub-tropical zones of the world. In New Guinea, species have been recorded in genera such as *Acrogenotheca*, *Antennatula*, *Asteridiella*, *Asterina*, *Capnodium*, *Chaetothyrium*, *Ciferrioxyphium*, *Grallomyces*, *Lembosia*, *Meliola*, *Meliolina* and *Schnifferula* (Johnston 1961; Shaw 1963a and unpublished), many of the identifications having been made by Hansford and in recent years by Hughes, including several new species. Hansford also described two new genera on New Guinea material which may be mentioned here. The first was *Brooksia*, based on *B. tropicalis* with the asexual state *Hiospora hendrixksii* (Hansford 1956; Deighton and Piroxynski 1966) which is now widely recorded on many different plants in many parts of the moist tropics. The second was *Shawiella*, based on *S. grevilleae* on *Grevillea robusta* (Hansford 1957; Sutton 1975) which is parasitic on the leaf hairs. It has not yet been recorded on any other host, nor has it been found outside New Guinea as yet, and therefore may have a narrow ecological niche.

Another group of fungi occurring on leaves of many plants, but with unpigmented mycelium, is the powdery mildews. These are obligate plant parasites with superficial mycelium attached to the host by haustoria, and colonizing little of the internal tissue, although there is greater internal colonization by *Leveillula taurica*. The powdery mildews are represented in New Guinea with records of the asexual state on many important, mainly introduced agricultural plants including *Carica papaya*, *Citrullus vulgaris*, *Cucumis melo* and *C. sativus*, *Cucurbita maxima*, *C. pepo* and other Cucurbitaceae, rubber (*Hevea brasiliensis*), *Hibiscus manihot* ("aibika"), *Psophocarpus tetragonolobus* and species of *Vigna*, as well as on other garden plants, weeds and miscellaneous herbs (Johnston 1961; Shaw 1963a and unpublished; Price 1977).

Another group of fungi well represented in New Guinea are the rusts, which are important plant pathogens. Except for some strains of *Puccinia graminis* which have now been grown in culture, the rusts are known as obligate parasites of many plants. Some of the earliest collections of rusts in New Guinea were by Mrs Mary Strong Clemens, a missionary who worked mainly in the Morobe District inland from Lae. Cummins identified many of these rusts, and described several new species. One new genus, *Corbulopsora*, was also described, but has since been absorbed in the genus *Miyagia*. The records include rusts on plants indigenous to New Guinea as well as on introduced species. All the rust records for Papua New Guinea, including those of Cummins, were listed by Shaw (1963a) and rust records since then are now being compiled. Altogether they include species in temperate and tropical genera such as *Aecidium*, *Bubakia*, *Cerotelium*, *Crossopsora*, *Hamaspora*, *Kuehneola*, *Maravalia*, *Miyagia*, *Phakopsora*, *Puccinia*, *Sphaerophragmium*, *Uredo* and *Uromyces*. They also include coffee rust, *Hemileia vastatrix*, which has drastically affected coffee growing in the commercially producing countries of the world. In Papua New Guinea the entire ecological niche of coffee rust was deliberately destroyed by the eradication of all species of *Coffea* (the only known host) in the infected area, at an estimated cost of Au$70,000 (Shaw 1968). If this fungus had not been eradicated, and it had proceeded to fully utilize its niche throughout the country, it might have brought about the collapse of the entire coffee industry in Papua New Guinea. Since the destruction of its ecological niche and its eradication in 1965, however, the value of coffee exported from the country has been over Au$300,000,000.

Smuts of various plants were also collected by Mrs Clemens and others, many of the identifications being made by Zundel. These records were listed by Shaw (1963a), and others are now being compiled. They include species in well known genera such as *Cintractia*, *Entyloma*, *Sphacelotheca* and *Ustilago* on local and introduced hosts (Fig. 1, B).

Many other plant pathogens have been recorded in Papua New Guinea, on both indigenous and introduced hosts. The records (listed by Johnston 1961; Tantera, unpublished; Shaw 1963a and b and unpublished) include fungal parasites and pathogens whose niche on their respective hosts may involve little or no macroscopically visible damage, to other cases where the host is completely killed. The symptoms of disease may vary from hardly perceptible discoloration, to leaf spots, blotches, streaks, malformations, tumors, blights, rots, wilts, dieback and death. In some cases the niche is very narrow, being confined to one host species, while in others the pathogen has a very wide host range. The pathogens themselves are classified in all the major groups of fungi.

Further discussion on the many plant pathogens recorded in New Guinea will be confined to a few examples illustrating the above, especially if on important subsistence food plants or agriculturally important hosts, or if the taxon has been described on New Guinea material. As Powell (1976) pointed out, most of the staples of importance in subsistence agriculture have been introduced. Sago (*Metroxylon sagu*) and some of the bananas (Australimusa Section) are indigenous; yams (*Dioscorea* spp.) are possibly all SE Asian in origin; taro (*Colocasia esculenta*) is of Indian or Indonesian origin, and the sweet potato (*Ipomoea batatas*) of South American origin, while cassava (*Manihot esculenta*) is also from the New World.

Sago (*Metroxylon sagu*) would seem to be an inhospitable niche for pathogenic fungi in New Guinea, judging from the paucity of records. In pathological terms, this plant may show considerable horizontal resistance.

Some yams (*Dioscorea* spp.) are very susceptible to *Glomerella cingulata*. This fungus causes leaf spots which may become confluent and lead to premature death of most of the leaves. In one case of reputed witchcraft investigated by Shaw, the yams had been nearly totally destroyed by this fungus.

One of the introduced staples, taro (*Colocasia esculenta*) is host to several fungi confined to this species, with *Xanthosoma* apparently being immune. *Phytophthora colocasiae*, for example, which causes taro blight and is well-known overseas, may be severe on this crop at lower altitudes, as in New Ireland, the Gazelle Peninsula and North Solomons. *Cladosporium colocasiae*, recorded in Papua New Guinea and other overseas countries, is also confined to *Colocasia*, as is *Johnstonia colocasiae*, described on New Guinea material (Ellis 1976), but now recorded in the Solomons, Sabah and W Samoa.

Sweet potato (*Ipomoea batatas*) is host to such fungi as *Ceratocystis fimbriata*, known overseas on a wide variety of plants. Sweet potato is also host to some pathogens, such as *Elsinoe batatas* (Fig. 2, A) confined to *Ipomoea*.

Fungi causing diseases of bananas in New Guinea include the *Mycosphaerella* complex and *Cordana musae*, which are widely distributed in the world. Two new species have been described on New Guinea bananas, viz., *Uredo musae*, a rust (Cummins 1941) and *Phyllachora musicola* (Booth and Shaw 1961) causing black cross disease. This latter fungus is now known to occur on other Pacific islands including Fiji and the Philippines on AAB and ABB types, whereas the AAA group are probably immune, as is abaca (*Musa textilis*) (Stover 1972).

Cassava (*Manihot esculenta*) is a narrow ecological niche for some fungi in New Guinea, such as *Cercosporidium henningsii* (Fig. 2, B) restricted to this genus, whereas others such as *Corticium salmonicolor*, causing pink disease, affect a wide range of hosts.

Sugar cane (*Saccharum officinarum*) is indigenous to New Guinea, as is *S. robustum*. The fungus *Deightoniella papuana* (Shaw 1956) was described on these species and on *S. spontaneum*. On *S. robustum* the lesions on the leaves may reach a length of 60 cm, and occur on plants from sea level to about 2000 m. This fungus was also recorded on *Saccharum* from the Solomons. Some fungi known overseas also occur on New Guinea sugar cane, but others, such as *Ustilago scitaminea*, have apparently been excluded by the quarantine barrier.

Species described in New Guinea on coconut (*Cocos nucifera*) include *Anthostomella cylindrospora* (Booth and Shaw 1967) which casues petiole lesions on palms weakened by drought, although its full pathogenic potential has not been determined. *A. fusispora*, described on coconut material from Malaysia (Booth and Shaw 1967) has now also been recorded in New Guinea (Sivanesan, Shaw and Brown 1976). These latter authors also described *Leiosphaerella longispora* on coconut petiole from Papua and New Ireland, which is at present not recorded outside New Guinea. *Clasterosporium cocoicola* was also described on material from New Guinea (Ellis 1959), this being a black superficial mould with hyphopodia on the leaf surface.

Other economically important cash crops in Papua New Guinea such as coffee, cacao, rubber, tea and oil palm, have been introduced, and rigorous quarantine restrictions have prevented entry of many fungi which cause important diseases on them overseas (Shaw 1978). Some of the fungi which have been recorded on them in New Guinea have very narrow host ranges, while others are very wide. The fungus *Cercospora coffeicola*, for example, which causes leaf spot of coffee, is restricted to that host, whereas *Corticium salmonicolor*, causing pink disease, has been recorded on a wide variety of both trees and herbaceous hosts of different families, including rubber, cacao and citrus.

Various fungi find their ecological niche in cacao (*Theobroma cacao*) in New Guinea, although again, less than in other countries where this host is indigenous or has been established for a long time. Some of these fungi like *Phytophthora palmivora* which causes pod rot, canker and chupon wilt, are known from other hosts and are widely distributed in the tropical areas of the world. One new genus described on cacao in New Guinea is *Oncobasidium theobromae* (Talbot and Keane 1971) which causes vascular streak dieback. So far no other host is known, although the fungus has now been recorded in Malaysia (Keane and Turner 1972), again on cacao. These latter authors considered that one explanation of its wide distribution in Malaysia is that *O. theobromae* may be able to grow independently of cacao, possibly on another

Fig. 2. A. Scabby malformation of two shoots of sweet potato (*Ipomoea batatas*) caused by the pathogenic fungus *Elsinoe batatas*. B. Leaf spot of cassava (*Manihot esculenta*) caused by the pathogenic fungus *Cercosporidium henningsii*.

host plant. It this is so, its ecological niche may be wider than at first thought.

Very few fungi have so far been recorded as causing diseases of tea and oil palm in New Guinea. Both plants form favorable ecological niches for some important pathogens overseas. However, fungi such as *Exobasidium vexans*, which causes blister blight of tea, and *Cercospora elaeidis*, which causes *Cercospora* spot of oil palm, have so far been excluded by the rigorous quarantine barrier.

Rubber (*Hevea brasiliensis*) in New Guinea provides a narrow ecological niche for some pathogenic fungi, such as *Oidium heveae*, causing powdery mildew, considered by most workers to be restricted to this host. On the other hand, rubber is only one of many hosts for pathogenic fungi such as *Corticium salmonicolor* and *Phytophthora palmivora*, as mentioned previously. Rubber is also only one of many hosts of some root rotting fungi important overseas and in New Guinea, such as *Phellinus noxius* and *Rigidoporus lignosus*, to mention 2 of the most common. These attack several of the tree crops such as cacao and coffee as well as rubber, and also one of the important lowland shade species, *Leucaena leucocephala*. These root rot fungi have also been found fructifying on decaying stumps and logs in the rain forest. Their ecological niche is very wide in comparison with some other plant pathogens. Recently Simpson (1978) recorded *Heterobasidion annosum*, the wood rot fungus well known overseas, on dead *Araucaria cunninghamii* in natural forests in New Guinea. No evidence of root rot by this species has so far been found in plantations of either *Araucaria* or introduced *Pinus*. Other fungi isolated from living trees with root rot have been listed by Shaw (1963a and unpublished) while the heart rots particularly are at present under study by Simpson (unpublished).

Other plant pathogens described on New Guinea material include *Mycovellosiella puerariae* on *Pueraria triloba* (P. lobata) (Shaw and Deighton 1970). This fungus also appears to have a very narrow niche, as it is apparently restricted to *P. triloba*, with *P. phaseoloides* apparently immune. It also occurs in Taiwan. Another fungus originally described in New Guinea, *Ramularia oryzae* (Deighton and Shaw 1960) causing white leaf streak of rice (*Oryza sativa*) occurs on other species of *Oryza* and is recorded in the Solomon Is, N Borneo and Africa.

Two other new genera have been described on New Guinea plant material. The genus *Verrucispora*, with *V. proteacearum* as the type, on *Finschia chloroxantha* (indigenous) also occurs on *Hakea florulenta* in Queensland (Shaw and Alcorn 1967). *Verrucispora* was a monotypic genus up till 1978, when *V. indica* was described parasitizing living leaves of *Smilax prolifera* in India (Kamal and Singh 1978). The other new genus is *Stromatothecia*, *with S. nothofagi* on *Nothofagus* sp. (indigenous) as the type (Shaw and Hawksworth 1971). At present, this fungus has not been recorded outside New Guinea. On the other hand, species of *Cyttaria*, recorded on species of *Nothofagus* in other countries, have not been recorded in New Guinea.

Many fungi have been recorded on vegetables and other plants (Fig. 3) used by the people of New Guinea, but are not listed here. However, it should be pointed out that apart from primary pathogens which have the ability to enter undamaged plants, many other fungi can parasitize living plants once the first

Fig. 3. Defoliation of a variety of groundnuts (*Arachis hypogaea*) susceptible to both *Puccinia arachidis* (rust) and *Cercosporidium personatum* (leaf spot), with more resistant varieties at back.

line of resistance has been broken by primary pathogens, or after such occurrences as wounding or bruising during harvest, transport or storage; or after lighting strike (Fig. 4, A) or insect damage or if the host has reached a certain stage of senescence or is suffering from certain mineral deficiences etc. These secondary species may be important in causing rotting in herbaceous and woody plants, and especially in fruit and vegetables, particularly if produce has been held in transport or in unrefrigerated storage. Not only do these fungi cause deterioration of the produce itself, but some species include strains which produce mycotoxins deleterious to man and other animals.

Another important niche is that occupied by fungi forming mycorrhizal associations with plants. This association has been defined as a symbiotic, non-pathogenic or feebly pathogenic association of fungi and the roots of plants. The two main groups of mycorrhiza are the ectotrophic and the endotrophic (Harley 1968). In the ectotrophs, the fungus is on the surface of the roots and forms a Hartig net. This is the characteristic mycorrhiza of forest trees with different basidiomycetes, especially agarics and boletes. In the endotrophs, the fungus invades the roots, where it is frequently confined to well-marked layers. An example of this is found in orchids and other plants, including a few ferns and mosses, the hyphae being septate. In other cases the fungal hyphae are aseptate, and these are very common in angiosperms, gymnosperms, ferns and mosses. Fruiting bodies of some of the fungi in these associations have been identified in New Guinea by various workers, and the group is at present under further study by Simpson (unpublished).

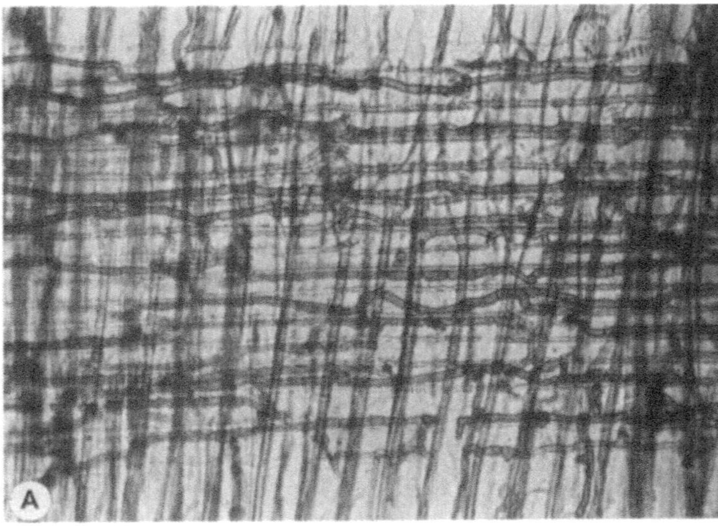

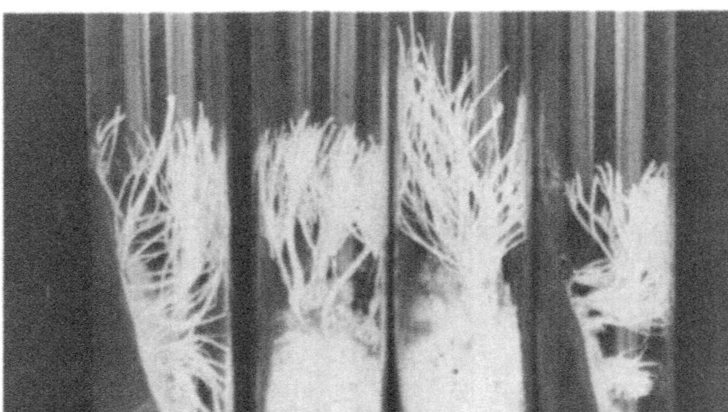

Fig. 4. A. Microscopic section through woody stem of *Leucaena leucocephala* after lightning strike showing cross cells invaded by fungal threads (hyphae) of the weak pathogen *Botryodiplodia theobromae*; vertical cells of a different type also visible but uninvaded. B. The fungus *Isaria cretacea* isolated from fecal pellets of the cuscus (*Phalanger maculatus*) growing in culture. Some species of *Isaria* are parasitic or saprophytic on insects.

4. Fungi in soil and on organic debris

Many species of fungi have been isolated from soils all over the world, including New Guinea. Some of the isolates are derived from propagules of species normally occurring above ground which have accidentally fallen into the soil. Some, like the truffles, are generally subterranean, while many others produce their spores above soil level. Species found in soil in New Guinea include the keratinophilous species, as already mentioned. Plant pathogens recorded in soil in New Guinea include species of *Phytophthora* and *Pythium* (Stamps, Shaw and Cartledge 1972; Arentz 1976) including *Ph. cinnamomi* (Shaw, Cartledge and Stamps 1972; Cartledge, Shaw and Stamps 1975; Arentz 1976 and unpublished). *Ph. cinnamomi* is well known overseas as an important cause of dieback and root rot of many herb and tree species. The main method used to isolate these fungi is by lupin radicle baiting of soil samples. Other important pathogens recorded in soil in New Guinea include species of *Fusarium*, *Rhizoctonia* and *Sclerotium* (Johnston 1961; Shaw 1963a; Shaw and Kula 1978 and other papers).

Some workers such as Park (1968) have considered the species spending some part of their life cycle in soil from the chemical state of the substrata utilized. The three main groups may be summarized as sugar users, cellulose users and lignin users. The sugar users occupy that niche provided by recently dead plants, which are quickly colonized. Plant sugars and other nutrients are used for quick production of spores. These fungi as quickly disappear, only to flare up again when the next batch of green manure or plant parts are added to the soil. Many of the Ascomycetes containing antibiotic and toxin producers are sugar users.

Once the sugar and other compounds disappear, fungi capable of cellulose decomposition take over. This group includes mushrooms (Fig. 5, A) which live on straw-type materials, not on fresh green plant parts. Some indigenous mushrooms are eaten by Papua New Guineans, and at least 2 other cultivated species have been introduced. The lignin users (Fig. 5, B) follow the cellulose decomposers. These species are especially present in forest floor litter, as well as in soil, and form a much more stable population than the sugar-users with their sudden flare ups.

Activities of the users of sugar, cellulose and lignin are of course essential for the breakdown of old biotic remains so as to enable recycling of nutrients to take place. The many species in all the above groups, involving diverse types of fungi, and including some species described on New Guinea material, are not listed here but will be found in papers by Aoshima, Berkeley, Bresadola, Cooke, Corner, Dennis, Doi, Frey, Gaudichaud, Heim, Hennings, Hongo, Horak, Imazeki, Kjoller, Kobayasi, Lloyd, Maas Geesteranus, Matsushima, Mueller, Nobles, Otani, Ramsbottom, Reid, Shepherd, Singer, Stevenson, Steyaert, Sydow, Takada, Udagawa, Walters, Yokoyama and others. Still other collections, especially recent ones by Horak, are still under study.

5. Fungi in other ecological niches

Man's artifacts also provide ecological niches for fungi. The species are of diverse groups and occur on other substrata as well. In New Guinea they have been recorded on a variety of objects such as tents, paper, books, household goods, linoleum, leather, mats, hairbrushes, and other wooden items, painted surfaces of houses and other structures. Some of these records arose during investigations of spoilage of fabric and equipment during World War II, as listed by Shaw (1963a) while other records are still unpublished.

Some timbers are particularly susceptible to wood rot fungi and to species which cause blue stain and other discolorations. In recent years a wood chip

Fig. 5. A. A cluster of mushroom-type fructifications of *Collybia reineckiana*. B. Fungal fans of a lignicolous species, *Schizophyllum commune*, growing on rotting log. The fans shrivel in dry weather but re-expand under moist conditions.

industry for export has been developed in Papua New Guinea, and the chip piles provide another niche for fungi. Studies carried out on experimental piles of mixed tropical hardwood chips by Greaves (1973, 1974 and 1975) showed that a range of fungi as well as other microorganisms was responsible for discoloration. Decay was mostly caused by soft rotting microfungi such as *Chaetomium globosum*, *C. thermophile*, *Humicola lanuginosa*, *Cephalosporium acremonium* and *Gliomastix subiculosa*. The fungi present included some thermophilic and thermotolerant species as would be expected, since a rapid rise in temperature is a characteristic of wood chip piles.

Dung or coprophilous fungi have a special ecological niche, in that their spores will only germinate if they are passed through the alimentary canal of grazing animals. They are voided with the feces and grow in succession on dung pads. Overseas Ingold (1971) recorded that mucoraceous fungi usually precede ascomycetous and discomycetous genera, which tend to appear before Pyrenomycetes while lastly a crop of agarics is produced usually consisting of small species of *Coprinus*. Ingold thought that a basic factor determining the successive appearance of the different fungi may be the time interval between germination and the production of spore-bearing structures. For example, on sterile pellets of rabbit dung, the interval was 2–4 days for the mucoraceous fungi, but 1–2 weeks for *Coprinus*. Ingold has shown that some of the species have what appear to be beautiful and parallel adaptations in the organization of their spore discharge, especially involving active discharge of the projectiles which enables the spores to reach the grass without the aid of wind. In New Guinea coprophilous species have been recorded by various authors, including Udagawa and Takada (1971) and Matsushima (1971). One fungus isolated from dung by Shaw is shown in Fig. 4, B.

As mentioned previously, some fungi are parasitic on other fungi. In other cases, some fungal species are often found associated with or over-growing certain fungal bodies or colonies. Udagawa and Horie (1971) recorded various ascomycetous and dematiaceous fungi isolated from some polyporaceous (bracket-forming) species and other fungi in New Guinea, although the degree of parasitism, if any, was not specified. Both hyperparasites and overgrowers have been recorded on various other fungi in New Guinea, particularly on plant pathogens such as the rusts or other fungi closely associated with plants, such as meliolaceous leaf parasites. The records of these include one for Irian Jaya (Johnston 1961) and others for Papua New Guinea by Shaw (1963a and unpublished) and include species in genera such as *Ampelomyces*, *Cryptodidymosphaeria*, *Eudarluca*, *Spiropes* and *Hansfordia*.

Some fungi are known to have "constitutive" dormancy (Sussman and Halvorson 1966) and will fail to develop unless provided with an activation "trigger" treatment. Various physical and chemical factors which serve as activators include temperature, light, moisture, and the composition of the medium. Sussman and Halvorson stated that a relatively small number of fungal spores need heat activation, and that the requirement for heat-shock appears to be characteristic of the Ascomycetes. Many of these are associated with dung, as mentioned previously, and it is possible that the requirement for heat is associated with their passage through the digestive tracts of animals.

Ascospores of some species of *Neurospora* also require heat-shock for germination. This condition is provided, for example, by fire during vegetation burns. The asexual state (*Monilia*), forming pinkish-orange colonies on a number of substrata such as burnt grass stems, logs and tree bark, and also on autoclaved soil in the glasshouse, is common in New Guinea (Shaw unpublished). Perkins, Turner and Barry (1976) investigated the mating types and species from nine collections in Papua by crossing with testers at Stanford University. Collections from eight sites proved to be *N. intermedia*, a species widely recorded in the East, Australia, Hawaii and the southern United States. Both mating types were present at three sites, and either one or the other mating type at five sites. The collection from the ninth site is at present unidentified, as it produced no fertile crosses with any tester.

Some fungi grow best at high temperatures, although only about 20 or so species are at present known in the world which are truly thermophilic with growth at 50°C or above, and these appear to have worldwide distribution (Emerson 1968). Some of these fungi, such as *Chaetomium thermophile*, *Humicola grisea* var. *thermoidea*, *H. lanuginosa*, *Malbrancea pulchella*, *Paecilomyces variotii*, *Talaromyces emersoni* and *Thermoascus auranticus* have already been recorded in New Guinea on wood chip piles (Greaves 1973, 1974, 1975).

Cold-loving species of fungi could be expected to be in a minority in an equatorial country such as New Guinea. However, lower temperatures are available, of course, on the high moutains of the main cordillera. Some fungi have been recorded on Mt Wilhelm by Kobayasi and Otani (1971) and Otani (1973), including 4 new species (*Cantharella alpinus*, *Herpotrichiella papuapteridis*, *Hygrocybe alpicola*, *Ompalina novoguineensis*) all collected in the zone 4300–4400 m.

Mention should also be made of aquatic fungi or water moulds. As Sparrow (1968) pointed out, these fungi are to be found not only in pools, ponds, lakes, streams and bogs, but also in muds on lake bottoms and in temporary bodies of water, as well as in marginal and strictly terrestrial habitats. These are the permanent inhabitants, as distinct from the transient terrestrial soil fungi washed into water bodies by rain action. Fungi also exist in the marine environment, both in pelagic and benthic zones as well as in estuaries (Johnson 1968). Within each of the above zones are many ecological niches varying with the mass of the water body and its duration, chemical composition and pH, amount of dissolved oxygen, depth and light, temperature, amount of turbulence, distance from shore, turbidity, substratum character and availability, seasonality and competition with other biota such as algae, etc. Some fungi are planktonic parasites, others parasitize fish, fish eggs and other anphibians, others infect damaged sea creatures, while still others are saprobic, living on the flotsam and jetsam such as dead biota of insects, pollen and other organic debris of the upper water mass, as well as on submerged branches, bark, twigs, fruit, leaves, wood etc. The fungi of the above niches are from all the main divisions of the Fungi, whereas species on the bottom, although diversified, are more limited. Sparrow (1968) pointed out that many aquatic fungi are ubiquitous, and studies by various workers overseas have shown that there is no appreciable difference between the aquatic fungi of the Hemispheres, and that even fungi from Central Pacific atolls have species readily identifiable with continental ones.

As yet, only a little work has been done in New Guinea in recording the fungal species present in its aquatic habitats. Fungi recorded to date by Tubaki (1965), Kobayasi and Konno (1971, 1973) and Shaw (unpublished) show that while a few new species have been described, most appear to be similar to those found in similar habitats overseas. Whether this will be invariable will not be known until many more investigations into the occupants of the various aquatic niches have been carried out.

Another important ecological niche concerning fungi is that of the lichens. These are commonly considered as a distinct group, though each is basically a self-supporting symbiotic association of a fungus and an alga. They are unique in that the resulting life form and its behavior differ markedly from that of the components. In most lichens, the fungal component, which is not recognizably free-living, is responsible for the overall form of the thallus and its fruiting bodies. Many of the algal components can be free living. The fungal components are mostly Ascomycetes, about half the Class being lichenized. Lichens are primarily colonizers in plant successions, occurring in all pioneer terrestrial habitats such as tree trunks, rock, soil, humus, from the world's frigid zones to the tropical areas (James and Hawksworth 1971). Lichens are abundant in New Guinea, many occurring as foliose growths on trees or as appressed colonies on trunks and leaves, and others in diverse situations.

6. Alterations to ecological niches

After his arrival in New Guinea, man at first probably had little effect on ecological niches affecting fungi. Later the felling of trees, burning of vegetation and the development of agricultural activities with more extensive planting of staples, would have brought about some alteration in niches. Shifting agriculture would have made certain niches less favorable for some fungal pathogens by increasing space and time between crops. Selection by indigenous growers of a better type of planting material (which in New Guinea often means undiseased or little diseased stock) is practised, and probably has been practised for some considerable time. This again may lessen the activities of some fungi on such crops. However, as there was no recognition of the causal agents of plant disease, or of their microscopic nature, in many cases fungal propagules would have carried over even on selected planting pieces.

During this century, especially in the last 40 years, man's activities have been bringing about accelerated changes in the New Guinea environment. Man is now deliberately providing niches advantageous to some fungi in New Guinea, for example, by providing green manure for enhanced soil microbiological acitivity and preparing straw beds and other substrata for edible mushroom cultivation. Some activities are deliberately designed to limit the extent of niches, or to favor certain fungi over others. For example, better land management and rotation of crops may decrease certain soil pathogens;

chemical treatments may be efficaceous against some plant and human pathogens and against fungi causing timber, wood chip and paint deterioration. In some cases he has deliberately destroyed the entire ecological niche, for example, by the eradication of all *Coffea* species in the area affected by coffee rust after the 1965 outbreak of *Hemileia vastatrix*.

Other changes carried out for purposes quite unrelated to man's deliberate control or cultivation of fungi include the extensive felling of forest, the introduction of animals such as cattle and deer; extension of acreages of staples due to population growth and of selected plant species such as coconut, coffee, cacao, rubber, tea, oil palm, tobacco, peanuts, shade and ground covers, pasture legumes and grasses, cereals, vegetables and timber species, etc.; storage of water in dams; increased run-off by clearing and road making altering stream flow and composition, and the concentration of sewage effluents at certain points, etc. These changes, affecting land and aquatic environments, would in turn affect fungal species occupying the various niches, either by decreasing or increasing numbers or altering species composition altogether.

7. Discussion

Balgooy (1976) pointed out that the distribution of flora and fauna groups in New Guinea has given rise to much controversy. Phytogeographers place New Guinea in the Indo-Malesian* Floral Region, whereas zoogeographers basing their case on the distribution of mammals and birds, place New Guinea in the Australian Region, while students of invertebrate groups tend to agree with the phytogeographer's standpoint.

Powell (1976) has stated that the natural flora of New Guinea today comprises widespread tropical, Malesian and Australian elements, and that most of the extant families of the mountain areas would have reached New Guinea by the end of the Tertiary, and those of the lowlands somewhat later, in the Quarternary. She considered that if this were so, then diversification and speciation of the flora must have occurred rapidly during the Quaternary to explain the present endemism at both genus and species level.

Balgooy (1976) considered that to evaluate the phytogeographic position of New Guinea we should have a complete list of species growing naturally on the island. To evaluate the mycogeographical position, we would need not only a complete list of naturally-growing species in New Guinea, but also complete lists for adjacent areas, especially for W Malesia and Australia.

Complete lists of fungi are not available as yet for New Guinea or the other areas. Records of species present in New Guinea which have been made (only a few of which are mentioned in this paper), show, however, that many are of worldwide distribution, while others are abundant in the tropics. Of the new genera closely associated with plants, *Brooksia* is still monotypic although *B. tropicalis* is now recorded on plants of many different families in many parts of

*"Malesia" in the botanical sense following van Steenis, comprising Indonesia, Malaysia, Philippines and New Guinea (Paijmans 1968).

the humid tropics. *Oncobasidium*, on the introduced *Theobroma cacao*, is still monotypic, but has now been recorded in Malaysia, also on *T. cacao*, although not recognized to date in the original habitat of the host genus. *Verrucispora* (*V. proteacearum*) is still known only on *Finschia chloroxantha* in New Guinea and on *Hakea florulenta* in Australia (both Proteaceae), but a new species (*V. indica*) has now been described in India on *Smilax* (Smilacaceae). *Shawiella* (*S. grevilleae*) on *Grevillea robusta* (introduced, Proteaceae) and *Stromatothecia* (*S. papuana*) on *Nothofagus* sp. (indigenous, Fagaceae) are still monotypic and known only from New Guinea.

Other pathogens described as new species on hosts of genera some species of which occur naturally in New Guinea, such as *Deightoniella papuana* and *Ramulispora sacchari* on *Saccharum*, *Phyllachora musicola* on *Musa* and *Ramularia oryzae* on *Oryza*, have now been recorded overseas. *Johnstonia colocasiae*, described as new on *Colocasia*, an introduced genus of the Araceae, while now recorded also in the Solomons and W Samoa, has not been recognized in the original habitat of *Colocasia*. Other pathogens described as new on hosts of genera some species of which occur naturally in New Guinea, such as *Phomopsis oryzae* on *Oryza*, *Phyllachora polytocae* on *Polytoca macrophylla*, *Mycovellosiella puerariae* on *Pueraria triloba* (*P. lobata*) and, to mention only one of the rusts, *Puccina finschiae* on the indigenous *Finschia chloroxantha*, have not as yet, to the writer's knowledge, been recorded elsewhere.

The apparent limited distribution of some of the fungi mentioned in the last two paragraphs may reflect the actual occurrence. In others, however, it may merely indicate that even the phytopathological fungi, which are often among the first to be studied (along with the macroscopically visible mushroom and bracket fungi) because of their economic importance, have not been fully catalogued in many parts of the tropical world at least, and of course, the same applies to New Guinea.

As mentioned previously, many species and genera in the diverse groups of fungi have already been recorded in New Guinea. It can already be said that some of these species are world wide in distribution, in some cases at least probably because their propagules are wind blown. In others, for example some aquatics of clear running streams, their world wide distributions are more difficult to explain. Other species are found in tropical habitats elsewhere. Whether New Guinea has also a unique endemic fungal flora will not be known until much more qualitative exploration and quantitative study of fungi in all the ecological niches have been carried out, not only in New Guinea, but in other regions as well.

References

Ainsworth, G. C. 1968. Fungal parasites of vertebrates. In The Fungi. An advanced treatise. Vol. III: The fungal population. Ed. G. C. Ainsworth and A. S. Sussman. Academic Press, New York and London. 738 pp. (pp. 211–226).

Arentz, F. 1976. The occurrence and distribution of Phytophthora species in Papua New Guinea. Science in New Guinea 4: 91–98.

Austwick, P. K. C. and J. W. Copland. 1974. Swamp cancer. Nature 250(5461): 84.

Balgooy, M. M. J. van 1976. Phytogeography. In New Guinea vegetation. Ed. K. Paijmans. Aust. Nat. Univ. Press. 213 pp. (pp. 1–22).
Booth, C. and D. E. Shaw. 1961. Black cross, caused by a new species of Phyllachora on banana. Papua New Guin. agric. J. 13: 157–159.
—— 1967. Anthostomella fusispora sp. nov. and A. cylindrospora sp. nov. on Cocos nucifera. Papua New Guin. agric. J. 19: 94–98.
Campbell, C. H. 1965. A case of sporotrichosis occurring in Papua. Med. J. Aust. 3: 23.
Cartledge, E. G., D. E. Shaw and D. J. Stamps. 1975. Studies in relation to dead patches of Nothofagus in Papua New Guinea. Dep. Agric. Stk Fish., P. Moresby, Res. Bull. No. 13: 1–25.
Champness, L. T. and J. K. A. Clezy. 1962. Torulosis in Papua and New Guinea. Med. J. Aust. April 14, p. 560.
Collins, D., J. H. Beard and P. Still. 1975. Aspergillus flavus and aflatoxin in groundnuts (Arachis hypogaea) from Papua New Guinea. Science New Guin. 3: 145–155.
Cummins, G. B. 1941. Uredinales of New Guinea. 3. Mycologia 33: 143–154.
—— 1961. A new rust on Finschia chloroxantha in New Guinea. Papua New Guin. agric. J. 14: 52.
Deighton, F. C. and K. A. Pirozynski. 1966. Microfungi II. Brooksia and Grallomyces; Acrogenotheca ornata sp. nov.; the genus Xenosporium. Mycol. Pap. 105: 1–35.
Deighton, F. C. and D. E. Shaw. 1960. White leaf-streak of rice caused by Ramularia oryzae sp. nov. Trans. Brit. mycol. Soc. 43: 516–518.
Duddington, C. L. 1968. Fungal parasites of invertebrates. 2. Predacious fungi. In The Fungi. An advanced treatise. Vol. III. The fungal population. Ed. G. C. Ainsworth and A. S. Sussman. Academic Press, New York and London. 738 pp. (pp. 239–251).
Egerton, J. R. and T. L. W. Rothwell 1964. The distribution of infectious and parasitic diseases of animals in Papua and New Guinea. Dep. Agric. Stk Fish., P. Moresby, Res. Bull. 1, Vet. Sci. Ser., 31 pp.
Ellis, M. B. 1959. Clasterosporium and some allied Demiatiaceae — Phragmosporae 2. Mycol. Pap. No. 72.
—— 1971. More dematiaceous Hyphomycetes. Commonw. Agric. Bur. and Commonw. Mycol. Inst., Kew, England. 507 pp.
Emerson, R. 1968. Thermophiles. In The Fungi. An advanced treatise. Vol. III. The fungal population. Ed. G. C. Ainsworth and A. S. Sussman. Academic Press, New York and London. 738 pp. (pp. 105–128).
Frey, D. 1959. Isolation of a new species of Aleurisma from soil in Australia and New Guinea. Mycologia 51: 641–646.
—— 1968. Catalogue of the Australian National Reference Laboratory in Medical Mycology. Inst. Med. Res., The Royal North Shore Hospital, Sydney. 39 pp.
Frey, D. and E. B. Durie. 1964. The isolation of Cryptococcus neoformans (Torula histolytica) from soil in New Guinea and pigeon droppings in Sydney, New South Wales. Med. J. Aust. 1: 947–949.
Frey, D. and D. M. Griffin. 1961. Ctenomyces serratus Eidam. Trans. Brit. mycol. Soc. 44: 449–452.
Greaves, H. 1973. Outside storage of tropical hardwood chips. III. Microbial ecology of chip piles after two and four months storage. Appita 27: 25–30.
—— 1974. Microbiology of wood chips stored under tropical conditions. Aust. Pl. Path. Soc. Newsl. 3: 36–37.
—— 1975. Microbiological aspects of wood chip storage in tropical environments. Aust. J. Biol. Sci., 28: 315–322.
Gregory, P. H. 1961. The microbiology of the atmosphere. Wiley (Interscience), New York. 251 pp.
Gressitt, J. L., J. Sedlacek and J. J. H. Szent-Ivany. 1965. Flora and fauna on backs of large Papuan moss-forest weevils. Science 150: 1833–1835.
Hansford, C. G. 1956. Australian fungi. Proc. Linn. Soc. N.S.W. 81: 23–51.
—— 1957. Australian fungi. IV. New records and revisions (cont.) Proc. Linn. Soc. N.S.W. 82: 200–229.
—— 1961. The Meliolineae. A monograph. Sydowia Beih. 2.
Harley, J. L. 1968. Mycorrhiza. In The Fungi. An advanced treatise. Vol. III. The fungi population. Ed. G. C. Ainsworth and A. S. Sussman. Academic Press, New York and London. 738 pp. (pp. 139–178).
Hart, D. 1970. Rainfall. In An atlas of Papua and New Guinea. Ed. R. G. Ward and D. A. M. Lea. Uni. Papua New Guin. and CollinLongman. 101 pp. (pp. 42–45).
Ingold, C. T. 1971. Fungal spores: their liberation and dispersal. Clarendon Press, Oxford. 302 pp.
James, P. W. and D. L. Hawksworth. 1971. Lichenes. In Dictionary of the fungi. Ed. G. C. Ainsworth. Commonw. Mycol. Inst., Kew, England. 663 pp.
Johnson, T. W., Jr. 1968. Saprobic marine fungi. In The fungi. An advanced treatise. Vol. III. The fungal population. Ed. G. C. Ainsworth and A. S. Sussman. Academic Press, New York and London. 738 pp. (pp. 95–104).

Johnston, A. 1961. A preliminary plant disease survey in Netherlands New Guinea. Landbouwkundige serie No. 4. 55 pp.
Kamal and Singh, R. P. 1978. A new species of Verrucispora (Hyphomycetes) from India. Can. J. Bot. 56: 2785–2786.
Keane, P. J. and P. D. Turner. 1972. Vascular-streak dieback of cocoa in West Malaysia. In Cocoa and coconuts in Malaysia. Ed. R. L. Wastie and D. A. Earp. The Incorporated Society of Planters, Kuala Lumpur. pp. 50–57.
Kobayasi, Y. and K. Konno. 1971. Lower Phycomycetes including Glaziella. Bull. Nat. Sci. Mus. Tokyo 14: 373–386.
—— 1973. Some additional water moulds from soil. Bull. Nat. Sci. Mus. Tokyo 16: 497–502.
Kobayasi, Y. and Y. Otani. 1971. Fungal flora of the alpine region of Mt Wilhelm. 1. Bull. Nat. Sci. Mus. Tokyo 14: 484–500.
Kobayasi, Y. and D. Shimizu. 1976. The genus Cordyceps and its allies from New Guinea. Bull. Nat. Sci. Mus. Ser. B. (Bot.) 2: 133–151.
McAlpine, J. R., G. Keig and K. Short. 1975. Climatic tables for Papua New Guinea. Commonw. Sci. Ind. Res. Org., Div. Land Use Res. Tech. Pap. No. 37. 177 pp.
MacLennan, R. and M. O'Keefe. 1975. Altitude and prevalence of tinea imbricata in New Guinea. Trans. Roy. Soc. Trop. Med. Hyg. 69: 91–93. (Abstr. Rev. Med. Vet. Myc., 10, Item No. 1687).
Madelin, M. F. 1968. Fungal parasites of invertebrates. 1. Entomogeneous fungi. In The Fungi. An advanced treatise. Vol. III. The fungal population. Ed. G. C. Ainsworth and A. S. Sussman. Academic Press, New York and London. 738 pp. (pp. 227–238).
Matsushima, T. 1971. Microfungi of the Solomon Islands and Papua-New Guinea. Published by the Author. Kobe. 78 pp. + figs. and plates.
Morahan, R. J. 1970. Microsporum gypseum in soil and skin lesions in New Guinea. Papua New Guin. Med. J. 13: 61.
Otani, Y. 1973. Fungal flora of the alpine region of Mt Wilhelm. 2. Bull. Nat. Sci. Mus. Tokyo 16: 477–492.
Paijmans, K. 1976. Vegetation. In New Guinea vegetation. Ed. K. Paijmans. Aust. Nat. Univer. Press. 213 pp. (pp. 23–105).
Park, D. 1968. The ecology of terrestrial fungi. In The Fungi. An advanced treatise. Vol. III. The fungal population. Ed. G. C. Ainsworth and A. S. Sussman. Academic Press, New York and London. 738 pp. (pp. 5–39).
Perkins, D. E., B. C. Turner and E. G. Barry. 1976. Strains of Neurospora collected from nature. Evolution 30: 281–313.
Peters, W. 1960. Tinea imbricata and malnutrition. Trans. Roy. Soc. Trop. Med. Hyg. 54: 197–198.
Powell, J. M. 1976. Ethnobotany. In New Guinea vegetation. Ed. K. Paijmans. Aust. Nat. Univ. Press. 213 pp. (pp. 106–199).
Price, T. V. 1977. Powdery mildew: a new disease of winged bean (Psophocarpus tetragonolobus) in Papua New Guinea. Pl. Dis. Reptr 61: 384–385.
Schofield, F. D., A. D. Parkinson and D. Jeffrey. 1963. Observations on the epidemiology, effects and treatment of tinea imbricata. Trans. Roy. Soc. Trop. Med. Hyg. 57: 214–227.
Shaw, D. E. 1956. A Deightoniella disease of Saccharum species. Papua New Guin. agric. J. 11: 77–81.
—— 1963a. Plant pathogens and other microorganisms in Papua and New Guinea. Dep. Agric. Stk Fish., P. Moresby, Res. Bull. 1: 1–78.
—— 1963b. Annotated list of references to plant pathogens and miscellaneous fungi in West New Guinea. Dep. Agric. Stk Fish., P. Moresby, Res. Bull. 1: 79–82.
—— 1968. Coffee rust outbreaks in Papua from 1892 to 1965 and the 1965 eradication campaign. Dep. Agric. Stk Fish., P. Moresby, Res. Bull., Pl. Path. Ser. 2: 20–52.
—— 1978. The plant quarantine barrier and some breaches in Papua New Guinea. Papua New Guin. agric. J. 27: 121–135.
Shaw, D. E. and J. L. Alcorn. 1967. The genus Verrucispora gen. nov. (Fungi Imperfecti) on Proteaceae in New Guinea and Queensland. Proc. Linn. Soc. N.S.W. 92: 171–173.
Shaw, D.E., J. S. Brown and A. Cunningham. 1972. Fungi isolated from stored and waste foodstuffs in Papua New Guinea. Papua New Guin. agric. J. 15: 108–112.
Shaw, D. E., G. E. Cartledge and D. J. Stamps. 1972. First records of Phytopthora cinnamomi in Papua New Guinea. Papua New Guin. agric. J. 23: 46–48.
Shaw, D. E. and F. C. Deighton. 1970. Yellow leaf mould of Pueraria lobata caused by Mycovellosiella puerariae sp. nov. Trans. Brit. mycol. Soc. 54: 326–330.
Shaw, D. E. and D. L. Hawksworth. 1971. Stromatothecia, a new genus and species of scolecospored Ascomycetes from Nothofagus in New Guinea. Papua New Guin. Sci. Soc. Proc. 1970 22: 21–25.

Shaw, D. E. and G. Kula. 1978. A disease of some legumes in Papua New Guinea caused by Sclerotium sp. Papua New Guin. agric. J. 29: 5–10.

Simpson, J. A. 1978. Heterobasidion annosum on Araucaria cunninghamii in Papua New Guinea. Pl. Dis. Reptr. 62: 404–405.

Silvanesan, A., D. E. Shaw and J. S. Brown. 1976. Leiosphaerella longispora sp. nov. on coconut petiole in Papua New Guinea. Trans. Brit. mycol. Soc. 67: 529–532.

Soneda, M. and S. Uchida. 1971. A survey on the yeasts. Bull. Nat. Sci. Mus. Tokyo 14: 438–459.

Sparrow, F. K. Jr. 1968. Ecology of freshwater fungi. In The Fungi. An advanced treatise. Vol. III. The fungal population. Ed. G. C. Ainsworth and A. S. Sussman. Academic Press, New York and London. 738 pp. (pp. 41–93).

Stamps, D. J., D. E. Shaw and E. G. Cartledge. 1972. Species of Phytophthora and Pythium in Papua New Guinea. Papua New Guin. agric. J. 23: 41–45.

Still, P. and P. E. Munro. 1978. Aflatoxin and wing beans in Papua New Guinea. In Programme and abstracts. International Conference "Mycotoxins", August 14–15, 1978, Munich. pp. 25–26.

Stover, R. H. 1972. Banana, plantain and abaca diseases. Commonw. Mycol. Inst. Kew, England. 316 pp.

Sutton, B. 1975. Eucalyptus microfungi. Satchmopsis gen. nov., and new species of Coniella, Coniothryrium and Harknessia. Nova Hedwigia 26: 1–16.

Sussman, A. S. and H. O. Halvorson. 1966. Spores: their dormancy and germination. Harper and Row, New York and London. 354 pp.

Talbot, P. H. P. and P. J. Keane. 1971. Oncobasidium: a new genus of Tulasnelloid fungi. Aust. J. Bot. 19: 203–206.

Tubaki, K. 1965. Short note on aquatic spora in East New Guinea. Trans. Myc. Soc. Japan 6: 11–14.

Udagawa, S. and Y. Horie. 1971. Isolation of mushroom-inhabiting fungi. Bull. Nat. Sci. Mus. Tokyo 14: 516–543.

Udagawo, S. and M. Takada. 1971. Soil and coprophilous microfungi. Bull. Nat. Sci. Mus. Tokyo 14: 501–515.

Wookey, S. J. 1972. Aflatoxin in foods grown in Papua New Guinea. Papua New Guin. Med. J. 15: 113–116.

11. Ecology and distribution of the seagrasses

I. M. Johnstone

1. Introduction

The history of seagrass studies in New Guinea is brief. O. Beccari in the 1870s collected 8 species from Irian Jayan waters, 7 from Sorong at the western end of the Vogelkop and 1 from Warbusi in Geelvink Bay on the N coast (den Hartog 1970). During the 1950s and '60s the distribution records for these 8 species were extended by a number of collectors in both Irian Jaya and Papua New Guinea, and 1 new species was added to the known seagrass flora for New Guinea (den Hartog 1970).

The visit of C. den Hartog to the Torres Strait in 1967 could be considered the turning point in the history of New Guinea seagrass exploration with his discovery of 12 seagrass species at Thursday I, making this locality one of the most species-rich seagrass habitats in the world, possibly only rivaled by the diverse meadows of Western Australia where temperate and tropical species mix. My recent unpublished surveys have shown that this high species diversity is not restricted to the Australian islands of the Torres Strait, but extends across the Strait to the New Guinea mainland. These species-rich seagrass meadows in the Torres Strait are of considerable scientific interest, as they will allow the testing of hypotheses relating to community structure and function, which were developed on species-poor systems in the temperate zone.

In recent years distributional data has continued to be collected for Papua New Guinea (Johnstone 1975, 1976a, 1978a), and a start has been made, at least in Papua New Guinea, on describing the ecological structure and function of the New Guinea seagrass ecosystems (Johnstone 1976b, 1978b, 1979a). At present it appears that there is no active seagrass related research in Irian Jaya and, thus, data for the W half of New Guinea remains inadequate.

Although the role of mangroves in the secondary productivity of coastal waters has been acknowledged since the pioneering work in Florida (Odum and Heald 1972), the role of seagrasses, particularly in the tropics, remains to a great extent unappreciated. The extensive areas of seagrass meadows around the New Guinea coast coupled with their continuous leaf biomass production (see section 5) suggests that seagrasses are as important detritus producers as mangroves in this part of the world. In view of the small research effort expended to date, the present paper must be considered primarily as an

Monographiae Biologicae, Vol. 42, ed. by J. L. Gressitt.
© *1982, Dr W. Junk Publishers, The Hague.*

introduction and invitation to future research, rather than as the final chapter in our understanding of the New Guinea seagrass story.

2. General distribution

Information on the distribution of seagrasses around New Guinea is constantly being updated. The presently known distributions are given in Table 1. The main localities mentioned in Table 1 and the text are shown in Fig. 1.

For some species it is difficult to know whether the information in Table 1 gives a relatively true picture of their distribution. For example, *Halophila spinulosa* is frequently overlooked by "low tide collectors" and by collectors using only snorkle gear because, except in habitats where the water has a low transparency (e.g., Torres Strait), this species is found only at depths greater than 3 m below ELWS and thus it is normally only discovered by scuba divers or by dredging. However, for most of the species, although many gaps exist (particularly for Irian Jaya), certain distribution patterns are evident.

Cymodocea rotundata, *C. serrulata*, *Enhalus acoroides*, *Halodule uninervis* (narrow and wide leaved forms), *Halophila ovalis*, *Haloph. ovata*, *Syringodium isoetifolium* and *Thalassia hemprichii* are all widely distributed. *Thalassodendron ciliatum* has a peculiar disjunct distribution — Sorong (Irian Jaya), Thursday I. (Queensland), Urasi Cove, Malaita I. (Solomons) and Seeadler Harbor, Manus (Papua New Guinea) (Fig. 2). The absence of this plant from many New Guinea shores may result from the lack of suitable habitat (rocky or hard coral bottom and no freshwater influence), but it would seem to be only a matter of time before more, at least isolated, populations are found. *Zostera capricorni* is known only from Daru (where it is relatively abundant). This is the only *Zostera* record for the whole of Malaysia (den Hartog pers. comm.). I have found no evidence of its presence east of Daru, but isolated populations may occur to the west, possibly as far as Frederik Hendrik I. (Fig. 2). *Halophila decipiens* is known only from the S coast (Fig. 2), while *Halodule pinifolia* has been collected from 2 localities on the S coast of New Guinea and also from the S coast of Manus I. Although both of these species are easily overlooked they are certainly absent from most Papua New Guinea seagrass meadows. *Halophila spinulosa*, as previously mentioned, is easily overlooked by collectors. At present this plant is only known from Warbusi on the N coast and Daru, and Daugo and Lolorua islands on the S coast, but future surveys will almost certainly show it to have a much more continuous distribution around the coast of New Guinea. *Cymodocea* sp. has been collected from the Fly Is and Tufi on the Solomon Sea coast of Papua New Guinea and Lolorua I. (between 5 and 10 m below ELWS) on the S coast. A similar form has also been collected from Thursday I. in the Torres Strait (den Hartog, pers. comm.), and it occurs at Groote I. in the Gulf of Carpentaria (pers. observation).

Biogeographical relationships of New Guinea seagrasses are all to the west with the Indian Ocean — SE Asian region except for *Zostera capricorni* whose population is centered to the S of New Guinea. It is tempting to consider the Torres Strait as a center of origin for seagrasses since it is the area of greatest

Table 1. Distribution of New Guinea seagrasses

Species	Irian Jaya[4]					Papua New Guinea[5]								
	Aru	Sorong	Warbusi	Batanta I.	Salawatti Japen I. Jayapura	Warrior Reefs	Daru	Central Prov.	Milne Bay Prov.	Northern Prov. (Tufi)	Fly Is	Rai Coast	Wewak and Raboin I.	Manus Prov.
Potamogetonaceae														
Cymodocea rotundata Ehrenb. and Hempr. ex Aschers	×	×					×	×	×	×	×		×	×
Cymodocea serrulata (R.Br.) Aschers. and Magnus	×	×					×	×	×	×		×	×	×
Cymodocea sp.[1]										×	×			
Halodule pinifolia (Miki) den Hartog														
normal form							×							×
long leaved form								×						
Halodule uninervis (Forsk.) Aschers.[2]	×	×												
narrow leaved form							×	×	×	×	×		×	×
wide leaved form						×	×	×	×	×	×	×	×	×
Syringodium isoetifolium (Aschers.) Dandy	×	×		×			×	×	×	×	×	×	×	×
Thalassodendron ciliatum (Forsk.) den Hartog		×												×
Zostera capricorni Aschers.[3]							×							
Hydrocharitaceae														
Enhalus acoroides (L.f.) Royle	×	×			×	×	×	×	×	×	×	×	×	×
Halophila decipiens Ostenfeld	×						×							
Halophila ovalis (R.Br.) Hook.	×	×				×	×	×	×	×	×		×	×
Halophila ovata Gaud.							×	×	×	×	×		×	×
Halophila spinulosa (R.Br.)Aschers			×				×	×						
Thalassia hemprichii (Ehremb.) Aschers	×	×				×	×	×	×	×	×	×	×	×

1. This *Cymodocea*, misidentified as *T. ciliatum* by Johnstone (1978a,b), is phenotypically the same as *C. serrulata* except that the leaves are borne on erect branches that are up to 50 cm long.
2. The narrow and wide leaved forms of *H. uninervis* are treated separately (except where the form is unknown) by me since they have separate distributions within and between communities. Their status as separate species clearly is still open to question.
3. The Daru *Zostera* is vegetatively identical to *Z. capricorni* except that the leaf tips of the Daru material are mucronate. Flowering material has not been found.
4. From den Hartog (1970).
5. Author's records, material housed in the U.P.N.G. herbarium (Tami Is collections by B. Johns, Bulolo Forestry College Herbarium).

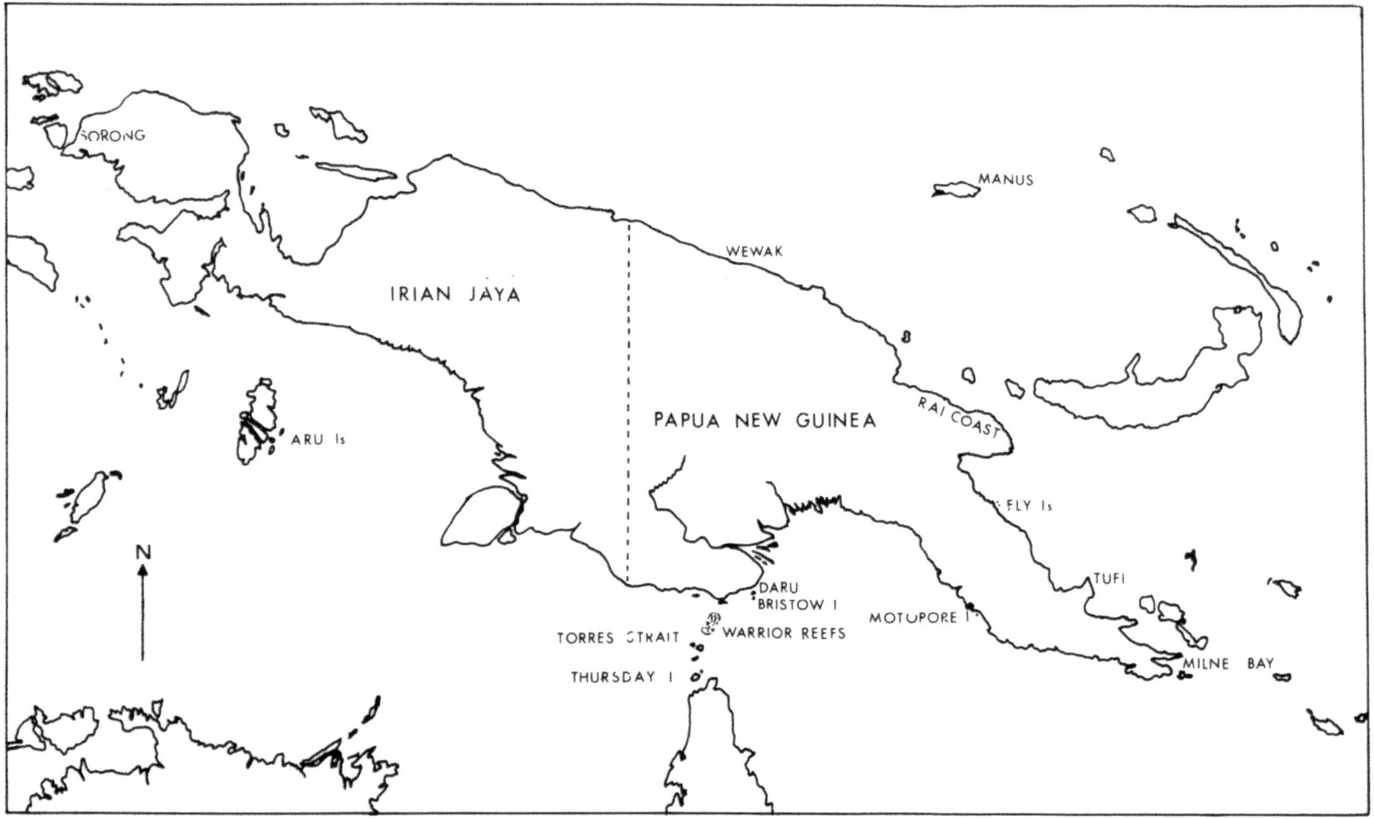

Fig. 1. New Guinea showing the main localities mentioned in the text.

species diversity, but this seems an unlikely proposition for several reasons. First, the Torres Strait is the present day geographical limit for at least one species (*Z. capricorni*). Secondly, my field observations suggest that interspecific competition in the most highly mixed stands at Daru is severe because population densities are low compared with other areas. This interspecific competition should, theoretically, lead to extinctions rather than speciation. That the competition can be tolerated must in part be a result of optimal environmental conditions in the area. In addition, continued coexistence of the Torres Strait assemblage may, in fact, depend on the maintenance of a disequilibrium caused by the high levels of dugong grazing which may prevent any particular seagrass from becoming dominant. Thirdly, the earliest known fossil records of seagrasses and their associated foraminifera are not from Australian-New Guinea deposits (Brasier 1975). The Torres Strait, therefore, appears to be an area which is within the dispersal range of most extant tropical Indo-Pacific seagrasses and is ecologically conducive to their coexistence.

Recent reports for the more remote Pacific islands enable one to consider the biogeography east of New Guinea (Fig. 3). The decrease in species diversity is similar to the classic area-diversity and distance-diversity patterns found as one

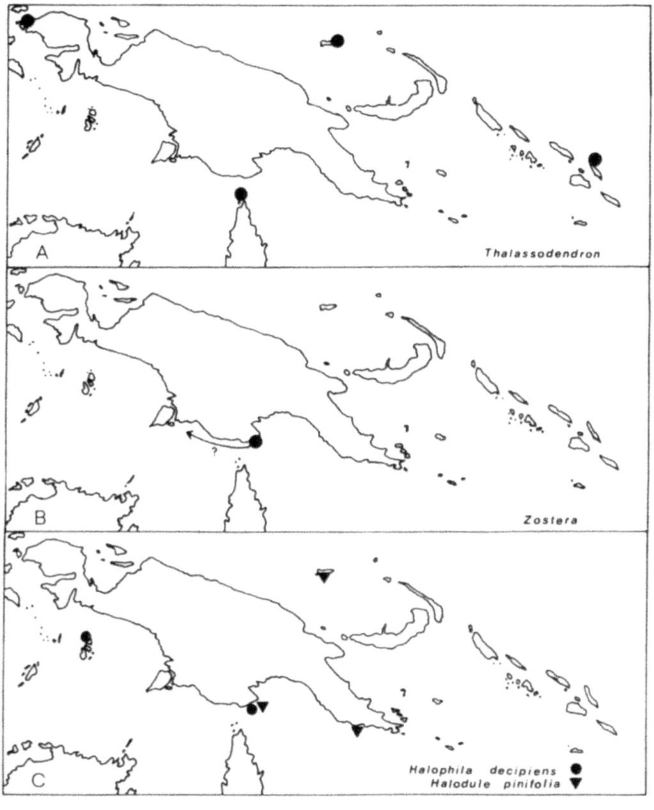

Fig. 2. New Guinea seagrass distributions. A: *T. ciliatum*; B: *Z. capricorni*; C: *H. decipiens* and *H. pinifolia*.

moves east of New Guinea as, for example, in birds (Diamond 1974) and weevils (Simpson 1965, p. 88; after Zimmerman).

Seagrass distributions in the Pacific do not appear to be stable as suggested by the evidently recent arrival of *Halodule uninervis* in Fiji (Ryland pers. comm.) and *Syringodium isoetifolium* at Truk (Koch and Tsuda 1978). Whether these are natural or man-assisted colonizations is unknown, but the colonization pattern of the Mediterranean, following the opening of the Suez Canal, by *Halophila stipulacea* strongly suggests that seagrass may be dispersed by ships (Lipkin 1975).

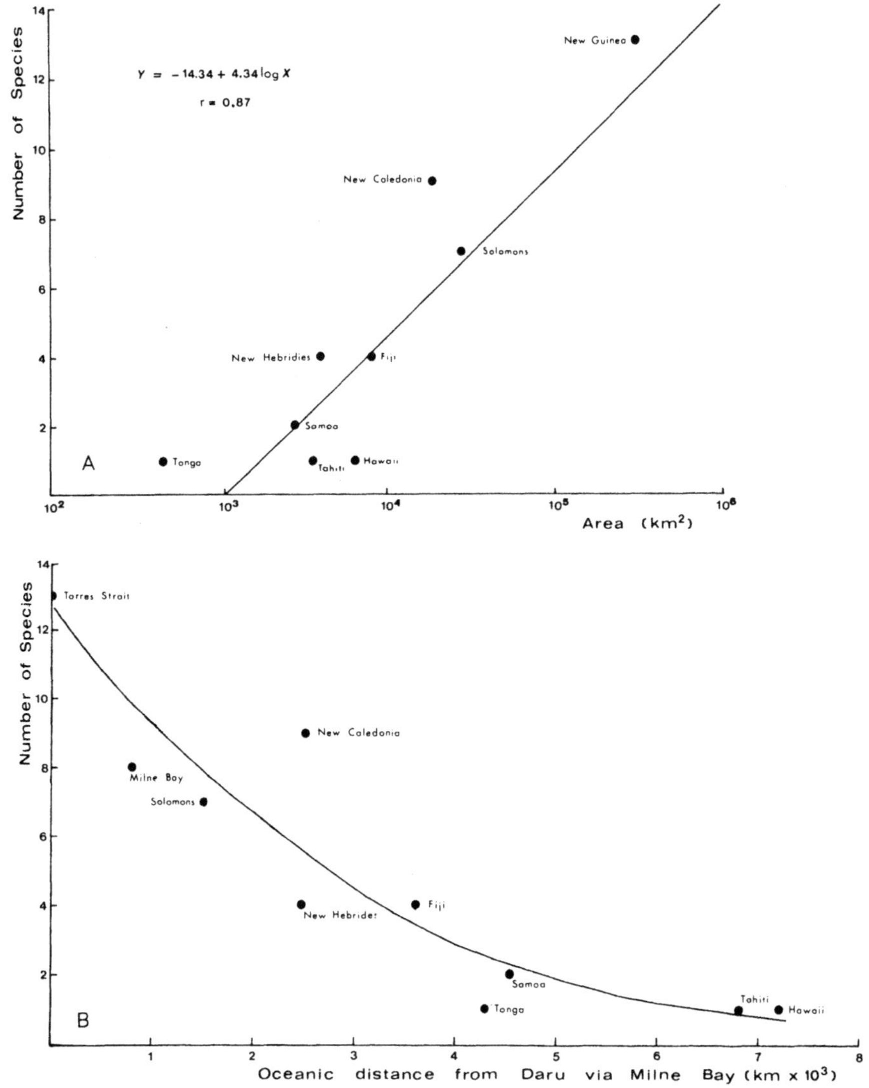

Fig. 3. Seagrass biogeographical relationships in the Pacific Ocean. A: area-diversity relationship; B: distance-diversity relationship. The curve in B was drawn in by eye. Non-New Guinea data from den Hartog (1970), Fosberg (1976), Parham (1976), Ryland (pers. comm.), Taylor (1978) and Womersley and Bailey (1970).

3. Zonation patterns

Zonation patterns of algae and seagrass in the littoral zone on New Guinea shores are considerably less obvious to the casual observer than on temperate shores. The main reason for this inconspicuous zonation is the almost complete absence of larger algae and seagrass in that part of the littoral uncovered by most tides (i.e., on the Papuan Coast, higher than about 0.5 m above ELWS; 0.6 m chart datum). On rocky shores in the temperate zone, algae in particular are found at levels subject to daily exposure and are zoned, to a large extent, on their ability to withstand desiccation. The general absence of seagrasses on New Guinea shores from that area of the littoral uncovered by most tides appears to be due to an inability of these plants to withstand exposure and the associated desiccation. During spring lows the exposed foliage of the seagrasses in the lower littoral dies if the low is during midday hours and it is a sunny day. The relationship between the number of daytime exposures experienced in a year and the upper limit of the seagrasses is shown in Table 2. The absence of a seagrass zonation based on exposure tolerance does not, however, mean an absence of zonation. Distributional analysis shows that in the lower littoral (i.e., below 0.6 m on the Papuan coast) and in the adjacent sublittoral the various seagrass species are present in relatively well defined zones. The lack of color and other phenotypic differences between seagrass species makes the zonation pattern relatively inconspicuous at most localities. Although direct experimental evidence has not been published, I believe that the zonation of seagrasses is mainly related to light intensity (i.e., water clarity and depth). For example, the main difference between zonation at Daru (muddy substrate) where underwater visibility is usually less than 3 m and Daugo I. (coral sand substrate) where visibility is usually more than 10 m is that the corresponding seagrass zones are simply at a shallower depth at Daru (Fig. 4).

For most New Guinea seagrass meadows the community can be divided into

Table 2. Tidal exposures with depth in the lower littoral, during 1979, at Motupore I. (from Port Moresby tide tables).

Tide height (m) (chart datum)	Exposures during hours 07:00–17:00	Cumulative total exposures hours 07:00–17:00	Exposures during hours 17:00–07:00	Seagrass upper limit
1.0	53	219	43	
0.9	32	166	44	
0.8	32	134	39	*H. uninervis* (narrow)
0.7	26	102	37	
0.6	35	76	29	*C. rotundata* *T. hemprichii* *E. acoroides*
0.5	22	41	18	*S. isoetifolium* *H. uninervis* (wide)
0.4	5	19	17	
0.3	7	14	14	*C. serrulata*
0.2	6	7	5	
0.1	1	1	1	
0.0	0	0	0	

4 main zones. In the following discussion these zones are named by reference to the most abundant species in the zone.

Zone 1: Halodule uninervis (narrow leaved) zone. The narrow leaved form of *H. uninervis* forms a pure, but sparse, stand in a number of localities. This zone is considerably above (with respect to tidal hight) the main seagrass meadow and it can be easily overlooked because it is frequently some distince away from the edge of the main seagrass meadow. A good example of this zone was at Ela Beach, Port Moresby, but this monospecific zone has recently completely disappeared from this beach. The reason for this disappearance is probably the heavy recreational use and mechanical cleaning operations that Ela Beach has been subject to in recent years. At Motupore I. the *H. uninervis* zone is found between 0.7 and 0.9 m above ELWS. The zone is only found where the sand is relatively stable and therefore is not included in the zonation diagram for Motupore (Johnstone 1979a, Fig. 1) which was drawn from a transect at the tip of a mobile sandspit. On the north coast at the Fly Is *H. uninervis* is replaced in this habitat by *Halophila ovata* (Johnstone 1978b).

Zone 2A: Cymodocea rotundata zone. This zone forms the upper fringe of the main seagrass bed. It is usually only 2 or 3 m wide, but on coral reef flats it can be several hundred m wide (e.g. Bristow I., Western Prov.). The main species associated with *C. rotundata* is *Halodule uninervis* (wide and narrow leaved forms), while *Syringodium isoetifolium*, *Halophila ovata*, *Haloph. ovalis*, *Thalassia hemprichii* and *Enhalus acoroides* are all usually present in small numbers.

Zone 2B: Halophila ovata-ovalis zone. In areas where scouring or active sand movement is taking place the landward edge of the *C. rotundata* zone may be replaced by stands of *H. ovata* and /or *H. ovalis* with only occasional plants of *C. rotundata* making up the assemblage.

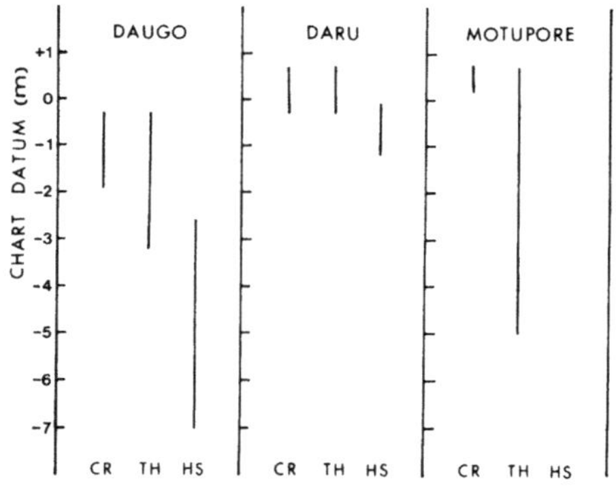

Fig. 4. Zonation comparison of 3 "zone indicator" species at Daugo, Daru and Motupore islands. The depth limits at Daru are considerably shallower than those at Daugo. At Daugo (N end of island) the seagrasses do not extend above −0.3 m (chart datum) due to wave generated beach erosion. Although the depth range is accurate for Daru, the absolute chart datum heights are possibly slightly higher or lower as no tide tables exist for Daru and the heights were estimated from the Port Moresby/Thursday I. tables. CR = *C. rotundata*; TH = *T. hemprichii*; HS = *H. spinulosa*.

Zone 3: Enhalus-Thalassia zone. This zone usually constitutes the bulk of the seagrass meadow and ranges from a few meters to several thousand m in width depending on the local topography. Both of the dominant species may be present (e.g., Motopure I.) or one may be absent. For example, *E. acoroides* is not present at Raboin I. (Johnstone 1978b) and Daugo I., while *T. hemprichii* is absent from the meadow at the Tufi wharf beach and from part of the meadow at Sehulea, Normanby I. (B. Hudson pers. comm.). The presence of other species in this zone is quite variable. Usually *Halophila ovalis*, *Halodule uninervis* (wide leaved form) and, where the substrate is sandy rather than muddy, *Syringodium isoetifolium* are present. *Cymodocea serrulata* is sometimes present in small numbers, but it can, especially when *Enhalus* is absent, contribute up to about 30% of the leaf area of this zone. At Daru *Halodule pinifolia*, *Halophila decipiens* and *Zostera capricorni* are also found in this zone.

Zone 4: Halophila spinulosa zone. This, the deepest zone, is relatively distinct and forms an open community growing below the deepest *Thalassia* or *Enhalus* plants. Only two other species (*C. serrulata* and *H. uninervis*-wide leaved form) are commonly found in this zone, while *H. ovalis* is sometimes found here, particularly if *H. spinulosa* is absent. All 4 species are usually not present at any one locality. In some localities (e.g. Loloata I.) the *H. spinulosa* assemblage is replaced by a monospecific meadow of *H. decipiens*.

The zones described above are for shores where there is adequate sand or mud for the seagrasses to remain anchored. *Thalassodendron ciliatum* can not be placed in any of the above zones because it normally forms monospecific stands on rocky or coral outcrops where other seagrass species can not invade (e.g., Paradise Beach, Manus; Fig. 5). At sites where a layer of sand overlies the rock, however, *T. ciliatum*, can occur in association with other seagrass species, notably *T. hemprichii*, *E. acoroides*, *C. serrulata* and *H. ovalis*. Such habitats occur both at Manus and Thursday I.

Exceptions to the zonation pattern described are fairly easy to discover. For example, on the weather coast of Bristow I. (Western Prov.) the seagrass growing closest to the shore, at the interface of the seagrass/mangrove communities, is *C. serrulata*. This shift from its normal position in the zonation pattern is due to the presence of tidal pools at the edge of the mangrove swamp in which the *C. serrulata* is growing.

In general the distribution of seagrass species with depth of New Guinea

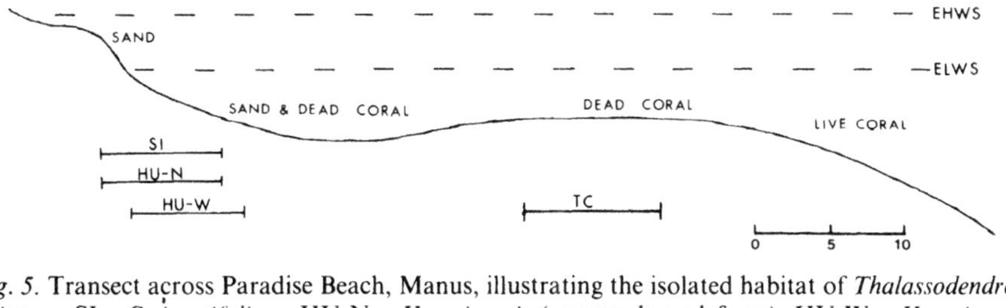

Fig. 5. Transect across Paradise Beach, Manus, illustrating the isolated habitat of *Thalassodendron ciliatum*. SI = *S. isoetifolium*; HU-N = *H. uninervis* (narrow leaved form); HU-W = *H. uninervis* (wide leaved form); TC = *T. ciliatum*. After Green and Sander (1979).

shores fits the generalized zonation diagram for seagrasses proposed by den Hartog (1977).

4. Community structure

The presence of zonation in New Guinea seagrass meadows implies at least a certain level of plant community structure. The literature on tropical seagrass communities provides no information as to the structure of these systems except for recording the particular combinations of seagrasses found at various sites. This present account, based mainly on the author's unpublished observations, is an attempt to provide a framework for future studies.

Because there is little evidence for an interaction in either time or space between the seagrass zones that have already been described, to attempt to relate them in a unifying theme clearly has little justification. However, in parallel with the work of ecologists on other systems one can apply terminology relating to succession. Such an approach, implying a time relationship in seagrass meadows, also fits in with the community concepts developed by C. P. McRoy and members of the SES group to explain the variations in chemical and physical parameters that these workers have observed in the species-poor seagrass communities of the north temperate zone (McRoy pers. comm.). If this is done one could call zones 1, 2A and 2B (as described in section 3) a "colonizing community", zone 3 a "climax community" and zone 4 a "senescent community".

A structural-functional classification of New Guinea seagrass meadows, whether based on succession or some other community process, clearly will only be justified when many meadows have been studied in detail. Many difficulties exist with a successional classification. For example, at Motupore I. the "senescent community" of *Halophila spinulosa* and its associated species is not present. Instead there is an open meadow of rather sick-looking *Thalassia* plants. Does this mean that the senescent phase is absent or that it is present and that, simply, the composition of the phase is different at Motupore Island? In bays where there is considerable freshwater influence it is common to find pure stands of *Enhalus*. Does this mean that in such situations *Enhalus* is a colonizer, climax species and senescent species depending on its position in the community? Do seagrass communities colonize landward or seaward? If seaward, then the *H. spinulosa* community becomes the "colonizing community" and zones 1 and 2 would be the "senescent communities". There is no evidence which could help clarify any of these problems at this time, nor is it impossible that *both* the landward and seaward fringes of the seagrass meadow are "colonizing."

If a successional approach is at all valid for New Guinea seagrass communities, then various community parameters should be correlated with the main stages. Limited data (Fig. 6) reveal that two measures of community structure, leaf area index and species diversity (as species number) are indeed least in the seaward and landward fringes and greatest at the center of the meadow. It is of some interest that the maximum leaf area index in both of the communities shown in Fig. 6 (max. LAI 2.2 Motupure; max. LAI 3.3 Daugo) are relatively low compared to figures reported for seagrass communities elsewhere (McRoy and McMillan 1977, Table 3).

5. Trophic relationships

The role of the seagrass community as an energy resource for the New Guinea shallow water coastal consumer community has two main aspects — the grazing food chain and the detritus food chain. The relative importance of these seagrass-based food chains depends both on the productivity of the seagrasses and the presence and abundance of organisms capable of exploiting this resource.

Primary productivity on a community basis has not as yet been determined for any New Guinea seagrass meadow. The only productivity data available is that for *Enhalus* (Johnstone 1979a). However, in the absence of any other data, the *Enhalus* leaf growth rates (net leaf biomass production) can be used to gain a rough idea of the productivity of the Motupore sandbar meadow if one assumes the productivity of the other species in the meadow is roughly similar to that of *Enhalus* on a leaf area basis. *Enhalus* comprises approximately 50% of the leaf area of the Motupore meadow (determined from the individual leaf area data used to construct Fig. 6). On the basis of this 50% the community productivity could be expected to be double that of *Enhalus* alone. This extrapolation gives a net productivity value of 0.64 g (epiphyte-free, ash-free, dry weight of leaf material)/m^2/day. Assuming this ash-free component is carbohydrate, this gives a value of 0.25 g C/m^2/day or 93.4 g C/m^2/year (i.e., using 0.4 as a conversion factor). The annual figure is derived by multiplying the daily rate by 365 because *Enhalus*, and almost certainly the other seagrasses of Motupure I., has a 365 day growing season (Fig. 7) in contrast to McRoy and McMillan's (1977, p. 73) suggestion that growth does not occur at all seasons in tropical waters.

This computed community primary productivity for the Motupore meadow is very low compared with levels reported even for single-species studies for other areas of the world, in which values of up to 16 g C/m^2/day have been given (Phillips 1978). The only relevant Indo-Pacific data available for comparison seems to be that for a *Thalassia*/*Cymodocea* bed at Kavaratti Atoll (Laccadives) for which a net production of 5.81 g C/m^2/day was computed from diurnal curves (Qasim and Bhattathiri 1971). Whether or not New Guinea seagrasses are remarkably unproductive in comparison to other areas with similar species or whether differences are due to methodology (leaf measurements vs oxygen production) clearly needs to be clarified.

Although 0.25 g C/m^2/day is only a quarter the productivity of a crop like corn (Phillips 1978) seagrasses on the New Guinea coast are still a significant energy resource. Furthermore, in comparison with mangroves, although the standing crop of seagrass is far less this does not necessarily mean that their contribution to the food supply is insignificant. This is because the turnover rate is probably the critical factor in determining the importance of primary producers (see Minshall 1978). A new *Enhalus* leaf is initiated every 27 days (my

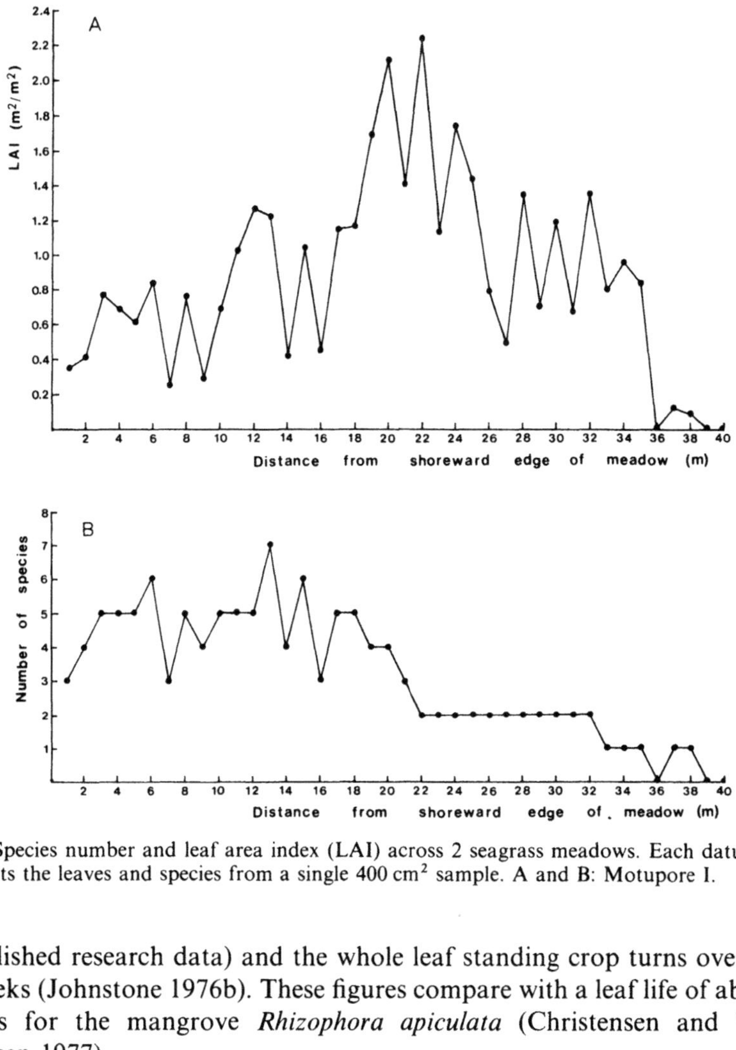

Fig. 6. Species number and leaf area index (LAI) across 2 seagrass meadows. Each datum point represents the leaves and species from a single 400 cm^2 sample. A and B: Motupore I.

unpublished research data) and the whole leaf standing crop turns over every 7.5 weeks (Johnstone 1976b). These figures compare with a leaf life of about 18 months for the mangrove *Rhizophora apiculata* (Christensen and Wium-Andersen 1977).

At Motupore as well as in most other New Guinea seagrass meadows familiar to me there is evidence for only minor amounts of direct grazing. One or 2 semi-circular bites, 0.5 cm^2 in area, which are probably due to parrotfish, are frequently found at the tips of *Enhalus* and *Thalassia* leaves. Both *Halophila ovalis* and *Haloph. ovata* frequently exhibit leaf damage; however, the organism responsible has not been identified. Other seagrass species usually show little or no leaf damage that could be attributed to herbivores.

Direct leaf consumption by herbivores, therefore, in New Guinea waters, seems to be mainly effected by turtles and dugongs. An instance has been

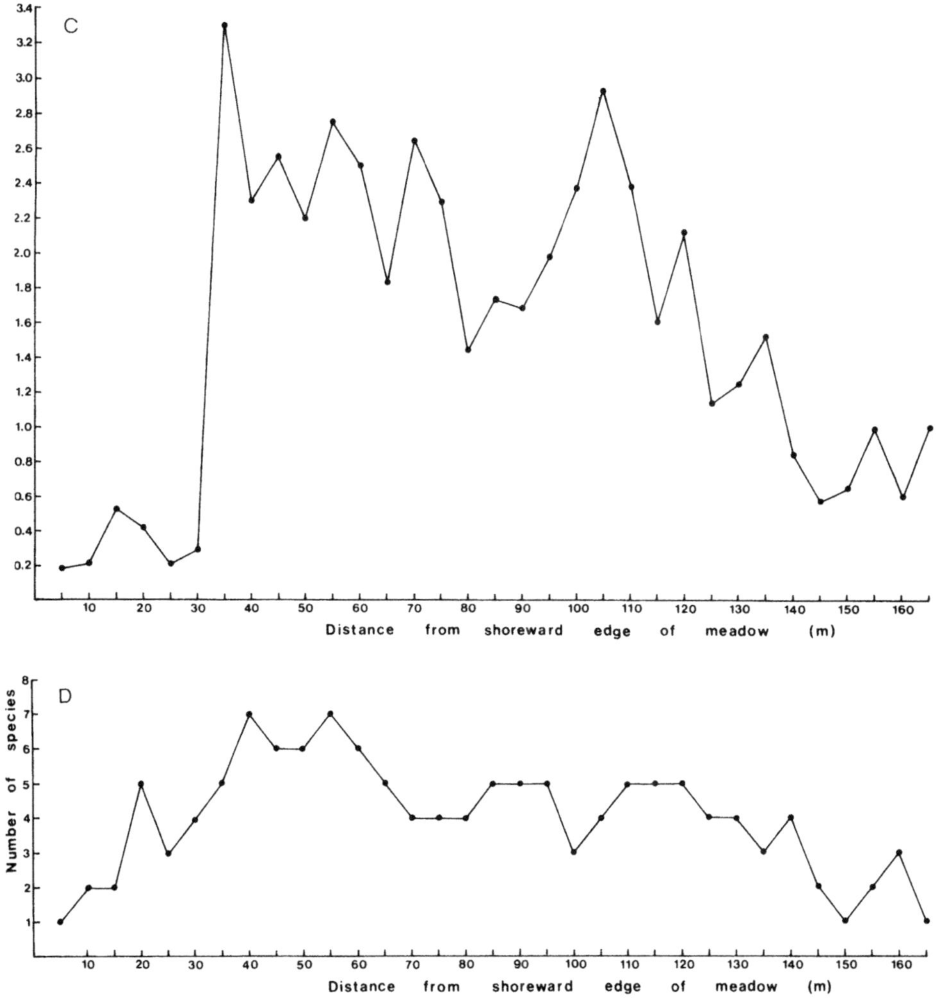

Fig. 6. C and D: Dango I.

reported where a meadow was, except for the shoreward fringe, completely grazed to a lawn-like height apparently by turtles (Johnstone 1978b). Dugongs, in contrast to the browsing action of turtles, "vacuum-clean" as they feed. Roots, rhizomes and leaves are all consumed as the dugong feeds and a trail of destruction is left behind. Such dugong feeding trails can be clearly seen from the air when flying over seagrass meadows in the Torres Strait area. According to Nietschmann (1977) "grazing pressure on seagrasses is high (*in the Torres Strait*) and several areas have been stripped of grass, primarily by dugongs." The dugong's feeding strategy results in a relatively mixed diet of seagrasses (Table 3). A simple survey technique has been developed to provide a quantitative estimate of the suitability of a meadow as a dugong or turtle feeding site (Johnstone 1979b).

In general the majority of New Guinea seagrass meadows seem underutilized by herbivores, and probably the main bulk of the seagrass biomass is entering a

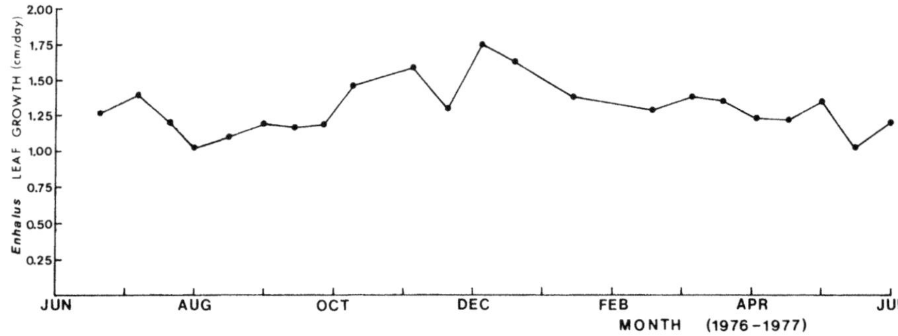

Fig. 7. Seasonal leaf growth rates of *Enhalus acoroides* determined from c. 14 day growth increments (n = 10 plants per analysis). Differences in growth rates are not significant.

detrital food chain. There appear to be no published studies on the seagrass detrital food chain with respect to the tropical Western Pacific-East Indian Ocean area. Therefore it is not possible to give a quantitative estimate of the value of seagrasses to the detritus based shallow water consumers or to commercially important coastal fish stock which depend on this food chain. However, a recent report provides a qualitative comparison between subsistence (village-level) fish catch and the presence of seagrass meadows (Green and Sander 1979). Their maps show some correlation between fishing areas/fish catch and the presence of seagrass meadows in Seeadler Harbor, Manus I., although the data clearly reveals that more fish are caught in mangrove, compared with seagrass, areas (Green and Sander 1979, Figs. 10 and 13).

6. Autecology

There is only one report in the literature on a single-species ecological study in New Guinea. The main aspects discussed are factors affecting leaf width and leaf growth rates of *Enhalus* (Johnstone 1979a).

Information on the phenological aspects of New Guinea seagrasses is particularly in need of documentation. Limited data collected by me on reproductive timing are given in Table 3 for all species (except *Enhalus*, which flowers at any season as long as the tide is low enough) that have been found with reproductive organs.

7. Human impacts

There is little information available on the role of human generated, density independent impacts on seagrasses. That which is available has been summarized by Zieman (1975). In Papua New Guinea only recreational use (e.g., Ela Beach, Port Moresby) and the destructive activities of overabundant subsistence level food/shell collectors present any problem at this time, and these are relatively minor.

Table 3. Flowering and fruiting times for seagrass in New Guinea waters.

Seagrass	Site	Date	Stage
C. rotundata	Bristow I	22.III.79	♂ and ♀ flowers and mature seeds
C. serrulata	Daru	9.XI.78	seeds
	Bristow I	22.III.79	♂ flowers
H. decipiens	Daru	9.XI.78	♀ flowers
H. ovata	Daru	9.XI.78	♀ flowers
H. ovalis	Lolorua I	23.IX.76	♀ flowers and germinating seeds
	Motupore I	27.IX.76	♂ and ♀ flowers
	Taurama Beach	12.XI.78	♀ flowers
	Motupore I	19.XI.78	♂ and ♀ flowers and germinating seeds
	Cape Suckling	9.VIII.79	♂ and ♀ flowers
	Lolorua I	24.VIII.79	♂ and ♀ flowers
S. isoetifolium	Papa	26.XI.75	flowers
	Daru	9.XI.78	flowers
	Daugo I	22.XI.78	flowers
T. hemprichii	Daugo I	22.XI.78	♀ flowers and fruit
	Seeadler Hr	14.II.79	fruit
	Bristow I	22.III.79	♀ flowers and fruit
	Taurama Beach	29.VIII.79	♂ and ♀ flowers.

8. Summary

A start has been made on the description and analysis of the complex seagrass meadows of New Guinea. However, uncertainties remain. It is still almost impossible to categorically give a precise answer to most questions relating to the New Guinea seagrass ecosystem. Whether or not our knowledge of this system will progress depends on not just the interest of single researchers, but also the activities of teams of workers due to the problems of adequate sampling in a patchy environment. As commercial and industrial development accelerates in both political halves of New Guinea it will become even more urgent that a greater scientific and political appreciation of this system develops.

The New Guinea seagrass flora consists of 13 species. (This number becomes 15 if the wide and narrow leaved forms of *Halodule uninervis* are considered as distinct species and the *Cymodocea* sp. with long leaf-bearing branches is not an ecotype of *C. serrulata*.) Eight species are widely distributed, while 5 species and *Cymodocea* sp. are restricted to a few localities. All the species except *Zostera capricorni* have biogeographical affinities with the Indian Ocean-SE Asian area. *Z. capricorni* just reaches New Guinea as a northern extension of the Australasian part of its range. The Torres Strait, between New Guinea and Australia, possesses meadows in which all New Guinea species are represented. These species-diverse communities in the Torres Strait region are believed to be maintained in part by the destructive feeding action of dugongs. East of New Guinea, seagrass diversity drops off rapidly as one moves further into the Pacific.

Seagrasses on New Guinea shores are zoned, possibly in relationship to light intensity. Distinct zones and, by implication, communities are recognizable.

Biomass production (0.25 g C/m^2/day), as estimated from a limited data set, is lower than that in most other temperate or tropical seagrass areas for which data has been reported. However, differences in methodology amongst seagrass scientists may explain the higher rates obtained elsewhere. Seagrasses are important energy sources for turtles and dugongs and, to an undefined extent, for grazing and detrital food chain consumers of the shallow water coastal communities.

References

Brasier, M. D. 1975. An outline history of seagrass communities. Palaeontology 18(4): 681–702.

Christensen, B. and S. Wium-Andersen. 1977. Seasonal growth of mangrove trees in southern Thailand. I. The phenology of Rhizophora apiculata Bl. Aq. Bot. 3: 281–286.

den Hartog, C. 1970. The Sea-grasses of the World. North-Holland Pub. Co., Amsterdam, 275 pp.

—— 1977. Structure, function and classification in seagrass communities. In: McRoy, C. P. and Helfferich, C. (Editors) Seagrass Ecosystems a scientific perspective. Marcel Dekker, Inc., New York, pp. 89–121.

Diamond, J. M. 1974. Colonization of exploded volcanic islands by birds: the Supertramp strategy. Science 184: 803–806.

Fosberg, F. R. 1976. Distributional extensions of marine spermatophytes. Micronesica 12(2): 317–318.

Green, W. and H. Sander. 1979. Environment Studies — proposed tuna canning operations Lombrum, Manus Province. Office of Environment and Conservation, Waigani, in press.

Johnstone, I. M. 1975. The Sea-grasses of the Port Moresby Region. U.P.N.G. Department of Biology, Occ. Paper 7, 38 pp.

—— 1976a. Additions to the sea-grass flora of Papua New Guinea. Science in New Guinea 4: 21–23.

—— 1976b. Seagrasses: productivity and pollution. Paper presented to PNG Botanical Society, 27 March 1976, Mimeo., 28 pp.

—— 1978a. The ecology and distribution of Papua New Guinea seagrasses. I. Additions to the seagrass flora of Papua New Guinea. Aq. Bot. 5: 229–233.

—— 1978b. The ecology and distribution of Papua New Guinea seagrasses. II. The Fly Islands and Raboin Island. Aq. Bot. 5: 235–243.

—— 1979a. Papua New Guinea seagrasses and aspects of the biology and growth of Enhalus acoroides (L.f) Royle. Aq. Bot., 7: 197–208.

—— 1979b. Survey methods for determining the potential of seagrass meadows as dugong and turtle habitat. U.P.N.G., Dept Biology, Occ. Paper 8, 42 pp.

Kock, R. L. and R. T. Tsuda. 1978. Seagrass assemblages of Yap, Micronesia. Aq. Bot. 5: 245–249.

Lipkin, Y. 1975. Halophila stipulacea, a review of a successful immigration. Aq. Bot. 1: 203–215.

McRoy, C. P. and C. McMillan. 1977. Production ecology and physiology of seagrass. In: McRoy, C. P. and Helfferich, C. (Editors) Seagrass Ecosystems a scientific perspective. Marcel Dekker, Inc., New York, pp. 53–87.

Nietschmann, B. 1977. Torres Strait islander hunters and environment. Work-in-progress seminar by the Department of Human Geography, Research School of Pacific Studies. Mimeo., 18 pp.

Odum, W. E. and E. J. Heald. 1972. Trophic analyses of an estuarine mangrove community. Bull. Marine Sci. 22: 671–738.

Parham, J. W. 1972. Plants of the Fiji Islands. Govt Printer, Suva,

Phillips, R. C. 1978. Seagrasses and the coastal marine environment. Oceanus 21(3): 30–40.

Qasim, S. Z. and P. M. A. Bhattahiri. 1971. Primary production of a seagrass bed on Kavaratti Atoll (Laccadives). Hydrobiologia 31: 29–38.

Simpson, G. G. 1965. The Geography of Evolution. Chilton Co., Philadelphia, 249 pp.

Taylor, F. J. 1978. Seagrasses in the New Hebrides. Aq. Bot. 4: 373–375.

Womersley, H. B. S. and A. Bailey. 1970. Marine algae of the Solomon Islands. Phil. Trans. R. Soc. London, B Biol. Sci. 259: 257–352.

Zieman, J. C. 1975. Tropical sea grass ecosystems and pollution. In: Ferguson Wood, E. J. and Johannes, R. E. (Editors) Tropical Marine Pollution. Elsevier Sci. Pub. Co., Amsterdam, pp. 63–74.

12. Mangroves of the Papuan Subregion

I. M. Johnstone & D. G. Frodin

1. Introduction

Although collections of mangrove species and their associates have been made in the New Guinea region from the days of Bougainville and Cook, serious studies of the mangal in this region only date from the second world war, which is rather later than in Java, the Malay Peninsula and Philippines. This late start may be explained, perhaps, by the almost complete absence of economic and other pressures on New Guinea mangroves, by the lack of local scientific units before World War II in New Guinea, and by the general emphasis on taxonomic collections rather than ecological studies before 1942.

With much of the collecting in New Guinea before 1942 in the hands of non-specialists, mangrove collections were few and scattered. However, by 1942 most widespread species had been collected, with useful contributions in the general papers by Hemsley (1885), Warburg (1891), Lauterbach (1928–30), Lane-Poole (1925), and in the Archbold reports (1936–42) by Brass and others. Collectors in the former Territory of Papua between the wars, especially Brass and Carr, paid particular attention to mangroves, and only then did the peculiarities of the south coast mangrove community begin to be realized. The types or paratypes of two key Coral/Arafura Sea taxa, *Ceriops tagal var. australis* and *Bruguiera exaristata* were both collected during this period from the Central Provincial coastline.

Large-scale resident collecting began after World War II in connection with increased overall involvement in New Guinea by the governing powers. By 1975 the extent and general composition of mangal in New Guinea was relatively well known, especially in the east, following extensive collecting and general surveys by governmental agencies over a 25 year period. These studies fittingly culminated in the publication by Percival and Womersley of "Floristics and Ecology of the Mangrove vegetation of Papua New Guinea" in 1975 and in the papers by Frodin *et al.* (1975) and Paijmans (1976).

Since 1975 most studies have become more ecological in their orientation, e.g., Paijmans and Rollet (1977); however, vast stretches of mangrove forests in both the eastern and western halves of New Guinea still remain to be studied even on a simple floristic level, while several genera (e.g., *Bruguiera*, *Avicenna*) still need to be biosystematically investigated.

Monographiae Biologicae, Vol. 42, ed. by J. L. Gressitt.

© 1982, Dr W. Junk Publishers, The Hague.

2. Biogeography

Mangroves occur on all coast lines and in all estuaries and embayments in the Papuan Subregion (New Guinea, nearby large islands and the Bismarck Archipelago) where suitable conditions permit. Areas most favorable to mangal development include deltaic areas, embayments and deep estuaries that are sheltered from excessive wave action and where mud and silt accumulation are taking place, together with high rainfall and no or only a feeble dry season. Tidal range as such (generally less than 3 m in the Papuan Subregion) does not affect the presence or absence of mangroves. In fact, even in areas of extreme tidal range, mangroves may still flourish, as for example the extra-Papuan Darwin coast (N Australia).

The largest stands of mangroves in the Papuan Subregion have been mapped by Percival and Womersley (1975), but smaller occurrences and thin fringing belts are considerably more widespread than is implied by their map.

Areas largely devoid of mangroves include exposed strand where sand accretion is taking place (*Casuarina equisetifolia* often being dominant) and sites where the coast is being eroded or is otherwise relatively abrupt, as, for example, along much of the Rai coast. Markedly seasonal rainfall does not in itself eliminate mangal, but the size of the plants in such regions may be considerably reduced as, for instance, in the Port Moresby coastal belt.

The actual number of mangrove species has long been a matter for debate; the range of species accepted differs considerably from Schimper (1891) through Watson (1928) to Percival and Womersley (1975). However, even taking the term mangrove in a narrow sense (MacNae 1968), for which one of us has used the term "obligate" (DGF, AIMS Mangrove Workshop 1979, unpublished), and limiting mangroves to erect woody dicotylodonous species, the New Guinea mangrove flora, with some 32 obligate taxa presently recognised, is one of the richest in the world (Table 1).

All species are not uniformly distributed throughout the region. Apart from restrictions imposed by ecological factors, there are floristic differences within the Papuan Subregion, most notably that between the north and south coasts of New Guinea.

The south coast mangal facing the Coral Sea, Torres Strait and Arafura Sea has close affinities with that of the extreme north of Australia. With increasing distance south from New Guinea, however, species drop out, and the close affinity is lost. Only one — and that doubtfully distinct within the *Avicennia marina*-complex (*A. resinifera* Forst.) — comes in going south rather than vice versa (Lear and Turner 1977).

The south coast mangal is more distinctive than that of the north coast. Those species not known so far from the north coast (except in some cases in the Vogelkop or Berau Peninsula) include *Aegialites annulata*, *Avicennia officinalis*, *Bruguiera exaristata*, *Camptostemon schultzii*, *Ceriops tagal* v. *australis*, *Osbornia octodonta*, *F.hizophora lamarckii* (one record from New Hanover), *Sonneratia caseolaris* (narrow leaved form) and *Sonneratia ovata*; while *Avicennia eucalyptifolia* and *Xylocarpus australasicus* are both more common on the southern coast. Several of these species, *A. annulata*, *C. schultzii*, *B.*

Table 1. Obligate mangrove species of the Papuan subregion and their distribution in New Guinea.

Mangrove species	Family	Distribution
Aegialites annulata R. Br.	Plumbaginaceae	South Coast
Aegiceras corniculatum (L) Blanco	Myrsinaceae	Ubiquitous
A. floridum Roem and Schultes	Myrsinaceae	Irian Jaya
Avicennia alba Bl.	Verbenaceae	North Coast
A. eucalyptifolia Zipp. ex Mold.	Verbenaceae	Ubiquitous
A. officinalis L.	Verbenaceae	South Coast
A. rumphiana Hall. f.	Verbenaceae	Ubiquitous
Bruguiera cylindrica (L.) Bl.	Rhizophoraceae	Ubiquitous
B. exaristata Hou	Rhizophoraceae	South Coast
B. gymnorrhiza (L.) Lamk.	Rhizophoraceae	Ubiquitous
B. parviflora (Roxb.) W. and A.	Rhizophoraceae	Ubiquitous
B. sexangula (Lour.) Poir	Rhizophoraceae	Ubiquitous
Camptostemon schultzii Mart	Bombacaceae	South Coast
Ceriops decandra (Griff.) Poir.	Rhizophoraceae	Ubiquitous
C. tagal (Perr.) B. B. Rob.		
var. australis C. T. White	Rhizophoraceae	South Coast
var. tagal	Rhizophoraceae	Ubiquitous
Excoecaria agallocha L.	Euphorbiaceae	Ubiquitous
Heritiera littoralis Ait.	Sterculiaceae	Ubiquitous
Lumnitzera littorea (Jack) Voigt	Combretaceae	Ubiquitous
L. racemosa Willd.	Combretaceae	Ubiquitous
Osbornia octodonta F. Muell.	Myrtaceae	South Coast
Rhizophora apiculata Bl.	Rhizophoraceae	Ubiquitous
R. lamarckii Montr.	Rhizophoraceae	South Coast and New Hanover
R. mucronata Lamk.	Rhizophoraceae	Ubiquitous
R. stylosa Griff.	Rhizophoraceae	Ubiquitous
Scyphiphora hydrophyllacea Gaertn.	Rubiaceae	Ubiquitous
Sonneratia alba J. E. Sm.	Sonneratiaceae	Ubiquitous
S. caseolaris (L.) Engl.	Sonneratiaceae	Ubiquitous
Narrow form "Claudie"	Sonneratiaceae	South Coast
Wide form "Tully"	Sonneratiaceae	North Coast
S. ovata Backer	Sonneratiaceae	South Coast (not East of Daru)
Xylocarpus australasicus Ridl. (*X. mekongensis* Pierre p.p.)	Meliaceae	Ubiquitous
X. granatum Koen.	Meliaceae	Ubiquitous

exaristata, *O. octodonta*, and to a lesser extent *C. tagal v. australis*, are characteristic Arafura Sea-Coral Sea centered taxa, with western limits in Timor and Maluku (the last named seems to extend to Mindanao, as does *A. eucalyptifolia*). *R. lamarckii* is centered around the Coral Sea, but also occurs sporadically in the Bismarcks and Solomons. However, this species is atypical and perhaps should be treated as a special case (Tomlinson 1978).

The mangal on the north coast, facing the Pacific Ocean and Solomon Sea, is more closely related to the general Indo-Pacific mangal with, in comparison to the south coast association, fewer distinctive elements. The only taxa apparently restricted to the north coast are *Avicennia alba* and *Sonneratia caseolaris* (broad leaved form), although the latter returns to the west a short distance along the south coast to the Abu District (Central Prov.). The mangroves of the Milne Bay islands exhibit a closer affinity to the north coast flora. *A. officinalis* extends along the north coast of the Berau Peninsula, and the Central Malesian

Aegiceras floridum is known from a few localities in the same area, but not further east. *Avicennia rumphiana* is more widely distributed on the north coast than on the south coast, where it is only known from one locality east of Daru (Bogoro Inlet, Central Prov.).

The sharp discontinuity reported by van Slooten (1937) for *Lumnitzera* in Java apparently does not occur in New Guinea. One of us (IMJ) has recently found *L. littorea* and *L. racemosa* growing side-by-side at Cape Moem (East Sepik Prov.) while mixed stands of the two species also occur in the Northern Province (J. Aruga, pers. comm.), and hybrids between these 2 species have recently been reported from Queensland (Tomlinson *et al.* 1978). Clearly further studies are needed on the supposed ecological separation of the two *Lumnitzera* species.

The biogeographic affinities of the vast majority of the mangrove species (plus the associated herbs, for example, *Acanthus* spp, *Acrostichum* spp) as well as *Nypa fruticans* are with the Indo-Malesian floral province. Thus, mangroves show the same geographic relationships as do the plants of the lowland forest (van Royen 1963). *Osbornia*, previously considered to be the only mangrove of Australian origin, has recently been transferred by Briggs and Johnson (1979) to the Myrtoideae. It appears to be related to the extra-Australia centered genera *Rhodomyrtus*, *Xanthomyrtus* and *Octamyrtus* and, thus, it is probably not fundamentally Australian. A few species (*C. schultzii*, *A. floridum*) are basically Central Malesian (van Slooten 1937). *Camptostemon* as a whole is restricted to Central Malesia (Troll 1933), but its affinities are unclear; although it is usually assigned to the Bombacaceae it seems to one of us (DGF) to bear greater overall resemblance to the Sterculiaceae-Sterculioideae. *A. rumphiana* is also most common in Central Malesia, but apart from New Guinea it also extends through Borneo to the east coast of the Malay Peninsula. The remaining species are either widely distributed in Indo-Malesia (some even extending to Africa), or are closely related to such species.

To the east, beginning with New Britain and New Ireland, species diversity decreases in the now classical pattern as seen in the terrestrial vascular flora (Balgooy 1971) and among seagrasses (Johnstone this volume). Ultimately all species drop out; no mangroves have ever reached the Hawaiian Islands naturally, although two or three species, including *Rhizophora mangle*, have been successfully introduced.

The affinities of mangrove species and genera to terrestrial forms within their families or orders remains to be critically studied. It is interesting, however, that many of the close inland relatives of mangroves are relatively hygrophilous, for example, *Carallia* (Rhizophoraceae) and *Duabanga* (Sonneratiaceae) (Bews 1927). The evolution of mangroves from such ancestors may still be continuing, as, for example, *Myristica hollrungii* which is both a freshwater swamp tree and estuarine mangrove; while speciation of "obligate" mangroves very likely continues to be an active process, as, for example, in the East-Asiatic and West and Central Malesian species *Kandelia candel* (not yet reported from New Guinea) which appears to be very closely related related to *Ceriops* and possibly originated from this genus in the recent past. Reflecting their diverse terrestrial ancestry, mangroves are a taxonomically heterogeneous group of plants

(Table 1). They are a classical example of convergent evolution in the plant kingdom with their pneumatophores or stilt roots; inconspicuous flowers; vivipary; water dispersed propagules; thick, entire leaves; and (except for *Xylocarpus*) evergreen habit.

The antiquity of many mangrove genera (Muller 1970) and their long existence in Malesia, coupled with the great diversity of the recent mangrove flora, suggests a possible "center of origin" (of the traditional Darwin/Willis kind) in Malesia, followed by migration. Another interpretation, however, which may be correlated with ecology, suggests that mangroves may have been part of the old Tethyan biota which was essentially a tropical assemblage (*sensu* Croizat *et al.* 1974) and thus their origin dated from the time of closer association of the continents (McCoy and Heck 1976). At the time of effective separation of the Americas from the old world, however, the full complement of generic development may not have yet taken place(Meggers *et al.* 1973). Further additions to the mangrove flora were subsequently of a regional nature; this later speciation being particularly important in the Indo-Malesian area possibly as a result of conducive habitat conditions. Land in the present New Guinea region, probably as islands, existed as part of an unstable belt running from southeast Asia towards the SW Pacific well before the present land mass came into existence in connection with the northward drift of the Australian continent and the associated interplate collision. Thus the extant mangrove flora of New Guinea most likely results from short distance dispersal and evolution within the Indo-Malesian region of which New Guinea is and was a part. We believe the presence of *Aegialites*, one of the oldest mangroves in the fossil record (Muller 1970), in the Papuan subregion to be of crucial importance in terms of the possible age and origins of the New Guinea mangrove flora, which we regard as entirely of Indo-Malesian rather than Australian affinity.

3. Zonation patterns

Where environmental gradients are present, mangrove species occur in characteristic patterns with respect to these gradients. Few studies of zonation patterns using transects have been done in New Guinea. Floyd (1977) illustrates a transect through a mangrove swamp in the Purari Delta, Green and Sander (1979) illustrate 3 transects from Manus I.; Johnstone (1978) describes the zonation pattern at Papa near Port Moresby and Percival and Womersley (1975) provide two zonation diagrams apparently based on transects. The two patterns described by Percival and Womersley, however, are somewhat unusual. For example, their Fig. 15 (as per erratum) shows *R. mucronata* where normally either *R. stylosa* or *S. alba* is present in the present authors' experience. Most authors have described mangrove "zonation" simply in terms of the main associations or plant communities found in the mangal. This latter approach was adopted in the CSIRO Land Research Series (Heyligers 1965, 1967, 1972; Paijmans 1967, 1969, 1971; Robbins 1968) and was also used by Paijmans and Rollet (1977). Paijmans (1976) briefly reviews these main mangrove communities. The discussions by Chapman (1976, 1977) are mainly based on the

CSIRO Land Research Series reports. Most of the CSIRO surveys were based on aerial photography, and the purpose of these studies was to develop a land classification rather than a study of the mangrove system *per se*. A general categorization of the mangal into several sub-communities is adequate for inventorying forests, but is inadequate for studying the ecology of individual species. A number of obscure patterns become obvious after an actual field-based quantitative survey; for example, the bimodal distribution of *A*. "*marina*" (i.e., *A. marina* sensu lato).

The traditional approach of Percival and Womersley (1975) in which the mangal is divided into seven zones is, in comparison to the CSIRO approach, of value only if its inherent limitations are recognized. These limitations can be illustrated by the variability that occurs in the "seaward zone." Depending on the environmental conditions, the species occupying the seaward fringe may include any of the following: *A. corniculatum*, *A. alba*, *A. annulata*, *A*. "*marina*," *A. officinalis*, *A. rumphiana*, *C. tagal* v. *tagal*, *O. octodonta*, *R. lamarckii*, *R. mucronata*, *R. stylosa*, *S. alba*, *S. caseolaris*, *X. mekongensis* and the palm *N. fructicans* (Fig. 1).

Field experience suggests that much of the apparent complexity in mangrove zonation is due to a limited set of factors which vary in their relative importance from site to site. Those factors which appear to be the most important in determining the pattern of zonation are discussed below.

(a) Inundation-Water Depth

Zonation patterns are most obviously associated with water depth (tidal regime). In areas with a perceptible slope there is always some zonation present (e.g., most open coast situations and certain large lagoons such as Hood Lagoon); at sites with little or no ground slope, zonation patterns are inconspicuous or

Fig. 1. R. stylosa occupying the seaward fringe at Daru (Western Province). At other sites on this island *R. stylosa* is replaced by *S. alba* or *R. apiculata* or *A. eucalypitfolia* as the most seaward mangrove.

absent over large areas (e.g., Galley Reach). The period of inundation is probably more important than water depth *per se*, as recognized by Watson, De Haan, and other authors many years ago (see Chapman 1976:124–125), although water depth almost certainly controls the seaward limit of the mangroves. However, inundation-water depth may not be the most important factor in regulating the internal patterns. Two reasons for suggesting this possibility are: (1) Recent transplant studies of seedlings have shown that various new world mangrove species will grow throughout the whole of the tidal zone normally occupied by mangroves (Rabinowitz 1978). There is little reason to think that old world mangroves should not show the same behavior (except for certain essentially drift-line species such as *Excoecaria agallocha* and *Scyphiphora hydrophyllacea* and the "fresh-water" mangroves such as *Myristica* and *Amoora*). (2) Many New Guinean mangroves have bimodal distributions, i.e., their population is densest on the landward and seaward fringes of the swamp (e.g., *A. "marina," A. corniculatum*), while several species are usually (but not always) distributed throughout the swamp with only a slight tendency towards a contagious distribution (e.g., *C. tagal v. tagal, R. stylosa*).

Inundation-water level, however, is probably responsible for the gross zonation of Papuan subregion mangal into a "true mangrove" zone and a "transition" (to land) zone as described by Paijmans and Rollet (1977) for Galley Reach. Although many species (e.g., *R. stylosa, A. "marina"*) occur in both zones, some species are almost always restricted to the true mangrove zone (e.g., *B. cylindrica, B. gymnorrhiza, R. mucronata, S. alba, S. caseolaris, X. granatum*), while many species are normally restricted to the transition zone (e.g., *A. rumphiana, E. agallocha, H. littoralis, L. racemosa*, and the herbs *A. aureum, A. ilicifolius*). The transition between these two zones is quite sharp (about the tidal level, going seaward, where *Bruguiera* knee roots are first encountered) and thus appears to be a tidally related discontinuity.

(b) Wave Action

Wave action is a well known modulator of algal species distribution on rocky shores (Lewis 1964; Seapy and Littler 1978). Mangroves also appear to have varying tolerances to the intensity of wave action (Jennings and Coventry 1973; Johnstone 1978). Most mangroves do not appear to tolerate high-energy localities, but this may simply be due to the absence of a suitable substrate in such situations rather than the waves themselves. However, irrespective of the ultimate reason, a survey of localities in the Port Moresby region has shown that distinct groupings can be made of species based on their tolerance to wave action (Johnstone 1978).

(c) Drainage (Soil Water and Water Table)

No direct experimental evidence relating to the effects of drainage is available for Papuan subregion mangals. However, drainage (facilitating aeration or modulating soil salinity?) probably plays a role that is separate from the inundation-water depth factor. Percival and Womersley (1975) consider drainage to be one of three controlling factors in zonation (the others being inundation and salinity), although authors in other geographic areas do not seem to give drainage such importance. However, there are a number of conditions in the opinion of the present authors which suggest that drainage conditions are quite important in the Papuan mangals, both in terms of zonation across the tidal gradient and in terms of general distribution (between-habitat) patterns.

(1) *The banks of mangrove creeks.* The banks of mangrove creeks and estuaries support a flora quite distinct to that of the mudflats less than 20 m from the bank. For example, *R. mucronata* and *A. officinalis* are most frequent on the banks of well drained creek banks, while *S. caseolaris* tends to occur on poorly drained banks (i.e., with little slope).

(2) *Lobster mounds.* The large (up to 2 m high) lobster (*Thalassina anomala*) mounds found in areas such as Galley Reach are usually occupied by species such as *H. littoralis*, *A. speciosum* and *A. ilicifolius*. The soil of these mounds is well drained, which could play a role in their colonization by transition zone species.

(3) *Coarse substrate habitats.* Sand, coral rubble and rocky-gravel foreshores retain considerably less interstitial water when the tide is out compared with habitats where the substrate is of a muddy-silty composition. The amplified drainage in such coarse substrate habitats probably partially accounts for the occurrence of certain mangroves in such situations. For example, *A. annulata* and *O. octodonta* are most common on rocky, coral rubble and sandy shores.

(4) *Undrained back mangal.* Due to terrestrial seepage and lack of ground slope many back mangrove areas remain with a cover of water even at low tides. Such areas are normally occupied by *A.* "*marina*" in low rainfall areas and *N. fructicans* in high rainfall areas.

(5) *Modified drainage situations.* Watson (1928) briefly commented on marked effects of both natural (loc. cit., p. 140) and unnatural (loc. cit., p. 220) changes to the drainage pattern of Malayan mangrove swamps. Similar observations may be made with respect to the swamps of the Papuan Subregion. The most spectacular changes result from road construction through mangals (e.g., Hood Lagoon). If soil and surface water movement is stopped or diverted a number of species (e.g., *Rhizophora* spp, *Bruguiera* spp, *Ceriops* spp) die and they are replaced by a scrub community including *A. ilicifolius*, *A. aureum*, *E. agallocha*, *A.* "*marina*" and, sometimes, *S. hydrophyllacea*, unless the environmental modification results in areas of ponded water, in which case there is no recolonization of the area. Naturally ponded areas as are found in the Galley Reach swamps are usually devoid of mangroves (Paijmans and Rollet 1977).

(d) Salinity-Freshwater Regime

In probably all mangrove habitats the distribution pattern of the plant species is greatly influenced by salinity levels. Salinity has been downrated in recent years, particularly by workers in the new world (see, Lugo and Snedaker 1974), but the distribution of a number of species in the Papuan subregion seems to be almost completely dependent on salinity levels. Most studies on the effects of salinity on mangroves have been short-term experimental studies (e.g., McMillan 1975), and until more long-term (seasonal) field data are available it seems premature to dismiss salinity. For example, Christensen and Wium-Andersen (1978) and Wium-Andersen and Christensen (1978) have recently shown that leaf production phenology of several SE Asian mangroves seems to be related to salinity. For New Guinea the only published salinity figures seem to be those of Paijmans and Rollet (1977) and these only indicate broad limits for the forest types of Galley Reach. Long-term field studies are urgently needed for New Guinea and elsewhere.

There are two conditions, based on our field observations, under which salinity is apparently important in New Guinea.

(1) *High salinity*. In low rainfall areas of the Papuan coast, salt pans develop at the back of the mangrove forest which are similar to those in N Australia (Saengar and Hopkins 1975). the last mangrove to drop out before the forest gives way to open ground is *Ceriops tagal v. australis*. In fact, this species seems to be completely restricted to high salt environments. Spot measurements by one of us (IMJ) during the dry season in *C. tagal v. australis* habitats around the Port Moresby harbor have revealed a mean salinity in these habitats of 70 mg Cl/g dry wt soil (c. 125 ‰ NaCl) in the top cm of soil and 30 mg Cl/g dry wt soil (c. 53.7 ‰ NaCl) at a depth of 30–35 cm. Under these very high salinity levels, salinity is almost certainly controlling the species distribution pattern at the landward fringe of the mangal. Both *L. racemosa* and *A. "marina"* may be present with *C. tagal v. australis* when the salinity drops to about 50 mg Cl/g dry wt soil, but most other mangroves appear to prefer a salinity equal to or less than that of seawater.

(2) *Low salinity*. Most mangroves can probably grow under essentially freshwater conditions and their absence from such habitats is probably not due to an obligate requirement for chloride, but most likely results from some form of interspecific competition (see below). On the other hand, many species of the Papuan Subregion mangal probably find elevated salinity levels toxic and simply can't invade more saline habitats. Included in this category would be species of brackish habitats such as *S. caseolaris*, and *N. fructicans* and the "freshwater" mangroves *Amoora cucullata*, *Brownlowia argentata* and *Myristica hollrungii*.

Not only does salinity appear to play a role in zonation of New Guinea mangroves, but it seems also to affect their biomass and growth because on high rainfall coasts the standing crop is far greater than that on coasts subject to low rainfall (unpublished data, IMJ).

(e) Substrate

As is well known most mangroves prefer a muddy habitat, but there are a number of important exceptions in the Papuan Subregion mangals and these exceptions contribute to the overall zonation patterns found in the field. *A. annulata* prefers rock platforms where it establishes in crevices. *O. octodonta* prefers coral platforms and coarse coral sand habitats as does *Pemphis acidula*; sometimes *A. corniculatum* and *S. hydrophyllacea* are also to be found growing vigorously under such conditions. Many species prefer a firm sand substrate, e.g., *A. "marina," H. littoralis*, *Ceriops* (*tagal* and *decandra*). *Lumnitzera* (*racemosa* and *littorea*), *A. rumphiana*, *S. hydrophyllacea* and *X. mekongensis*. However, the last two also grow vigorously in very muddy habitats.

Species found in the softest mud situations usually belong to those genera whose propagules have long viviparously produced hypocotyls. Presumably this adaptation is advantageous in such situations since there would be less chance of the seedling sinking completely into the mud.

(f) Biota

Little attention has been given to the role of biotic interactions in mangrove distributions. The fauna probably plays a minor role. For example, Paijmans and Rollet (1977) suggest that the mud lobster (*Thalassina anomola*) may influence species distribution as a result of its burrowing activities. The destruction of propagules of species such as *R. mucronata* by crabs may also play a minor role (S. Wium-Anderson pers. comm.). Interactions between the flora, on the other hand, seem to have been almost completely ignored both in New Guinea and elsewhere. This is surprising because the first thing one notices about a mangrove forest is the "crowded" nature of the system. The canopy in natural stands is normally continuous, with root systems intertwining. Competition for space must be "fierce". Numerous discussions exist in the literature on mangrove succession (see Chapman 1976), but the possibility that plant-plant interactions play any role in the determination of zonation-succession patterns seems to have been an alien idea to the authors.

In the species-rich swamps of New Guinea, interspecific competition is possibly the factor responsible for the ultimate distribution patterns present. Direct field evidence is lacking, but the reasoning behind this suggested emphasis on competition is simple. Physio-chemical factors such as already discussed interact to create preferred habitats for each species, but exclusion from other habitats does not seem to be on a physio-chemical basis in the majority of cases. Transplant experiments have shown that mangroves will grow away from their "normal" zone (Rabinowtz 1978). In addition, and perhaps more importantly, field observation by the present authors show that individual plants are frequently found growing away from their "preferred" habitat or zone. For example, a thorough search will usually turn up one or two *A. annulata* plants in the middle of the muddy *Bruguiera-Rhizophora* zone. Aerial photographs always show a scattering of individuals of the "colonizer" *A. "marina"* throughout a swamp, a fact readily confirmed by ground based observation. Numerous other examples exist. The implication, then, is that competitive exclusion must play a powerful role in the zonation patterns of mangals (see Johnstone 1980).

The most obvious examples of competition are those for light (for example, *A. annulata* is generally not found under a *Rhizophora-Bruguiera* canopy because there is not enough light — the individuals that do establish there are strongly etiolated) and for space (for example, *Bruguiera* tends to replace *Rhizophora* because the surface root system of *Bruguiera* reduces the possibility of *Rhizophora* or other mangrove propagules lodging in the ground). However, many interactions are presumably far more subtle and remain to be identified.

This emphasis on competition is novel, but the idea certainly is not; for example, MacNae (1966) recognized that shading by *Rhizophora* prevents the establishment of *Avicennia* seedlings under *Rhizophora*.

As suggested by the foregoing discussion no single factor controls the zonation patterns of mangals, while, in reality, the zonation patterns vary from site to site and, thus, one cannot give a specific example as representative of some general pattern. Either one can describe each mangrove habitat in great detail as has been done for Australia recently (Galloway 1979) or describe a few patterns which are relatively representative. Using this latter approach two patterns — the "open coast pattern" and the "estuarine pattern" are described.

The open coast pattern is represented by the sequence found on the NW side of Hood Lagoon (Fig. 2). The zones from the sea are *A. "marina"* and *S. alba*; *R. stylosa*; *R. apiculata*; *B. cylindrica*; *B. gymnorrhiza*; *C. tagal v. tagal*; *A. "marina."* These species all form complete canopies within the area of their maximal density, and, although many other mangrove species are present at Hood Lagood, none exhibit this degree of dominance.

The estuarine pattern is represented by that found at Galley Reach. The zonation here is more complex and depends on the prevailing salinity. The estuary-river bank will be occupied by some combination of the following species — *A. officinalis*, *N. fructicans*, *R. mucronata*, *S. caseolaris* and *X. granatum* where the water is relatively fresh and *A. "marina,"* *S. alba*, *R. apiculata*, *R. mucronata* and *R. stylosa* in more salty situations. On the frequently inundated flood-plain immediately behind the "bank-species" one finds either a forest of *B. gymnorrhiza*, *B. parviflora* and *R. apiculata* where the salinity is relatively high or *N. fructicans* where the salinity is lower. Moving into the infrequently inundated (or transition) zone one encounters, in the transition to dry land, *B. cylindrica*, then *C. tagal v. tagal*, then *H. littoralis* and finally *C. tagal v. australis* or *A. "marina"* or *S. hydrophyllacea*. The transition to freshwater swamp, in contrast to dry land, is gradual and *Nypa* slowly gives way to freshwater swamp plants such as *Phragmites*, *Saccharum*, *Terminalia*, *Nauclea*, *Octomeles*, *Artocarpus*, and, on slightly higher ground, *Melaleuca*.

In general, the zonation of mangroves in the Papuan subregion is similar to that in the rest of Malesia and in northern Australia (Watson 1928; van Steenis 1958; Chai 1975; Saenger and Hopkins 1975; MacNae 1967).

4. Community structure and development

4.1. Vertical Stratification

New Guinea mangrove forests are mostly single storey forests although emergents are a conspicuous feature of nipa palm swamps (see Paijmans 1976, pl 7), sometimes remnant (?) *R. stylosa* and *R. apiculata* are emergents in *B. cylindrica* and *B. exaristata* forest, and, similarly, *A. "marina"* in *C. tagal v. tagal* forest. In higher rainfall areas a dense ground layer or understory can develop towards the rear of the swamp. This understory usually consists of *A. ilicifolius*, *Acrostichum speciosum*, *E. agallocha*, *S. hydrophyllacea*, *Dalbergia candenatensis* and "occasionals" such as *Maytenus emarginata* and the various lianes and scrambling vines. In open *A. "marina"* woodland a ground cover of *Tecticornia* is frequently present.

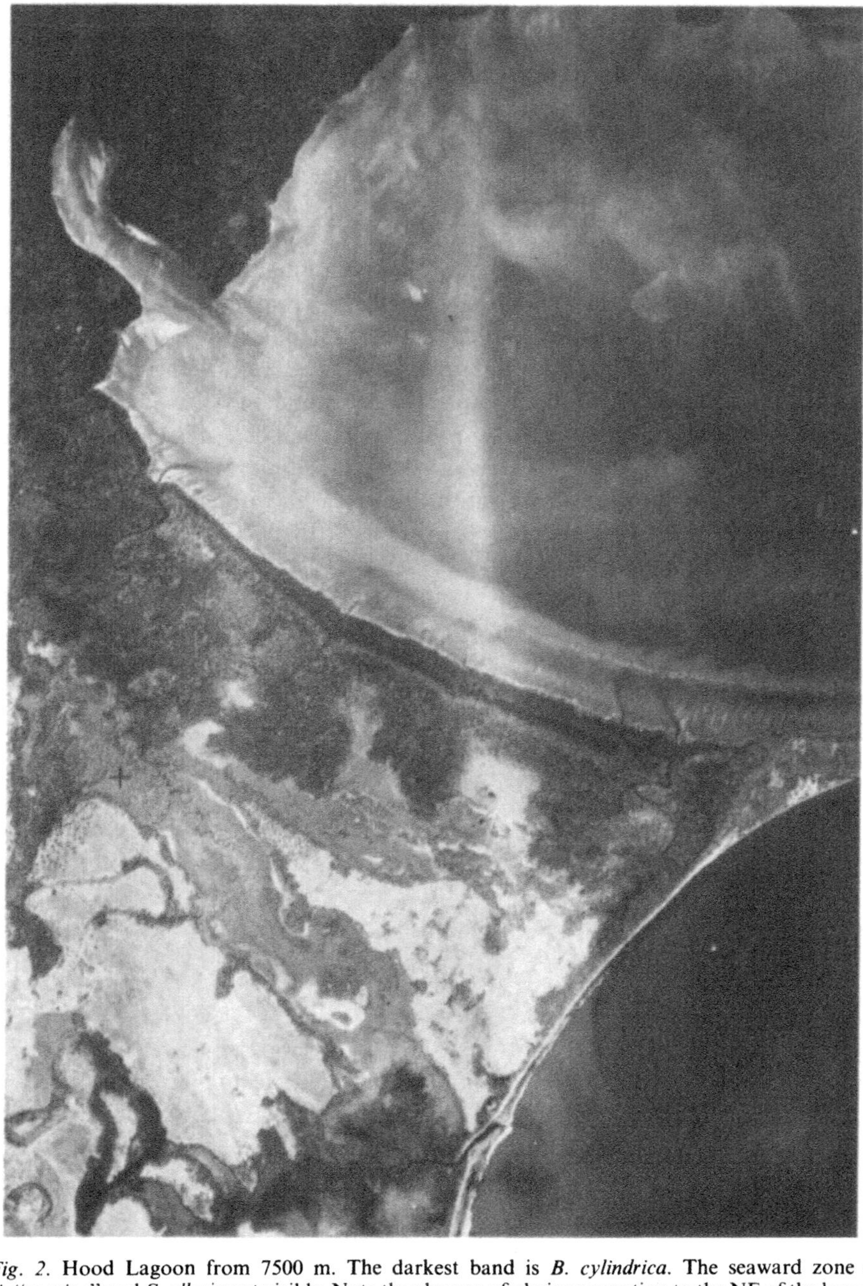

Fig. 2. Hood Lagoon from 7500 m. The darkest band is *B. cylindrica*. The seaward zone of *A.* "*marina*" and *S. alba* is not visible. Note the absence of obvious zonation to the NE of the lagoon (N is to the left).

Some authors (see Lugo and Snedaker 1974) have suggested there is no ecological succession in mangroves. Rather, they believe, the zones are static entities which move seaward or landward depending on sea-level fluctuations. An ecosystem without succession seems a contradiction and is not an acceptable concept to us. Instead, we believe succession occurs, but due to the strong environmental gradients across the mangal the whole swamp is prevented from developing into a homogenous climax community. Thus, it seems that in all except the climax zone succession has been curtailed by one or more environmental factors. A series of terms could be coined to describe these environmentally maintained stages, e.g., haloclimax, flucticlimax, fossaclimax, etc., if the actual controlling factor in each case had been described.

Since the precise interactions controlling the zonation patterns are not yet properly understood, it seems premature to describe succession in any detail. However, there are clearly two possible approaches which are briefly described with reference to the Hood Lagoon system (only the main species in each zone/sere are mentioned (Fig. 3)).

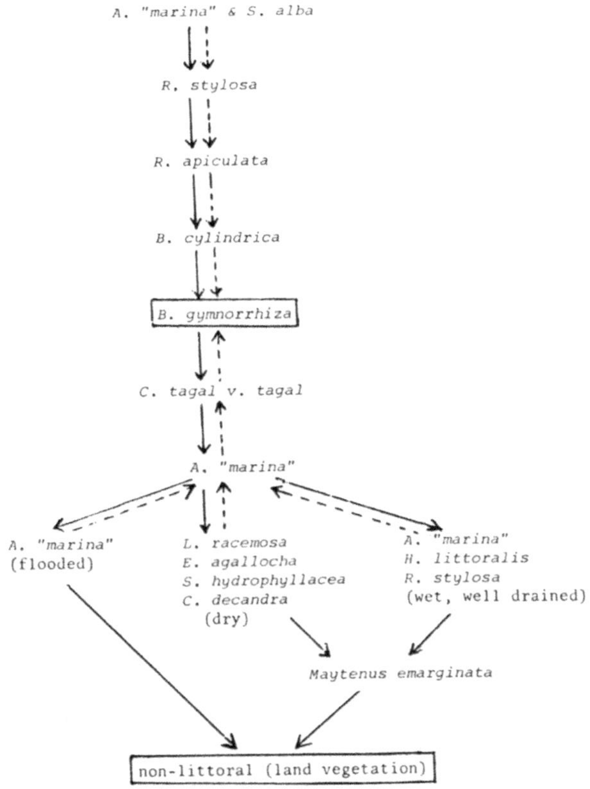

Fig. 3. Succession at Hood Lagoon. Solid lines illustrate succession towards classical terrestrial climax, broken lines indicate succession to proposed within-mangrove climax (Johnstone 1980).

With reference to Fig. 3, the classical scheme with the dry land community as the climax is well known (see Chapman 1976 for numerous additional examples). The second scheme follows from field work mainly at Hood Lagoon (Johnstone 1980). This scheme implies that the mangal is a community in its own right with both colonizing and climax elements.

5. Miscellaneous aspects

5.1. Trophic relationships and dependent biota

No ecological studies in New Guinea have been made of either the herbivore or detritus food chains which originate with the mangroves. Recently Johnstone (1981) has completed a study of leaf damage by herbivores. No obvious correlations were found in this study. Redfield (1979) has reviewed the state of knowledge on trophic pathways based on mangroves in Australia, and much of the Australian data probably is applicable to the New Guinea situation. However, work is currently being carried out in the extensive mangals of the Purari Delta which will have a bearing on these aspects (Haines 1979; Petr 1979). Liem and Haines (1977) have listed the fauna of the Purari mangrove system.

5.2. Conservation and utilization

Human generated impacts and the utilization of mangroves in Papua New Guinea have been discussed by Percival and Womersley (1975), Liem and Haines (1977) and Petr (1977). However, in the short term, the only major direct significance the mangroves will have to the indigenous population will continue to be in terms of subsistence fisheries (Green and Sander 1979; Haines 1979), and as local sources of fuel and timber, with increasingly noticeable effects near Port Moresby, Wewak and Lae.

6. Summary

The Papuan subregion possesses possibly the last remaining extensive, unmodified mangals in the world. They are highly diverse. Not only are these important as a scientific resource, but their commercial significance is also rapidly becoming appreciated. To maintain the scientific integrity of these forests in the face of inevitable "development" will be difficult. A convincing data base needs to be rapidly built so that development projects likely to have an effect on the stability and survival of the mangals may be planned within a context of maximum conservation. Serious studies of mangroves in the Papuan Subregion have only been in progress since World War II and most ecological publications have only appeared during the past decade. The south coast mangrove flora of New Guinea is more distinctive than that of the north coast.

Palynological evidence (from Indonesia) and the presence of the genus *Aegialites* in New Guinea suggest an early arrival or evolution of mangrove in the Papuan Subregion. Zonation and succession are examined. Evidence discussed suggests that the mangal of the Papuan Subregion is not just a colonizing assemblage but, rather, a system in its own right.

References

Archbold Reports. 1936–42. Bull. Amer. Mus. Nat. Hist.

Balgooy, M. van. 1971. Plant Geography of the Pacific. Blumea (suppl.) 6. Leiden.

Bews, J. W. 1927. Studies in the ecological evolution of angiosperms. New Phytol. (Reprint) 16. London.

Briggs, B. G. and L. A. S. Johnson. 1979. Evolution in the Myrtaceae — Evidence from infloresence structure. Proc. Linn. Soc. (N.S.W.) 102: 157–265.

Chai, P. P. K. 1975. Mangrove forests in Sarawak. Malay. For. 38: 108–134.

Chapman, V. J. 1976. Mangrove Vegetation. J. Cramer, Vaduz. 447 pp.

—— 1977. Wet coastal formations of Indo-Malesia and Papua-New Guinea. In: V. J. Chapman (Ed) Wet Coastal Ecosystems. Elsevier, Amsterdam, pp. 261–270.

Christensen, B. and Wium-Andersen, S. 1978. Seasonal growth of mangrove trees in southern Thailand. I. The phenology of Rhizophora apiculata Bl. Ag. Bot. 3: 281–286.

Croizat, L., G. Nelson and D. E. Rosen. 1974. Centers of origin and related concepts. Syst. Zool. 23: 265–287.

Floyd, A. G. 1977. Ecology of the Tidal Forests in the Kikori-Romilly Sound area Gulf of Papua. Ecol. Rept. 4, Office of Forests, Lae, 59 pp.

Frodin, D. G., C. R. Huxley and K. W. Kirina. 1975. Mangroves of the Port Moresby Region. Occ. Pap. 3, Biology Dept., UPNG, Waigani, 53 pp.

Galloway, R. W. 1979. Distribution and Patterns of Australian Mangals. National Mangrove Workshop: Position Papers, A.I.M.S., Townsville.

Green, W. and H. Sander. 1979. Manus Province Tuna Cannery Environmental Study. Office of Environment and Conservation, Waigani, 123 pp.

Haines, A. K. 1979. The subsistence fishery of the Purari Delta. Science in New Guinea 6: 80–95.

Hemsley, W. B. 1885. The Botany of the Southeastern Moluccas. Challenger Exp. Bot. Vol. 1, part 3.

Heyligers, P. C. 1965. Vegetation and ecology of the Port Moresby-Kairuku area. CSIRO Aust. Land Res. Ser. 14: 146–173.

—— 1967. Vegetation and ecology of Bougainville and Buka Islands. CSIRO Aust. Land Res. Ser. 20: 121–145.

—— 1972. Vegetation and ecology of the Aitape-Ambunti area. CSIRO Aust. Land Res. Ser. 30; 73–99.

Jennings, J. N. and R. J. Coventry. 1973. Structure and texture of a gravelly barrier island in Fitzroy estuary, Western Australia, and the role of mangroves in shore dynamics. Marine Geol. 15: 45–167.

Johnstone, I. M. 1978. The mangroves and mangrove ecosystems of Papua New Guinea. 2nd Int. Congr. Ecol., mimeogr., 26 pp.

—— 1980. Succession in zoned mangrove communities: where is the climax? 2nd Int. Symp. Biol. Manag. Mangroves & Trop. Shallow Water Comm. (Port Moresby), mimeogr., 18 pp.

—— 1981. Consumption of mangrove leaves by herbivores in mixed mangrove stands: search for a correlation. Biotropica. In press.

Lane-Poole, C. E. 1925. The forest resources of the territories of Papua and New Guinea. Reports Parliam. Commonw. Australia 1925, II, Melbourne.

Lauterbach, C. 1928–30. Die Pflanzen Formationen einiger Gebiete Nordost-Neu-Guineas und des Bismarck Archipels. I-IV. Bot. Jarhb. 62: 284–304, 550–564; 63: 1–28, 419–476.

Lear, R. and T. Turner. 1977. Mangroves of Australia. Univ. Qld Press, St. Lucia.

Liem, D. S. and A. K. Haines. 1977. Purari River (Wabo) Hydro-electric Scheme Environmental Studies. Vol. 3. The ecological significance and economic importance of the mangroves and estuarine communities of the Gulf Province, Papua New Guinea. Office of Environment and Conservation, C.G.O., Waigani, 35 pp.

Lewis, J. R. 1964. The ecology of rocky shores. English Universities Press, London. 323 pp.

Lugo, A. E. and S. C. Snedaker. 1974. The ecology of mangroves. Ann. Rev. Ecol. Syst. 5: 39–64.
Macnae, W. 1966. Mangroves in eastern and southern Australia. Aust. J. Bot. 14: 67–104.
—— 1967. Zonation within mangroves associated with estuaries in North Queensland. In: G. H. Lauff (Ed) Estuaries. A.A.A.S., Washington, pp. 432–441.
—— 1968. A general account of the fauna and flora of mangrove swamps and forests in the Indo-West-Pacific region. Advances Marine Biol. 6: 73–270.
McCoy, E. D. and K. L. Heck. 1976. Biogeography of corals, seagrasses, and mangroves: an alternative to the center of origin concept. Syst. Zool. 25: 201–210.
McMillan, C. 1975. Interaction of soil texture with salinity tolerances of black mangrove (Avicennia) and white mangrove (Laguncularia) from North America. Proc. Int. Symp. Biol. and Manag. Mangroves (Honolulu) 2: 561–566.
Meggers, B. J., E. S. Ayensu and W. D. Duckworth. 1973. Tropical Forest Ecosystems in Africa and South America: a comparative review. Smithsonian Inst. Press., Washington.
Muller, J. 1970. Palynological evidence on early differentiation of angiosperms. Biol. Rev. 45: 417–450.
Paijmans, K. 1967. Vegetation of the Safia-Pongani area. CSIRO Aust. Land Res. Ser. 17: 142–167.
—— 1969. Vegetation and ecology of the Kerema-Vailala area. CSIRO Aust. Land Res. Ser. 23: 95–116.
—— 1971. Vegetation, forest resources and ecology of the Morehead-Kiunga area. CSIRO Land Res. Ser. 29: 88–113.
—— 1976. Vegetation. In: K. Paijmans (Ed) New Guinea Vegetation. Aust. Nat. Uni. Press, Canberra, pp. 23–105.
Paijmans, K. and B. Rollet. 1977. The mangroves of Galley Reach, Papua New Guinea. Forest Ecology and Management 1: 119–140.
Percival, M. and J. S. Womersley. 1975. Floristics and Ecology of the Mangrove Vegetation of Papua New Guinea. Bot. Bull. 8, Dept of Forests, Lae 96 pp.
Petr, T. 1977. Mangroves of Papua New Guinea. Paper presented at the IPFC/UNEP/Philippine workshop on mangrove and estuarine area development in the Indo-Pacific region, Manila 14–19 Nov. 1977. 10 pp.
—— 1979. The Purari River Hydroelectric Development at Wabo and its environmental impact: an assessment of a scheme in the planning stage. Science in New Guinea 6: 105–116.
Rabinowitz, D. 1978. Early growth of mangrove seedlings in Panama, and an hypothesis concerning the relationship of dispersal and zonation. J. Biog. 5: 113–133.
Redfield, J. A. 1979. Trophic relationships in mangrove communities. In: National Mangrove Workshop: Position Papers. A.I.M.S., Townsville.
Robbins, R. G. 1968. Vegetation of the Wewak-Lower Sepik area. CSIRO Aust. Land Res. Ser. 22: 109–124.
Royen, P. van. 1963. The vegetation of the islands of New Guinea. 70 pp. (cyclostyled) Department of Forests, Lae.
Saengar, P. and M. S. Hopkins. 1975. Observations on the mangroves of the southeastern Gulf of Carpentaria, Australia. Proc. Int. Symp. Biol. Manag. Mangroves (Honolulu) 1: 126–136.
Schimper, A. F. W. 1891. Die Indo-Malayische Strandflora. Bot. Mitteil. Trop. 3. Fischer, Jena 204 p.
Seapy, R. R. and M. M. Littler. 1978. The distribution, abundance, community structure, and primary productivity of macroorganisms from two central California rocky intertidal habitats. Pacific Science 32: 293–314.
Slooten, D. V. van. 1937. Die Verbreitung von Lumnitzera und anderen Mangrovegewachsen. Blumea Suppl. 1: 162–175.
Steenis, C. G. G. J. van. 1958. Ecology (mangrove species). Flora Malesiana Ser. I, 5: 431–441.
Tomlinson, P. B. 1978. Rhizophora in Australasia — some clarification of taxonomy and distribution. J. Arnold Arb. 59: 156–169.
Tomlinson, P. B., J. B. Bunt, R. B. Primack and N. C. Duke. 1978. Lumnitzera rosea (Combretaceae) — its status and floral morphology. J. Arnold Arb. 59: 342–351.
Troll, W. 1933. Camptostemon schultzii Mart. und C. philippensis Becc. als neue Vertreter der australasiatischen Mangrove-vegetation. Flora (N.S.) 28: 348–352.
Warburg, O. 1891. Beitrage zur Kenntnis der Papuanischen Flora. Bot. Jahrb. 13: 230–455.
Watson, J. G. 1928. Mangrove forests of the Malay Peninsula. Malayan Forest Records 6: 1–275.
Wium-Andersen, S. and B. Christensen. 1978. Seasonal growth of mangrove trees in southern Thailand. II. Phenology of Bruguiera cylindrica, Ceriops tagal, Lumnitzera littorea and Avicennia marina. Aq. Bot. 5: 383–390.

13. Fly pollination in *Drimys* (Winteraceae), a primitive angiosperm

Leonard B. Thien

The primitive angiosperms are an assemblage of relictual plants considered to be representative of the earliest flowering plants (Stebbins 1974). These living fossils are distributed throughout the temperate and tropical regions of the world (Takhtajan 1969); an exceptionally large number of extant primitive angiosperms are concentrated in New Guinea, New Caledonia and NE Australia (Thorne, 1965, 1969).

The Winteraceae retain the largest number of primitive vegetative and floral characteristics of any family of angiosperms (Stebbins 1974; Thorne 1974). this group of shrubs and small trees contains approximately 100 species in six genera: *Drimys*, *Belliolum*, *Pseudowintera*, *Zygogynum*, *Bubbia* and *Exospernum* (Smith 1943a). Prominent features include vesselless wood, stomata occluded by waxy deposits, small actinomorphic flowers and styleless conduplicate carpels with papillate stigmatic crests (Thorne 1974).

The genus *Drimys* is found in the Old and New World tropics with the vast majority occurring in New Guinea (over 30 species). In the Old World it occurs in Australia and Tasmania, New Guinea, the Philippines, Borneo, Celebes and Ambon (Smith 1943a). The New World species have bisexual flowers and a chromosome number of $n = 43$; Old World species are usually unisexual (dioecious) with some entities possessing separate plants with male, female or bisexual flowers and a chromosome number of $n = 13$ (Vink 1970; Smith 1943b). Smith (1943b) originally divided *Drimys* into two sections, *Drimys* (New World) and *Tasmannia* (Old World). Subsequently Smith (1969) raised both sections to generic status but Vink (1970) in the latest monograph considers the taxon to be *Drimys* with two sections.

On Mt Kaindi, Papua New Guinea (see Gressitt and Nadkarni 1978), *Drimys piperita* entity *polymera* occurs as an understory shrub in the *Nothofagus* forest. The bisexual, female and male flowers are produced on separate plants (8:1:1 ratio respectively). All the flowers are small (1–1.5 cm dia.), with 2–5 white petals and produce a faint sweet odor. An individual flower never closes once open and remains functional for 10–12 days. The flowers are pollinated primarily by various species of Diptera during daylight hours (Table 1).

Bisexual flowers. Flies are attracted to large buds prior to opening and move their proboscides over the surface (Fig. 1). This behavior was observed on numerous occasions although no attraction mechanism or food could be

Monographiae Biologicae, Vol. 42, ed. by J. L. Gressitt.

© *1982, Dr W. Junk Publishers, The Hague.*

detected. Upon opening, the stigmatic crests of the carpels secrete large drops of nectar which collect at the bases of the carpels. A large variety of Diptera visits the flowers to consume the nectar (fig. 1; table 1).

The connectives of the stamens also exude nectar commencing on the first day of anthesis with the innermost series; each day a subsequent series of stamens

Fig. 1. A, B and D. *Drimys piperita* entity *polymera*; C. *Drimys piperita* entity *cordata*. A. Arrow shows fly on bud just prior to opening. ×0.65. B. Fly with proboscis on surface of large bud. ×0.65. C. A newly opened bisexual flower with fly feeding on nectar secreted by carpels. ×0.65. D. Fly feeding on nectar and pollen secreted by stamens. ×0.65.

Table 1. Pollinators and visitors captured from flowers of *Drimys piperita* entity *polymera* on Mt Kaindi, Papua New Guinea

Insects	Number of insects ♂	♀	Average pollen tetrads; number ( ) of individuals
Diptera			
DROSOPHILIDAE			12(12)
Drosophila sp.	3		
Liodrosophila sp.	3		
Chymomyza sp.	1		
MUSCIDAE			23(8)
Dichaetomyia 6 spp. (A-F)	8	6	
Helina sp.	1	1	
Limnophora sp.	2	3	
LAUXANIIDAE			7(10)
Homoneura decemseriata	6	5	
H. sexseriata	4	8	
New genus	5	3	
DOLICHOPODIDAE			
New genus	3	2	
SIMULIIDAE			
Simulium sp.		1	
TIPULIDAE			
New genus?	1		
CERATOPOGONIDAE			
SCIARIDAE	65		1(30)
Hymenoptera			
HYLAEINAE			
Hylaeus sp.	2	4	4(3)
Unidentified genera	10		
Coleoptera			8(1)
CURCULIONIDAE	4		
CHRYSOMELIDAE	1		
DERMESTIDAE	3		

produces a mixture of pollen (tetrads) and nectar. The dehisced anthers continue to shed pollen throughout the life of the flower. In the process of feeding from the stamens and carpels the insects become coated with nectar and pollen and presumably pollinate bisexual or female flowers.

Unisexual flowers. The unisexual flowers tend to have a lower number of petals (2–3) than bisexual flowers. Nectar production is similar except female flowers offer only nectar whereas male and bisexual flowers offer pollen mixed with nectar.

Pollinators. Insect visits are diurnal with peak activity on bright days. Only one individual, a cockcroach, was observed on the flowers at night. Many of the species of flies that feed from the flowers during the day, rest on the lower side of the leaves at night. The Lauxaniidae, Muscidae and Drosophilidae are the prime pollinators based upon observations of entity *Polymera* (Table 1). A large number of Sciaridae was captured but these are minor pollinators because of the low pollen loads carried by individuals and their behavior. Sciarids consistently fed from nectar secreted by the carpels and were rarely captured or sighted on the stamens. Species of *Hylaeus* (a solitary bee) visit the flowers for pollen but rarely feed from the carpels; the Coleptera also prefer to eat the pollen and are infrequent visitors.

Crossing. On Mt Kaindi 27 female buds on 5 plants were bagged to deny insects access to the developing flowers — 26 fruits were formed (one bud was lost); 35 bisexual flowers on 6 plants were bagged and no fruits developed. Six bisexual flowers on 3 different plants were self-pollinated and none set fruit. In addition, 100 plants with bisexual flowers were surveyed for fruits; none were found. Thus it is not known how much sexual reproduction occurs in the populations of entity *polymera* but bisexual flowers function as males. Seedlings are rarely found in nature and apprently the plants spread by runners and rooting at nodes on fallen plants (Vink 1970). Apparently little sexual reproduction is occurring in the entity *polymera* populations on Mt Kaindi.

Discussion. In addition to *Drimys piperita* entity *polymera*, entities *heteromera*, *montis-wilhelmi* and *cordata* were observed on Mt Wilhelm, Papua New Guinea and are also pollinated by a wide variety of Diptera. the myophilous flowers of *Drimys* are small, sweet scented, open only once, are long lived, lack massive stamens and secrete nectar as the primary food for insects. The flowers are obviously not anemophilous as suggested by Pervukhima (1967) nor pollinated by beetles as proposed by Gottsberger (1940).

The flowers of *Drimys granadensis* var. *mexicana* on Mt Tzontehuitz, Chiapas, Mexico are pollinated by a large variety of Diptera, Hymenoptera and Coleoptera. Beetles, however, comprised less than 1% of the insect visitors but carried larger pollen loads (Thien 1977).

Cantharophilous flowers of extant primitive angiosperms present yet another mode of pollination. The flowers tend to be large, short lived (2–4 days), possess numerous leaf-like stamens that provide large amounts of pollen and produce strong fragrances. A typical beetle-pollinated flower within the extant primitive angiosperms is *Magnolia* (Thien 1974).

It is commonly held by many biologists that the early angiosperm flowers were pollinated by beetles (Diels 1916; Grant 1950; van der Pijl 1960). This concept is based on early observations of extant primitive angiosperms pollinated by beetles; plus the fact that beetles are very abundant as fossils of the Cretaceous (Riek 1973) when the angiosperms arose.

Downes (1973, 1974) suggests that honeydew, the partially digested sugar-rich plant sap liberated by small Homoptera, has probably been available on leaves since the Permian. Sugar feeding insects would readily be attracted to nectaries of reproductive shoots and perhaps Diptera have been associated with the evolution of the early angiosperm flower.

Yet another mode of pollination besides beetles and flies exists within the primitive angiosperms. In New Caledonia, *Taeniothrips novocaledonensis*, an endemic thrips, pollinates 3 species of *Belliolum* (a vesselless primitive angiosperm with about 4 species in New Caledonia and 4 species in the Solomons). The insects ingest nectar and pollen from the protogynous flowers and apparently their life cycle revolves about the plants. Large numbers of larvae were captured from the flowers of all species and the insect-plant relationship may be similar to that of thrips and *Calluna* in the Faeroes (Hagerup 1950). The

Solomon Is taxon of *Belliolum* desperately needs study to correlate with observations made in New Caledonia.

The flowers of *Bubbia*, a genus closely related to *Drimys* (many species in New Guinea), may also be pollinated by Diptera, as the flowers exhibit a similar syndrome of pollination characters. The highland region of New Guinea holds many insect-plant relationships in a primeval group of plants.

References

Diels, L. 1916. Käferblumen bei den Ranales und ihre Bedeutung für die Philogenese der Angiospermen. Ber. Dtsch. Bot. Ges. 34: 758–774.

Downes, J.A. 1973. Endopterygote insects and the origin of the angiosperm flower. 1st. Int. Congr. of Syst. and Evol. Biol. Abst.

—— 1974. The feeding habits of adult Chironomidae. Ent. Tidskr. 95: 80–84.

Gottsberger, G. 1974. The structure and function of the primitive angiosperm flower — a discussion. Acta. bot. Neerl. 23: 461–471.

—— 1976. Some aspects of beetle pollination in the evolution of flowering plants. In K. Kubitzki (ed.) Flowering plants evolution and classification of higher categories. Symp., Hamburg, 1976. Springer-Verlag, Wien-New York.

Grant, V. 1950. The pollination of Calycanthus occidentalis. Amer. J. Bot. 37: 294–297.

Gressitt, J. L. and N. Nadkarni. 1978. Guide to Mt Kaindi: background to montane New Guinea ecology. Wau Ecology Inst. Hdbk 5: 1–135.

Hagerup, O. 1950. Thrips pollination in Calluna. Det. Kgl. Danske Videnskabernes Selskab. Biologiske Meddelser. 18(4): 1–16.

Pervukhina, N. V. 1967. The pollination of the primary angiosperms and the evolution of modes of pollination (Russ.) Bot. Zh. (Moskow) 52: 157–188.

Riek, E. F. 1973. The Insects of Australia. Melbourne Univ. Press.

Smith, A. C. 1943a. Taxonomic notes on the Old World species of Winteraceae. J. Arnold Arb. 24: 119–164.

—— 1943b. The American species of Drimys. J. Arnold Arb. 24(1): 1–29.

—— 1969. A reconsideration of the genus Tasmannia (Winteraceae). Taxon. 18: 286–290.

Stebbins, G. L. 1974. Flowering plants — Evolution above the species level. The Belknap Press, Harvard University Press, Cambridge.

Takhtajan, A. 1969. Flowering plants — Origin and dispersal. Oliver and Boyd, Edinburgh.

Thien, L. B. 1974. Floral biology of Magnolia. Amer. J. Bot. 61(10): 1037–1045.

—— 1977. Personal observation.

—— 1980. Patterns of pollination in the primitive angiosperms. Biotropica. 12(1): 1–13.

Thorne, R. F. 1965. Floristic relationships of New Caledonia. The Univ. of Iowa Studies in Natural History. 20(7): 1–14.

—— 1969. Floristic relationships between New Caledonia and the Solomon Islands. Phil. Trans. Roy. Soc. B 255: 595–602.

—— 1974. A phylogenetic classification of the Annoniflorae. Aliso 8(2): 147–209.

van der Pijl, L. 1960. Ecological aspects of flower evolution. I. Evolution 14: 403–416.

Vink, W. 1970. The Winteraceae of the Old World I. Pseudowintera and Drimys — morphology and taxonomy. Blumea 23: 219–250.

GPSR Compliance

The European Union's (EU) General Product Safety Regulation (GPSR) is a set of rules that requires consumer products to be safe and our obligations to ensure this.

If you have any concerns about our products, you can contact us on ProductSafety@springernature.com

In case Publisher is established outside the EU, the EU authorized representative is:

Springer Nature Customer Service Center GmbH
Europaplatz 3
69115 Heidelberg, Germany

Batch number: 10823775

Printed by Printforce, the Netherlands